Edited by
Thomas Wirth

Organoselenium Chemistry

Related Titles

Toru, T., Bolm, C. (eds.)

Organosulfur Chemistry in Asymmetric Synthesis

2008
Hardcover
ISBN: 978-3-527-31854-4

Steinborn, D.

Fundamentals of Organometallic Catalysis

2011
Hardcover: ISBN: 978-3-527-32716-4
Softcover: ISBN: 978-3-527-32717-1

Song, C. E. (ed.)

Cinchona Alkaloids in Synthesis and Catalysis

Ligands, Immobilization and Organocatalysis

2009
Hardcover
ISBN: 978-3-527-32416-3

Zecchina, A., Bordiga, S., Groppo, E. (eds.)

Selective Nanocatalysts and Nanoscience

Concepts for Heterogeneous and Homogeneous Catalysis

2011
Hardcover
ISBN: 978-3-527-32271-8

Cordova, A. (ed.)

Catalytic Asymmetric Conjugate Reactions

2010
Hardcover
ISBN: 978-3-527-32411-8

Bandini, M., Umani-Ronchi, A. (eds.)

Catalytic Asymmetric Friedel-Crafts Alkylations

2009
Hardcover
ISBN: 978-3-527-32380-7

Ackermann, L. (ed.)

Modern Arylation Methods

2009
Hardcover
ISBN: 978-3-527-31937-4

Carreira, E. M., Kvaerno, L.

Classics in Stereoselective Synthesis

2009
Hardcover: ISBN: 978-3-527-32452-1
Softcover: ISBN: 978-3-527-29966-9

Ding, K., Dai, L.-X. (eds.)

Organic Chemistry – Breakthroughs and Perspectives

2012
Softcover
ISBN: 978-3-527-32963-2

Edited by Thomas Wirth

Organoselenium Chemistry

Synthesis and Reactions

WILEY-VCH Verlag GmbH & Co. KGaA

The Editor

Prof. Dr. Thomas Wirth
Cardiff University
School of Chemistry
Park Place Main Building
Cardiff CF10 3AT
United Kingdom

All books published by **Wiley-VCH** are carefully produced. Nevertheless, authors, editors, and publisher do not warrant the information contained in these books, including this book, to be free of errors. Readers are advised to keep in mind that statements, data, illustrations, procedural details or other items may inadvertently be inaccurate.

Library of Congress Card No.: applied for

British Library Cataloguing-in-Publication Data
A catalogue record for this book is available from the British Library.

Bibliographic information published by the Deutsche Nationalbibliothek
The Deutsche Nationalbibliothek lists this publication in the Deutsche Nationalbibliografie; detailed bibliographic data are available on the Internet at <http://dnb.d-nb.de>.

© 2012 Wiley-VCH Verlag & Co. KGaA, Boschstr. 12, 69469 Weinheim, Germany

All rights reserved (including those of translation into other languages). No part of this book may be reproduced in any form – by photoprinting, microfilm, or any other means – nor transmitted or translated into a machine language without written permission from the publishers. Registered names, trademarks, etc. used in this book, even when not specifically marked as such, are not to be considered unprotected by law.

Typesetting Toppan Best-set Premedia Limited, Hong Kong
Printing and Binding Fabulous Printers Pte Ltd, Singapore
Cover Design Grafik-Design Schulz, Fußgönheim

Printed in Singapore
Printed on acid-free paper

Print ISBN: 978-3-527-32944-1

ePDF ISBN: 978-3-527-64196-3

oBook ISBN: 978-3-527-64194-9

ePub ISBN: 978-3-527-64195-6

mobi ISBN: 978-3-527-64197-0

Contents

Preface *XI*
List of Contributor *XIII*

1 Electrophilic Selenium *1*
Claudio Santi and Stefano Santoro
1.1 General Introduction *1*
1.1.1 Synthesis of Electrophilic Selenium Reagents *3*
1.1.2 Reactivity and Properties *7*
1.2 Addition Reactions to Double Bonds *11*
1.2.1 Addition Reaction Involving Oxygen-Centered Nucleophiles *11*
1.2.2 Addition Reaction Involving Nitrogen-Centered Nucleophiles *22*
1.2.3 Addition Reactions Involving Carbon-Centered Nucleophiles *26*
1.2.4 Addition Reaction Involving Chiral Nucleophiles or Chiral Substrates *28*
1.3 Selenocyclizations *30*
1.3.1 Oxygen Nucleophiles *31*
1.3.2 Nitrogen Nucleophiles *35*
1.3.3 Competition between Oxygen and Nitrogen Nucleophiles *40*
1.3.4 Carbon Nucleophiles *42*
1.3.5 Double Cyclization Reactions *44*
References *45*

2 Nucleophilic Selenium *53*
Michio Iwaoka
2.1 Introduction *53*
2.1.1 Development of Nucleophilic Selenium Reagents *53*
2.1.2 Examples of Recent Applications *54*
2.2 Properties of Selenols and Selenolates *56*
2.2.1 Electronegativity of Selenium *56*
2.2.2 Tautomerism of Selenols *57*
2.2.3 Nucleophilicity of Selenolates *58*
2.3 Inorganic Nucleophilic Selenium Reagents *59*
2.3.1 Conventional Reagents *59*

2.3.2	New Reagents 61
2.4	Organic Nucleophilic Selenium Reagents 65
2.4.1	Preparation 65
2.4.2	Structure 66
2.4.3	Ammonium Selenolates (NH_4^+) 67
2.4.4	Selenolates of Group 1 Elements (Li, Na, K, and Cs) 67
2.4.5	Selenolates of Group 2 Elements (Mg, Ca, and Ba) 70
2.4.6	Selenolates of Group 3 Elements (Sm, Ce, Pr, Nb, and U) 71
2.4.7	Selenolates of Group 4 Elements (Ti, Zr, and Hf) 73
2.4.8	Selenolates of Group 5 Elements (V, Nb, and Ta) 74
2.4.9	Selenolates of Group 6 Elements (Mo and W) 75
2.4.10	Selenolates of Group 7 Elements (Mn and Re) 76
2.4.11	Selenolates of Group 8 Elements (Fe, Ru, and Os) 78
2.4.12	Selenolates of Group 9 Elements (Co, Rh, and Ir) 81
2.4.13	Selenolates of Group 10 Elements (Ni, Pd, and Pt) 84
2.4.14	Selenolates of Group 11 Elements (Cu, Ag, and Au) 90
2.4.15	Selenolates of Group 12 Elements (Zn, Cd, and Hg) 92
2.4.16	Selenolates of Group 13 Elements (B, Al, Ga, and In) 95
2.4.17	Selenolates of Group 14 Elements (Si, Ge, Sn, and Pb) 97
2.4.18	Selenolates of Group 15 Elements (P, As, Sb, and Bi) 100
	References 102
3	**Selenium Compounds in Radical Reactions** 111
	W. Russell Bowman
3.1	Homolytic Substitution at Selenium to Generate Radical Precursors 111
3.1.1	Bimolecular S_H2 Reactions: Synthetic Considerations 112
3.1.1.1	Radical Reagents 115
3.1.2	Alkyl Radicals from Selenide Precursors 115
3.1.3	Acyl Radicals from Acyl Selenide Precursors 119
3.1.4	Imidoyl Radicals from Imidoyl Selenides 123
3.1.5	Other Radicals from Selenide Precursors 125
3.2	Selenide Building Blocks 126
3.3	Solid-Phase Synthesis 128
3.4	Selenide Precursors in Radical Domino Reactions 130
3.5	Homolytic Substitution at Selenium for the Synthesis of Se-Containing Products 132
3.5.1	Intermolecular S_H2 onto Se 132
3.5.2	Intramolecular S_H2: Cyclization onto Se 132
3.6	Seleno Group Transfer onto Alkenes and Alkynes 134
3.6.1	Seleno-Selenation 135
3.6.2	Seleno-Sulfonation 136
3.6.3	Seleno-Alkylation 137
3.7	PhSeH in Radical Reactions 138
3.7.1	Radical Clock Reactions 138

3.7.2	Problem of Unwanted Trapping of Intermediate Radicals	138
3.7.3	Catalysis of Stannane-Mediated Reactions	139
3.8	Selenium Radical Anions, $S_{RN}1$ Substitutions	141
	References	143
4	**Selenium-Stabilized Carbanions**	**147**
	João V. Comasseto, Alcindo A. Dos Santos, and Edison P. Wendler	
4.1	Introduction	147
4.2	Preparation of Selenium-Stabilized Carbanions	149
4.2.1	Deprotonation of Selenides	149
4.2.2	Element-Lithium Exchange	154
4.2.3	Conjugate Addition of Organometallics to Vinyl- and Alkynylselenides	158
4.3	Reactivity of the Selenium-Stabilized Carbanions with Electrophiles and Synthetic Transformations of the Products	161
4.3.1	Reaction of Selenium-Stabilized Carbanions with Electrophiles	166
4.3.2	Selenium-Based Transformations on the Reaction Products of Selenium-Stabilized Carbanions with Electrophiles	167
4.4	Stereochemical Aspects	168
4.4.1	Cyclic Selenium-Stabilized Carbanions	173
4.4.2	Acyclic Selenium-Stabilized Carbanions	176
4.5	Application of Selenium-Stabilized Carbanions in Total Synthesis	176
4.5.1	Examples Using Alkylation Reactions of Selenium-Stabilized Carbanions	177
4.5.2	Examples Using the Addition of Selenium-Stabilized Carbanions to Carbonyl Compounds	180
4.5.3	Examples Using 1,4-Addition of Selenium-Stabilized Carbanions to α,β-Unsaturated Carbonyl Compounds	184
4.6	Conclusion	186
	References	187
5	**Selenium Compounds with Valency Higher than Two**	**191**
	Józef Drabowicz, Jarosław Lewkowski, and Jacek Ścianowski	
5.1	Introduction	191
5.2	Trivalent, Dicoordinated Selenonium Salts	192
5.3	Trivalent, Tricoordinated Derivatives	194
5.4	Tetravalent, Dicoordinated Derivatives	211
5.5	Tetravalent, Tricoordinated Derivatives	225
5.6	Pentavalent Derivatives	239
5.7	Hexavalent, Tetracoordinated Derivatives	240
5.8	Hypervalent Derivatives	244
5.8.1	Selenuranes	244
5.8.2	Selenurane Oxides	249
5.8.3	Perselenuranes	250
	Acknowledgment	251
	References	251

6	**Selenocarbonyls** *257*
	Toshiaki Murai
6.1	Overview *257*
6.2	Theoretical Aspects of Selenocarbonyls *259*
6.3	Molecular Structure of Selenocarbonyls *261*
6.4	Synthetic Procedures of Selenocarbonyls *261*
6.5	Manipulation of Selenocarbonyls *270*
6.6	Metal Complexes of Selenocarbonyls *278*
6.7	Future Aspects *280*
	References *281*

7	**Selenoxide Elimination and [2,3]-Sigmatropic Rearrangement** *287*
	Yoshiaki Nishibayashi and Sakae Uemura
7.1	Introduction *287*
7.2	Preparation and Properties of Chiral Selenoxides *288*
7.3	Selenoxide Elimination *292*
7.3.1	Enantioselective Selenoxide Elimination Producing Chiral Allenes and α,β-Unsaturated Ketones *293*
7.3.2	Diastereoselective Selenoxide Elimination Producing Chiral Allenecarboxylic Esters *295*
7.4	[2,3]-Sigmatropic Rearrangement via Allylic Selenoxides *297*
7.4.1	Enantioselective [2,3]-Sigmatropic Rearrangement Producing Chiral Allylic Alcohols *297*
7.4.2	Diastereoselective [2,3]-Sigmatropic Rearrangement Producing Chiral Allylic Alcohols *299*
7.5	[2,3]-Sigmatropic Rearrangement via Allylic Selenimides *305*
7.5.1	Preparation and Properties of Chiral Selenimides *307*
7.5.2	Enantioselective [2,3]-Sigmatropic Rearrangement Producing Chiral Allylic Amines *309*
7.5.3	Diastereoselective [2,3]-Sigmatropic Rearrangements Producing Chiral Allylic Amines *310*
7.6	[2,3]-Sigmatropic Rearrangement via Allylic Selenium Ylides *311*
7.6.1	Preparation and Properties of Optically Active Selenium Ylides *312*
7.6.2	Enantioselective [2,3]-Sigmatropic Rearrangements via Allylic Selenium Ylides *313*
7.6.3	Diastereoselective [2,3]-Sigmatropic Rearrangement via Allylic Selenium Ylides *315*
7.7	Summary *317*
	References *317*

8	**Selenium Compounds as Ligands and Catalysts** *321*
	Fateh V. Singh and Thomas Wirth
8.1	Introduction *321*
8.2	Selenium-Catalyzed Reactions *321*
8.2.1	Stereoselective Addition of Diorganozinc Reagents to Aldehydes *322*

8.2.1.1	Diethylzinc Addition	*322*
8.2.1.2	Diphenylzinc Addition	*323*
8.2.2	Selenium-Ligated Transition Metal-Catalyzed Reactions	*324*
8.2.2.1	Selenium-Ligated Stereoselective Hydrosilylation of Ketones	*324*
8.2.2.2	Selenium-Ligated Copper-Catalyzed Addition of Organometallic Reagents to Enones	*325*
8.2.2.3	Selenium-Ligated Palladium-Catalyzed Asymmetric Allylic Alkylation	*326*
8.2.2.4	Selenium-Ligands in Palladium-Catalyzed Mizoroki–Heck Reactions	*328*
8.2.2.5	Selenium-Ligands in Palladium-Catalyzed Phenylselenenylation of Organohalides	*330*
8.2.2.6	Selenium-Ligands in Palladium-Catalyzed Substitution Reactions	*331*
8.2.2.7	Selenium-Ligands in the Palladium-Catalyzed Allylation of Aldehydes	*331*
8.2.2.8	Selenium-Ligands in Palladium-Catalyzed Condensation Reactions	*332*
8.2.2.9	Ruthenium-Catalyzed Substitution Reactions	*333*
8.2.2.10	Selenium-Ligands in Zinc-Catalyzed Intramolecular Hydroaminations	*334*
8.2.3	Selenium-Ligands in Organocatalytic Asymmetric Aldol Reactions	*334*
8.2.4	Selenium-Ligands in Stereoselective Darzens Reactions	*334*
8.2.5	Selenium-Catalyzed Carbonylation Reactions	*335*
8.2.6	Selective Reduction of α,β-Unsaturated Carbonyl Compounds	*336*
8.2.7	Selenium-Catalyzed Halogenations and Halocyclizations	*336*
8.2.8	Selenium-Catalyzed Staudinger–Vilarrasa Reaction	*337*
8.2.9	Selenium-Catalyzed Elimination Reactions of Diols	*338*
8.2.10	Selenium-Catalyzed Hydrostannylation of Alkenes	*339*
8.2.11	Selenium-Catalyzed Radical Chain Reactions	*340*
8.2.12	Selenium-Catalyzed Oxidation Reactions	*342*
8.2.12.1	Selenium-Catalyzed Epoxidation of Alkenes	*342*
8.2.12.2	Selenium-Catalyzed Dihydroxylation of Alkenes	*344*
8.2.12.3	Selenium-Catalyzed Oxidation of Alcohols	*346*
8.2.12.4	Baeyer–Villiger Oxidation	*347*
8.2.12.5	Selenium-Catalyzed Allylic Oxidation of Alkenes	*349*
8.2.12.6	Selenium-Catalyzed Oxidation of Aryl Alkyl Ketones	*350*
8.2.12.7	Selenium-Catalyzed Oxidation of Primary Aromatic Amines	*350*
8.2.12.8	Selenium-Catalyzed Oxidation of Alkynes	*351*
8.2.12.9	Selenium-Catalyzed Oxidation of Halide Anions	*352*
8.2.13	Stereoselective Catalytic Selenenylation–Elimination Reactions	*353*
8.2.14	Selenium-Catalyzed Diels–Alder Reactions	*355*
8.2.15	Selenium-Catalyzed Synthesis of Thioacetals	*355*

8.2.16	Selenium-Catalyzed Baylis–Hillman Reaction 356	
	References 356	
9	**Biological and Biochemical Aspects of Selenium Compounds** *361*	
	Bhaskar J. Bhuyan and Govindasamy Mugesh	
9.1	Introduction 361	
9.2	Biological Importance of Selenium 361	
9.3	Selenocysteine: The 21st Amino Acid 362	
9.4	Biosynthesis of Selenocysteine 363	
9.5	Chemical Synthesis of Selenocysteine 366	
9.6	Chemical Synthesis of Sec-Containing Proteins and Peptides 367	
9.7	Selenoenzymes 369	
9.7.1	Glutathione Peroxidases 369	
9.7.2	Iodothyronine Deiodinase 379	
9.7.3	Synthetic Mimics of IDs 384	
9.7.4	Thioredoxin Reductase 387	
9.8	Summary 389	
	References 392	

^{77}Se NMR Values *397*

Index *435*

Preface

Selenium was discovered in 1818 – almost 200 years ago – by the Swedish chemist Jöns Jacob Berzelius. Selenium is a common companion of sulfur, but was named after the goddess of the moon *Selene*. This indicates the chemical relation to tellurium, named after the Greek word for earth, *tellus*. Berzelius studied the element selenium and its inorganic compounds in detail. Nowadays, selenium is obtained in the electrolytic refining of copper and its production reaches several thousand tons per year. The first organoselenium derivative (ethyl selenol) was published in 1847 by Wöhler and Siemens. The use of selenium dioxide as an oxidant was described in a patent in 1929. Since that time, selenium and its derivatives have appeared as reagents in organic synthesis. Organoselenium chemistry started blossoming in the early seventies with the discovery of the selenoxide elimination for the introduction of double bonds under mild reaction conditions. Since 1971, the chemistry of selenium and tellurium is also regularly promoted in a conference series (ICCST, International Conference on the Chemistry of Selenium and Tellurium). Several monographs and many review articles have appeared during the last decades. This book summarizes the latest developments with a strong emphasis on the last decade providing an updated picture of the many facets of organoselenium chemistry.

I am very grateful to all the distinguished scientists who have contributed to this book with their time, knowledge, and expertise. I hope that all chapters will not only be a rich source of information, but also a source of inspiration to students and colleagues.

Cardiff, July 2011 *Thomas Wirth*

List of Contributors

Bhaskar J. Bhuyan
Indian Institute of Science
Department of Inorganic and Physical Chemistry
Bangalore 560 012
India

W. Russell Bowman
Loughborough University
Department of Chemistry
Loughborough, Leics LE11 3TU
UK

João V. Comasseto
Universidade de São Paulo
Instituto de Química
Av. Prof. Lineu Prestes, 748
05508-000 São Paulo-SP
Brazil

Alcindo A. Dos Santos
Universidade de São Paulo
Instituto de Química
Av. Prof. Lineu Prestes, 748
05508-000 São Paulo-SP
Brazil

Józef Drabowicz
Polish Academy of Sciences
Department of Heteroorganic Chemistry
Center of Molecular and Macromolecular Studies
Sienkiewicza 112
Łódź 90–363
Poland
and
Jan Długosz University
Institute of Chemistry and Environmental Protection
Armii Krajowej 13/15
Częstochowa 42–200
Poland

Michio Iwaoka
Tokai University
School of Science
Department of Chemistry
Kanagawa 259-1292
Japan

Jarosław Lewkowski
University of Łódź
Faculty of Chemistry
Department of Organic Chemistry
Tamka 12; 91-403 Łódź
Poland

Toshiaki Murai
Gifu University
Faculty of Engineering
Department of Chemistry
Yanagido, Gifu 501-1193
Japan

Govindasamy Mugesh
Indian Institute of Science
Department of Inorganic and Physical Chemistry
Bangalore 560 012
India

Yoshiaki Nishibayashi
The University of Tokyo
School of Engineering
Bunkyo-ku
Tokyo 113-8656
Japan

Claudio Santi
University of Perugia
Dipartimento di Chimica e Tecnologia del Farmaco
Perugia 06123
Italy

Stefano Santoro
University of Perugia
Dipartimento di Chimica e Tecnologia del Farmaco
Perugia 06123
Italy

Jacek Ścianowski
Nicolaus Copernicus University
Faculty of Chemistry
Department of Organic Chemistry
Gagarina 7
Toruń 87–100
Poland

Fateh V. Singh
Cardiff University
School of Chemistry
Park Place
Cardiff CF10 3AT
UK

Sakae Uemura
Okayama University of Science
Faculty of Engineering
Okayama 700-0803
Japan

Edison P. Wendler
Universidade de São Paulo
Instituto de Química
Av. Prof. Lineu Prestes, 748
05508-000 São Paulo-SP
Brazil

Thomas Wirth
Cardiff University
School of Chemistry
Park Place
Cardiff CF10 3AT
UK

1
Electrophilic Selenium
Claudio Santi and Stefano Santoro

1.1
General Introduction

During the last few decades, organoselenium compounds have emerged as important reagents and intermediates in organic synthesis.

Selenium can be introduced as an electrophile, as a nucleophile, or as a radical and generally it combines chemo-, regio-, and stereoselectivity with mild experimental conditions. Once incorporated, it can be directly converted into different functional groups or it can be employed for further manipulation of the molecule.

Since the discovery in the late 1950s that species of type RSeX add stereospecifically to simple alkenes [1], electrophilic organoselenium compounds provided the synthetic chemist with useful and powerful reagents and the selenofunctionalization of olefins represents an important method for the rapid introduction of vicinal functional groups, often with concomitant formation of rings and stereocenters (Scheme 1.1a and b).

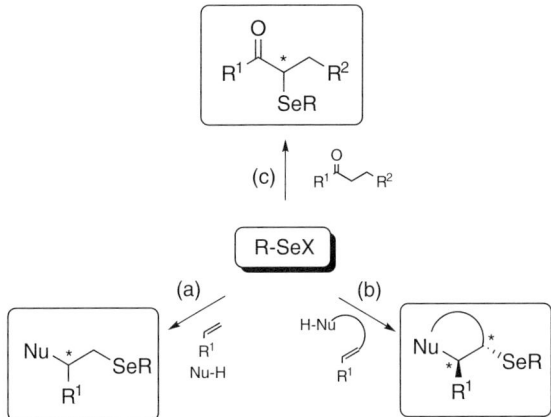

Scheme 1.1 The reactivity of electrophilic selenium reagents.

Organoselenium Chemistry: Synthesis and Reactions, First Edition. Edited by Thomas Wirth.
© 2012 Wiley-VCH Verlag GmbH & Co. KGaA. Published 2012 by Wiley-VCH Verlag GmbH & Co. KGaA.

1 Electrophilic Selenium

In addition, electrophilic selenium reagents can be also used for the α-selenenylation of carbonyl compounds (Scheme 1.1c) affording useful intermediates for the synthesis of α,β-unsaturated [2] derivatives or 1,2-diketones through a seleno-Pummerer reaction [3].

Oxidation of selenides to the corresponding selenoxide for the synthesis of α,β-unsaturated compounds represents a current topic in organic chemistry and has been used successfully also in structurally complex product synthesis. An example has been very recently reported in which the electrophilic selenenylation followed by an oxidative elimination represent a crucial step in the total synthesis of heptemerone G, a diterpenoid fungi-derived with interesting antibacterial activity (Scheme 1.2) [4].

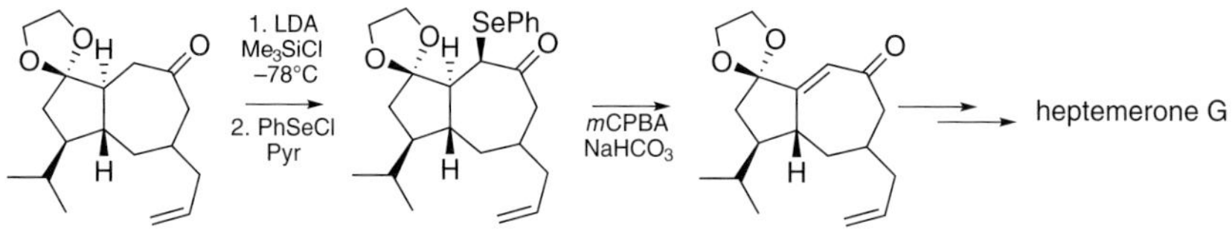

Scheme 1.2 Electrophilic selenium reagent in the total synthesis of heptemerone G.

The kinetic lithium enolate **1**, trapped as trimethysilyl derivatives, reacts with PhSeCl affording the selenide **2** that, after oxidation with metachloroperbenzoic acid, is converted into the enone **3** from which the heptemerone G can be prepared in some additional steps.

The treatment of selenides with tin hydrides, in the presence of AIBN, produces the homolytic cleavage of the carbon–selenium bond generating a carbon radical and opening the way for interesting radical reactions.

An elegant application was reported for the total synthesis of (+)-Samin (Scheme 1.3). The selenide **4** was subjected to radical deselenenylation conditions affording the tetrahydrofurane derivative **5** following a 5-*exo-trig* radical cyclization mechanism. From **5**, (+)-Samin was obtained through a few classical steps [5].

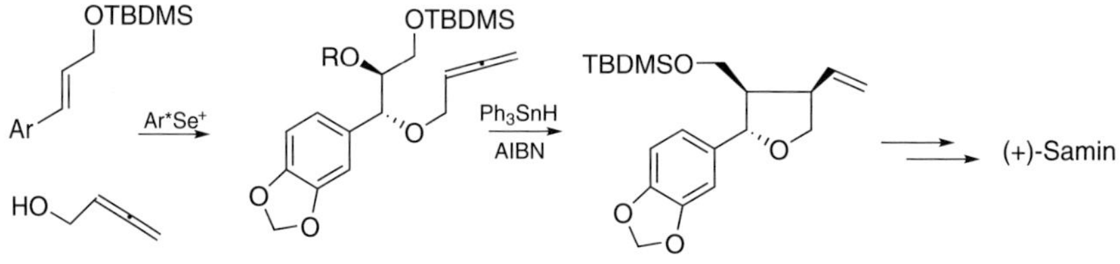

Scheme 1.3 Electrophilic selenium reagent in the total synthesis of (+)-Samin.

The main aspects of organoselenium chemistry have been described in a series of books [6] and review articles and, in recent times, the synthesis of chiral selenium electrophiles as well as their applications in asymmetric synthesis represents a very interesting field of interests for many research groups [7].

In this chapter, we take in consideration some general aspects of the chemistry promoted by electrophilic selenium reagents by reporting selected examples and some more recent and innovative applications.

1.1.1
Synthesis of Electrophilic Selenium Reagents

Some phenylselenenyl derivatives such as chloride, bromide, and N-phenylselenophthalimide [8] are nowadays commercially available and represent the most common electrophilic reagents used to introduce selenium into organic molecules. Otherwise, in a more general procedure, very versatile precursors for the preparation of various electrophilic selenium species are the corresponding diselenides **6**. They can be easily converted into selenenyl halides **7**, **8** by treatment with sulfuryl chloride or chlorine in hexane and bromine in tetrahydrofuran, respectively (Scheme 1.4).

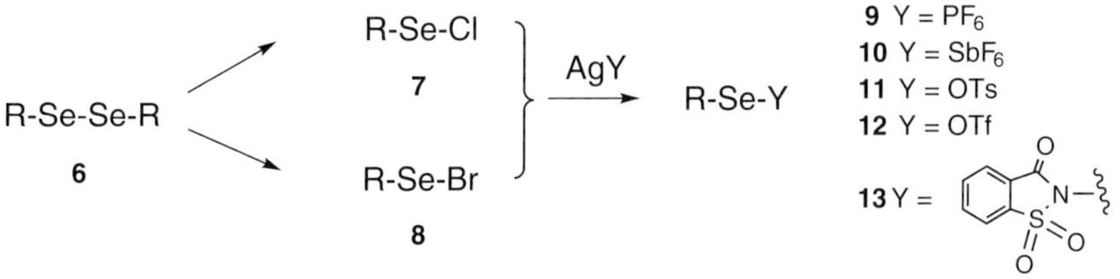

Scheme 1.4 Electrophilic selenium reagents.

The use of halides in synthesis often gives rise to side processes due to the nucleophilicity of the halide anions. For this reason, a series of new selenenylating agents with nonhalide counterions have been reported.

Some of them were directly prepared starting from the appropriate selenenyl halide with silver salts such as hexafluorophosphate **9** [9], hexafluoroantimoniate **10** [10], tolylsulfonate **11** [11], and triflate **12** [12].

This latter is probably the most commonly used electrophilic selenium reagent even if, in many cases, the stoichiometric amount of trifluoromethanesulfonic acid formed is not compatible with the stability of the substrates and/or of the products. More recently, Tingoli reported a similar procedure to prepare the N-saccharin derivatives **13** containing a sulfonamide anion that is scarcely nucleophilic and generating saccharin that is a very weak acidic species [13].

In other cases, the electrophilic reagent can be more conveniently produced by the *in situ* oxidation of **6** with several inorganic reagents: KNO_3 [14], $CuSO_4$ [15], $Ce(NH_4)_2(NO_3)_6$ [16], $Mn(OAc)$ [17], or nitrogen dioxide [18]. Among these, starting from diphenyl diselenide, $(NH_4)_2S_2O_8$ [19] produces the strongly electrophilic phenylselenenyl sulfate (PSS) **14** through a mechanism that reasonably involves an electron transfer or an S_N2 reaction. A product derived from a single electron transfer has been proposed also as an intermediate in the reaction of diphenyl diselenide with 1,2-dicyanonaphthalene [20] that leads to the formation of the phenylselenenyl cation **15** as depicted in Scheme 1.5.

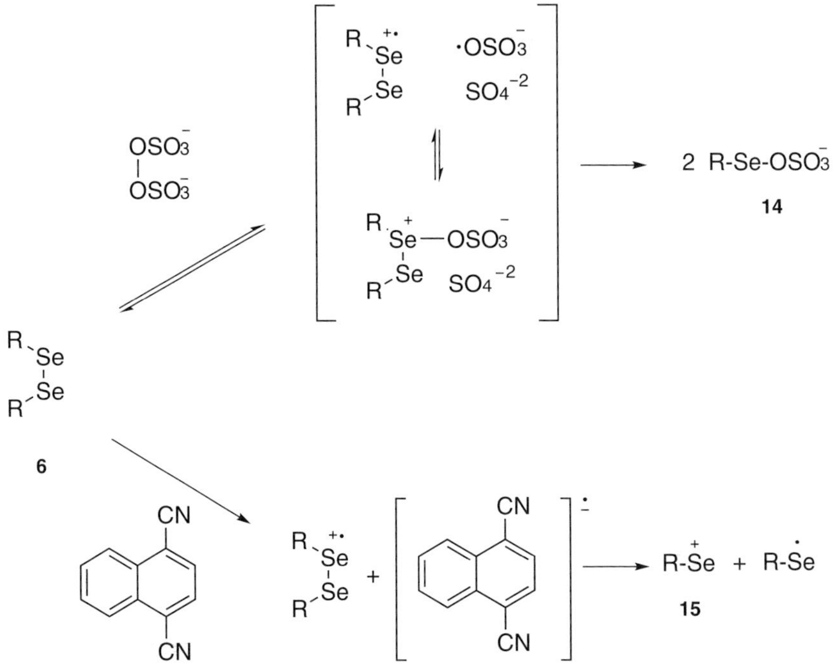

Scheme 1.5 Electrophilic selenium reagents through a single electron transfer mechanism.

Some other organic oxidizing agents such as *m*-nitrobenzenesulfonyl peroxide [21], (bis[trifluoroacetoxy] iodo)benzene [22], (diacethoxy iodo)benzene [23], and 2,3-dichloro-5,6-dicyano-1,4-benzoquinone (DDQ) [24] have been also used in some cases. The choice of the best reagent strongly depends on the chemical susceptibility of the substrate and its functional groups and is mainly dictated by the requirements of the addition reaction to be carried out.

In recent years, polymer-supported reagents have also attracted interest because they can provide attractive and practical methods for combinatorial chemistry and solid-phase synthesis.

Polymer-supported selenium reagents represent an interesting improvement for synthetic organic chemists because of their facile handling without the formation of toxic and odorous by-products. Some electrophilic selenium-based approaches for solid-phase chemistry have been reported by different groups and the use of these reagents allows easy purification and recycling of the reagent for a next reaction. In addition, it represents useful strategies especially for constructing libraries of heterocyclic derivatives [25]. Wirth and coworkers compared the efficiency of polystyrene, TentaGel, and mesopouros silica as a solid support for enantioselective electrophilic addition reactions [26].

In a recent application, polystyrene-supported selenenylbromide was reacted with methyl acrylate and a primary amine to afford in a *one-pot* procedure a resin that has been used to prepare libraries of 2-pyridones, 1,4-diazepines, 1,4 oxazepines [27], and other nitrogen heterocycles [28].

Even if the mild reaction conditions usually required for the selenenylation of unsatured substrates represent an attractive aspect for this chemistry, some of these conversions suffer the drawback that the selenium reagent must be used in stoichiometric amounts. However, addition–elimination sequences using catalytic amount of diselenides in the presence of an excess of oxidizing reagent have been reported using peroxydisulfates [29] as well as hypervalent iodine compounds [30].

An electrochemical procedure to generate a selenium electrophile starting from diphenyl diselenide involves the use of tetraethylammonium bromide as redox catalyst and as electrolyte. The anodic oxidation of bromide to bromine initiates the reaction producing the electrophilic phenylselenenyl bromide from the diselenide [31].

During the last 10 years, several research groups devoted their efforts to the preparation of different optically active diselenides that have been used as electrophilic selenenylating agents precursors [32].

Since the first binaphthyl-based diselenides developed by Tomoda and Iwaoka [33], a series of interesting chiral scaffolds have been proposed and evaluated as chiral sources in asymmetric electrophilic addition and cyclofunctionalization reactions. Selected examples are collected in Scheme 1.6.

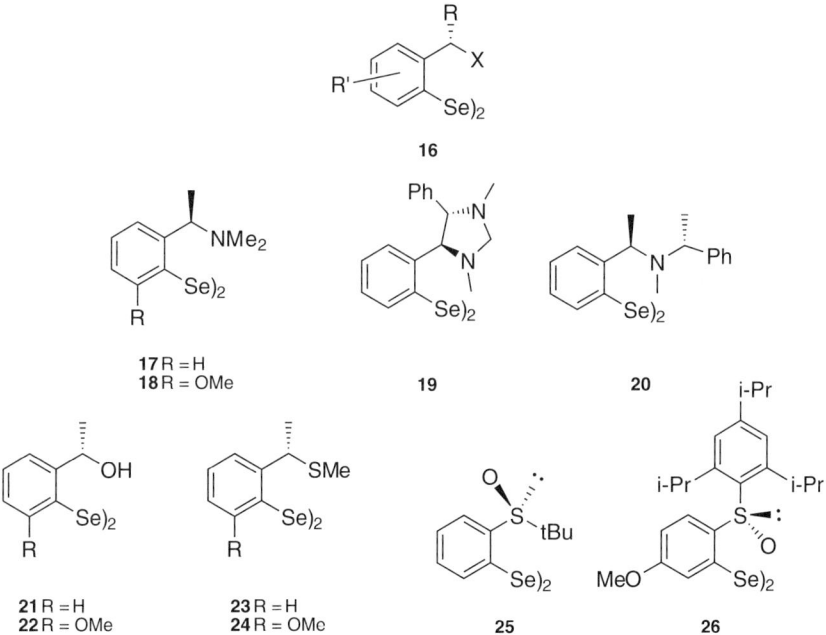

Scheme 1.6 Aromatic chiral diselenides.

Wirth and coworkers reported that easily accessible diselenides having the general structure **10** can be conveniently used to prepare electrophilic selenium reagents and promote asymmetric electrophilic addition reactions [34].

Starting from readily available chiral precursors, diselenides **17–24** can be prepared in one-step syntheses consisting of *ortho*-lithiation, reaction with elemental selenium and air oxidation in generally good overall yields. All these compounds have as common characteristic the presence of heteroatom able to interact with the nearby electrophilic selenium forcing the chiral moiety to come close to the reaction center and leading, at the same time, to a more rigid transition state. Both these conditions have been proposed to play an important role in the enantioselective addition to double bonds and we demonstrated that the substitution of the heteroatom with a methylene group determines a complete lost of diastereoselectivity [35].

The existence of nonbonding interaction like that depicted in structure **27** has been demonstrated by several authors for a large number of organochalcogen systems using a combination of techniques such as X-ray crystallography, NMR, DFT calculations, and has been object of a recent review [36].

The major factor contributing toward nonbonding interactions in these compounds is identified as arising from the orbital interaction.

We first described a Se···S interaction in the selenenyl halides **28** [37] and **29** [38] prepared starting from diselenide **23**. For these reagents, we reported spectroscopic and chemical evidences for the existence of this interaction. X-ray analysis evidenced a T-shaped coordination geometry around the selenium atom with a distance between Se and S (2.497[7] Å for **29** and 2.344[2] Å for **28**) that is significantly shorter than the sum of the van der Waals radii (3.7 Å) [35].

The shorter distance observed in **28** compared to **29** seems to indicate a stronger interaction when the counterion is a chloride.

Comparison of nuclear Overhauser effect (n.O.e.) of arylselenyl halides **28** and **29** to the corresponding aryl methylselenide **30** proves a greater conformational rigidity reasonably deriving from the Se···S interaction atom, which is not only present in the crystal form but also in CDCl$_3$ solution (Scheme 1.7).

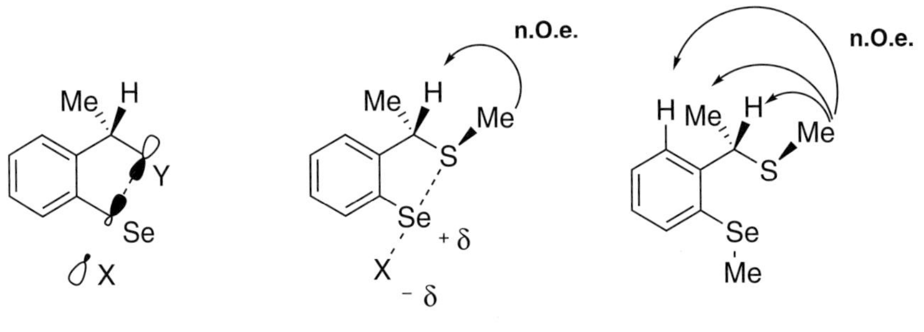

Scheme 1.7 Selenium–heteroatom nonbonding interaction.

Organoselenium compounds exhibiting intramolecular Se···N and Se···O interactions are particularly interesting since these interactions are expected to modulate the biological activity of the selenium compounds. Selenocyanates are commonly used as intermediates in the synthesis of diorganoselenates and are investigated as antitumor drugs. Jones, Mugesh, and du-Mont determined the X-ray structure of methyl-2-selenocyanatobenzoate in which the Se···O contact is sensibly shorter in comparison with the Se···N one [39].

More recently starting from (S)-ethyllactate and (−)-(1R)-para-toluensulfinate, Wirth and coworkers reported the syntheses of two new optically active sulfoxide containing diselenides **25** and **26** [40]. These were used to prepare new selenium electrophiles that have been successfully used for stereoselective functionalization of alkenes.

X-ray comparison between diselenides **25** and **22** evidences that the oxygen attached to the chiral center in compound **25** has no interaction with the selenium atom in the solid state.

Convenient methods for the synthesis of optically active nonfunctionalized or functionalized selenium reagents from mono- and bicyclic terpenes have also been developed in the recent period; Scheme 1.8 summarizes some of the most representative structures (**31–37**) [41].

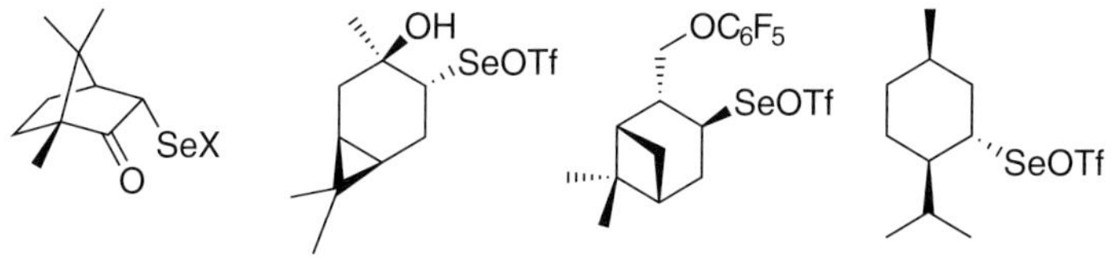

31 X = Cl
32 X = Br
33 X = OTf
34 X = OSO$_3$H

35

36

37

Scheme 1.8 Terpene-based electrophilc selenium reagents.

In some of these cases, X-ray analysis and DFT calculations showed the existence of an intramolecular heteroatom-selenium interaction, which seems to be an important factor for the chirality transfer in the transition state of the addition reactions.

Concerning this class of diselenides, it is possible to generalize that the facial selectivity produced by the aliphatic electrophilic selenium reagent is usually lower in respect to those obtained using an aromatic core.

1.1.2
Reactivity and Properties

In many aspects, the properties of organoselenium compounds are similar to those of the better-known sulfur analogues. However, the introduction of the heteroatom, the manipulation of the resulting molecules, and, in particular, the removal of the selenium-containing functions occurs under much simpler and milder conditions than those required for the corresponding sulfur compounds.

The reaction of selenium electrophiles with alkenes consists in a stereospecific *anti*-addition that involves the initial formation of a seleniranium ion intermediate. This is rapidly opened in the presence of a nucleophile that can be external, leading to the addition products, or internal, giving the corresponding cyclized derivative.

The intermediate seleniranium ion can be ring-opened to afford two different regioisomers. The regiochemistry usually follows the thermodynamically favored Markovnikov orientation even if examples of *anti*-Markovnikov addition were reported, as a consequence of the coordinating effect of a hydroxyl group in the allylic position.

When the reagent is chiral, a differentiation between the two faces of unsymmetrically substituted alkenes can be observed (Scheme 1.9). Depending on the reaction conditions, the formation of the seleniranium ion can be reversible and at low temperatures the reaction is under kinetic control.

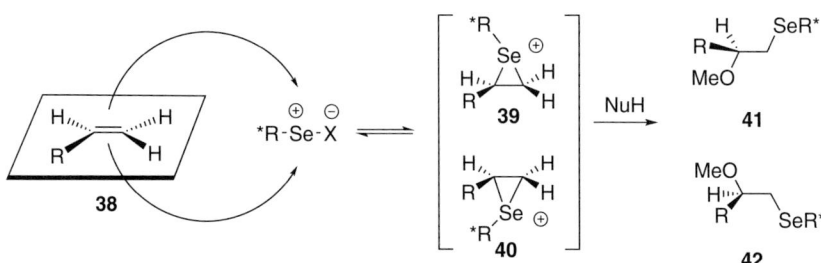

Scheme 1.9 Stereoselective addition.

Steric and electronic effects control the formation of the diastereoisomeric intermediates **39** and **40** and the ratio reflects the different stability between them. Subsequent reaction with a nucleophile affords the diastereoisomers **41** and **42** derived from the stereospecific ring opening of **39** and **40**, respectively.

The presence of an equilibrium between the starting materials and the seleniranium intermediates has been chemically demonstrated starting from the hydroxyselenides **43** and **44**, easily obtained by the nucleophilic ring opening of the corresponding optically pure (*R*)- and (*S*)-styrene epoxide (Scheme 1.10) [42].

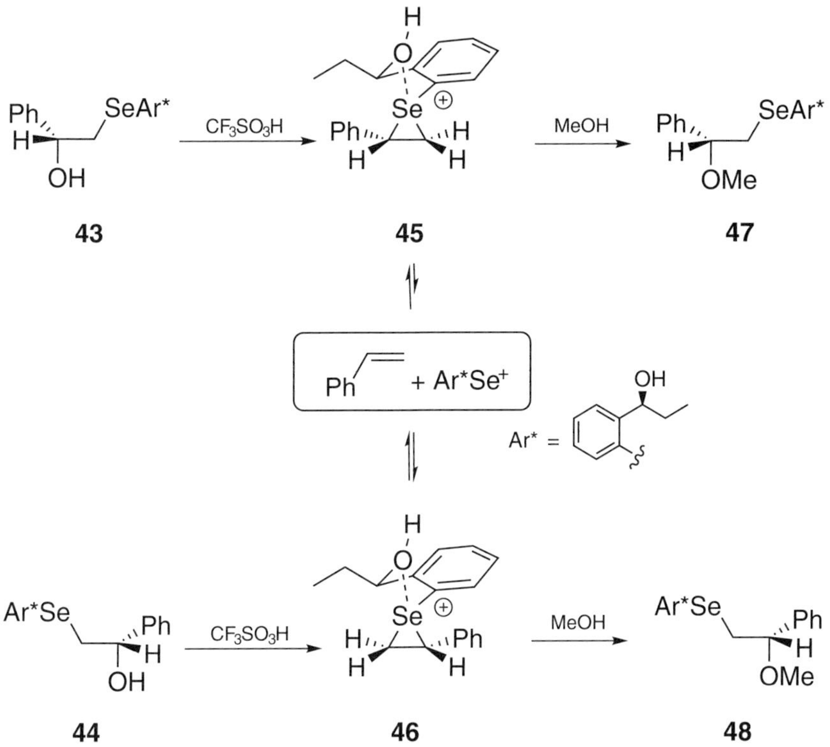

1.1 General Introduction

Scheme 1.10 Mechanism of the enantioselective addition reactions.

Treatment of **43** and **44** with trifluoromethane sulfonic acid generates selectively the corresponding seleniranium ions **45** and **46** according to an intramolecular S_N2 displacement of a molecule of water. In the case of **43**, only the seleniranium ion **45** is formed and the subsequent treatment with methanol affords the adduct **47** corresponding to the Re-attack of the selenium electrophile to the styrene double bond. *Ab initio* calculations on the stability of the intermediates indicated that **45** is more stable than **46**.

Under the same experimental conditions, the hydroxyselenide **44** produces the less-stable seleniranium ion **46**. After reaction with methanol, the formation of both isomers **47** and **48** in a 3:1 ratio clearly indicates a decomplexation–complexation mechanism that is involved in the above-mentioned equilibrium [43].

Even if the central tenet of this class of reaction is the formation of the high reactive seleniranium ions, only few examples were reported in which they have been independently synthesized and analyzed. Many studies have been performed with the assumption that the putative intermediate seleniranium ions are responsible for the observed products.

After seminal computational studies in which the enthalpic activation barrier for direct, intramolecular thiiranium–olefin and seleniranium–olefin transfer were compared for alkylthiiranium, arylthiiranium, and arylseleniranium ions [44] Denmark et al. reported a crossover experiment as the first direct observation of the transfer of a selenonium cation from one olefin to another [45].

On the basis of these observations, the authors suggest that the rapid olefin-to-olefin transfer represents one of the most likely pathways for racemization of the enantiomerically enriched seleniranium ions (Scheme 1.11).

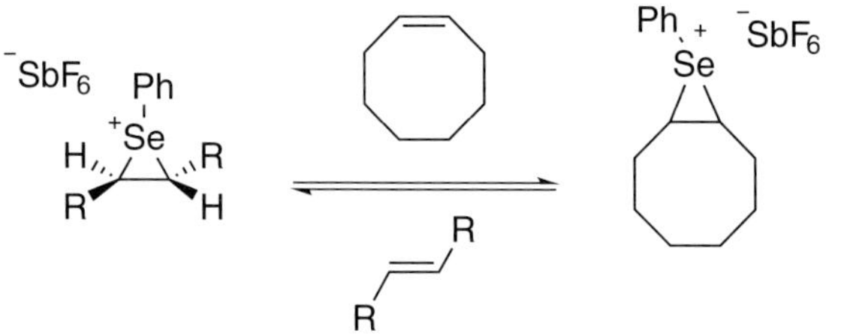

Scheme 1.11 Crossover experiment.

Our recent NMR investigations on the haloselenenylation of styrene derivatives [46], according to other evidence previously reported by Garrat [47], demonstrate the presence of a more complicated equilibrium when the nucleophile is chlorine or a bromine. The initial formation of the Markovnicov's adduct is very fast and rapidly reaches the equilibrium with the anti-Markovnicov regioisomer. NMR analysis of the equilibrium mixture, combined with kinetic investigation, evidenced the presence of two different intermediates in the formation of the Markovnicov and anti-Markovnicov product that were assigned to the episelenurane **49** and seleniranium ion **50**, respectively (Scheme 1.12).

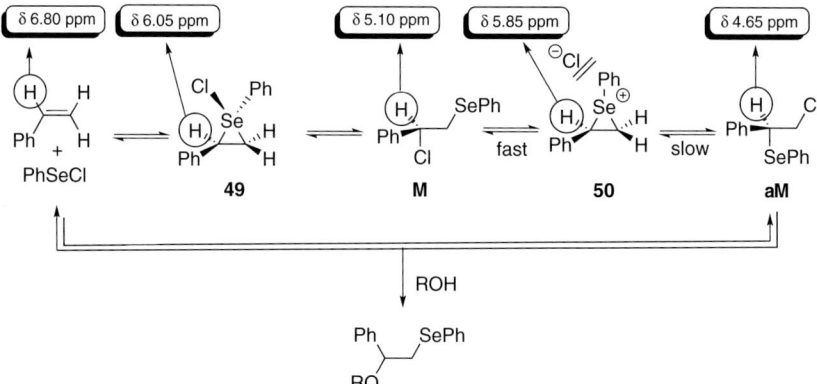

Scheme 1.12 Mechanism of the chloroselenenylation reaction.

When the anion is chloride, the treatment of the above-described equilibrium mixture with an oxygen-containing nucleophile such as methanol affords quantitatively the Markovnikov alkoxyselenide.

After 60 years since the first stereospecific selenoaddition reaction, some mechanistic aspects are still under investigation and a clear explanation of the species involved in these reactions should represent an important aspect in the development of more efficient and selective reactions.

Electrophilic selenenylation of dienes has not been extensively used and investigated. The first example of conjugated diene selenenylation reported that the reaction with PhSeCl gives either an allylic alcohol or an enone depending on the experimental procedure [48].

More recently, treatment of dienes with aryl selenenamides in the presence of phosphorus(V)oxyhalides has been proposed as a useful method to effect the 1,4-haloselenenylation of conjugated dienes. Similar experimental conditions applied to nonconjugated dienes afforded only mono-haloselenides (Scheme 1.13) [49].

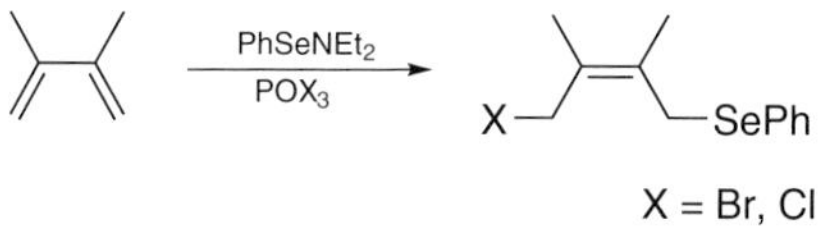

Scheme 1.13 Conjugated electrophilic addition to dienes.

Furthermore, the reaction of divinylsulfide with selenium dibromide prepared from selenium and bromine in carbon tetrachloride gives, in near quantitative yield, the six-membered heterocyclic compound 2,6-dibromo-1,4-thiaselano that underwent spontaneous rearrangement to a five-membered thioselenolan-derivative [50].

1.2
Addition Reactions to Double Bonds

1.2.1
Addition Reaction Involving Oxygen-Centered Nucleophiles

Addition reactions to double bonds promoted by electrophilic selenium reagents are usually carried out in the presence of a solvent that acts as external nucleophile. Simple and efficient procedures to introduce oxygen and nitrogen nucleophiles have been reported and are currently employed in the functionalization of olefins. In the first case, the process is named oxyselenenylation and leads to the introduction of hydroxy-, alkoxy-, or acetoxy groups.

PhSeBr and acetic acid in the presence of acetic anhydride and KNO_3 promote the acetoxyselenenylation of alkenes, but this methodology suffers of considerably low regioselectivity in the case of terminal olefins [51]. An alternative and more efficient procedure involves the oxidation of diphenyl diselenide with (diacetoxy iodo) benzene in acetonitrile [23].

Currently the most relevant developments in this field of research concern the use of optically pure electrophilic reagents in the asymmetric synthesis of alkoxy- and hydroxy derivatives. Methoxyselenenylation of styrene has been used by several groups as a test reaction to compare the diastereoselectivity induced by different chiral selenenylating reagents and to compare the effect of the different experimental conditions as well as the structural features of the reagents.

Selected examples are collected in Table 1.1 and, even if in all the cases the diastereoselectivity is usually good, some general considerations on the ability of various electrophilic reagents to transfer the chirality to the newly generated center can be attempted.

Considering the above-mentioned role of the heteroatom on coordinating to the electrophilic selenium atom, from the data reported in Table 1.1, sulfur atom (entries 10–11), in respect to oxygen and nitrogen atom (entries 1,2,8,9), seems to be more effective, leading to higher diasteromeric excesses at higher reaction temperatures. The results reported in entries 3–7 also suggest that the presence of a second chiral center could produce a positive effect in terms of diastereoselectivity. Using the electrophilic reagent derived from the oxidation of diselenide **20**, a match/mismatch effect related to the relative configuration of the two chiral centers has been described [52].

As shown in entries 8–9 and 10–11 reagents, with a methoxy group in the *ortho*-position to the selenium electrophile shows higher selectivities than unsubstituted one. Probably this arises from a different coordination of the side chain, even if detailed calculations [53] showed that the strength of coordination does not correlate with the trend of the experimentally observed selectivities. Probably the situation is different when the chiral moiety is a sulfoxide. In this case in fact, as already reported, no selenium heteroatom interaction was observed and the methoxy selenenylation of styrene proceeds with moderate diasteroselectivity (d.r. 6:1). For this reagent, the introduction of a substituent (MeO-) in the *ortho*-position to the electrophilic selenium atom dramatically reduces the yield (24%) as well as the diasteromeric excess (2:1).

The information gained from these experiments can be used as preliminary indications even if a strong dependence on the alkene structure as well as on the nucleophile and the solvent must be taken in consideration.

As an illustrative example when the conditions reported on entry 12 were applied to effect the methoxyselenenylation of 2-chlorostyrene or β-methyl styrene, the selectivity increases affording a diasteromeric ratios of 11:1 in dichloromethane. Other solvents such as THF, diethyl ether, and chloroform give considerably lower selectivity. Concerning the nucleophile, slight differences in selectivity have been observed using different alcohols, and this suggests that probably different nucleophiles coordinate in different ways to the selenium electrophile. Using the selenenyl triflates generated from oxygen-containing diselenides (**21**,

Table 1.1 Asymmetric methoxyselenenylation of styrene.

Entry	R*	X	T (°C)	de (%)	Yield (%)	Reference
1	*o*-(CH(NMe₂))C₆H₄–	TfO⁻	0	35	64	[34c]
2		HSO₄⁻	25	65	80	[34c]
3	imidazolidine (Ph, N)	TfO⁻	−100	86	98	[34f]
4		HSO₄⁻	−15	86	74	[34f]
5		HSO₄⁻	25	76	95	[34f]
6	*o*-(CH(N(Me)Ph))C₆H₄–	TfO⁻	−78	90	40	[51]
7		HSO₄⁻	25	94	70	[51]
8	*o*-(CH(OH))C₆H₄–	TfO⁻	−100	83	67	[34e]
9	2-OMe,6-CH(OH)-C₆H₃–	TfO⁻	−100	96	55	[34e]
10	*o*-(CH(SMe))C₆H₄–	TfO⁻	−78	92	80	[36]
11	2-OMe,6-CH(SMe)-C₆H₃–	HSO₄⁻	−30	96	72	[37]
12	*o*-(S(O)tBu)C₆H₄–	TfO⁻	−78	70	52	[38]

22), the selenenylation of styrene with methanol is more selective than with acetic acid.

Scianowsky et al. reported several examples of methoxyselenenylation mediated by terpene-based electrophilic selenium reagents [41f–n]. A large screening of p-menthane derivatives indicates that the diastereoselectivity is strongly correlated to the stability of the chair conformers of the terpeneselenenyl bromide as calculated using density functional theory [41h].

The electrophilic reagent **32** proceeds with the methoxyselenenylation of styrene at −78 °C in 86% yield with only 40% de [41a–e]. The results obtained with **35** and **36** indicate that a more rigid bicyclic structure and the presence of a coordinating heteroatom increase the diastereomeric excess at the price of a reduced reactivity (using **35**: 54% yield, 80% de; using **36**: 53% yield, 72% de) [41m].

An important role, even if not yet completely understood, is played in the selenenylation reaction by the counterion. On the basis of some results reported by Tomoda [32p], it can be suggested that a decrease in the nucleophilicity of the counterion reasonably correspond to an increase in the electrophilicity of the selenium reagent, and hence to an increased selectivity which is reflected in higher diastereomeric excesses.

The reaction of methyl cinnamate with different camphor-derived selenenyl reagents (**31–34**) at room temperature in methanol follows a similar trend (see Table 1.2). It is interesting to note that in the case of the sulfate **34**, the addition occurs with the opposite facial selectivity in respect to that observed with other anions [54].

Similarly, the selectivity changes passing from the bromide **32** to the sulfate **34** for some linear olefins like (E)-β-methylstyrene, (E)-4-octene, and methyl-(E)-3-hexenoate, while for Z-symmetric alkenes, such as cyclohexene and cycloctene, the facial selectivity resulted to be unchanged. In this latter case, a different mechanism for the stereoselection can be envisioned: the electrophile attacks the (Z)-olefins leading to the formation of a *meso*-seleniranium ion. Consequently the

Table 1.2 Effect of the counterion on the facial selectivity [54].

X	T (°C)	Time (h)	Yield (%)	d.r.
Cl (31)	0	4	49	40:60
Br (32)	0	2	56	35:65
OTf (33)	0	5	77	42:58
OSO$_3$H (34)	25	24	94	85:15

1.2 Addition Reactions to Double Bonds

ring-opening reaction mediated by the nucleophile becomes the stereoselective reaction step. Since this occurs far from the chiral environment of the electrophilic reagent, this can explain the generally lower facial selectivity observed for these substrates (Scheme 1.14).

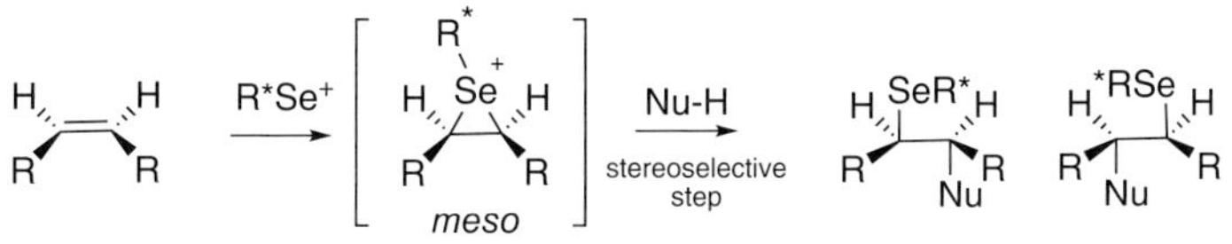

Scheme 1.14 Mechanism for asymmetric addition on Z-olefins.

Enantiomerically pure electrophilic selenium reagents prepared from diselenide **24** can be used to perform the kinetic resolution of allylic alcohols. The methoxyselenenylation of **51** effect using 0.5 equivalents of selenenylating reagent affords the formation of the corresponding addition products (e.g., **52**) with a facial selectivity ranging from 95:5 to 98:2. The reaction is regiospecific and the resolution resulted to be more efficient when aryl-substituted olefins were employed as substrates. From the crude, the unreacted alcohol can be recovered in optically enriched form with 90–94% ee.

In order to complete the resolution, the arylseleno and the methoxy groups were eliminated from **52** by treatment with a catalytic amount of CF_3SO_3H in the presence of styrene. The regio- and stereoselectivity observed on different substrates have been correlated to a stabilizing interaction between the allylic oxygen and the selenium atom (Scheme 1.15) [55].

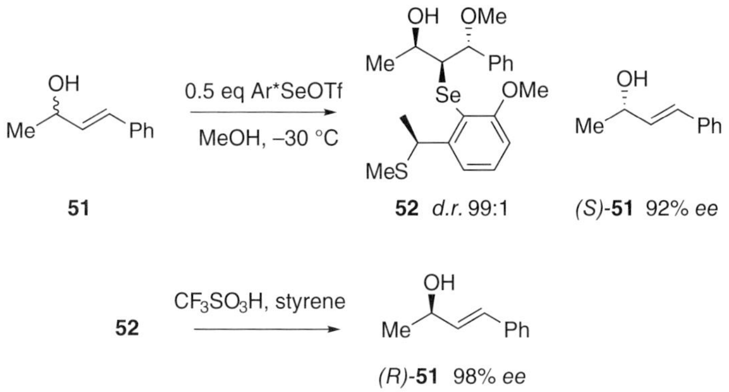

Scheme 1.15 Kinetic resolution of allylic alcohols.

Methoxyselenenylations of α,β-unsaturated aldehydes were first reported by Paulmier in 1991 using PhSeCl in MeOH at −30 °C [56]. The reaction proceeds with poor regio- and stereoselectivity through the formation of a dimethylacetal

1 Electrophilic Selenium

intermediate in which the double bond is a better nucleophile because it is no longer stabilized by conjugation and undergoes electrophilic addition via the seleniranium ion intermediate.

Very recently, the asymmetric version has been proposed using the chiral reagent **28** in the presence of MgSO$_4$–methanol at 0 °C. The reaction in this case resulted in the stereoselective formation of mainly one of the four possible isomers (Scheme 1.16) [57].

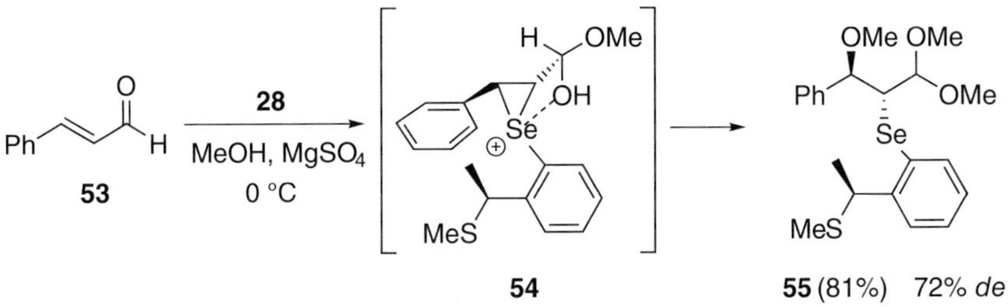

Scheme 1.16 Methoxyselenenylation of α,β-unsaturated aldehydes.

A series of evidences demonstrated that probably in this case, the formation of the dimethylacetal is not the first reaction step and the methoxyselenenylation occurs on the corresponding hemiacetals **54**, in which a nonbonding interaction between seleniranium ion and the hydroxyl group in the allylic position controls the regiochemistry as well as the stereochemistry of the process.

When the electrophilic selenium addition is effected using water instead of methanol as external nucleophile, the introduction of a hydroxyl group and the synthesis of alcohols can be easily achieved.

Chiral selenenyl sulfates prepared by oxidation of the corresponding diselenides with ammonium persulfate react with alkenes in acetonitrile in the presence of traces of water affording the hydroxyselenenylation products, generally in good yields and with a diastereoselectivity that mainly depends on the combination of reagents and substrates.

In order to effect hydroxyselenenylations reactions, the sulfate counteranion presents two important advantages: (1) it can be conveniently prepared in aqueous media, and (2) the reaction promoted by selenenyl sulfate derivatives can be performed at higher temperatures with respect to those required by other counteranions (e.g., from −30°/+40° against −100°/−78° for triflate analogues).

Selected results are summarized in Table 1.3 in order to compare the reactivity and the selectivity of three different chiral scaffolds. The hydroxyselenenylation of mono- and disubstituted olefins affords regiospecifically a couple of hydroxyselenides deriving from an *anti*-stereospecific addition, with good yields and moderate to excellent facial selectivity.

Camphor selenenyl sulfate (**34**) showed better selectivity for the alkyl-substituted substrates in contrast to sulfur-containing aryl selenenyl sulfate (**56**) that resulted

Table 1.3 Asymmetric hydroxyselenenylation.

			34 [58]	56 [36, 37]	57 [53]
			d.e. (%) (yield (%))	d.e. (%) (yield (%))	d.e. (%) (yield (%))
a)	PhCH=CH$_2$ → PhCH(OH)CH$_2$SeR* (R*SeOSO$_3$H, MeCN/H$_2$O)		30 (68)	96 (65)	90 (40)
b)	PhCH=CHMe → PhCH(OH)CH(Me)SeR* (R*SeOSO$_3$H, MeCN/H$_2$O)		30 (62)	96 (70)	62 (82)
c)	cyclohexene → trans-2-(SeR*)-cyclohexanol (R*SeOSO$_3$H, MeCN/H$_2$O)		56 (68)	44 (60)	–
d)	C$_4$H$_9$CH=CHC$_4$H$_9$ → C$_4$H$_9$CH(OH)CH(C$_4$H$_9$)SeR* (R*SeOSO$_3$H, MeCN/H$_2$O)		88 (89)	76 (79)	–

in higher selectivities when the carbon–carbon double bond is substituted with a phenyl group. Concerning the sulfur-containing reagent, it was demonstrated that the presence of the methoxy group in the *ortho*-position in respect to the electrophilic selenium, as well as for the methoxyselenenylation reaction, effects a positive influence on the selectivity giving higher diastereomeric ratios.

Starting from cyclohexene is not surprising to observe that a low selectivity is produced by both reagents. In this case, as previously outlined, the first step leads to a *meso*-seleniranium ion and the stereoselection arises from the subsequent attack of a molecule of water, responsible for the seleniranium ring-opening reaction. This latter event occurs far from the chiral environment of the intermediate and it is reasonably less sterically and electronically controlled.

The reaction temperature is usually chosen based on the reactivity of the reagent, looking for the best compromise between reaction time and selectivity. Comparing sulfate derivatives, the sulfur-containing electrophilic selenium reagent showed the best reactivity: the reactions are fast even at −30 °C affording the best diastereomeric ratio. On the contrary, when the coordinating heteroatom is a nitrogen atom (**57**), the reactivity is strongly reduced and the reaction occurs only at 25 °C. These experimental conditions are enough to affect the hydroxyselenenylation of styrene derivatives but not of aliphatic olefins and the higher temperature is probably responsible for the reduced selectivity.

A promising development of these methodologies is certainly represented by the possibility to affect selenenylation and deselenenylation in *one-pot* using an excess of a suitable oxidant. This approach has received high attention during the last years from several research groups.

The first example was reported by Torii who effected the catalytic conversion of alkenes into allylic ethers or esters using catalytic amounts of diphenyl diselenide [59].

The strategy is to introduce a new functionality in a substrate after electrophilic activation of a carbon–carbon double bond and subsequently remove the organoselenium function accordingly with a two-step *one-pot* procedure.

The first step, depending on the experimental conditions, consists of α-methoxyselenenylation or hydroxyselenenylation and leads to an α-alkoxy or hydroxy selenide. In the second step, the selenium moiety is transformed into a good leaving group which in turn is removed through α,β-elimination or, less frequently, through direct nucleophilic substitution.

During the second step, the electrophilic selenenylating reagent is regenerated, allowing the use of a catalytic amount of the selenium-containing reagent (Scheme 1.17).

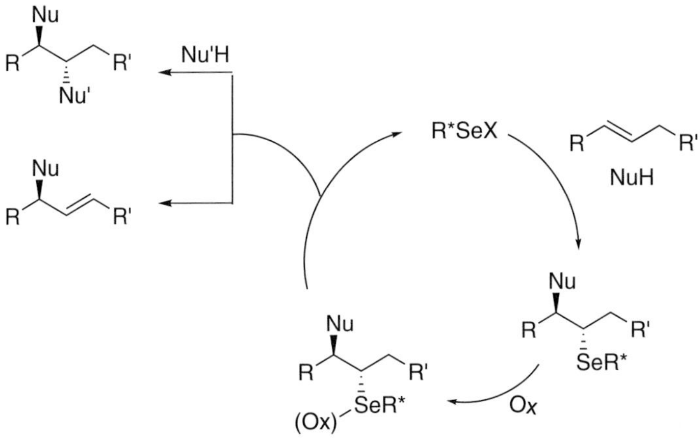

Scheme 1.17 One-pot selenenylation–deselenenylation sequence.

Several research groups have recently investigated the asymmetric version of the catalytic *one-pot* selenenylation–deselenenylation sequence. The asymmetric version of this experiment can be simply effected by replacing diphenyl diselenide with an optically pure selenenylating reagent. All these aspects will be fully treated in Chapter 9 and have been very recently object of some reviews [7h, i]. Here just some general concepts will be shortly introduced as an example of electrophilic selenium reagents.

Tiecco and coworkers reported that ammonium persulfate was the best oxidant for the *one-pot* selenenylation–deselenenylation reaction and applied this strategy to effect the conversion of β,γ-unsaturated esters, amides, and nitriles into γ-alkoxy or γ-hydroxy α,β-unsaturated derivates, also using a catalytic amount of diphenyl diselenide [60].

The selenenyl sulfate is produced *in situ* from the reaction of the corresponding diselenide with ammonium persulfate. It reacts with the alkenes to give alkoxy- or hydroxyselenenylation products. The reaction of these addition products with ammonium persulfate in excess is suggested to generate a selenonium ion which affords, by elimination, the observed reaction products and regenerates the electrophilic reagent.

An electron-withdrawing group in the allylic position is essential for the success of the reaction since simple unsubstituted alkenes gave rise to product mixtures. It was supposed that the driving force of the process is the stabilization by conjugation of the double bond formed during the deselenenylation.

With chiral diselenides, asymmetric catalytic reactions can be performed. Table 1.4 reports some results that put in evidence that passing from camphor selenenyl

Table 1.4 Asymmetric *one-pot* methoxyselenenylation deselenenylation sequence.

Catalyst	% cat	R	EWG	R'	Yield (%)	ee (%)	Ref.
34 (SeOSO$_3$H, camphor)	100	Ph	COOMe	Me	72	65	[61]
	100	Et	COOMe	Me	71	86	[61]
	100	Me	COOMe	Me	60	70	[61]
	100	Me	CN	Me	60	70	[61]
56 (N-Ph, SeOSO$_3$H)	10	Ph	COOMe	Me	12	94	[53]
	100	Ph	COOMe	Me	50	94	[53]
57 (SMe, SeOSO$_3$H, OMe)	5	Ph	COOMe	Me	98	78	[37]
	5	Et	COOMe	Me	98	68	[37]
	5	Ph	COOMe	H	98	55	[37]

Reaction scheme: R–CH=CH–EWG + (R*Se)$_2$ / (NH$_4$)$_2$S$_2$O$_8$ / R'OH → R–CH(OR')–CH=EWG

sulfate **34** [61] to aromatic selenenyl analogues **56** [36, 37] and **57** [53], probably the mechanism for the catalytic stereoselection is different. In the first case, in fact, to obtain a good facial selectivity a stoichiometric amount of selenenylating reagent is necessary even if, at the end of the reaction, camphoryl diselenide can be recovered quantitatively.

Regarding the aromatic reagents **57** and **56**, as already observed for the addition reactions, the nitrogen-containing derivative **57** is less reactive than those in which the electrophilic selenium is coordinated by a sulfur (**56**). In the first case, a stoichimetric amount of reagent is required in order to accelerate a very slow reaction but the amount of selenenylating agent did not affect the diastereoselectivity. In the latter one, using as electrophilic reagent **56**, moderate to good enantiomeric excesses were obtained with only 5% of the catalyst even at 20 °C.

A second possibility than a β-elimination of the oxidized selenium moiety from the selenenylation intermediate is the direct nucleophilic substitution resulting in a double functionalization of the double bond. In a very recent example, the first direct selenium-mediated dihydroxylation of olefins has been achieved with this approach [62]. The oxidant employed is a persulfate and in the proposed mechanism, the oxidized selenium moiety is substituted by a molecule of water. The diastereoselectivity (*syn* vs *anti*) is strongly dependent on the nature of the substrate and the method proved to be efficient even when a sulfur-containing chiral diselenide was used instead of diphenyl diselenide, affording the corresponding diols with an enantiomeric excess that strongly depends on the amount of catalyst. A more efficient and eco-friendly procedure has been also reported using hydrogen peroxide as oxidant but a completely different reaction mechanism is involved in this case [63].

Similarly diphenyl diselenide and ammonium persulfate have been also successfully used to oxidize carbon–carbon triple bonds (**58**) [64]. The reactions (see Scheme 1.18), performed in aqueous media, lead to 1,2-unprotected dicarbonyl

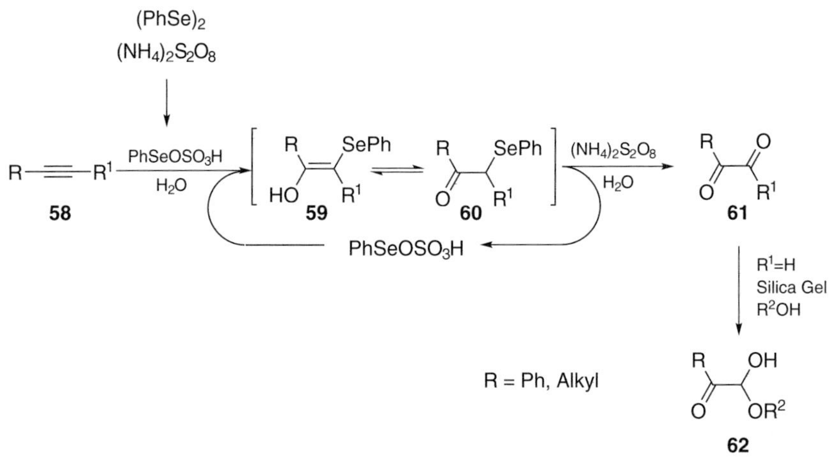

Scheme 1.18 One-pot oxidation of alkynes.

derivatives **61** starting from internal triple bonds and to α-keto hemiacetals **62** when the substrate contained a terminal alkyne. The reaction proceeds in *one-pot* starting with the hydroxyselenenylation of the triple bond and the formation of enol **59** that exists in equilibrium with the corresponding α-phenylseleno ketone **60**. The latter, in the presence of an excess ammonium persulfate and water, undergoes the nucleophilic substitution of the phenylseleno group with an hydroxyl group that is subsequently oxidized to a ketone or aldehyde, depending on the alkyne **58** (Scheme 1.18).

During the second step, the PhSe–sulfate is regenerated and the reaction can be performed using a catalytic amount of diselenide in the presence of an excess ammonium persulfate.

Electrophilic selenium reagents can also be used to functionalize allenic systems and usually the reaction is highly regio- and stereoselective.

The 2,3-allenoates **63** treated with PhSeCl in MeCN gave the corresponding 3-phenylseleno-4-oxo-2(*E*)-alkenoates. The addition of Li$_2$CO$_3$ as base improved the yield and the selectivity of the reaction. A possible mechanism involves a decomposition of selenate esters, as shown in Scheme 1.19. The electrophile PhSeCl reacted with the relatively electronrich C=C bond in 2,3-allenoates **63** forming the seleniranium ion intermediates **64**, which may be attacked by a molecule of water in a process facilitated by the presence of Li$_2$CO$_3$. The hydroxyselenenylation intermediate **65** subsequently reacts with another molecule of PhSeCl in the presence of Li$_2$CO$_3$ to produce H$_2$O and the selenate esters **66**, which may decompose to form the ketone functionality in 3-phenylseleno-4-oxo-2(*E*)-alkenoates **67**.

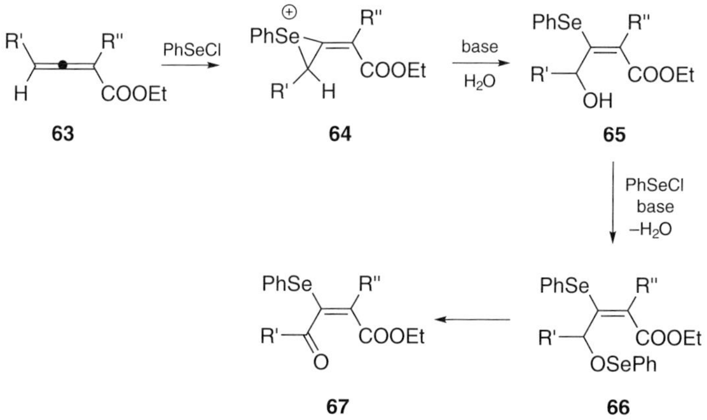

Scheme 1.19 Hydroxyselenenylation of allenes.

By increasing the amount of water, a competitive cyclization to the corresponding butenolides has been observed and represents an important criticism for this procedure.

The observed stereoselectivity may be explained by the favorable *trans*-arrangement of the phenylseleno and the ester group [65].

More recently, Ma and coworkers deeply investigated halo- and selenohydroxylations of allenyl phosphane oxides [66].

Their results confirmed that the electrophilic moiety adds to the central carbon atom of the allene and the hydroxyl group always attacks in the 3-position respect to the phosphorus atom.

All these reactions are characterized by an excellent regio- and (*E*)-stereoselectivity that should be correlated to the neighboring-group participation ability of the phosphane oxide [67].

High regio- and (*Z*)-stereoselectivity was obtained by the same authors in the chloro selenenylation of (1- and 3- monosubstituted 1,2-allenyl) phosphonates when the reaction is carried out using PhSeCl as selenenylating reagent.

1.2.2
Addition Reaction Involving Nitrogen-Centered Nucleophiles

In consideration of the biological and pharmaceutical relevance of nitrogen-containing compounds, the addition reactions of an electrophilic selenium reagent and a nitrogen nucleophile to unsaturated substrates represent synthetically important processes with potential practical applications.

Toshimitsu and Uemura first reported a Ritter-type amide synthesis [68]. Treatment of an olefin with PhSeCl in acetonitrile and water in the presence of trifluoromethane sulfonic acid affords the simultaneous insertion of a phenylselenenyl and of an acylamino group to a carbon–carbon double bond. The reaction proceeds in poor yield on styrene and electron-rich olefins. In these cases, better yields were achieved starting from the corresponding hydroxy- or methoxyselenenylated products by treatment with water, trifluoromethansulfonic acid, a stoichiometric amount of nitriles, and using 2-pyridylselenyl chloride as selenenylating agent [69]. Similar reactions can be effected using different nitriles, like propionitrile, butyronitrile, benzonitrile, and ethyl cyanoacetate affording the corresponding selenoamides.

An asymmetric variant of this procedure has been proposed using camphorselenenyl sulfate as selenenylating agent. Even if the amidoselenenylation reaction occurs with moderate facial selectivity, the two diastereomeric addition products **68** and **69** can be completely separated by chromatography and obtained in optically pure form (Scheme 1.20) [70].

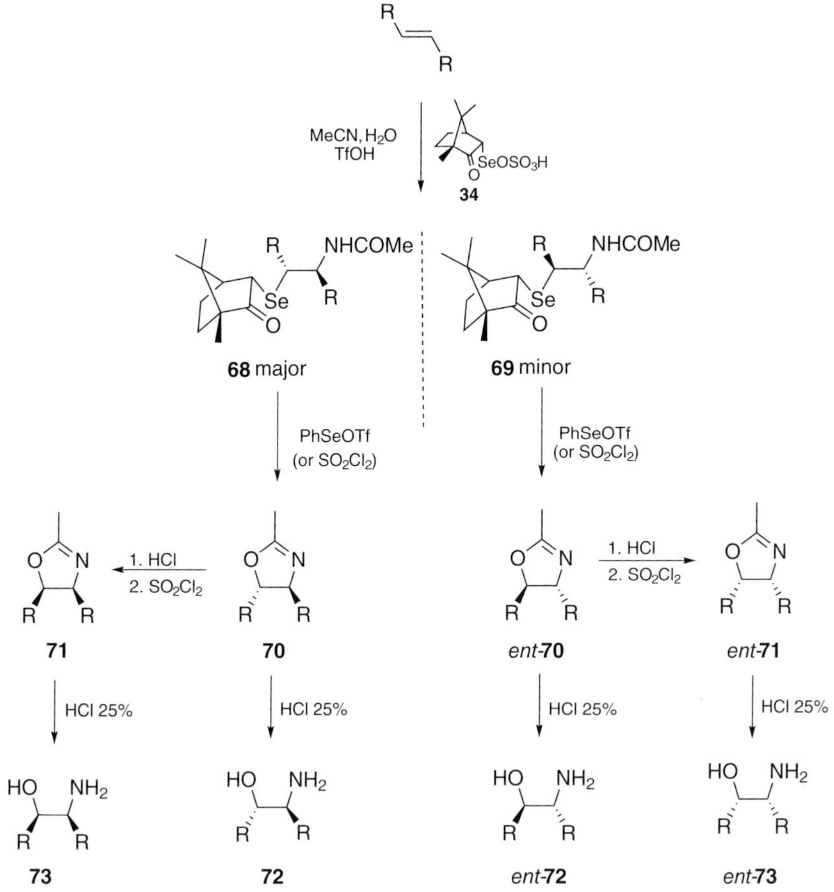

Scheme 1.20 Asymmetric amidoselenenylation and further stereospecific manipulations.

These compounds, subjected to deselenenylation, afforded enantiomerically pure derivatives.

The deselenenylation consists of the activation of the selenium moiety as a leaving group by transforming the selenide into selenonium ions by treatment with electrophilic selenenylating reagents or with SO_2Cl_2.

In these activated substrates, the presence of a suitably positioned nucleophilic substituent allows an intramolecular nucleophilic displacement that affords the corresponding heterocyclic derivatives. This intramolecular substitution occurs with inversion of configuration at the carbon atom bearing the selenium atom leading to the formation of the 4,5-*trans* oxazolines (**70**, *ent*-**70**); the hydrolysis and further treatment with thionyl chloride gave the corresponding *cis*-isomers (**71**, ent-**71**). Further manipulation consisted in the acidic hydrolysis of the four optically pure oxazolines in order to prepare the corresponding enantiomerically pure β-amino-alcohols. (**72**, ent-**72**, **73**, ent-**73**).

Enantiomerically pure thiazolines were prepared using a similar procedure. The mixture of diastereoisomeric amido selenides, treated with Lawesson's reagent, afforded the corresponding thioamido derivatives that were subjected to deselenenylation with the formation of a thiazoline structure [71].

Among the addition reactions which involve nitrogen nucleophiles, the azidoselenenylation of alkenes is one of the most important because both azido and phenylseleno group can give rise to several useful transformations. In addition to the already described manipulations of the selenenyl derivatives, the azido group can react with both nucleophilic and electrophilic reagents and can be used in 1,3-dipolar cycloaddition reactions. Krief first reported the azidoselenenylation of alkenes using a two-step *one-pot* procedure: a reaction with PhSeBr followed by the addition of sodium azide in DMF [72]. A similar reaction can be effected using PhSeCl, NaN$_3$ in DMSO or PhSe-pthalimide, azidotrimethylsilane in dichloromethane, even if the reactions carried out under these conditions suffer from very low regioselectivity [73]. Some years ago, Tiecco and coworkers demonstrated that the use of PhSeOTf and sodium azide in acetonitrile allows the synthesis of β-phenylselenoazides. The reaction is a stereospecific *anti*-addition in every case and the regiochemistry is determined by the structure of the starting alkenes. Complete regio- and stereoselectivity was observed in the azidoselenenylation of phenyl-substituted alkenes **74** when the reaction was carried out at 0 °C, obtaining the formation of the Markovnikov isomer **75** (Scheme 1.21) [74].

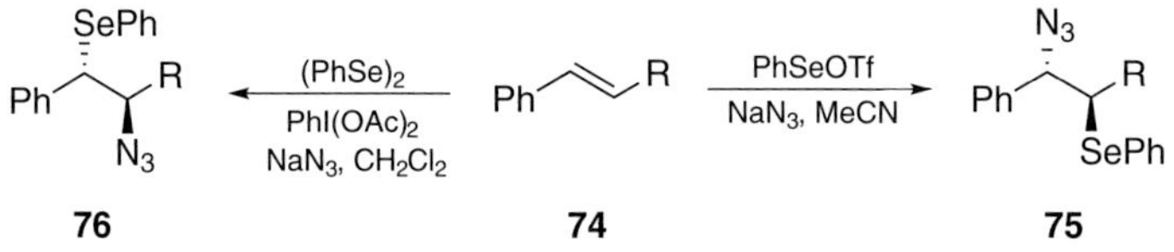

Scheme 1.21 Regiospecific azidoselenenylation reactions.

The *anti*-Markovnikov regioisomer **76** can be obtained by reacting the alkenes with PhSeSePh, NaN$_3$, and PhI(OAc)$_2$. In this case, the formation of the β-phenylseleno azides is the result of a radical addition initiated by the azido radical which is not stereospecific [75].

The electrophilic azidoselenenylation of *tert*-butyl cinnamyl ether has been used to prepare an intermediate that could be employed in the synthesis of the taxol side chain.

1.2 Addition Reactions to Double Bonds | 25

Asymmetric azidoselenenylation reactions were proposed using the selenenyl triflate prepared from the optically pure di-2-methoxy-6-([1S]-1-[methylthio]ethyl) phenyl diselenide (**24**) [76].

This reaction is stereo- and regiospecific and it is remarkable that it occurs with a high level of facial selectivity affording compounds **77a–e** in diasteromeric excess of up to 90% (Scheme 1.22).

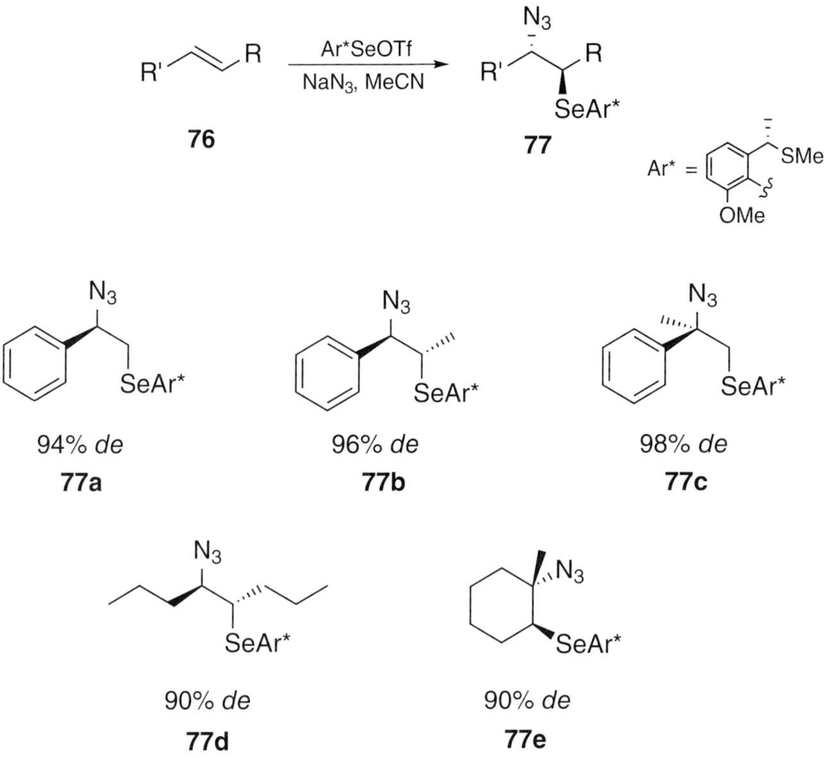

Scheme 1.22 Asymmetric electrophilic azidoselenenylations.

The synthetic relevance of the enantiomerically enriched β-arylselenoazides has been shown using these compounds as substrates for a subsequent synthesis of a series of optically active nitrogen-containing compounds such as amides **78**, aziridines **79**, oxazolines **80**, and triazoles **81–82** (Scheme 1.23).

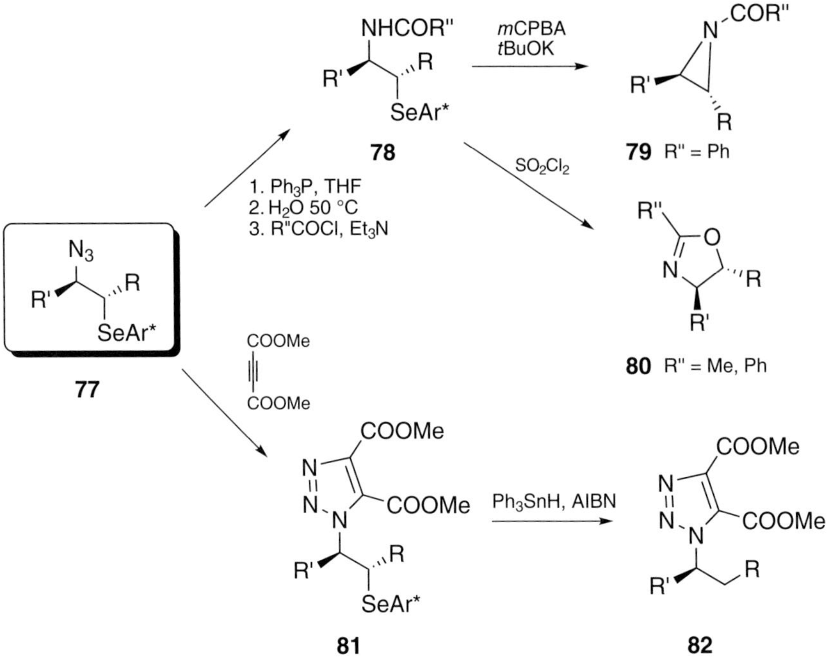

1 Electrophilic Selenium

Scheme 1.23 Stereospecific manipulation of optically pure azides.

The azido group can be reduced to the corresponding amines by treatment with Ph$_3$P and H$_2$O at 50 °C and subsequently by treatment *in situ* with a suitable acid chloride into the corresponding amido selenides **78**. Subsequent oxidation with *m*CPBA afforded the corresponding selenoxide that underwent spontaneous deselenenylation with the formation of the optically active aziridine **79**.

In a different way, acetamido and benzamido selenides can be deselenenylated by treatment with SO$_2$Cl$_2$; in this case the reaction leads to the formation of oxazolines **80**.

Enantiomerically enriched azides can also be employed in 1,3-dipolar cycloadditions to allow the synthesis of triazol **81**. After the cycloaddition reaction, the deselenenylation can be effected with triphenyltinhydride and AIBN affording **82**.

1.2.3
Addition Reactions Involving Carbon-Centered Nucleophiles

Toshimitsu and coworkers reported in 1996 the first stereospecific carbon–carbon bond formation by the reaction of a chiral seleniranium ion **84** with carbon-centered nucleophiles such as alkenyl silyl ethers, trimethylsilyl cyanide, allyltrimethylsilane by the use of a 2,4,6-tri-*tert*-butylphenyl (TTBPSe) group bound to the selenium atom (Scheme 1.24) [77].

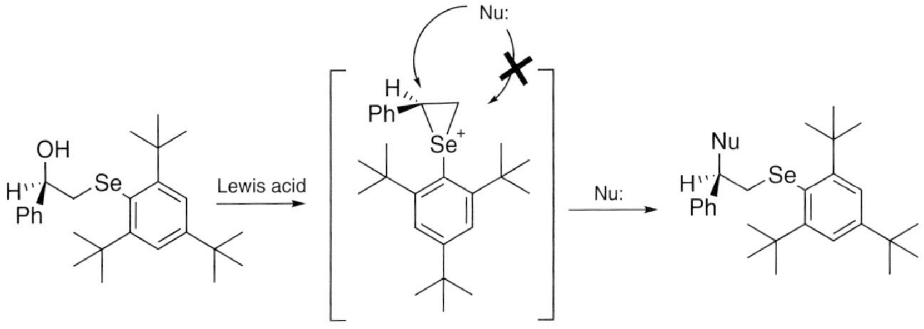

Scheme 1.24 Stereospecific carbon–carbon bond formation.

The steric protection of the selenium atom, deriving from the presence of the *tert*-butyl groups, has been proposed as necessary to prevent both the racemization of the stereogenic carbon atom and the attack of carbon-centered nucleophiles on the selenium atom.

Later, also aromatic compounds were used as nucleophiles and it was discovered that in this case the carbon–carbon bond forming reaction proceeds even without the steric protection of the selenium atom, and when a 2-pyridyl group is employed as the substituent, the stereospecificity of the reaction depends greatly on the reactivity of the aromatic compounds (Scheme 1.25) [78].

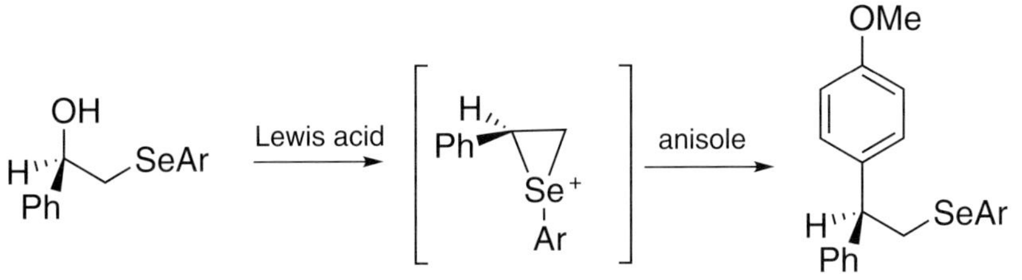

a Ar = Ph   85% yield, 2% ee
b Ar = 2-pyridyl   89% yield, 95% ee
c Ar = 2,4,6-t-Bu$_3$C$_6$H$_2$ 85% yield, 98% ee

Scheme 1.25 Aromatic carbon-centered nucleophile.

For example, starting from optically active hydroxyselenide **83**, prepared by the nucleophilic ring-opening reaction of the corresponding enantiomerically pure epoxide, the seleniranium ion **84** can be selectively generated by treatment with BF$_3$·OEt$_2$. When the nucleophile is anisole, selenophilic attack is rarely observed and the attack of **84** produces the addition product **86**. The selectivity in the addition reaction increases when a bulky substituent is introduced in the *ortho*-position of the aryl group attached to the selenium atom **86c** or when an electron-deficient aryl group is directly bound to the selenium atom **86b**.

Asymmetric carboselenenylation reactions of olefins can be formed in good yields and high diastereoselectivity using the electrophilic selenenyl triflate **87** characterized by a C2-symmetric structure (Scheme 1.26) [79].

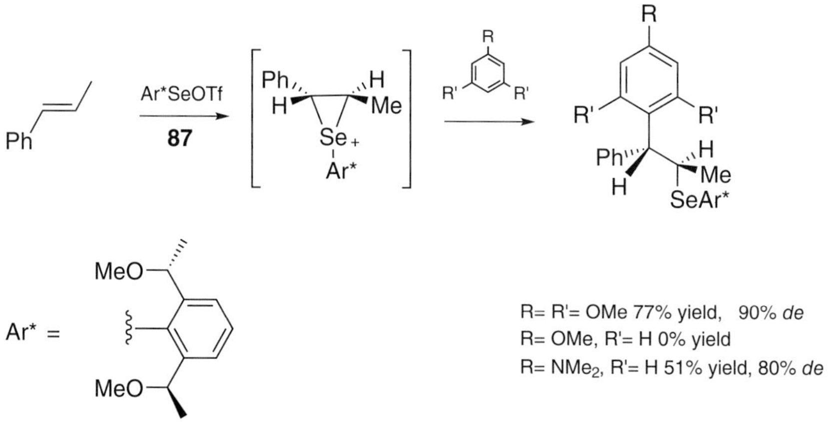

R= R'= OMe 77% yield, 90% de
R= OMe, R'= H 0% yield
R= NMe$_2$, R'= H 51% yield, 80% de

Scheme 1.26 Asymmetric carboselenenylation.

Not only heteroaromatic compounds but also electron-rich benzene derivatives such as *N,N*-dimethylaniline can be used as nucleophiles with selectivities ranging from 78% to 90% diasteromeric excess.

1.2.4
Addition Reaction Involving Chiral Nucleophiles or Chiral Substrates

In the previous section, we described several examples regarding the use of chiral organoselenium reagents to affect asymmetric synthesis. Here we are going to take in consideration that other strategies could be used to control the absolute configuration of the stereocenters generated by the addition process.

The selenenylation reaction of alkenes in the presence of enantiomerically pure nucleophiles has been employed in the preparation of cyclitols [80] as well as in the synthesis of carbohydrate derivatives (Scheme 1.27) [81].

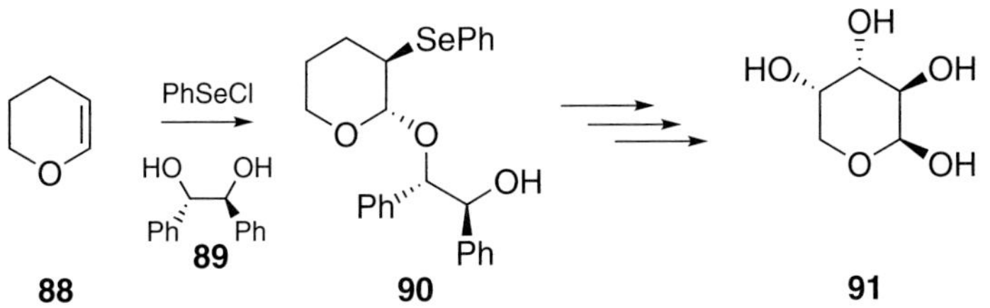

Scheme 1.27 Synthesis of L-Arabinose.

Oxyselenenylation reaction of 3,4-dihydro-2H-pyrane **88** with (S,S)-hydrobenzoin **89** affords a 1.2:1 mixture of (1S,2R)-oxyselenide **90** and its (1R,2S) diastereomer. After chromatographic separation, **90** was subjected to a series of stereoselective transformations affording enantiopure L-Arabinose **91**. In the same manner, D-Arabinose can be synthesized starting with the (1R,2S)-oxyselenide.

In another example, the regio- and stereospecific *anti*-addition of the enantiomerically pure diols **89**, **92**, **93** to a substituted alkene **94** promoted by N-(phenylseleno)phthalimide afforded enantiomerically pure diasteromeric selenoethers **95** and **96**.

These were separately transformed into the corresponding allylic ethers and finally into the four possible optically pure 1,4-dioxanes **97** by an intramolecular conjugated addition promoted by NaH (Scheme 1.28) [82].

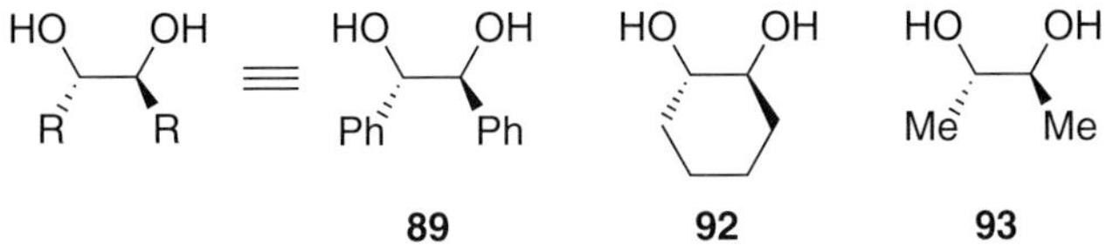

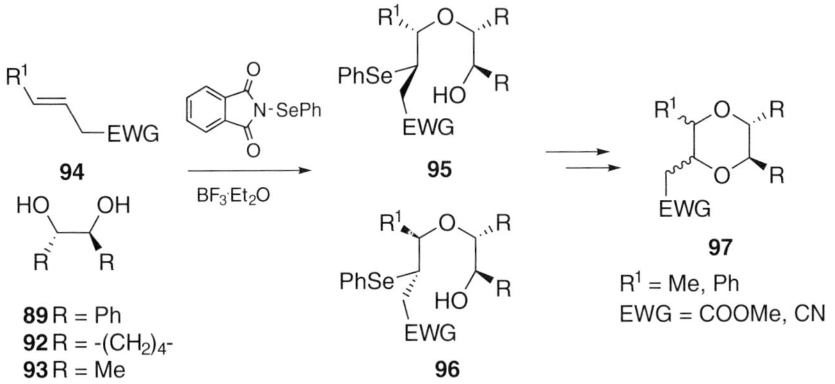

Scheme 1.28 Synthesis of the enantiomerically pure 1,4 dioxanes.

With N-Boc protected (R)-phenyl glycinol as nucleophile, enantiomerically pure 2,3,4-trisubstituted morpholines were prepared [83]. In this case, it is noteworthy that the addition reaction promoted by the electrophilic selenium reagent is completely chemiospecific and no traces of amidoselenenylation products were detected.

More recently, according to the "chiral pool approach" in asymmetric synthesis, mono- and bicyclic terpenes have been used as chiral substrates for azido- and hydroxyselenenylation reactions affording enantiomerically pure derivatives that represent interesting chiral-chelating auxiliaries for metal catalyzed conversions [61, 84].

The use of chiral auxiliaries covalently attached to the alkene has been also reported in the methoxyselenenylation of 2-vinylperhydro-1,3-benzoxazines **98**. The reaction afforded the α-methoxy selenides **99** and **100** with complete regio- and stereoselectivity and with a facial discrimination that, as well as the yields, are strongly dependent on the reaction conditions (Scheme 1.29) [85]. The stereoselection is strongly influenced by an intramolecular selenium–heteroatom coordination during the formation of the seleniranium intermediate.

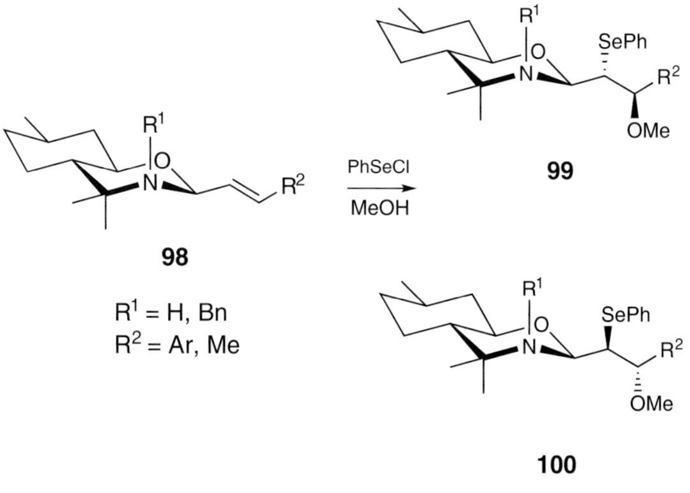

Scheme 1.29 Methoxyselenenylation of 2-vinylperhydro-1,3-benzoxazines **98**.

A preferential chelation with oxygen rather than with nitrogen atom has been proven and it should be a consequence of the axial position of the substituent at the nitrogen atom, assuming that the selenium atom is coordinated by the axial rather than the equatorial lone pair of the oxygen.

1.3
Selenocyclizations

As already mentioned in Section 1.1 of this chapter, when an internal nucleophile is present in a suitable position of the alkene, the reaction with the electrophilic selenium species results in a selenocyclization. This class of transformations is particularly interesting due to the high importance of heterocyclic compounds both as final products and intermediates in synthetic strategies. Part of the success of these conversions resides in the ease of manipulation of the introduced selenium moiety which, as discussed earlier in this chapter, can be removed or substituted in different ways. Among others, the most useful approaches are radical deselenenylation or oxidation followed by β-elimination. As the other mechanisms involved in the addition of an electrophilic selenium reagent to a carbon–carbon double bond, selenocyclizations are also stereospecific *anti*-processes, since in this case also the reaction occurs through the initial formation of a seleniranium intermediate.

The course of the cyclization depends on many factors, that is, the structure of the selenium electrophile, its counterion, solvent, and the nature of external additives which can coordinate to the selenium atom [86]. Detailed mechanistic studies have been reported very recently. Denmark *et al.* [87] studied the selenocyclization of β,γ-unsaturated acids and derivatives promoted by PhSeCl or PhSeBr, showing that the process is actually more complex that usually believed, involving a kinetic formation of haloselenenylation intermediates which afterward convert into the thermodynamically favored cyclization products.

In most of the examples reported in the literature, selenocyclizations afford five- or six-membered ring products by means of *5-endo-trig*, *5-exo-trig*, or *6-endo-trig* cyclizations. However, the synthesis of larger membered heterocyclic compounds is also possible.

In most cases, the internal nucleophile attacking the seleniranium intermediate is an oxygen or a nitrogen nucleophile. Very important, even if less represented in the literature, are selenocyclizations with an internal carbon nucleophile. In these cases, the final product is a carbocycle.

An overview of all these kinds of selenocyclizations will be reported in the following sections, focusing on the most recent examples.

1.3.1
Oxygen Nucleophiles

Selenocyclizations of alkenes bearing an internal oxygen nucleophile are the most thoroughly studied selenium-promoted cyclizations, due to the importance of oxygen-containing heterocycles. The most important oxygen nucleophiles involved in these processes are carboxylic acids and alcohols. The reactions originate lactones or cyclic ethers and can be named selenolactonizations and selenoetherifications, respectively. General examples are reported in Scheme 1.30a [88] and b [89]. The factors governing the outcome of these reactions have been recently investigated by synthesizing compound **111**, bearing both an hydroxy-group and a carboxylic acid in suitable positions to undergo these processes, and subjecting it to selenocyclization conditions, with different reactants and in the presence of different additives (Scheme 1.31) [90].

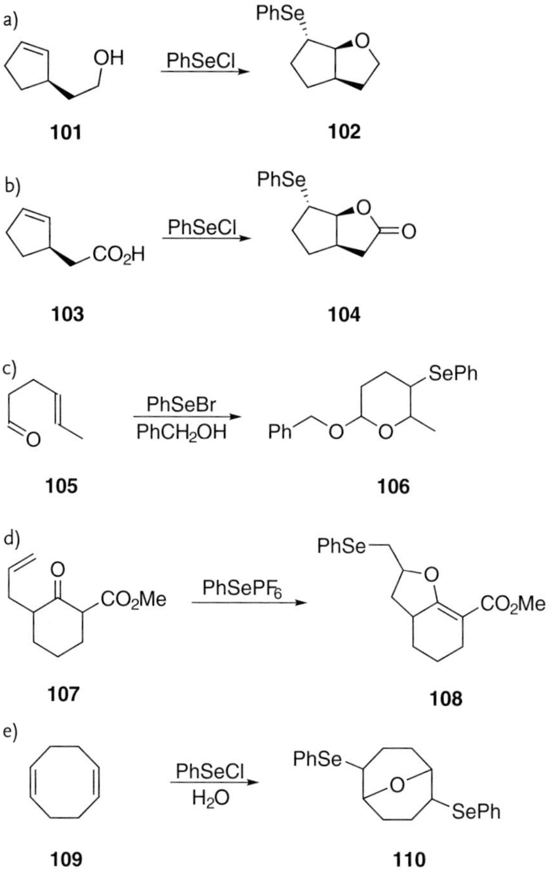

32 | *1 Electrophilic Selenium*

Scheme 1.30 Selenolactonizations and selenoetherifications.

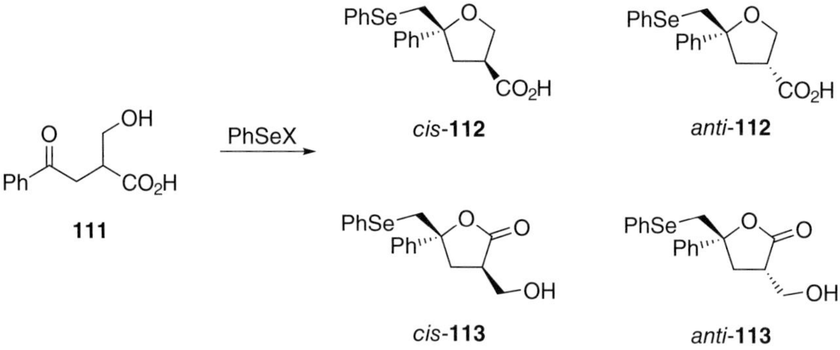

Scheme 1.31 Selenolactonizations versus selenoetherifications.

In situ generated hydroxyl groups can also participate in selenocyclizations. This strategy has been successfully applied in the synthesis of polycyclic compounds. An example in which the addition of an alcohol to an aldehyde generates *in situ* the hydroxyl group undergoing a 6-*endo-trig* cyclization is shown in Scheme 1.30c [91]. Another possibility for the *in situ* generation of the nucleophile comes from the tautomerization of β-dicarbonyl systems. The resulting vinylic oxygen can participate in selenocyclization processes, thus offering a valid strategy for the syntesis of cyclic vinyl ethers (see Scheme 1.30d for an example) [92]. Dienes can undergo sequential hydroxyselenenylation–selenoetherifications if the hydroxyl group generated from the addition oh PhSeOH to one of the olefins can attack the seleniranium formed on the other double bond. In Scheme 1.30e, the conversion of 1,5-cyclooctadiene **109** into a functionalized cyclic ether (**110**) is shown as an example [93].

Alkenyl nitrones can also undergo selenocyclizations. These processes generate reactive cyclic iminium salts which, depending on the substitution pattern, can undergo reduction by NaBH$_4$ and/or methanolysis. In both the cases, the product is a 1,2-oxazine [94].

2,3-Allenoates have been reported to undergo selenocyclizations in the presence of PhSeCl and water (Scheme 1.32) [95]. The reaction is proposed to occur through the formation of a seleniranium ion on the more electrophilic double bond (**115**) and the subsequent attack of the carbonylic oxygen. The formed carbocation (**116**) reacts then with a water molecule to afford the final product **117**, after an intramolecular proton transfer and the elimination of a molecule of alcohol.

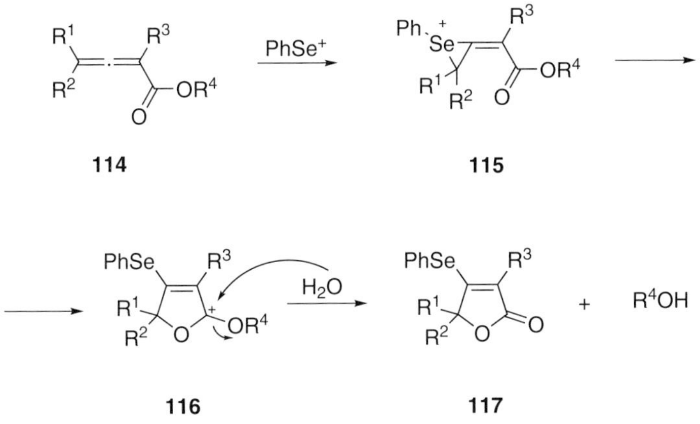

Scheme 1.32 Seleocyclizations of allenes.

Selenocyclizations have also been widely used in total syntheses. In a recent report by Danishefsky and coworkers, a selenocyclization with an *in situ* formed nucleophilic hydroxyl group (cf. Scheme 1.30c) is used in a key step of the synthesis of salinosporamide A (Scheme 1.33) [96].

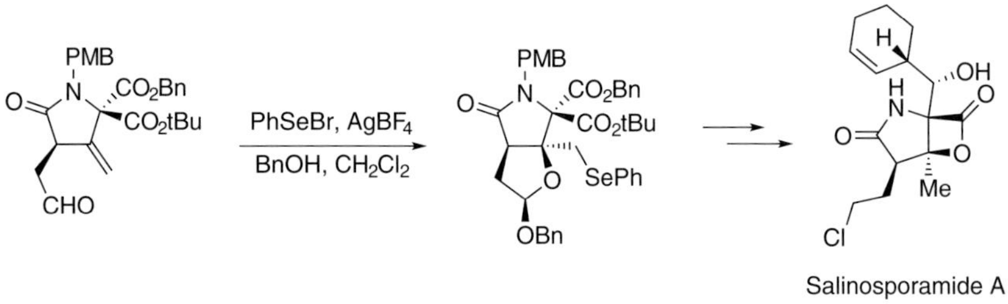

Scheme 1.33 Synthesis of salinosporamide A.

A selenocyclization is also used to synthesize the *erythro*-bis(2,2′)-tetrahydrofuran core of trilobacin [97]. In the key step of this total synthesis, PhSeCl promotes a nucleophilic transannular attack by an etheric oxygen to form the oxonium intermediate **121**. A subsequent nucleophilic attack of Cl$^-$ on C22 originates intermediate **122**, which in a few steps is converted to target product trilobacin (**123**) (Scheme 1.34).

Asymmetric selenocyclization of unsaturated alcohols and carboxylic acids has also been widely explored. Since selenocyclizations are stereospecific processes and they usually occur with a good regioselectivity and facial selectivity (when a chiral nonracemic electrophilic selenium reagent is employed), they represent an interesting way to access enantioenriched heterocycles.

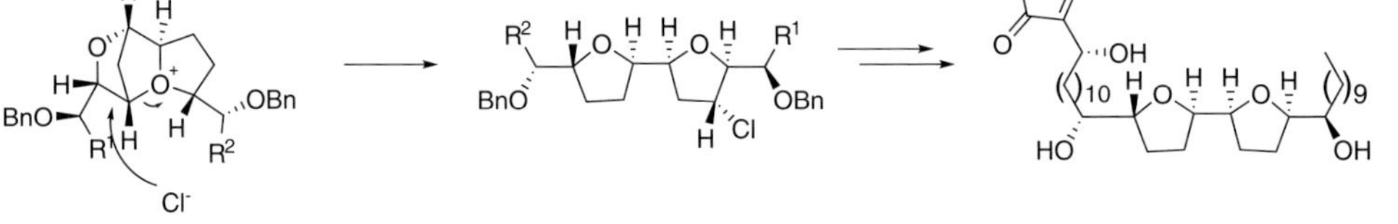

Scheme 1.34 Synthesis of trilobacin.

Almost all chiral selenium reagents synthetized have been tested in selenocyclization, together with selenofunctionalizations. Selected examples are reported in Scheme 1.35 [98]. The reaction is of general applicability and usually proceeds with good stereosectivities. The stereochemical outcome of the reaction is similar to that described for intermolecular oxyselenenylation processes.

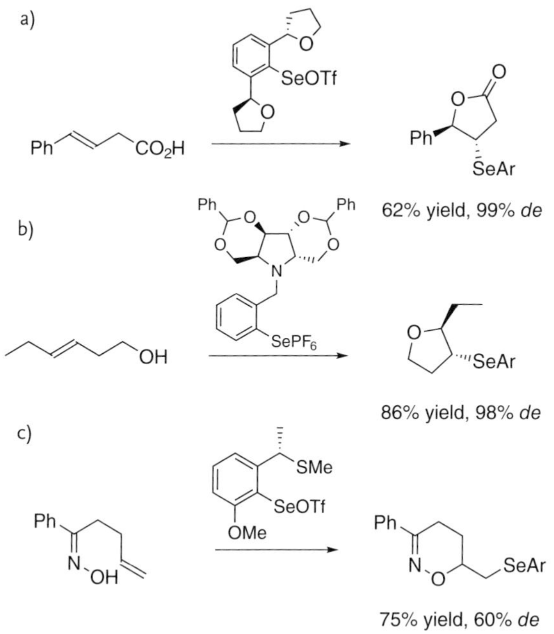

Scheme 1.35 Asymmetric selenocyclofunctionalizations [99].

1.3.2
Nitrogen Nucleophiles

Selenocyclization with nitrogen nucleophiles have received great attention due to their high potential in the synthesis of nitrogen-containing heterocycles and thus in the total synthesis of alkaloids. The first examples in this area were reported by Clive and coworkers [100], who showed that primary amines are poorly reactive in selenocyclization conditions, while amines functionalized with an electron withdrawing group smoothly cyclize to afford the desired nitrogen-containing heterocycle [101]. Most of the early literature reports use unsaturated carbamates as substrates for selenocyclizations. Some selected examples are shown in Scheme 1.36.

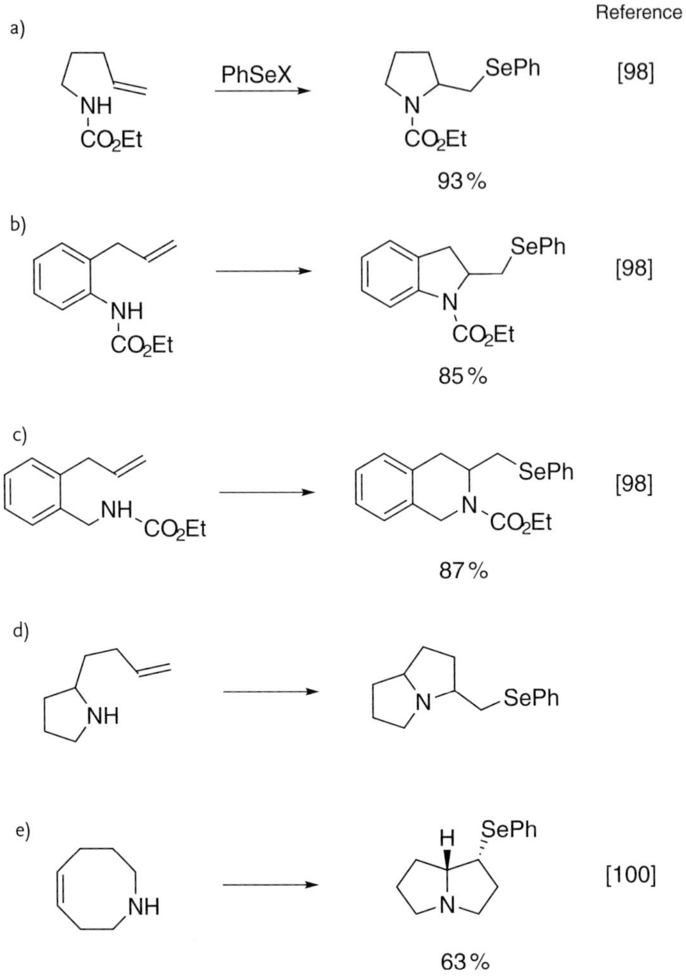

Scheme 1.36 Cyclofunctionalization by nitrogen nucleophilic attack.

Secondary substituted amines are more reactive than primary amines and afford the expected products, as shown in Scheme 1.36d and e [102]. In both examples, substituted pyrrolizidines are obtained, but the method is general in scope.

O-allyl hydroxylamines **124** have also been shown to react in the presence of PSS to give substituted isoxazolines **125** (Scheme 1.37). Two diastereoisomers of **125** were obtained [103].

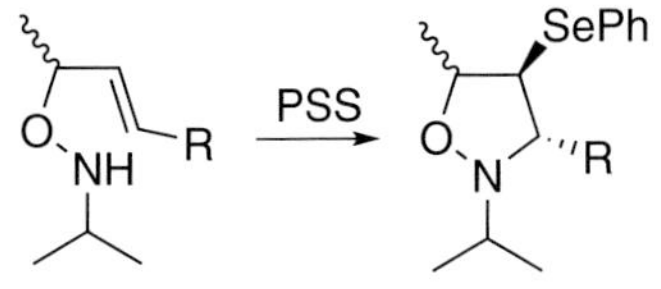

Scheme 1.37 Selenocyclization of O-allyl hydroxylamines.

Similarly to the previous example, O-allyl oximes **126** smoothly react in generating an intermediate iminium salt **127**, which can be easily converted into the final products either by reduction with NaBH$_4$ or by hydrolysis (Scheme 1.38) [104]. In the first case, N-alkyl oxazolines **128** are accessed, as in the example reported in Scheme 1.37; in the second one the hydrolysis afforded oxazolines **129** in very good yields.

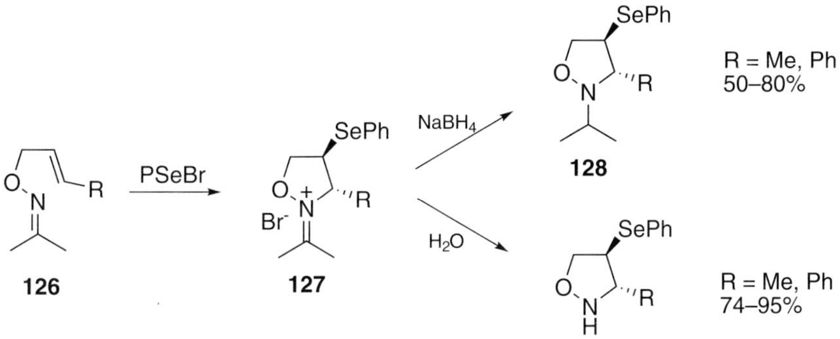

Scheme 1.38 Selenocyclization of O-allyl oximes via iminium salts.

The selenenyl triflate generated from the reaction of di-2-([1S]-1-[methylthio]-ethyl) phenyl diselenide (**23**) with silver triflate with various substituted O-allyl oximes afforded optically active isoxazolidines in high yields and with good diastereoselectivities up to 93:7. These heterocyclic compounds can be easily converted into the corresponding optically enriched 1,3-amino alcohols by N–O bond cleavage effected by treatment with zinc in aqueous acetic acid [105].

1 Electrophilic Selenium

The imino nitrogen can also react with seleniranium intermediates to afford nitrogen-containing heterocycles. Some interesting examples are shown in Scheme 1.39.

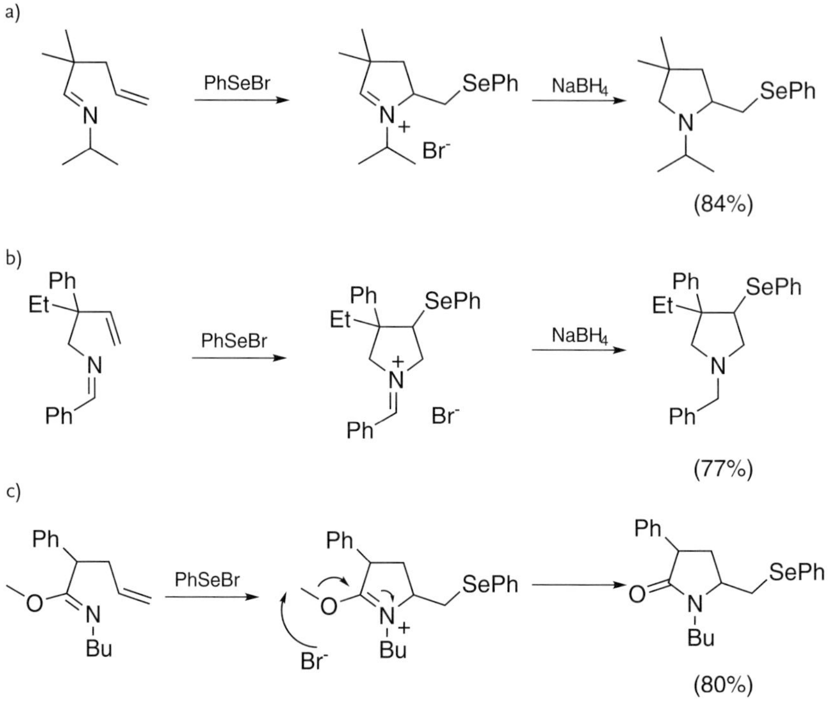

Scheme 1.39 Selenocyclization of imines.

The imine can be either internal (Scheme 1.39a) or external (Scheme 1.39b) to the formed heterocycle [106]. In all the cases, an iminium intermediate is formed, which can be reduced to an amine (a pyrrolidine in the presented examples) by treatment with NaBH$_4$. In Scheme 1.39c, the conversion of an imidate into a γ-lactam is shown. In this case, the PhSeBr cannot be substituted by other electrophilic selenium reagents since the bromide anion is necessary for the conversion of the iminium salt intermediate to the final product, through a substitution occurring on the methyl group and having the product as the leaving group [107].

An interesting recent application of electrophilic selenium to promote a cyclization with an internal nitrogen nucleophile (**130**→**131**) has been reported by Yao and coworkers in the total synthesis of chloptosin **132**, a dimeric cyclopeptide which showed potent apoptosis-inducing properties (Scheme 1.40) [108].

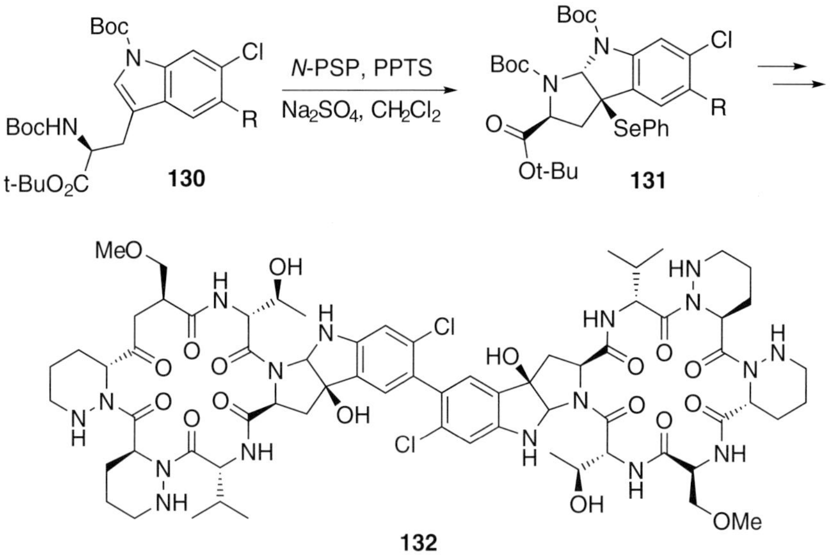

Scheme 1.40 Synthesis of chloptosin **132**.

In some cases, more than a nucleophilic nitrogen is present in the substrate and a chemoselectivity issue arises. When the alkenyl hydrazine **133** (Scheme 1.41a) is reacted with PSS, in principle both the nitrogen atoms can act as nucleophiles. Tiecco and coworkers demonstrated that the reaction outcome depends on the substituent R on the nitrogen. When R = Ph, a 6-*exo-trig* cyclization occurs affording pyridazine **134**. On the contrary, when an electron-withdrawing group is bound to the nitrogen, this is no more nucleophilic enough to react with the seleniranium intermediate, and the reaction occurs with the other nitrogen affording the pyrrolidineamine **135** as the product [109].

Some other interesting examples are reported in Scheme 1.41b and c. In this case, both an imino and amine nitrogen are present in the substrate, an alkenyl phenylhydrazone. Under the reaction conditions, no isomerization of the hydrazone is observed and the reaction outcome only depends on the geometry of the hydrazone in the starting material. Thus substrate **136** reacts with the imino nitrogen affording the intermediate **137** which can be reduced to pyrrolidineamine **138** (Scheme 1.41b). The other geometric isomer (**139**) reacts with the amino nitrogen to give tetrahydropyridazine **140** (Scheme 1.41c) [110].

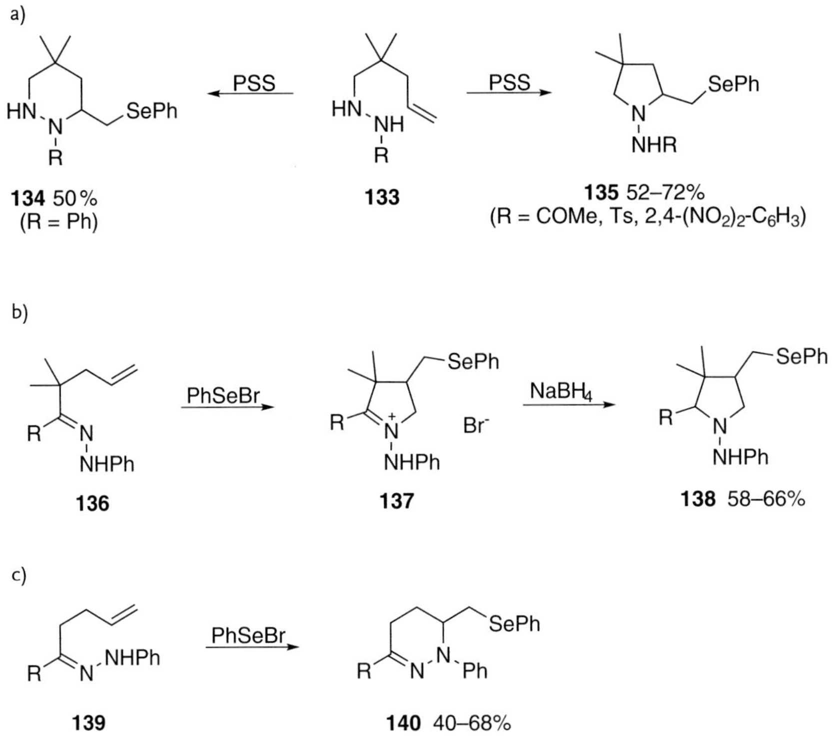

a)

134 50%
(R = Ph)

133

135 52–72%
(R = COMe, Ts, 2,4-(NO$_2$)$_2$-C$_6$H$_3$)

b)

136

137

138 58–66%

c)

139

140 40–68%

Scheme 1.41 Chemoselectivity in the cyclofunctionalization of hydrazine derivatives.

Recently RajanBabu and coworkers reported that N-aryl-N-tosylamides undergo a detosylative cyclization promoted by electrophilic selenium reagent such as phenylselenenyl bromide to afford 5- or 6-membered nitrogen-containing heterocycles (Scheme 1.42) [111].

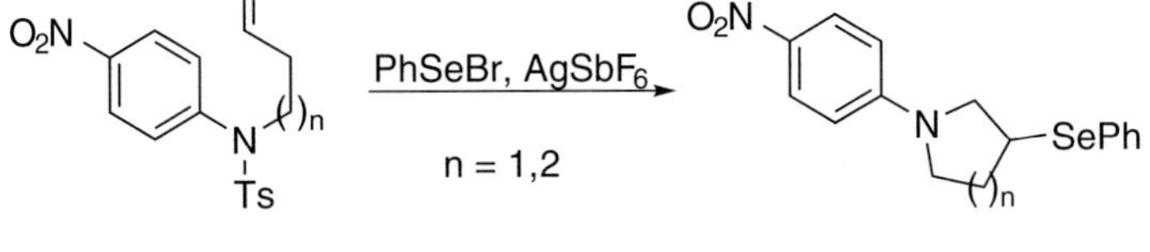

Scheme 1.42 Detosylative cyclization.

1.3.3
Competition between Oxygen and Nitrogen Nucleophiles

A more general issue about chemoselectivity arises when different nucleophiles, generally nitrogen and oxygen, are present in the same substrate and in principle both can react with the seleniranium intermediate to give different cyclization products. This problem has been extensively studied by different research groups.

Generally the observed outcome of the reaction only depends on the structure of the substrate, but a few cases have been reported in which a change in the experimental conditions is reflected in chemoselectivity of the process.

The first example of a competition between nitrogen and oxygen nucleophiles in selenocyclization was reported by Toshimitsu, who investigated this process in unsaturated amides (Scheme 1.43) [112]. When compound **141** (R = H) is allowed to react with an electrophilic selenium reagent, only the formation of product **142**, deriving from an oxygen attack on the seleniranium intermediate, was observed. The product is converted into a lactone during the work-up. On the other hand, compound **143** (R = Ph) affords the product of a 6-endo-trig cyclization **144**, with the nitrogen acting as a nucleophile. The reason for this switching both in the regiochemistry and in the nucleophile is not clear. It should be considered, however, that amides react more often with oxygen in selenocyclizations. The reaction with nitrogen as a nucleophile can be promoted by the introduction of electron-withdrawing groups on the nitrogen and by the use of a base.

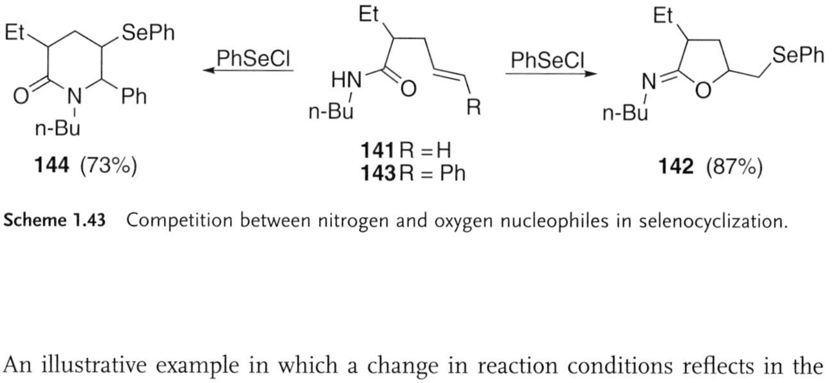

Scheme 1.43 Competition between nitrogen and oxygen nucleophiles in selenocyclization.

An illustrative example in which a change in reaction conditions reflects in the formation of different products is shown in Scheme 1.44. From the hydroxy carbamate **145** at low temperatures, the formation of product **146**, deriving from an oxygen attack on seleniranium **146**, is observed. At higher temperatures, the exclusive formation of the product derived from a nitrogen attack (**148**) is observed. The isolated product **147** can be completely converted to product **148** upon heating. This clearly shows that in these cyclizations different products can be originated by kinetic or thermodynamic control of the reaction [113].

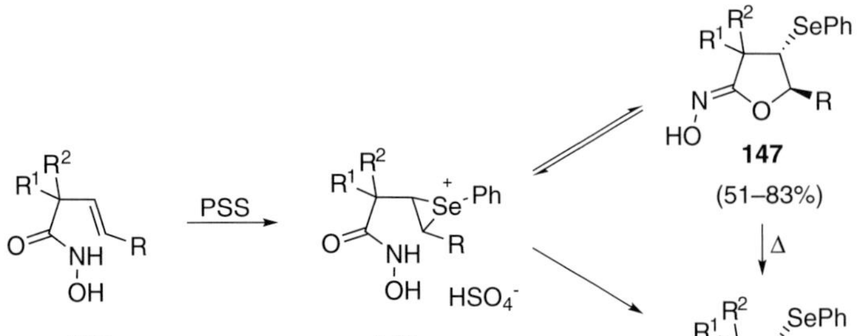

145
R = Et, Me, Ph
R¹ = H, Me
R² = H, Me

146

147

148

Scheme 1.44 Selenocyclization of hydroxy carbamic acids.

1.3.4
Carbon Nucleophiles

Carbocyclization occurs when a nucleophilic carbon, usually an olefin, a carbon in the α-position to a carbonyl group or an aromatic group attacks the seleniranium intermediate.

The first example has been reported by Clive and coworkers (Scheme 1.45) [114]. Diene **149** reacts in the presence of PhSeCl and acetic acid to afford bicyclic compound **150**. The initial attack of the olefin on the seleniranium intermediate leads to the formation of the carbocycle and a carbocation which can then be captured by a molecule of acetic acid.

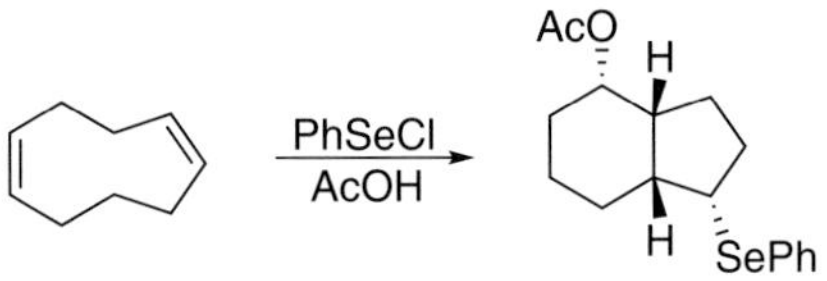

149 150

Scheme 1.45 Carbocyclization of diene **149**.

Toshimitsu and coworkers showed that diphenyl diselenide in the presence of molecular iodine in acetonitrile is effective for the carbocyclization of dienes. In this case, an acetamido group is attached to the product due to the solvolysis of the carbocationic intermediate [115].

The seleniranium intermediate can also be generated by reacting a β-hydroxyselenide with a strong acid (usually trifluoroacetic acid). This strategy is exemplified by the reaction reported in Scheme 1.46 [116].

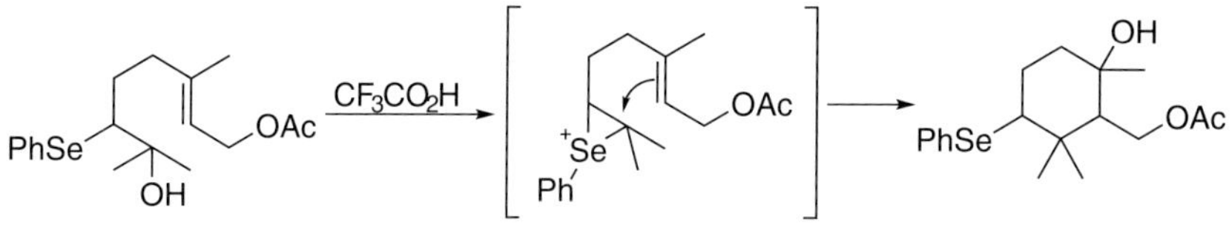

Scheme 1.46 Carbocyclization of hydroxyselenide.

Olefinic enolic bonds can also act as nucleophiles in selenium-promoted carbocyclizations. Ley and coworkers reported that β-dicarbonyl compound **151** reacts with N-PSP in the presence of a Lewis acid (ZnI_2, $AlCl_3$, or $SnCl_4$) to afford intermediate **152** (Scheme 1.47). Enolic oxygen is a better nucleophile and the initial formation of the selenoetherification product **153** is observed. Longer reaction times and the presence of strong acids lead to the formation of the thermodynamic carbocyclization product **154** [117].

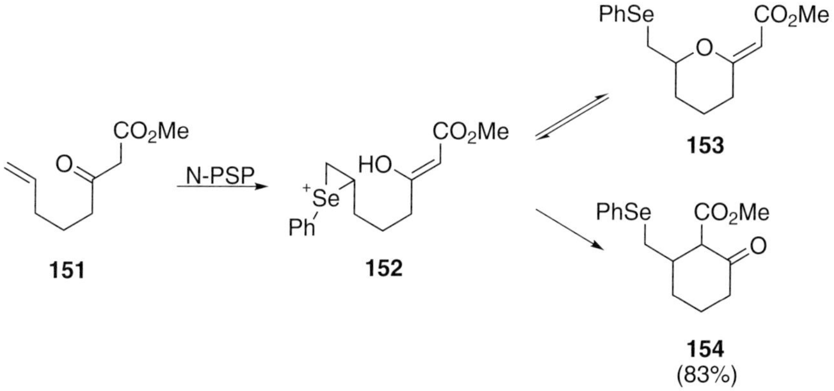

Scheme 1.47 Carbocyclization of β-dicarbonyl compound **151**.

α-Phenylseleno alkenyl ketones **155** can react with $TiCl_4$ in a carbocyclization process (Scheme 1.48). The reaction occurs through the formation of a titanium enolate **156**, while the phenylseleno group attacks the olefin [118].

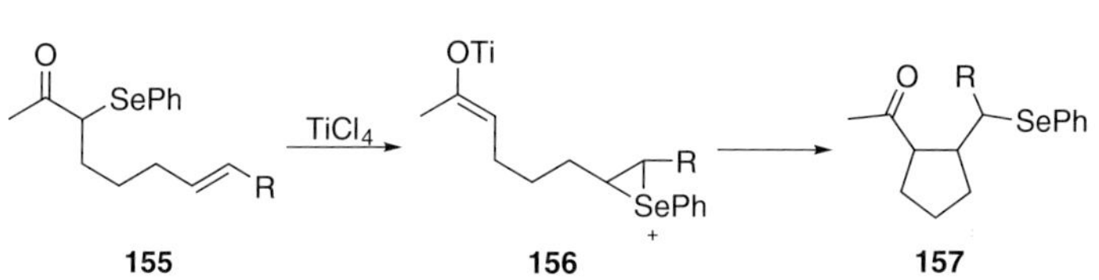

Scheme 1.48 Carbocyclization of enolates.

Wirth and coworkers reported that β-keto ester substituted stilbene derivatives undergo a cyclization and a subsequent 1,2-rearrangement of an aryl group when reacted with PhSeCl in the presence of Lewis acid (Scheme 1.49) [119].

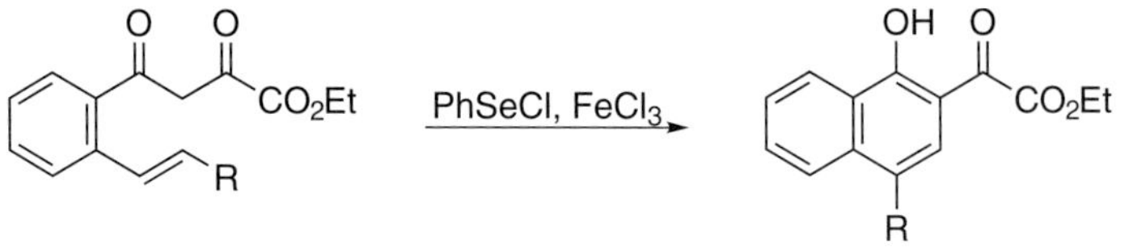

Scheme 1.49 Carbocyclization of a β-keto ester.

Aromatic carbons can also act as nucleophiles in electrophilic selenium promoted carbocyclizations. Deziel and coworkers reported that compound **158** reacts with a chiral arylselenenyl triflate derived from diselenide **159** in dichloromethane and methanol to afford an equimolar mixture of the methoxyselenenylation product **160** and carbocyclization product **161**. The former can be converted to the latter by treatment with trifluoroacetic acid. The carbocyclization product has been obtained in 70% yield and with 98% diastereoisomeric excess (Scheme 1.50) [120].

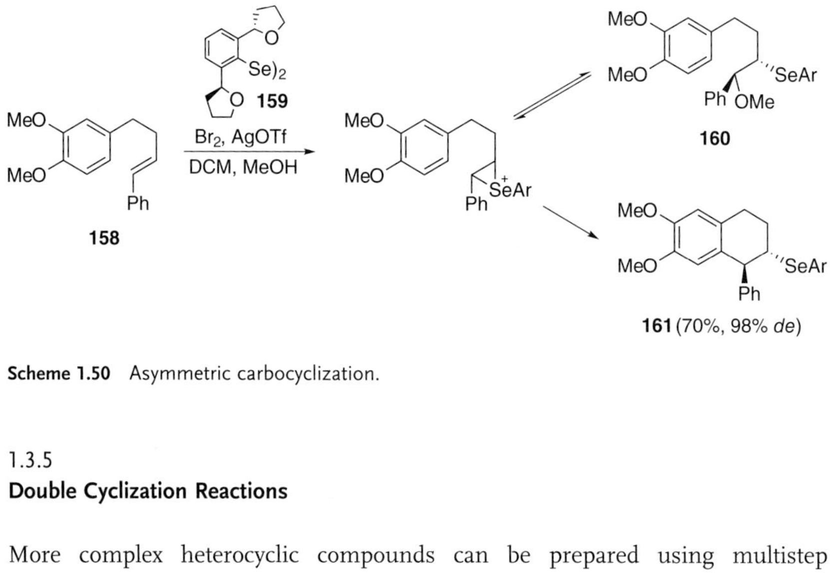

Scheme 1.50 Asymmetric carbocyclization.

1.3.5
Double Cyclization Reactions

More complex heterocyclic compounds can be prepared using multistep sequences as well as a direct double cyclization reaction. In the first example starting from the commercially available (R)-(+)-2,2-dimethyl-1,3-dioxolane-4-carboxaldehyde, by means of two consecutive selenium-promoted cyclizations, the enantiomerically pure compounds **162**, **163**, and **164** can be obtained (Scheme 1.51a) [121].

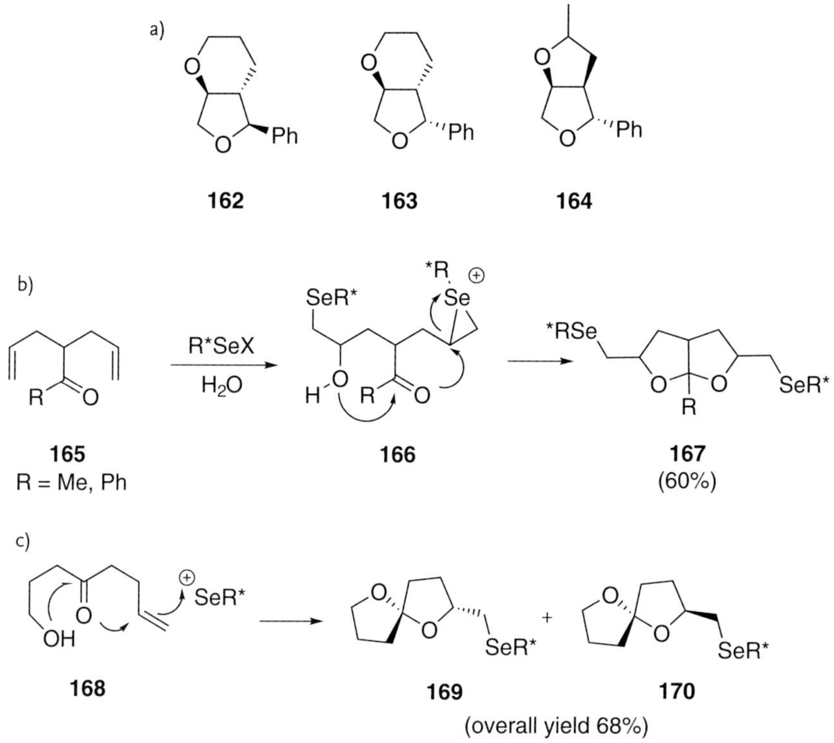

Scheme 1.51 Synthesis of bicyclic compounds.

More interestingly, the synthesis of related perhydrofuro[2,3-b]furans **167** has been achieved by the double cyclization of bisalkenylketones **165** promoted by camphoryl selenenyl sulfate in the presence of water. The mechanism consists of a selenohydroxylation leading to the intermediate **166** that subsequently underwent a double cyclization affording the final bicyclic compound **167** (Scheme 1.51b) [122].

Using a similar procedure and starting from 1-hydroxyoct-7-en-4-one **168**, enantiomerically pure 1,6-dioxaspiro[4,4]nonanes **169** and **170** can be rapidly prepared (Scheme 1.51c) [123].

References

1 Hölzle, G. and Jenny, W. (1958) *Helv. Chim. Acta*, **41**, 593–603.
2 Reich, H.J. and Wollowitz, S. (1993) *Org. React.*, **44**, 1–296.
3 (a) Marshall, J.A. and Royce, R.D. (1982) *J. Org. Chem.*, **47**, 693–698; (b) Schreiber, S.L. and Santini, C. (1984) *J. Am. Chem. Soc.*, **106**, 4038–4039.
4 Michalak, K., Michalak, M., and Wicha, J. (2010) *Tetrahedron Lett.*, **51**, 4344–4346.
5 Wirth, T., Kulicke, K.J., and Fragale, G. (1996) *J. Org. Chem.*, **61**, 2686–2689.

6 (a) Paulmier, C. (1986) *Selenium Reagents and Intermediates in Organic Synthesis*, Pergamon Press, Oxford; (b) Liotta, D.C. (1987) *Organoselenium Chemistry*, John Wiley & Sons, Inc., New York; (c) Wirth, T. (ed.) (2000) *Organoselenium Chemistry: Modern Developments in Organic Synthesis*, Top. Curr. Chem.; Springer, Berlin; (d) Back, T.G. (ed.) (1999) *Organoselenium Chemistry: A Practical Approach*, Oxford University Press, Oxford.

7 (a) Wirth, T. (1999) *Tetrahedron*, **55**, 1–28; (b) Wirth, T. (2000) *Angew. Chem. Int. Ed.*, **39**, 3740–3749; (c) Wirth, T. (2006) *Comprehensive Organometallic Chemistry III*, vol. 9 (eds R.H. Crabtree and D.M.P. Mingos), Elsevier, Oxford, pp. 457–500; (d) Braga, A.L., Ludtke, D.S., Vargas, F., and Braga, R.C. (2006) *Synlett*, 1453–1466; (e) Browne, D.M. and Wirth, T. (2006) *Curr. Org. Chem.*, **10**, 1893–1903; (f) Freudendahl, D.M., Shahzad, S.A., and Wirth, T. (2009) *Eur. J. Org. Chem.*, **11**, 1649–1664; (g) Freudendahl, D.M., Santoro, S., Shahzad, S.A., Santi, C., and Wirth, T. (2009) *Angew. Chem. Int. Ed.*, **48**, 8409–8411; (h) Santi, C., Santoro, S., and Battistelli, B. (2010) *Curr. Org. Chem.*, **14**, 2442–2462.

8 Nicolaou, K.C., Petasis, N.A., and Claremon, D.A. (1985) *Tetrahedron*, **41**, 4835–4841.

9 Jackson, W.P., Ley, S.V., and Whittle, A.J. (1980) *J. Chem. Soc., Chem. Commun.*, 1173–1174.

10 Davis, F.A., Zhou, P., and Sankar, L.G. (1990) *Tetrahedron Lett.*, **31**, 1653–1656.

11 Back, T.G. and Muralidharan, K.R. (1991) *J. Org. Chem.*, **56**, 2781–2787.

12 (a) Murata, S. and Suzuki, T. (1987) *Chem. Lett.*, 849–852; (b) Murata, S. and Suzuki, T. (1987) *Tetrahedron Lett.*, **28**, 4297–4298; (c) Murata, S. and Suzuki, T. (1987) *Tetrahedron Lett.*, **28**, 4415–4416.

13 Tingoli, M., Diana, M., and Panunzi, B. (2006) *Tetrahedron Lett.*, **47**, 7529–7531.

14 Tiecco, M., Testaferri, L., Tingoli, M., Chianelli, D., and Bartoli, D. (1988) *Tetrahedron*, **44**, 2273–2282.

15 Tiecco, M., Testaferri, L., Tingoli, M., Bagnoli, L., and Marini, F. (1993) *J. Chem. Soc., Perkin Trans. 1*, 1989–1993.

16 Bosman, C., D'Annibale, A., Resta, S., and Trogolo, C. (1994) *Tetrahedron Lett.*, **35**, 6525–6528.

17 Lee, D.H. and Kim, Y.H. (1995) *Synlett*, 349–350.

18 Han, L.B. and Tanaka, M. (1996) *J. Chem. Soc., Chem. Commun.*, 475–476.

19 Tiecco, M., Testaferri, L., Tingoli, M., and Bartoli, D. (1989) *Tetrahedron Lett.*, **30**, 1417–1420.

20 (a) Pandey, G., Rao, V.J., and Bhalerao, U.T. (1989) *J. Chem. Soc., Chem. Commun.*, 416–417; (b) Pandey, G., and Sekhar, B.B.V.S. (1993) *J. Chem. Soc., Chem. Commun.*, 780–782.

21 (a) Yoshida, M., Satoh, N., and Kamigata, N. (1989) *Chem. Lett.*, 1433–1436; (b) Yoshida, M., Sasage, S., Kawamura, K., Suzuki, T., and Kamigata, N. (1991) *Bull. Chem. Soc. Jpn.*, **64**, 416–422.

22 Roh, K.R., Chang, H.K., and Kim, Y.H. (1998) *Heterocycles*, **48**, 437–441.

23 Tingoli, M., Tiecco, M., Testaferri, L., and Temperini, A. (1998) *Synth. Commun.*, **28**, 1769–1778.

24 Tiecco, M., Testaferri, L., Temperini, A., Bagnoli, L., Marini, F., and Santi, C. (2001) *Synlett*, 1767–1771.

25 (a) Nicolaou, K.C., Pastor, J., Barluenga, S., and Winssinger, N. (1998) *Chem. Commun.*, 1947–1948; (b) Ruhland, T., Andersen, K., and Pedersen, H. (1998) *J. Org. Chem.*, **63**, 9204–9211; (c) Cohen, R.J., Fox, D.L., and Salvatore, R.N. (2004) *J. Org. Chem.*, **69**, 4265–4268; (d) Fujita, K., Hashimoto, S., Oishi, A., and Taguchi, Y. (2003) *Tetrahedron Lett.*, **44**, 3793–3795; (e) Nicolaou, K.C., Pfefferkorn, J.A., Mitchell, H.J., Roecker, A.J., Barluenga, S., Cao, G.Q., Affleck, R.L., and Lillig, J.E. (2000) *J. Am. Chem. Soc.*, **122**, 9954–9967; (f) Nicolaou, K.C., Pfefferkorn, J.A., Barluenga, S., Mitchell, H.J., Roecker, A.J., and Cao, G.Q. (2000) *J. Am. Chem. Soc.*, **122**, 9968–9976; (g) Nicolaou, K.C., Pfefferkorn, J.A., Roecker, A.J., Cao, G.Q., Barluenga, S., and Mitchell, H.J. (2000) *J. Am. Chem. Soc.*, **122**, 9939–9953; (h) Nicolaou, K.C., Roecker, A.J., Pfefferkorn, J.A., and Cao, G.Q. (2000) *J. Am. Chem. Soc.*, **122**, 2966–2967; (i) Nicolaou, K.C., Winssinger, N., Hughes,

R., Smethurst, C., and Cho, S.Y. (2000) *Angew. Chem. Int. Ed.*, **112**, 1126–1130; (j) Nicolaou, K.C., Pfefferkorn, J.A., and Cao, G.Q. (2000) *Angew. Chem. Int. Ed.*, **112**, 750–755; (k) Laura, A.M., Rosemary, A.M., and David, J.P. (2005) *Tetrahedron*, **61**, 11527–11576.

26 (a) Uehlin, L. and Wirth, T. (2001) *Chimia*, **55**, 65–67; (b) Uehlin, L. and Wirth, T. (2001) *Org. Lett.*, **3**, 2931–2933.

27 Xu, J.F. and Huang, X. (2009) *J. Comb. Chem.*, **11**, 938–942.

28 Xu, J.F. and Huang, X. (2009) *J. Comb. Chem.*, **11**, 350–354.

29 (a) Wirth, T., Häuptli, S., and Leuenberger, M. (1998) *Tetrahedron Asymmetry*, **9**, 547–550; (b) Iwaoka, M., and Tomoda, S. (1992) *J. Chem. Soc., Chem. Commun.*, 1165–1166; (c) Tiecco, M., Testaferri, L., Tingoli, M., and Bagnoli, S. (1993) *Synlett*, 798–800; (d) Fujita, K., Iwaoka, M., and Tomoda, S. (1994) *Chem. Lett.*, 923–926; (e) Fukuzawa, S., Takahashi, K., Kato, H., and Yamazi, H. (1997) *J. Org. Chem.*, **62**, 7711–7716; (f) Tiecco, M., Testaferri, L., and Santi, C. (1999) *Eur. J. Org. Chem.*, **4**, 797–803.

30 Browne, D.M., Niyomura, O., and Wirth, T. (2007) *Org. Lett.*, **9**, 3169–3171.

31 Niyomura, O., Cox, M., and Wirth, T. (2006) *Synlett*, 251–254.

32 (a) Tomoda, S. and Iwaoka, M. (1988) *J. Chem. Soc., Chem. Commun.*, 1283–1284; (b) Tomoda, S., Fujita, K., and Iwaoka, M. (1990) *J. Chem. Soc., Chem. Commun.*, 129–131; (c) Tomoda, S., Fujita, K., and Iwaoka, M. (1992) *Chem. Lett.*, 1123–1124; (d) Tomoda, S., Fujita, K., and Iwaoka, M. (1997) *Phosphorus Sulfur*, **67**, 247–252; (e) Déziel, R., Goulet, S., Grenier, L., Bordeleau, J., and Bernier, J. (1993) *J. Org. Chem.*, **58**, 3619–3621; (f) Déziel, R. and Malenfant, E. (1995) *J. Org. Chem.*, **60**, 4660–4662; (g) Déziel, R., Malenfant, E., and Bélanger, G. (1996) *J. Org. Chem.*, **61**, 1875–1876; (h) Déziel, R., Malenfant, E., Thibault, C., Fréchette, S., and Gravel, M. (1997) *Tetrahedron Lett.*, **38**, 4753–4756; (i) Nishibayashi, Y., Singh, J.D., Uemura, S., and Fukuzawa, S. (1994) *Tetrahedron Lett.*, **35**, 3115–3118; (j) Nishibayashi, Y., Srivastava, S.K., Takada, H., Fukuzawa, S., and Uemura, S. (1995) *J. Chem. Soc., Chem. Commun.*, 2321–2322; (k) Nishibayashi, Y., Singh, J.D., Fukuzawa, S., and Uemura, S. (1995) *J. Org. Chem.*, **60**, 4114–4120; (l) Fukuzawa, S., Takahashi, K., Kato, H., and Yamazaki, H. (1997) *J. Org. Chem.*, **62**, 7711–7716; (m) Fujita, K., Iwaoka, M., and Tomoda, S. (1994) *Chem. Lett.*, 923–926; (n) Fujita, K., Murata, K., Iwaoka, M., and Tomoda, S. (1995) *J. Chem. Soc., Chem. Commun.*, 1641–1642; (o) Fujita, K., Murata, K., Iwaoka, M., and Tomoda, S. (1995) *Tetrahedron Lett.*, **36**, 5219–5222; (p) Fujita, K., Murata, K., Iwaoka, M., and Tomoda, S. (1997) *Tetrahedron*, **53**, 2029–2048; (q) Iwaoka, M. and Tomoda, S. (1996) *J. Am. Chem. Soc.*, **118**, 8077–8084; (r) Komatsu, H., Iwaoka, M., and Tomoda, S. (1999) *J. Chem. Soc., Chem. Commun.*, 205–206.

33 (a) Tomoda, S. and Iwaoka, M. (1988) *Chem. Lett.*, 1895–1898; (b) Tomoda, S., Iwaoka, M., Yakushi, K., Kawamoto, A., and Tanaka, J. (1988) *J. Phys. Org. Chem.*, **1**, 179–184.

34 (a) Wirth, T. (1995) *Tetrahedron*, **36**, 7849–7852; (b) Wirth, T. (1995) *Angew. Chem. Int. Ed. Engl.*, **34**, 1726–1728; (c) Wirth, T. and Fragale, G. (1997) *Eur. J. Org. Chem.*, **3**, 1894–1902; (d) Wirth, T. (1997) *Liebigs Ann./Recueil*, **11**, 2189–2196; (e) Fragale, G., Neuburger, M. and Wirth, T. (1998) *Chem. Commun.*, 1867–1868; (f) Santi, C. and Wirth, T. (1999) *Tetrahedron Asymmetry*, **10**, 1019–1023; (g) Santi, C., Fragale, G., and Wirth, T. (1998) *Tetrahedron Asymmetry*, **9**, 3625–3628.

35 Tiecco, M., Testaferri, L., Santi, C., Tomasini, C., Santoro, S., Marini, F., Bagnoli, L., Temperini, A., and Costantino, F. (2006) *Eur. J. Org. Chem.*, **21**, 4867–4873.

36 Mukherjee, A.J., Zade, S.S., Singh, H.B., and Sunoj, R.B. (2010) *Chem. Rev.*, **110**, 4357–4416.

37 Tiecco, M., Testaferri, L., Bagnoli, L., Marini, F., Temperini, A., Tomassini, C., and Santi, C. (2000) *Tetrahedron Lett.*, **41**, 3241–3245.

38 Tiecco, M., Testaferri, L., Santi, C., Tomassini, C., Marini, F., Bagnoli, L.,

and Temperini, A. (2002) *Chem. Eur. J.*, **8**, 1118–1124.
39 Jones, P.G., Wisimach, C., Mugesh, G., and du Mont, W.-W. (2002) *Acta Cryst.*, **E58**, 1298–1300.
40 Freudendahl, D.M., Iwaoka, M., and Wirth, T. (2010) *Eur. J. Org. Chem.*, **20**, 3934–3944.
41 (a) Back, T.G., Dyck, B.P., and Parvez, M. (1994) *J. Chem. Soc., Chem. Commun.*, 515–516; (b) Back, T.G., Dyck, B.P., and Parvez, M. (1995) *J. Org. Chem.*, **60**, 703–710; (c) Back, T.G. and Dyck, B.P. (1996) *J. Chem. Soc., Chem. Commun.*, 2567–2568; (d) Back, T.G. and Nan, S. (1998) *J. Chem. Soc., Chem. Commun.*, 3123–3138; (e) Back, T.G., Dyck, B.P. and Nan, S. (1999) *Tetrahedron*, **55**, 3191–3194; (f) Scianowski, J. (2005) *Tetrahedron Lett.*, **46**, 3331–3334; (g) Scianowski, J., Rafinski, Z., and Wojtczak, A. (2006) *Eur. J. Org. Chem.*, **14**, 3216–3225; (h) Rafinski, Z., Scianowski, J., and Wojtczak, A. (2008) *Tetrahedron Asymmetry*, **19**, 223–230; (i) Rafinski, Z. and Scianowski, J. (2008) *Tetrahedron Asymmetry*, **19**, 1237–1244; (j) Scianowski, J. and Welniak, M. (2009) *Phosphorus Sulfur*, **184**, 1440–1447; (k) Ciechanska, M., Jozwiak, A., and Scianowski, J. (2009) *Phosphorus Sulfur*, **184**, 1502–1507; (l) Rafinski, Z., Scianowski, J., and Wojtczak, A. (2009) *Lett. Org. Chem.*, **6**, 321–328; (m) Scianowski, J., Rafinski, Z., Szuniewicz, A., and Wojtczak, A. (2009) *Tetrahedron*, **65**, 10162–10174; (n) Scianowski, J., Rafinski, Z., Wojtczak, A., and Burczynski, K. (2009) *Tetrahedron Asymmetry*, **20**, 2871–2879.
42 Fukuzawa, S.I., Kasugahara, Y., and Uemura, S. (1994) *Tetrahedron Lett.*, **39**, 9403–9406.
43 (a) Wirth, T., Fragale, G., and Spichty, M. (1998) *J. Am. Chem. Soc.*, **120**, 3376–3381; (b) Wang, X., Houk, K.N., Spichty, M., and Wirth, T. (1999) *J. Am. Chem. Soc.*, **121**, 8567–8576.
44 (a) Borodkin, G., Chernyak, E.I., Shakirov, M.M., and Shubin, V.G. (1997) *Russ. J. Org. Chem.*, **33**, 418–419; (b) Borodkin, G., Chernyak, E.I., Shakirov, M.M., and Shubin, V.G. (1998) *Russ. J. Org. Chem.*, **34**, 1563–1568.
45 Denmark, S.E., Collins, W.R., and Cullen, M.D. (2009) *J. Am. Chem. Soc.*, **131**, 3490–3492.
46 46 Santi, C. Mechanism of haloselenenylation reactions: an NMR evaluation. Unpublished results.
47 (a) Garratt, D.G. and Kabo, A. (1980) *Can. J. Chem.*, **58**, 1030–1041; (b) Shmid, G.H. and Garratt, D.G. (1983) *J. Org. Chem.*, **48**, 4169–4172.
48 Brown, R.S., Eyley, S.C., and Parson, P. (1984) *J. Chem. Soc., Chem. Commun.*, 438–439.
49 Antipin, R.A., Bleloglazkina, E.K., Zyc, N.V., and Zefrov, N.S. (2007) *Terahedron Lett.*, **48**, 729–731.
50 Amosova, S.V., Penzik, M.V., Albanov, A.I., and Potapov, V.A. (2009) *Terahedron Lett.*, **50**, 306–308.
51 Engman, L. (1989) *J. Org. Chem.*, **54**, 884–889.
52 Tiecco, M., Testaferri, L., Santi, C., Tomassini, C., Bagnoli, L., Marini, F., and Temperini, A. (2000) *Tetrahedron Asymmetry*, **11**, 4645–4650.
53 Spichty, M., Fragale, G., and Wirth, T. (2000) *J. Am. Chem. Soc.*, **122**, 10914–10916.
54 Tiecco, M., Testaferri, L., Santi, C., Marini, F., Bagnoli, L., and Temperini, A. (1998) *Tetrahedon Lett.*, **39**, 2809–2812.
55 Tiecco, M., Testaferri, L., Santi, C., Tomassini, C., Bonini, R., Bagnoli, L., Marini, F., and Temperini, A. (2004) *Org. Lett.*, **6**, 4751–4753.
56 Hout, J.F., Outurquin, F., and Paulmier, C. (1991) *Chem. Lett.*, 1599–1602.
57 Santi, C., Santoro, S., Tomassini, C., Pascolini, F., Testaferri, L., and Tiecco, M. (2009) *Synlett*, 743–746.
58 Tiecco, M., Testaferri, L., Santi, C., Bagnoli, L., Marini, F., Temperini, A., and Tomassini, C. (1998) *Eur. J. Org. Chem.*, **11**, 2275–2277.
59 Torii, S., Uneyama, K., Ono, M., and Bannou, T. (1981) *J. Am. Chem. Soc.*, **103**, 4606–4608.
60 Tiecco, M., Testaferri, L., Tingoli, M., Bagnoli, L., and Santi, C. (1993) *J. Chem. Soc., Chem. Commun.*, 637–639.

References

61 Tiecco, M., Testaferri, L., Marini, F., Santi, C., Bagnoli, L., and Temperini, A. (1999) *Tetrahedron Asymmetry*, **10**, 747–757.

62 Santi, C., Tiecco, M., Testaferri, L., Tomassini, C., Santoro, S., and Bizzoca, G. (2008) *Phosphorus Sulfur*, **183**, 956–960.

63 Santoro, S., Santi, C., Sabatini, M., Testaferri, L., and Tiecco, M. (2008) *Adv. Synth. Catal.*, **350**, 2881–2884.

64 Santoro, S., Battistelli, B., Gjoka, B., Si, C.W., Testaferri, L., Tiecco, M., and Santi, C. (2010) *Synlett*, 1402–1406.

65 Chen, G., Fu, C., and Ma, S. (2006) *J. Org. Chem*, **71**, 9877–9879.

66 He, G., Yu, Y., Fu, C., and Ma, S. (2010) *Eur. J. Org. Chem.*, **1**, 101–110.

67 (a) Guo, H., Quian, R., Guo, Y., and Ma, S. (2008) *J. Org. Chem.*, **73**, 7934–7938; (b) He, G., Guo, H., Quian, R., Guo, Y., Fu, C., and Ma, S. (2009) *Tetrahedron*, **65**, 4877–4889.

68 (a) Toshimitsu, A., Aoai, T., Uemura, S., and Okano, M. (1980) *J. Chem. Soc., Chem. Commun.*, 1041–1042; (b) Toshimitsu, A., Aoai, T., Owada, H., Uemura, S., and Okano, M. (1981) *J. Org. Chem.*, **46**, 4727–4733.

69 Toshimitsu, A., Hayashi, G., Terao, K., and Uemura, S. (1986) *J. Chem. Soc., Perkin Trans. 1*, 343–347.

70 Tiecco, M., Testaferri, L., Santi, C., Tomassini, C., Marini, F., Bagnoli, L., and Temperini, A. (2002) *Tetrahedron Asymmetry*, **13**, 429–435.

71 Tiecco, M., Testaferri, L., Santi, C., Tomassini, C., Marini, F., Bagnoli, L., and Temperini, A. (2000) *Eur. J. Org. Chem.*, **20**, 3451–3457.

72 Denis, J.N., Vincens, J., and Krief, A. (1979) *Tetrahedron Lett.*, **20**, 2697–2700.

73 (a) Hassner, A. and Amarasekara, A. (1987) *Tetrahedron Lett.*, **28**, 5185–5188; (b) Giuliano, R.M. and Duarte, F. (1991) *Synlett*, 419–421.

74 Tiecco, M., Testaferri, L., Temperini, A., Marini, F., Bagnoli, L., and Santi, C. (1998) *Synth. Commun.*, **28**, 2167–2179.

75 Tingoli, M., Tiecco, M., Chianelli, D., Balducci, D., and Temperini, A. (1991) *J. Org. Chem.*, **56**, 6809.

76 Tiecco, M., Testaferri, L., Santi, C., Tomassini, C., Marini, F., Bagnoli, L., and Temperini, A. (2003) *Angew. Chem. Int. Ed.*, **42**, 3131–3133.

77 (a) Toshimitsu, A., Nakano, K., Mukai, T., and Tamao, K. (1996) *J. Am. Chem. Soc.*, **118**, 2756–2757; (b) Toshimitsu, A., Terada, M., and Tamao, K. (1997) *Chem. Lett.*, 733–734.

78 Okamoto, K., Nishibayashi, Y., Uemura, S., and Toshimitsu, A. (2004) *Tetrahedron Lett.*, **45**, 6137–6139.

79 Okamoto, K., Nishibayashi, Y., Uemura, S., and Toshimitsu, A. (2005) *Angew. Chem. Int. Ed.*, **44**, 3588–3591.

80 (a) Kim, K.S., Park, J.I., Moon, H.K., and Yi, H. (1998) *J. Chem. Soc., Chem. Commun.*, 1945–1946; (b) Kim, K.S., Park, J.I., and Ding, P. (1998) *Tetrahedron Lett.*, **39**, 6471–6474.

81 Kim, K.S., Moon, C.W., Park, J.I., and Han, S.H. (2000) *J. Chem. Soc., Perkin Trans. 1*, 1341–1343.

82 Tiecco, M., Testaferri, L., Marini, F., Sternativo, S., Santi, C., Bagnoli, L., and Temperini, A. (2003) *Tetrahedron Asymmetry*, **14**, 1095–1102.

83 Tiecco, M., Testaferri, L., Marini, F., Sternativo, S., Santi, C., Bagnoli, L., and Temperini, A. (2003) *Tetrahedron Asymmetry*, **14**, 2651–2657.

84 Tiecco, M., Testaferri, L., Santi, C., Tomassini, C., Santoro, S., Marini, F., Bagnoli, L., and Temperini, A. (2007) *Tetrahedron*, **63**, 12373–12378.

85 Pedrosa, R., Andrès, C., Arias, R., Mendiguchìa, P., and Nieto, J. (2006) *J. Org. Chem.*, **71**, 2424–2428.

86 (a) Tiecco, M., Testaferri, L., Temperini, A., Bagnoli, L., Marini, F., and Santi, C. (2003) *Synlett*, 655–658; (b) Aprile, C., Gruttadauria, M., Amato, M.E., Anna, F.D., Meo, P.L., Riela, S., and Noto, R. (2003) *Tetrahedron*, **59**, 2241–2251; (c) Petragnani, N., Stefani, H.A., and Valduga, C.J. (2001) *Tetrahedron*, **57**, 1411–1448.

87 Denmark, S.E. and Edwards, M.G. (2006) *J. Org. Chem.*, **71**, 7293–7306.

88 Clive, D.L.J., Chittattu, G., and Wong, C.K. (1987) *Can. J. Chem.*, **55**, 3894–3897.

89 (a) Nicolaou, K.C. and Lysenko, Z. (1977) *J. Am. Chem. Soc.*, **99**, 3185–3187; (b) Nicolaou, K.C., Seitz, S.P., Sipio, W.J., and Blount, J.F. (1979) *J. Am. Chem. Soc.*, **101**, 3884–3893.

90 (a) Khokhar, S.S. and Wirth, T. (2004) *Angew. Chem. Int. Ed.*, **43**, 631–633; (b) Khokhar, S.S. and Wirth, T. (2004) *Eur. J. Org. Chem.*, **22**, 4567–4581.

91 Current, S. and Sharpless, K.B. (1978) *Tetrahedron Lett.*, **19**, 5075–5078.

92 Jackson, W.P., Ley, S.V., and Whittle, A.J. (1980) *J. Chem. Soc., Chem. Commun.*, 1173–1174.

93 Uemura, S., Toshimitsu, A., Aoai, T., and Okano, M. (1980) *Tetrahedron Lett.*, **21**, 1533–1536.

94 Tiecco, M., Testaferri, L., and Bagnoli, L. (1996) *Tetrahedron*, **52**, 6811–6822.

95 Chen, G., Fu, C., and Ma, S. (2006) *Tetrahedron*, **62**, 4444–4452.

96 Endo, A. and Danishefsky, S.J. (2005) *J. Am. Chem. Soc.*, **127**, 8298–8299.

97 Sohn, T.I., Kim, M.J., and Kim, D. (2010) *J. Am. Chem. Soc.*, **132**, 12226–12227.

98 (a) Déziel, R., Malenfant, E., Thibault, C., Fréchette, S., and Gravel, M. (1997) *Tetrahedron Lett.*, **38**, 4753–4756; (b) Fujita, K., Murata, K., Iwaoka, M., and Tomoda, S. (1997) *Tetrahedron*, **53**, 2029–2048; (c) Tiecco, M., Testaferri, L., Bagnoli, L., Purgatorio, V., Temperini, A., Marini, F., and Santi, C. (2001) *Tetrahedron Asymmetry*, **12**, 3297–3304.

99 Tiecco, M., Testaferri, L., Bagnoli, L., Marini, F., Temperini, A., Scarponi, C., Sternativo, S., Terlizzi, R., and Tomassini, C. (2006) *Arkivoc*, **7**, 186–206 and references therein cited.

100 (a) Clive, D.L.J., Wong, C.K., Kiel, W.A., and Menchen, S.M. (1978) *J. Chem. Soc., Chem. Commun.*, 379–380; (b) Clive, D.J., Farina, V., Singh, A., Wong, C.K., Kiel, W.A., and Menchen, S.M. (1980) *J. Org. Chem.*, **45**, 2120–2126.

101 (a) Tiecco, M., Testaferri, L., Tingoli, M., Bartoli, D., and Balducci, R. (1990) *J. Org. Chem.*, **55**, 429–434; (b) Webb, R.R., II and Danishefsky, S. (1983) *Tetrahedron Lett.*, **24**, 1357–1360; (c) Toshimitsu, A., Terao, K., and Uemura, S. (1986) *J. Org. Chem.*, **51**, 1724–1730; (d) Cooper, M.A. and Ward, A.D. (1994) *Tetrahedron Lett.*, **35**, 5065–5068; (e) Jones, A.D., Knight, D.W., Redfern, A.L., and Gilmore, J. (1999) *Tetrahedron Lett.*, **40**, 3267–3270.

102 Wilson, S.R. and Sawicki, R.A. (1979) *J. Org. Chem.*, **44**, 287–291.

103 Tiecco, M., Testaferri, L., Tingoli, M., and Santi, C. (1995) *Tetrahedron Lett.*, **36**, 163–166.

104 (a) Tiecco, M., Testaferri, L., Tingoli, M., Bagnoli, L., and Santi, C. (1995) *Tetrahedron*, **51**, 1277–1284; (b) Tiecco, M., Testaferri, L., Tingoli, M., and Bagnoli, L. (1995) *J. Chem. Soc., Chem. Commun.*, 235–236.

105 Tiecco, M., Testaferri, L., Marini, F., Sternativo, S., Santi, C., Bagnoli, L., and Temperini, A. (2001) *Tetrahedron Asymmetry*, **12**, 3053–3059.

106 (a) De Kimpe, N. and Boelens, M. (1993) *J. Chem. Soc., Chem. Commun.*, 916–918; (b) De Smaele, D. and De Kimpe, N. (1995) *J. Chem. Soc., Chem. Commun.*, 2029–2030.

107 (a) Toshimitsu, A., Terao, K., and Uemura, S. (1986) *J. Chem. Soc., Chem. Commun.*, 530–531; (b) Terao, K., Toshimitsu, A., and Uemura, S. (1986) *J. Chem. Soc., Perkin Trans. 1*, 1837–1843; (c) Toshimitsu, A., Terao, K., and Uemura, S. (1987) *J. Org. Chem.*, **52**, 2018–2026.

108 Yu, S.-M., Hong, W.-X., Wu, Y., Zhong, C.-L., and Yao, Z.-J. (2010) *Org. Lett.*, **12**, 1124–1127.

109 Tiecco, M., Testaferri, L., Marini, F., Santi, C., Bagnoli, L., and Temperini, A. (1997) *Tetrahedron*, **53**, 10591–10592.

110 Tiecco, M., Testaferri, L., Marini, F., Santi, C., Bagnoli, L., and Temperini, A. (1997) *Tetrahedron*, **53**, 7311–7318.

111 Lim, H.J. and RajanBabu, T.V. (2009) *Org. Lett.*, **11**, 2924–2927.

112 Toshimitsu, A., Terao, K., and Uemura, S. (1987) *J. Org. Chem.*, **52**, 2018–2026.

113 Tiecco, M., Testaferri, L., Tingoli, M., and Marini, F. (1994) *J. Chem. Soc., Chem. Commun.*, 221–222.

114 Clive, D.L.J., Chittatu, G., and Wong, C.K. (1978) *J. Chem. Soc., Chem. Commun.*, 1128–1129.

115 Toshimitsu, A., Uemura, S., and Okano, M. (1982) *J. Chem. Soc., Chem. Commun.*, 87–89.

116 Kametani, T., Suzuki, K., Kurobe, H., and Nemoto, H. (1979) *J. Chem. Soc., Chem. Commun.*, 1128–1129.

117 (a) Jackson, W.P., Ley, S.V., and Morton, J.A. (1980) *J. Chem. Soc., Chem. Commun.*, 1028–1029; (b) Jackson, W.P., Ley, S.V. and Morton, J.A. (1980) *J. Chem. Soc., Chem. Commun.*, 1173–1174; (c) Jackson, W.P., Ley, S.V., and Morton, J.A. (1981) *Tetrahedron Lett.*, **22**, 2601–2604; (d) Ley, S.V., Lygo, B., Molines, H., and Morton, J.A. (1982) *J. Chem. Soc., Chem. Commun.*, 1251–1252; (e) Cunat, A.C., Diez-Martin, D., Ley, S.V., and Montgomery, F.J. (1996) *J. Chem. Soc., Perkin Trans. 1*, 611–620.

118 Toru, T., Kawai, S., and Ueno, J. (1996) *Synlett*, 539–541.

119 Shahzad, S.A., Vivant, C., and Wirth, T. (2010) *Org. Lett.*, **12**, 1364–1367.

120 Deziel, R., Malenfant, E., and Thibault, C. (1998) *Tetrahedron Lett.*, **39**, 5493–5496.

121 Tiecco, M., Testaferri, L., Bagnoli, L., Terlizi, R., Temperini, A., Marini, F., Santi, C., and Scarponi, C. (2004) *Tetrahedron Asymmetry*, **15**, 1949–1955.

122 Tiecco, M., Testaferri, L., Bagnoli, L., Scarponi, C., Purgatorio, V., Temperini, A., Marini, F., and Santi, C. (2005) *Tetrahedron Asymmetry*, **16**, 2429–2435.

123 Tiecco, M., Testaferri, L., Bagnoli, L., Scarponi, C., Temperini, A., Marini, F., and Santi, C. (2006) *Tetrahedron Asymmetry*, **17**, 2768–2774.

2
Nucleophilic Selenium
Michio Iwaoka

2.1
Introduction

2.1.1
Development of Nucleophilic Selenium Reagents

The efficient utilization of nucleophilic selenium reagents in organic synthesis was discovered in the 1970s by several research groups [1–3]. Since then, various applications of selenium nucleophiles (RSe$^-$) in organic and inorganic syntheses have been eagerly investigated [4–6]. Many types of functional-group transformation by using a nucleophilic selenium reagent were well established till the end of the last millennium [7, 8]. These included the nucleophilic addition reactions to epoxides (Scheme 2.1) and the nucleophilic substitution reactions to halides or activated alcohols (Scheme 2.2). The reaction shown in Scheme 2.1 was further extended to asymmetric ring-opening reactions by using a chiral auxiliary attached to the selenium atom [9–11] and those catalyzed by a chiral gallium–titanium salen complex [12]. Thus, a wide range of organic selenium compounds with a variety of carbon skeletons and functional groups was synthesized, and the obtained compounds have been utilized for the fundamental physical organic chemistry research, that is, structures, interactions, and reactions of organic selenium compounds, as well as the applications to the fields of biological chemistry and material science.

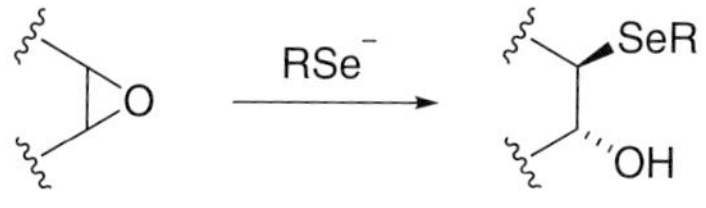

Scheme 2.1

Organoselenium Chemistry: Synthesis and Reactions, First Edition. Edited by Thomas Wirth.
© 2012 Wiley-VCH Verlag GmbH & Co. KGaA. Published 2012 by Wiley-VCH Verlag GmbH & Co. KGaA.

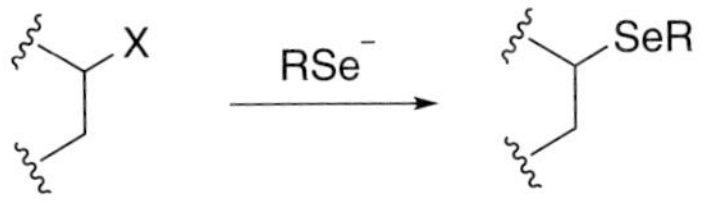

2 Nucleophilic Selenium

[Structure: R-X + RSe⁻ → R-SeR]

(X = leaving group)

Scheme 2.2

Selenium has enhanced redox reactivity compared to sulfur. Therefore, selenium compounds usually exhibit higher chemical and biological activities than sulfur compounds. More than 20 selenoenzymes that incorporate a selenium atom in the active site have been characterized [13], and most of them are oxidoreductases catalyzing biologically important redox processes. Accordingly, there is an increasing interest in development of small selenium molecules that mimic redox functions of the selenoenzymes in biological systems. Nucleophilic introduction of a selenium moiety into a desired carbon skeleton is a common and facile strategy toward the synthesis of such organoselenium compounds. Halides and epoxides are most widely applied as electrophilic organic substrates in those reactions.

Synthesis and reactions of nucleophilic selenium reagents have already been reviewed in two books in early 2000s [7, 8]. Herein, new features of the synthesis and reactions of nucleophilic selenium reagents are surveyed from the literature between 1999 and 2010. The results are summarized along with traditional reagents. After a brief introduction of some recent unique applications of nucleophilic selenium reagents (Section 2.1.2), properties of selenols and selenolates (Section 2.2), inorganic reagents (Section 2.3), and organic reagents with a variety of Se–M (M = metal) chemical bonds (Section 2.4) are described.

$$R^1SeSeR^1 + R^2X \xrightarrow[\text{THF-Et}_3\text{N}]{\substack{\text{cat. RhCl(PPh}_3)_3 \\ H_2}} R^1\text{-Se-}R^2$$

Scheme 2.3

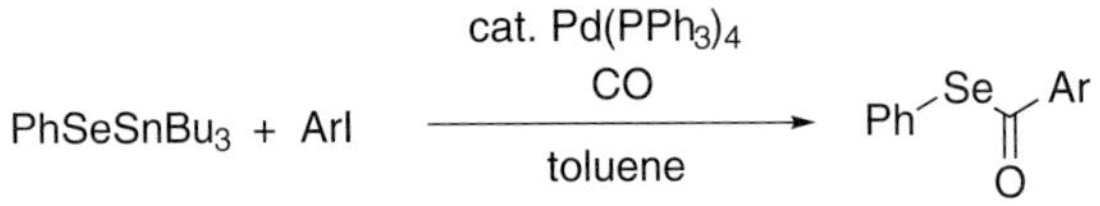

$$PhSeSnBu_3 + ArI \xrightarrow[\text{toluene}]{\substack{\text{cat. Pd(PPh}_3)_4 \\ CO}} Ph\text{-Se-C(=O)-Ar}$$

Scheme 2.4

2.1.2
Examples of Recent Applications

Nucleophilic selenium reagents are useful for the synthesis of various organoselenium compounds and inorganic metal complexes with a selenolate ligand, which

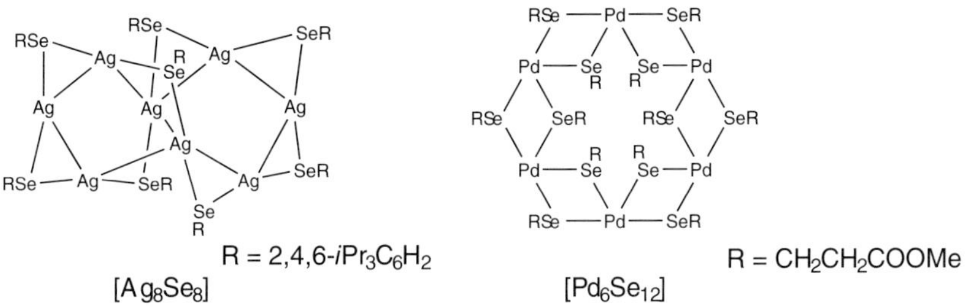

Figure 2.1 Examples of metal clusters containing bridging selenolate ligands.

can be applied as drugs, mimics of biomolecules, catalysts, advanced materials, etc. in various fields of chemistry. One interesting recent application of this chemistry is development of the coupling reaction between nucleophilic selenium reagents, such as benzeneselenol (PhSeH) and phenyl tributylstannyl selenide (PhSeSnBu$_3$), and aryl or alkyl halides catalyzed by transition metal complexes [14–17]. Examples are shown in Schemes 2.3 and 2.4. In these reactions, an involvement of metal selenolate intermediates during the catalytic cycle was suggested.

Another example is the synthesis of metal clusters containing bridging selenolate ligands by the reaction of nucleophilic selenium reagents with various metal compounds. For example, an octanuclear silver cluster (Ag$_8$Se$_8$) [18] and a hexanuclear palladium cluster (Pd$_6$Se$_{12}$) [19] were synthesized (Figure 2.1), and their molecular structures were determined by X-ray crystallographic analysis. These clusters would be useful as optical limiting materials, catalysts, or advanced electronic devices. The latter palladium complex was further transformed to (Pd$_{17}$Se$_{15}$) nanoparticle upon heating [19]. Similarly, ZnSe, CdSe, and HgSe nanoparticles have been prepared from metal selenolates M(SeR)$_2$ (M = Zn, Cd, and Hg) [20–22]. Structural diversity of metal selenolate complexes attracts much interest in the fields of inorganic chemistry and material science recently. Nucleophilic selenolates have also been utilized as an anchor to gold surface instead of conventional organic thiolates [23].

Among numerous applications of nucleophilic selenium reagents to biochemical research, the synthesis of radioactive organoselenium compounds is a unique example [24]. Radioactive compounds are useful to trace metabolic processes of sulfur-containing drugs and selenium compounds themselves. For these purposes, ^{73}Se (half-life 7.1 h) and ^{75}Se (half-life 120.4 d) isotope-labeled compounds were derivatized in a short period starting from radioactive selenite (SeO$_3^{2-}$) (Scheme 2.5). After conversion of the selenite to selenourea **1** [25, 26] or lithium methaneselenolate (MeSeLi) [26], the resulting nucleophilic selenium species were applied to the synthesis of asymmetric selenoethers, such as selenomethionine **2** [25], homoselenocysteine, and the selenolactone [26]. These isotope-labeled selenium compounds are then utilized for pharmaceutical research. The metabolic process of radioactive SeO$_3^{2-}$ was investigated by the direct injection to rats [27].

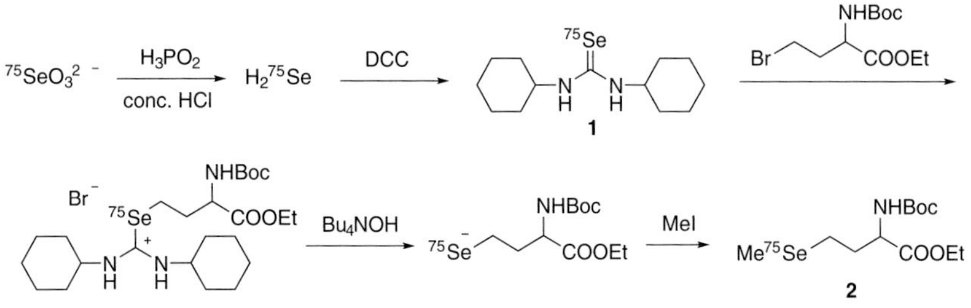

Scheme 2.5

The selenium metabolites are targeted by the metal ions that are involved at the active sites of various metalloenzymes [28].

Another interesting biological application of nucleophilic selenium reagents is extension of the native chemical ligation methodology mediated by cysteine to selenium chemistry (Scheme 2.6) [29–31]. Since selenol has a lower pK_a value than thiol (vide infra) and the corresponding selenolate anion is more nucleophilic than the thiolate anion, the coupling between a C-terminal thioester group of a polypeptide and a selenocysteine residue at the N-terminal of another polypeptide proceeds efficiently to produce an elongated peptide molecule having a selenocysteine residue at the coupling point. This selenocysteine-mediated native chemical ligation method is useful for the preparation of selenopeptides or selenoproteins that have selenocysteine residues at the desired positions.

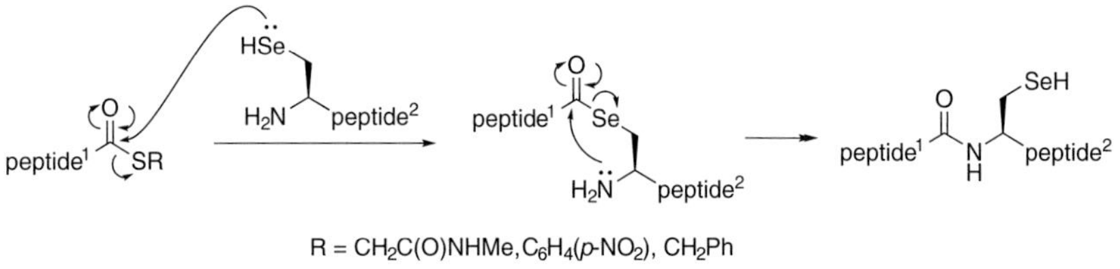

Scheme 2.6

2.2
Properties of Selenols and Selenolates

2.2.1
Electronegativity of Selenium

Since selenium is located at the right-hand region on the periodic table of elements, the chemical bonds with all metallic atoms (M) and a hydrogen atom (H) are polarized with a partial negative charge on the selenium atom (Se). Such a selenium atom can be utilized as a nucleophile toward various electrophilic substrates. Nucleophilicity of the selenium atom is controlled by the extent of polarization of the Se–M bond, which should depend on the electronegativity of the metallic center: the more electropositive the metal atom, the more nucleophilic the selenium atom. Most of all the possible Se–M chemical bonds are already known and have been characterized in the solid state by X-ray analysis. Some of these are already summarized in recent reviews [32–37].

2.2.2
Tautomerism of Selenols

The acid dissociation constant in water for hydrogen selenide (H_2Se; pK_a = 3.89) is lower than that for hydrogen sulfide (H_2S; pK_a = 7.05), indicating that H_2Se is a stronger acid. Analogously, organic selenols (RSeH) are stronger acids than thiols (RSH). The pK_a values for $HSeCH_2CO_2H$ and C_6H_5SeH are 4.7 [38] and 5.9 [39], respectively, whereas those for CH_3SH, $H_2NCH(CO_2H)CH_2SH$ (L-cysteine), and C_6H_5SH are 10.33, 8.7, and 6.62 [40], respectively, indicating that selenols are dominantly deprotonated to selenolate ions in neutral water.

Selenols are usually unstable under air atmosphere because they are easily oxidized to diselenides by oxygen. Therefore, selenols must be handled under inert atmosphere. Various organic selenols were synthesized previously, including not only common alkyl and aryl selenols but also vinyl [41], allyl [42], propargyl [42], allenyl [43], and alkynyl [43] selenols. When a selenol (SeH) group is attached to an unsaturated C=C bond, slight bond-shortening of the Se–C bond was observed, suggesting the significant conjugation between the lone pair of the selenium atom and the π-electrons of the adjacent double bond [44]. On the other hand, the presence of an intramolecular weak hydrogen bond between an SeH group and distant π-electrons was reported for 2-propene-1-selenol ($H_2C{=}CHCH_2SeH$) [45].

In solution, the SeH group attached to an unsaturated bond tautomerizes to the corresponding selone (or selenoketone) form having a C=Se bond. The selenol and selone forms can exist as discrete species like enol/ketone tautomerism, and this was clearly demonstrated for pyridine-2-selenol **4** [46]. The reaction of 2-bromopyridine **3** with Na_2Se_2 yielded **4** as a 1 : 6 mixture of the selenol **4a** and selone **4b** forms (Scheme 2.7). For the case of selenoimidazole **5**, however, it was proven that the selone tautomer **5b** is dominant (Scheme 2.8) [47, 48], which is highly polarized with a resonance contribution from the zwitterionic form **5c**. Indeed, significant nucleophilicity of the selenium atom in the selone tautomer **5b** was demonstrated in the reactions with diselenides [47]. Selenocarbonyl

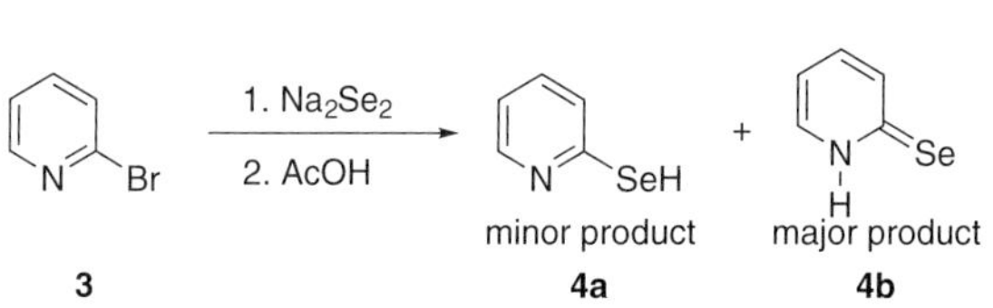

Scheme 2.7

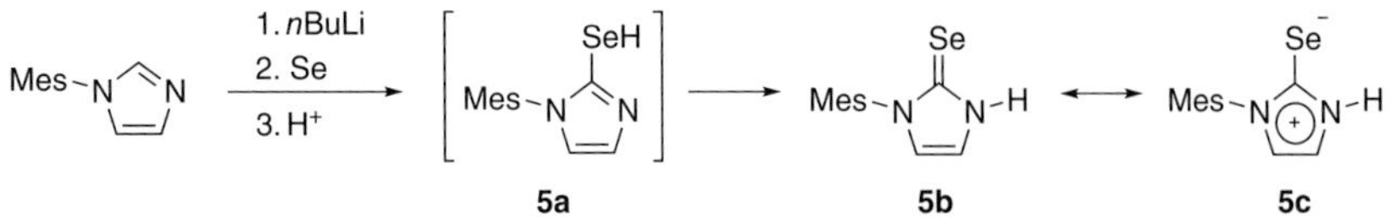

Scheme 2.8

compounds, such as selenoaldehydes (RC[=Se]H) and selenoketones (RC[=Se]R′), are generally unstable unless they are sterically protected from reactions with other molecules. However, selenoamides (RC[=Se]NR′R″) and selenoureas (RR′NC[=Se]NRR′) are fairly stable due to resonance delocalization of the nitrogen lone-pair electrons.

Polyselenols (RSe$_n$H) and the corresponding polyselenolates (RSe$_n^-$) are unstable and can be obtained only in solution. The reaction of two equivalents of elemental selenium with n-butyl lithium afforded lithium n-butyldiselenolate **6** in situ, which reacted with n-butyl bromide to produce di(n-butyl) diselenide **7** in good yield [49]. n-Butyldiselenol **8** was formed by the addition of hydrochloric acid to the solution of **6**, but it readily disproportionated to n-butyl selenol **9** with precipitation of gray selenium (Scheme 2.9) [49].

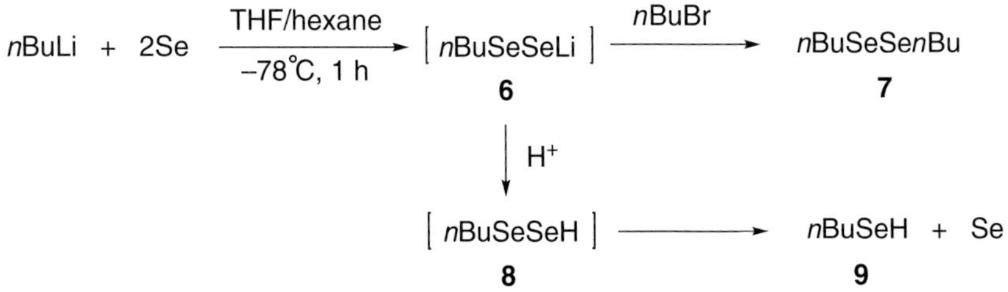

Scheme 2.9

Paris and coworkers studied polyselenolate dianions (Se$_n^{2-}$) produced by electrochemical reduction of selenium deposited on a gold grid electrode [50]. The Se$_8^{2-}$, Se$_6^{2-}$, and Se$_4^{2-}$ species could be characterized by absorption spectroscopy, but formation of Se$_2^{2-}$ was suspected: Se$_2^{2-}$ was suggested to be highly unstable and easily disproportionate to Se$_3^{2-}$ and Se^{2-}. On the other hand, formation of polyselenolates ions, PhSe$_n^-$ and PhCH$_2$Se$_n^-$ (n = 2–4), by the reaction of the corresponding monoselenolate (PhSe$^-$ and PhCH$_2$Se$^-$) with elemental selenium in dimethylacetamide was observed in the absorption spectra [50]. Arenepolyselenolate ArSe$_n^-$ was useful for the synthesis of diaryl triselenides (ArSe$_3$Ar) [51].

2.2.3
Nucleophilicity of Selenolates

Selenolates are highly nucleophilic and much more easily oxidized with various oxidants, such as oxygen and peroxides, than selenols. The oxidation path of

2.3 Inorganic Nucleophilic Selenium Reagents

methaneselenolate (MeSe⁻) with hydrogen peroxide (H$_2$O$_2$) was studied theoretically at high levels of theory, such as QCIST(D) and MP2, including solvent models [52]. The results indicated that the oxidation proceeds through a single transition state with a close contact between the Se atom and one HO fragment of H$_2$O$_2$. Although selenolates are more easily oxidized than the corresponding thiolates, the activation energy for MeSe⁻ in water (G_s = 65.9 kJ mol^{-1}) was not significantly different from the energy calculated for the corresponding thiolate substrate.

Bare selenolate species without any close contacts between the negatively charged selenium atom and the counter cation have been isolated and characterized in the solid state [46, 53, 54]. For example, the reaction of bis(2-pyridyl) diselenide **10** with KB(sBu)$_3$H in the presence of 18-crown-6 in THF afforded the corresponding free selenolate anion **11** (Scheme 2.10) [46]. A significant low field shift of the ^{77}Se NMR signal (δ = 441.8 ppm) suggested a contribution from the selone form. A free 2,4,6-tri-*tert*-butylbenzeneselenolate anion **12** was obtained by the treatment of selenol with potassium hydride in the presence of 18-crown-6 (Scheme 2.11) [54]. Such separated selenolates would be more reactive than the selenolates having a strong contact with a counter cation.

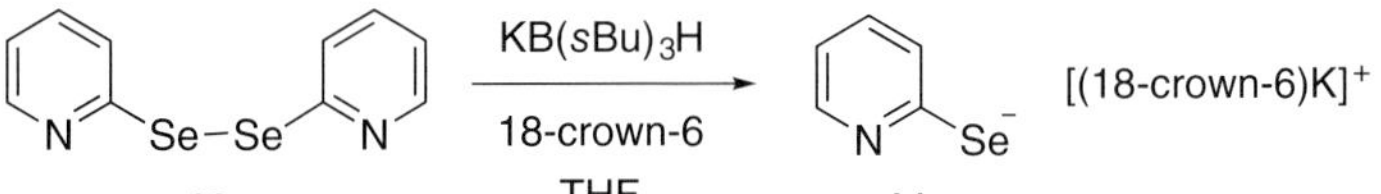

Scheme 2.10

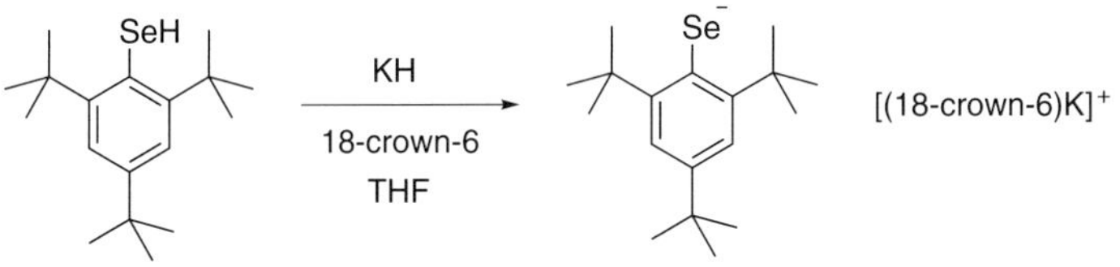

Scheme 2.11

2.3 Inorganic Nucleophilic Selenium Reagents

2.3.1 Conventional Reagents

There are many choices of inorganic selenium nucleophiles, such as Li$_2$Se$_2$, NaHSe, Na$_2$Se, Na$_2$Se$_2$, and KSeCN, in order to introduce a selenium atom into organic skeletons. These reagents are easily prepared *in situ* (Scheme 2.12) or available from commercial sources in the case of KSeCN. Among them, NaHSe

2 Nucleophilic Selenium

$$Se + NaBH_4 \xrightarrow[\text{solvent}]{} NaHSe \text{ or } Na_2Se_2$$

(solvent = H$_2$O, MeOH, EtOH, DMF)

$$Se + 2Na \xrightarrow[\text{liq. NH}_3]{} Na_2Se$$

$$Se + SmI_2 \longrightarrow Se^{2-} \text{ or } Se_2^{2-}$$

Scheme 2.12

and Na$_2$Se$_2$ obtained by the reaction of elemental selenium with an appropriate amount of NaBH$_4$ in a solvent, such as water [55], methanol [55], ethanol [55], or DMF [56], are most useful and easy-to-handle reagents. The reaction of selenium with sodium in liquid ammonia yields Na$_2$Se, which, after exchanging the solvent, is also frequently applied for various synthetic reactions [57]. The reaction of elemental selenium with samarium diiodide (SmI$_2$) affords nucleophilic selenium species, Se^{2-} or Se$_2^{2-}$, depending on the reaction stoichiometry [58]. These nucleophilic selenium reagents have been widely applied to the synthesis of various selenides and diselenides by reacting with halides, epoxides, activated alcohols, etc. [7, 8].

Bis(trimethylsilyl) selenide (Se[SiMe$_3$]$_2$) [59] is another frequently used inorganic selenium nucleophile [32]. In the recent applications, the reagent was used for the preparation of monomeric copper and silver complexes with a silylselenolate ligand, (RR′$_2$P)$_3$MSeSiMe$_3$ **13** (M=Cu) and **14** (M=Ag), by the reaction with CuOAc or AgOAc in the presence of a phosphine ligand (Scheme 2.13) [60]. The reaction of Se(SiMe$_3$)$_2$ with ([tmeda]Zn[OAc]$_2$, where tmeda = N,N,N′,N′-tetramethylethylenediamine) produced ([tmeda]Zn[SeSiMe$_3$]$_2$) **15** (Scheme 2.14),

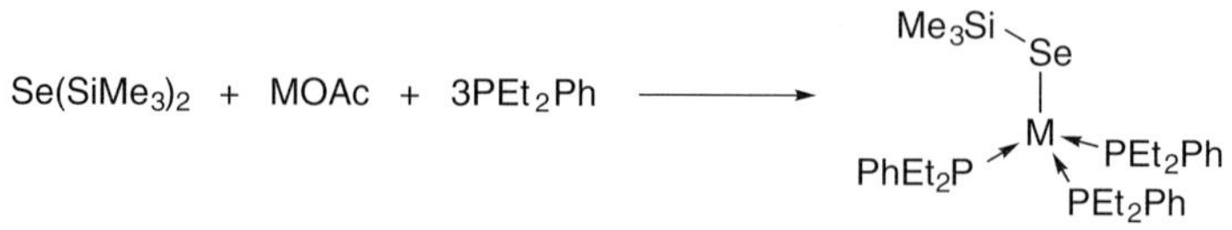

Se(SiMe$_3$)$_2$ + MOAc + 3PEt$_2$Ph ⟶ [Me$_3$Si–Se–M(PEt$_2$Ph)$_3$ complex] **13** M = Cu **14** M = Ag

Scheme 2.13

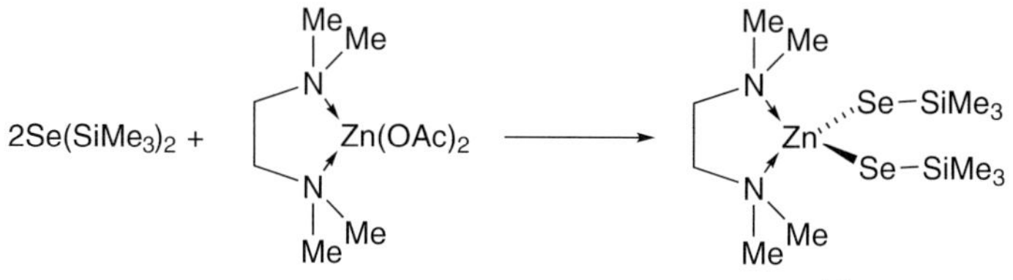

2Se(SiMe$_3$)$_2$ + [tmeda]Zn(OAc)$_2$ ⟶ [tmeda]Zn(Se–SiMe$_3$)$_2$ **15**

Scheme 2.14

which was then converted by the reaction with a cadmium compound to a complex cluster having Zn–Se–Cd linkages [61]. Thus, Se(SiMe$_3$)$_2$ can be used for the stepwise preparation of heteronuclear metal complexes with μ-Se ligands. Similar disilanyl selenides ([Me$_3$Si]Me$_2$Si)$_2$Se, ([Me$_3$Si]$_2$MeSi)$_2$Se, and ([Me$_3$Si]$_3$Si)$_2$Se were also synthesized, and their structures were characterized by NMR [62].

2.3.2
New Reagents

In addition to traditional inorganic nucleophilic selenium reagents, new nucleophilic selenium reagents have been developed recently. LiAlHSeH **16** was prepared by the reaction of gray selenium with LiAlH$_4$ in THF (Scheme 2.15) [63]. This highly nucleophilic selenium reagent was used for the synthesis of various selenium compounds (Scheme 2.16). For example, acyl chlorides were converted to diacyl selenides **17** or diacyl diselenides **18** in good yields [63]. The reaction of **16** with cyanamides in the presence of HCl yielded N,N-disubstituted selenoureas **19** [64]. Similarly, N-alkyl-Se-alkylselenocarbamates **20** [65] and selenothiocarbamates **21** or diselenocarbamates **22** [66] were synthesized from isocyanates and dichloromethylenedimethyliminium chloride, respectively. Reagent **16** was also applied as a key reagent for the coupling reaction between carboxylic acids, including amino acids and peptides, and azides to yield amide products **23** (Scheme 2.17) [67].

$$\text{LiAlH}_4 + \text{Se} \xrightarrow[\text{0°C, 0.5 h}]{\text{THF}} \text{LiAlHSeH} + \text{H}_2$$
16

Scheme 2.15

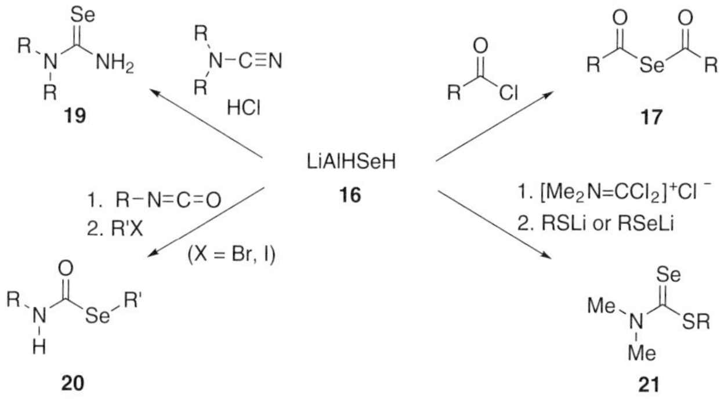

Scheme 2.16

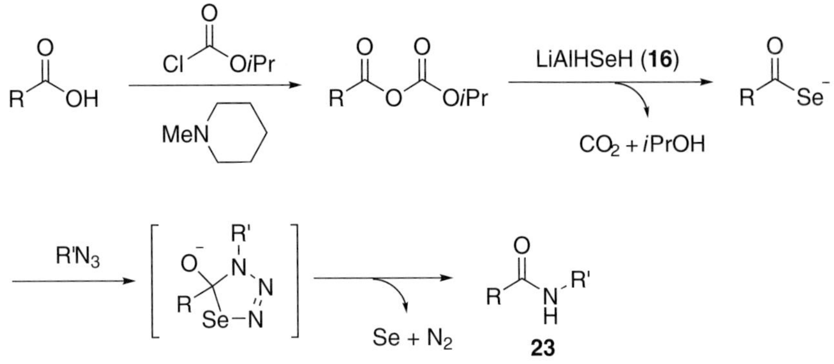

Scheme 2.17

Tetraethylammonium tetraselenotungstate ([Et$_4$N]$_2$WSe$_4$) **24** [68], which was prepared from K$_2$Se$_3$ by the reaction with W(CO)$_6$ and then Et$_4$NBr [69, 70], was applied as a nucleophilic selenium reagent to convert various halides and activated alcohols to the corresponding selenolates. Thus, various organic diselenides **25** [70], selenoamino acid derivatives **26** [71], and seleno-sugar compounds **27** [70, 72] were synthesized by the reaction with the corresponding halides or tosylates (Scheme 2.18). Amides were also converted to the corresponding selenoamides **28** by the reaction with **24** via a chloroiminium intermediate **29** (Scheme 2.19) [73]. Treatment of **24** with dimethyl acetylenedicarboxylate afforded unique alkyne and selenide complexes **30** and **31** (Scheme 2.20) [74].

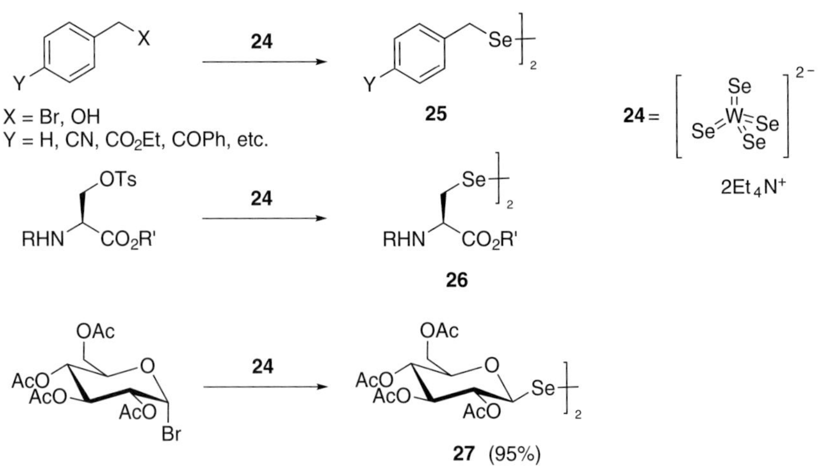

Scheme 2.18

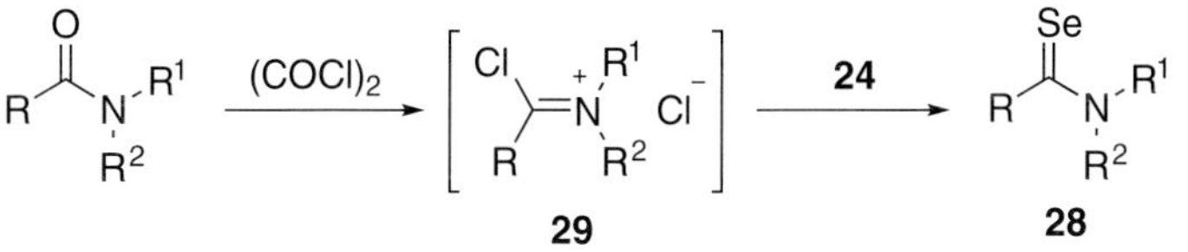

2.3 Inorganic Nucleophilic Selenium Reagents

Scheme 2.19

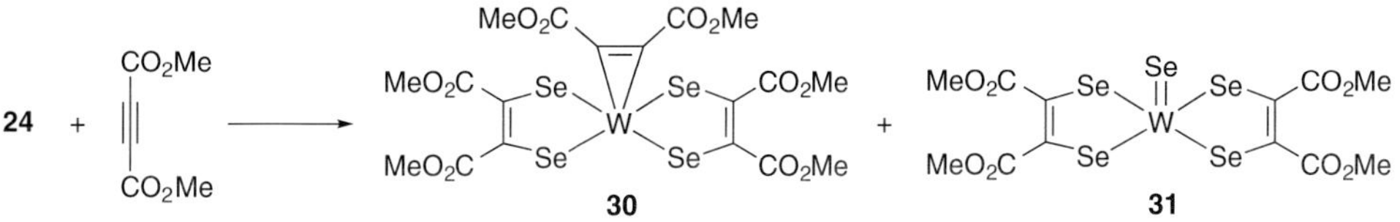

Scheme 2.20

Phosphorus–selenium compounds are also useful reagents to convert organic electrophiles to various selenium compounds. Tetraphosphorus decaselenide (P_4Se_{10}), prepared from gray selenium and red phosphorus [75], is a reagent frequently used for the conversion of a carbonyl (C=O) group to a selenoketone (C=Se) group [76]. For example, optically active selenoamides **32** and **33** were synthesized from the corresponding amides by the reaction with P_4Se_{10} in the presence of barium carbonate (Scheme 2.21), and their chiroptical behaviors were investigated [77]. Another important phosphorus–selenium reagent has been developed by Woollins (Scheme 2.22) [78]. Woollins' reagent (W.R.) **34**, like analogous Lawesson's reagent with two μ_2-sulfur atoms [79], can be used for the introduction of a chalcogen atom into various organic and inorganic substrates. Amides, carboxylic acids, and furans can be converted into selenoamides **35** [80], selenocarboxylic acids **36** [81], and selenophenes **37** [82], respectively, by using W.R.

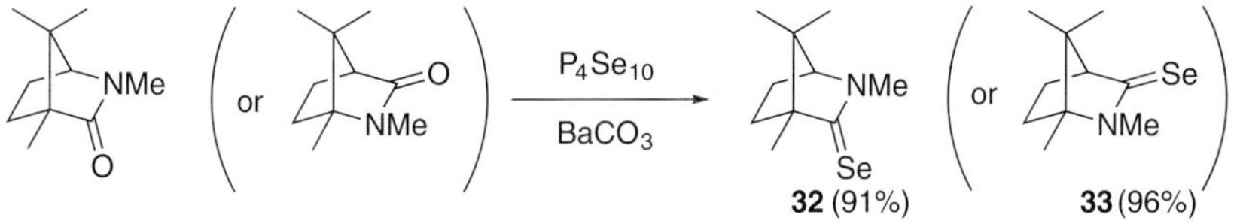

Scheme 2.21

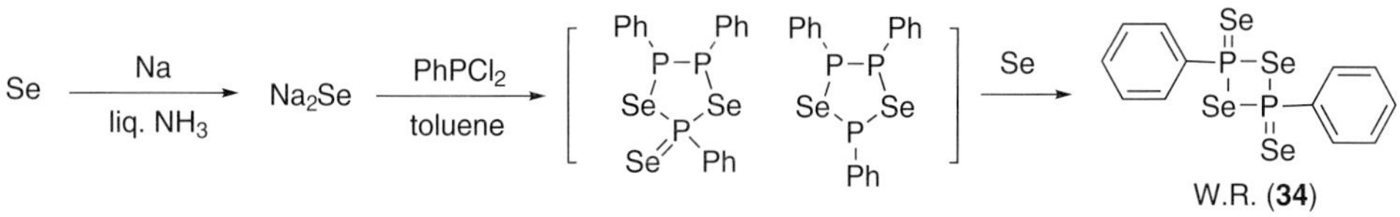

Scheme 2.22

(Scheme 2.23). W.R. was also applied to the synthesis of selenocysteine derivatives **38** from serine derivatives **39** by nucleophilic substitution probably via oxazoline intermediates **40** (Scheme 2.24) [83]. Nucleophilic ring opening of W.R. with metal salts, such as KOtBu, NaOAc, and CuOAc, produced various types of organometallic coordination polymers with P–Se–M (M = K, Na, Cu) linkages [84].

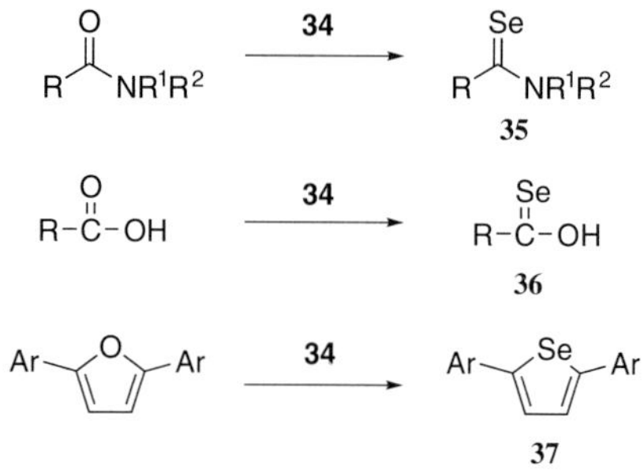

Scheme 2.23

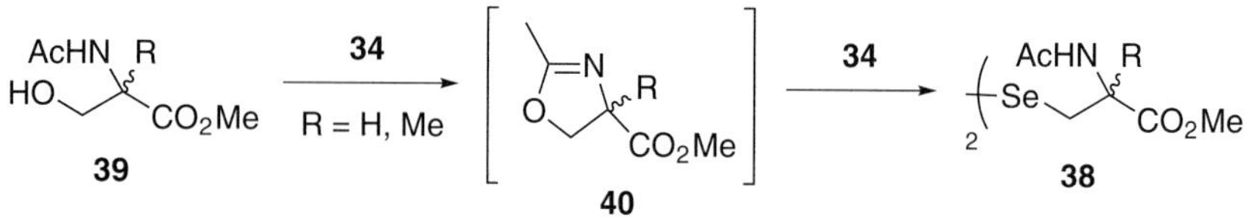

Scheme 2.24

The Wurtz reaction of dimeric ([Cl][Se=]P[μ-NtBu])$_2$ **41**, which was prepared by the reaction of (ClP[μ-N-tBu])$_2$ and Se, with sodium metal gave inorganic macrocycle ([Se=]P[μ-NtBu]$_2$P[μ-Se])$_6$ **42** via a selenium-centered nucleophilic intermediate **43** (Scheme 2.25) [85]. The inorganic macrocyclic backbone contains alternating P(III) and P(V) centers and possesses a cavity with a 8.2 Å diameter.

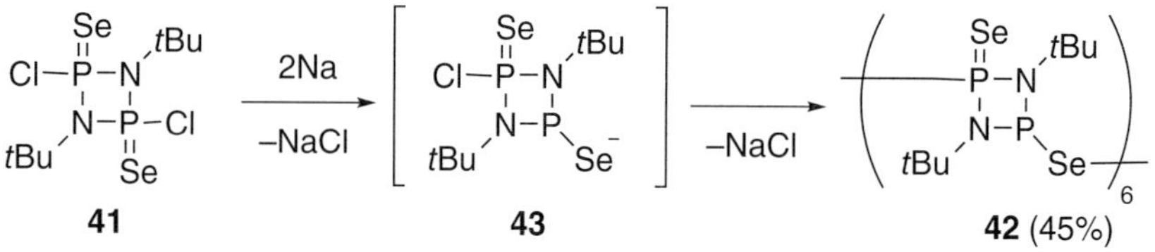

Scheme 2.25

The use of CuO nanopowder as a catalyst for the preparation of diselenides was reported by Braga and coworkers [86]. When a selenium nucleophile (Se^{2-}) **44**, which was generated by the reaction of elemental selenium with potassium hydroxide, was treated with aryl, alkyl, and heteroaryl iodides **45** in the presence of CuO nanopowder, the corresponding symmetrical diselenides **46** were obtained in good to excellent yields (Scheme 2.26) [86].

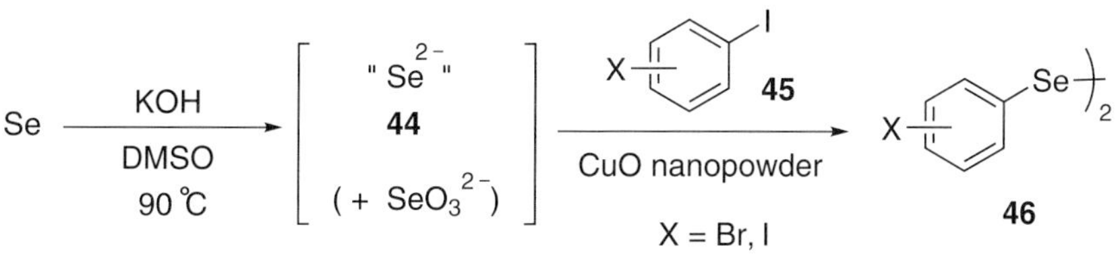

$$Se \xrightarrow[\text{DMSO}]{\text{KOH}} [\text{"Se}^{2-}\text{"} \quad \textbf{44} \quad (+ SeO_3^{2-})] \xrightarrow[\text{CuO nanopowder}]{X-C_6H_4-I \quad \textbf{45}} (X-C_6H_4-Se)_2 \quad \textbf{46}$$

X = Br, I

Scheme 2.26

2.4
Organic Nucleophilic Selenium Reagents

2.4.1
Preparation

There are several useful methods to prepare organic nucleophilic selenium reagents. Insertion of a selenium atom into a metal–carbon (M–C) bond (Scheme 2.27) and reductive cleavage of the Se–Se bond of diselenides (Scheme 2.28) are most frequently applied [7, 8]. Various reducing agents such as Na, NaH, NaBH$_4$, LiBEt$_3$H, LiAlH$_4$, Bu$_3$P–NaOH, iBu$_2$AlH, and SmI$_2$ have been used to cleave the Se–Se bond. In addition to these traditional reagents, a number of new reduction conditions were reported recently. For example, diaryl or dibenzyl diselenides (RSeSeR) were reduced with Zn/ZrCl$_4$ [87], Zn/AlCl$_3$ [88–90], or Zn/RuCl$_3$ [91] system to produce zinc selenolate ([RSe]$_2$Zn). Indium(I) iodide and indium(I)

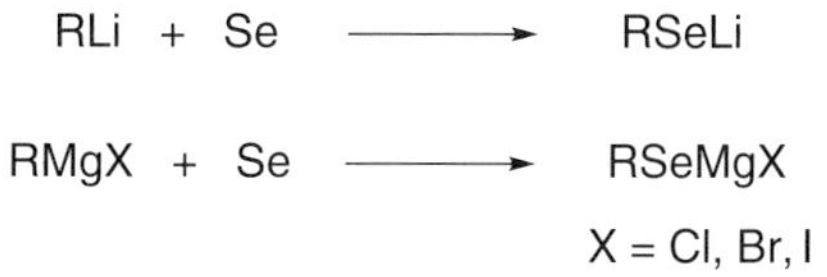

RLi + Se ⟶ RSeLi

RMgX + Se ⟶ RSeMgX

X = Cl, Br, I

Scheme 2.27

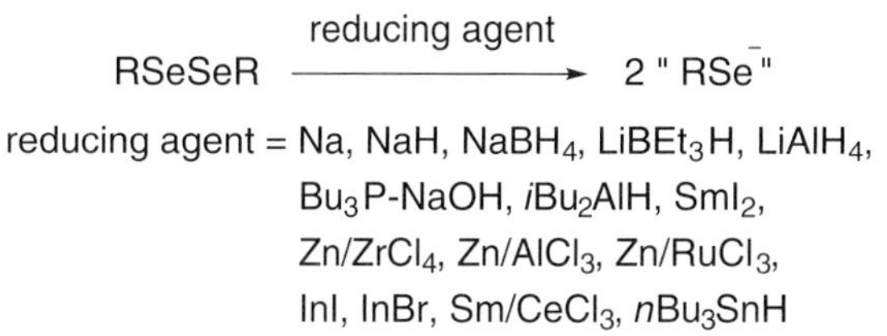

RSeSeR $\xrightarrow{\text{reducing agent}}$ 2 "RSe$^-$"

reducing agent = Na, NaH, NaBH$_4$, LiBEt$_3$H, LiAlH$_4$,
Bu$_3$P-NaOH, iBu$_2$AlH, SmI$_2$,
Zn/ZrCl$_4$, Zn/AlCl$_3$, Zn/RuCl$_3$,
InI, InBr, Sm/CeCl$_3$, nBu$_3$SnH

Scheme 2.28

bromide were applied to reduction of PhSeSePh to indium(III) bisselenolates (PhSe)$_2$InX (X = I or Br) [92, 93]. Sm/CeCl$_3$ could also reduce diaryl diselenides to selenolate anions ArSe$^-$, which was subsequently reacted with α,β-unsaturated esters or nitriles to afford β-selenoesters or β-selenonitriles, respectively [94]. Allylic and propargylic diselenides were efficiently reduced with nBu$_3$SnH in tetraglyme to yield the corresponding selenols [42]. Selenolates, prepared by reduction of selenocyanates (RSeCN) with LiAlH$_4$, were converted to selenols by treating with succinic acid [44].

The organic selenium nucleophiles thus obtained can be used not only for the preparation of various organoselenium compounds by reacting with various organic electrophiles, such as halide and epoxides but also for the synthesis of various metal complexes having a selenolate ligand by exchanging the metal ion with other metal ions. Thus, numerous selenolate complexes, which have a variety of selenium–metal (Se–M) bonds, have been prepared. Furthermore, various metal selenolate complexes with more than two different metal ions in the core, such as AgSeCo clusters [95], SnSeMn clusters [96], RhSeNi and RhSePd clusters [97], CuSeIn clusters [98, 99], and CuSeFe clusters [100], have also been synthesized.

2.4.2
Structure

A wide range of molecular structures for metal selenolates has been determined in the solid state by X-ray crystallography. The coordination patterns around the selenium atom can be classified to several categories (Figure 2.2). Free selenolates, which do not have any strong contacts to the neighboring atoms except for the covalently bonded carbon atom, have rarely been observed [46, 53, 54]. For most selenolate complexes, the selenium atom has close interaction with the counter cation(s). Usually, the selenolate ligand has a single contact to the metallic center to build a terminal selenolate complex (M–SeR). However, μ$_2$-selenolate complexes with a bridging selenolate ligand (M–Se[R]–M) are also observed frequently in larger metallic clusters. Complexes with μ$_3$- and μ$_4$-selenolate ligands are possible, but they are seldom observed. Recent examples of such selenolate compounds are categorized below according to the group of the elements involved as a counter cation.

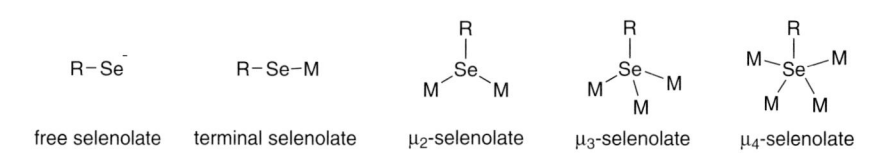

Figure 2.2 Coordination patterns around a selenium atom for metal selenolates.

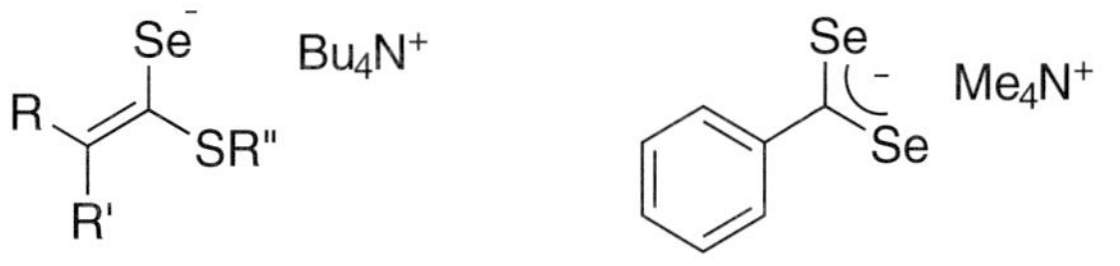

Figure 2.3 Examples of stable ammonium selenolates.

2.4.3
Ammonium Selenolates (NH$_4^+$)

Ammonium selenolates are unstable, but when the selenium atom is bonded to a sp^2 carbon atom, the resonance between the lone pair of the anionic selenium atom and the π-electrons would stabilize the ammonium salts. Thus, tetrabutylammonium eneselenolates **47** [101] and tetramethylammonium aromatic diselenolates **48** [102] were characterized in solution by ^{1}H, ^{13}C, and ^{77}Se NMR as well as in the solid state by X-ray analysis (Figure 2.3). N-Methylmorpholinium [103] and tetraphenyl phosphonium [53] salts were also reported.

2.4.4
Selenolates of Group 1 Elements (Li, Na, K, and Cs)

Lithium and sodium selenolates are the most common organic selenium nucleophiles. Lithium selenolates are usually synthesized *in situ* by the insertion of Se to the C–Li bond of organolithium reagents (Scheme 2.27). Reduction of diselenides with a lithium reagent, such as LiBEt$_3$H, LiAlH$_4$, can also be applied for the preparation of lithium selenolates. For example, PhSeLi was obtained by the reduction of PhSeSePh with methyl lithium (Scheme 2.29) [104]. On the other hand, sodium selenolates are easily prepared by the reduction of diselenides with various sodium reagents, such as metallic Na, NaH, and NaBH$_4$ (Scheme 2.28). Other alkaline metal selenolates are also known [34].

PhSeSePh + MeLi ⟶ PhSeLi + PhSeMe

Scheme 2.29

The molecular structure of dimeric lithium alkylselenolate (*n*BuSeLi·tmeda)$_2$ **49** was determined by X-ray analysis [105]. The selenolate formed a Li$_2$Se$_2$ cluster in the solid state with the Se–Li bond lengths in a range of 2.544 to 2.592 Å (Figure 2.4). On the other hand, selenolate **49** was in equilibrium between the dimeric and tetrameric forms in solution. The ^{77}Se NMR was observed at $\delta = -660$ ppm, suggesting that a large negative charge is localized on the Se atom. On the other hand, lithium 1-heptynylselenolate **50** showed the ^{77}Se NMR signal at −114.6 ppm in

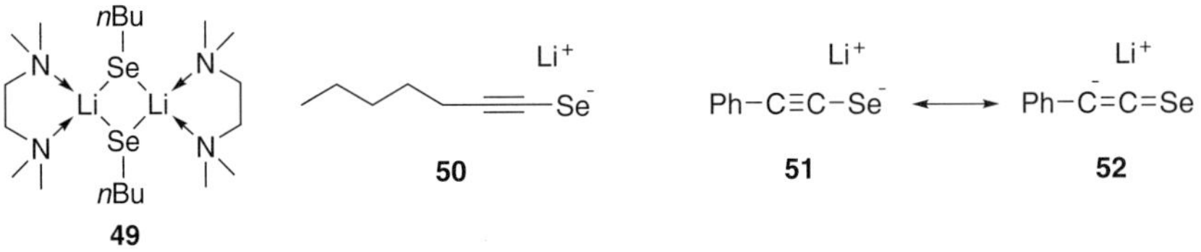

Figure 2.4 Examples of lithium selenolates.

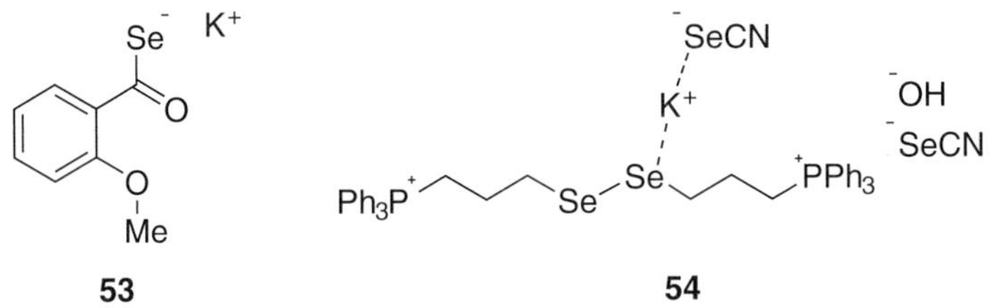

Figure 2.5 Examples of potassium selenolates.

THF, which shifted to −15.1 ppm in benzene. The downfield shift suggested that the selenolate and the lithium ion are separated in a polar solvent [106]. The ^{77}Se NMR chemical shift of lithium phenylalkynylselenolate **51** was observed at −59.3 ppm, showing importance of a selenoketene resonance structure **52** [106]. Similar behavior was also reported for potassium alkynylselenolates [107].

The solid-state structure of potassium selenolates was reported for selenocarboxylate **53**, in which the Se⋯K atomic distances were in a range of 3.309 to 3.625 Å [108]. A similar Se⋯K nonbonded distance (3.459 Å) was observed for $(Ph_3P^+[CH_2]_3Se)_2(SeCN^-)_2 \cdot KOH$ complex **54** in the solid state (Figure 2.5) [109].

Alkaline metal selenolates react with various electrophiles, such as halides [110], epoxides [111], and aziridines [111], as shown in Schemes 2.1 and 2.2. Besides these typical substrates, the reactions with α,β-unsaturated esters and alkynes were recently investigated. The reaction of PhSeLi with t-butyl acrylates in the presence of aldehydes gave Michael/Aldol tandem adducts **55** in moderate to good yields (Scheme 2.30) [104]. On the other hand, lithium eneselenolates, prepared from selenoamides by the reaction with LDA, reacted with various aldehydes, including α,β-unsaturated aldehydes, to give aldol-type condensation products **56** (Scheme

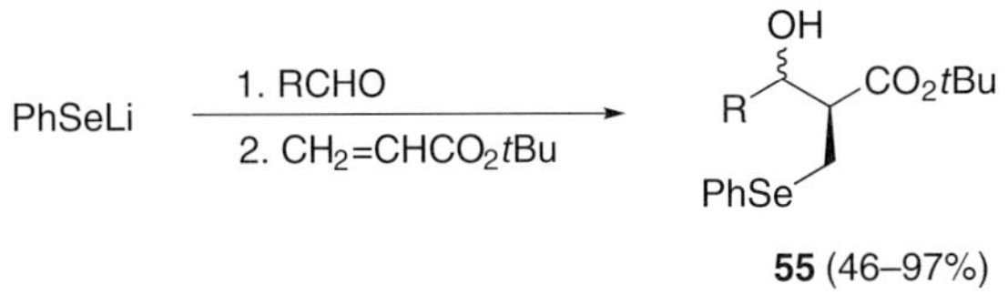

Scheme 2.30

2.4 Organic Nucleophilic Selenium Reagents | 69

2.31) [112]. Lithium alkynylselenolates (**51** ↔ **52**) exhibited ambident nucleophilic behavior depending on the solvent polarity and the electronic nature of the adjacent substituent [106, 113]. Intramolecular addition of a lithium selenolate moiety to a C≡C triple bond was used for the synthesis of selenophene-based heteroacenes **57** (Scheme 2.32) [114].

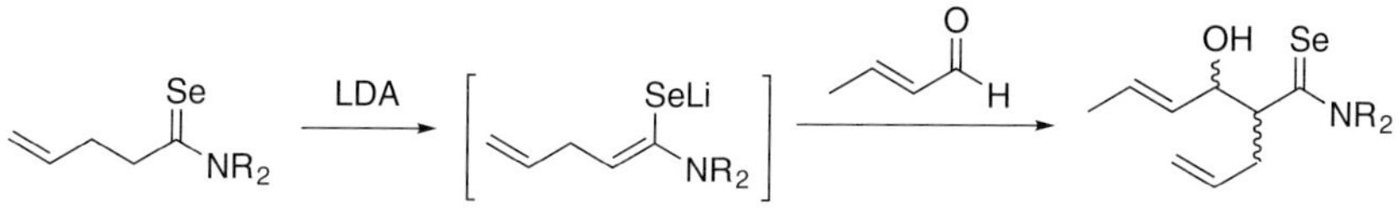

Scheme 2.31

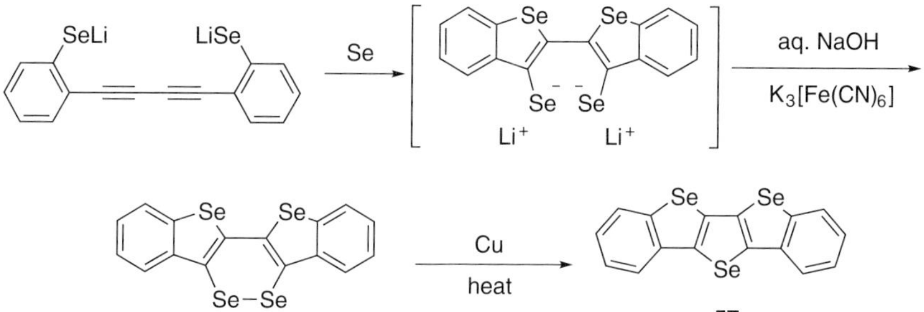

Scheme 2.32

Sodium benzeneselenolates (PhSeNa) was added to substituted alkynes to afford vinyl selenides **58** stereoselectively (Scheme 2.33) [115, 116]. Cesium selenolate (RSeCs) was suggested to be involved as an activated intermediate in the coupling reaction between Se-acyl selenoglycosides **59** and various halides in the presence of a secondary amine and Cs_2CO_3 (Scheme 2.34) [117].

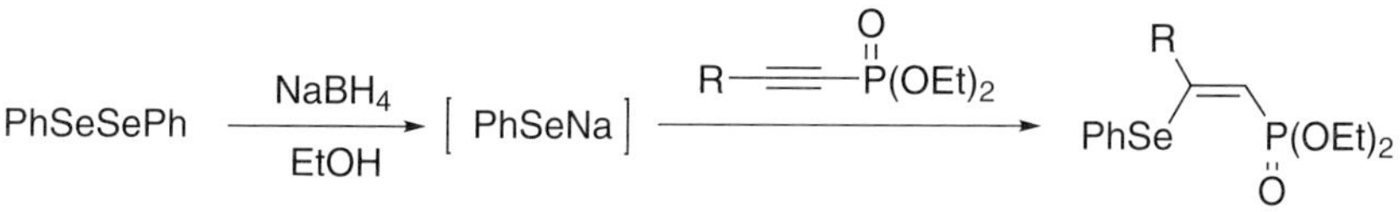

Scheme 2.33

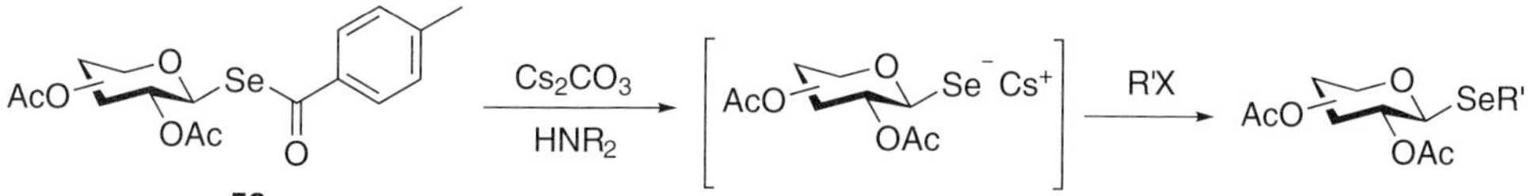

Scheme 2.34

2.4.5
Selenolates of Group 2 Elements (Mg, Ca, and Ba)

Magnesium selenolates (RSeMgX), which are useful nucleophilic selenium reagents in organic synthesis, can be obtained by reacting Grignard reagents with elemental selenium, as shown in Scheme 2.27. PhSeMgBr was alternatively synthesized by the reaction of PhSeSePh with MeMgBr, although it afforded PhSeMe as a side product (Scheme 2.35) [118]. The reagent was applied for the preparation of Michael/Aldol tandem adducts **60** by the reaction with β-substituted-α,β-unsaturated esters and aldehydes (Scheme 2.36) [118, 119].

PhSeSePh + MeMgBr ⟶ PhSeMgBr + PhSeMe

Scheme 2.35

PhSeMgBr + R–CH=CH–CO$_2$tBu + R'CHO ⟶ product **60** (PhSe, OH, R, R', CO$_2$tBu)

Scheme 2.36

In contrast to common magnesium reagents, selenolates of the other alkaline earth metals are rarely reported. Calcium selenolates and barium selenolates were synthesized by the reaction of bis(2,4,6-tri-*tert*-butylphenyl) diselenide **61** with calcium or barium in liquid ammonia. The selenolates were obtained as ion contact pairs **62** or separated ions **63** in the solid state, depending on the coexisting ligands (Scheme 2.37) [120, 121]. Interaction of a calcium ion with selenouracils **64** was studied in the gas phase by density functional theory calculation [122]. The result suggested that a calcium ion favors to associate with the oxygen and nitrogen atoms rather than the selenium atom and the association has a clear effect on the tautomerization process (Scheme 2.38).

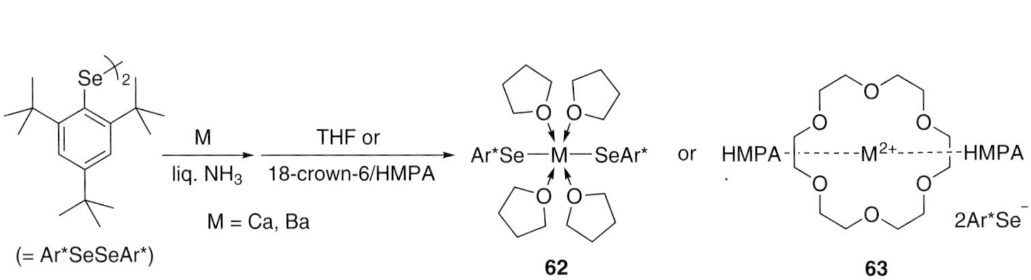

61 (= Ar*SeSeAr*) + M (M = Ca, Ba) → THF or liq. NH$_3$ / 18-crown-6/HMPA → Ar*Se—M—SeAr* (**62**) or HMPA---M^{2+}---HMPA (**63**) 2Ar*Se$^-$

Scheme 2.37

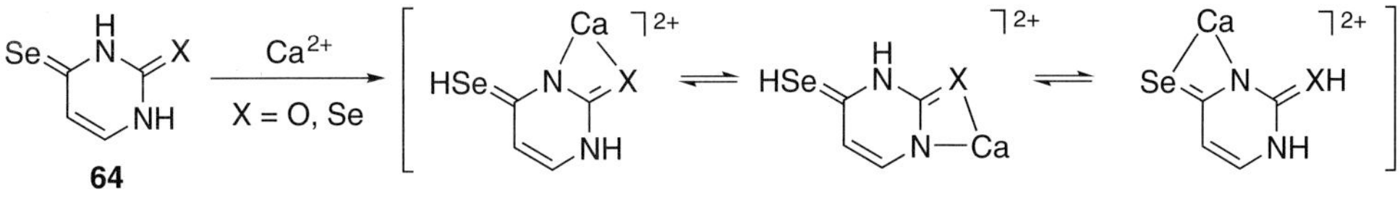

2.4 Organic Nucleophilic Selenium Reagents | 71

Scheme 2.38

2.4.6
Selenolates of Group 3 Elements (Sm, Ce, Pr, Nb, and U)

Samarium selenolates are most extensively studied among this class of selenolates. Reductive cleavage of the Se–Se bond of organic diselenides (RSeSeR) by the Sm/$HgCl_2$ system produced samarium selenolates (RSeSmCl$_2$) **65**, which were subsequently reacted with acyl chlorides, methyl chloroformats, halides, etc. to afford selenoesters **66**, selenoformates **67**, and unsymmetrical selenides **68** in good yields under mild and neutral reaction conditions (Scheme 2.39) [123]. On the other hand, diaryl diselenides (ArSeSeAr) reacted with samarium tris-pyrazolylborates **69** to give samarium selenolate complexes **70** (Scheme 2.40), whose structures were determined by X-ray analysis [124]. The reaction of PhSeSePh with metallic samarium in THF gave the unstable divalent samarium selenolate Sm(SePh)$_2$ **71**,

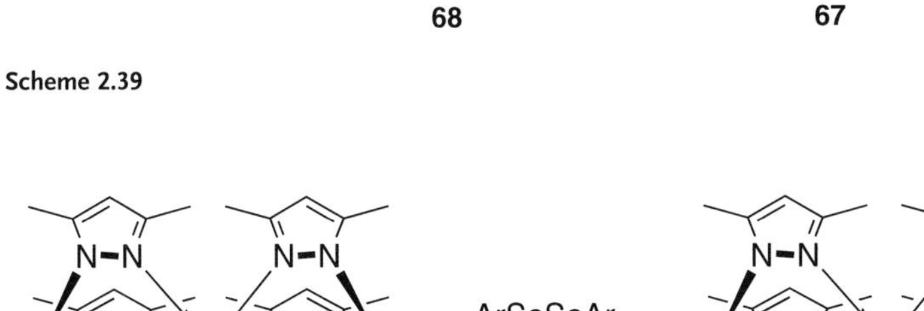

Scheme 2.39

Scheme 2.40

which could be isolated in the presence of Zn(SePh)$_2$ as polymeric heteronuclear complex **72** with bridging selenolate ligands (Scheme 2.41) [125].

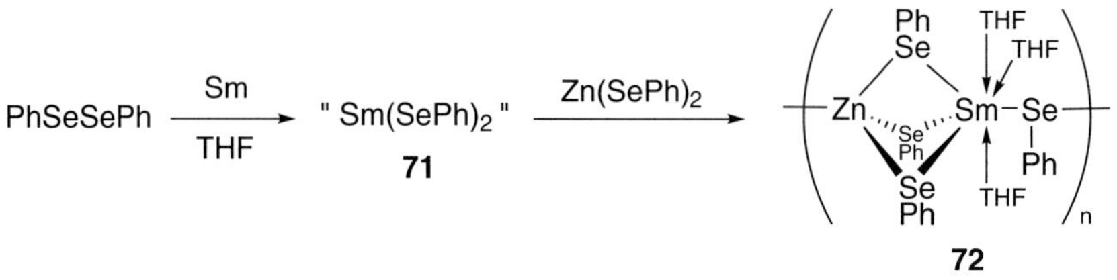

Scheme 2.41

The reaction of diselenides with samarium diiodide (SmI$_2$) in THF afforded samarium selenolate RSeSmI$_2$ [126]. The reactive nucleophilic selenium species was applied to the synthesis of selenosulfides **73** by reacting with sodium alkyl thiosulfonates or phenylsulfenyl chloride [127] and the synthesis of selenocarbamates **74** by reacting with chlorocarbonyl compounds (Scheme 2.42) [128].

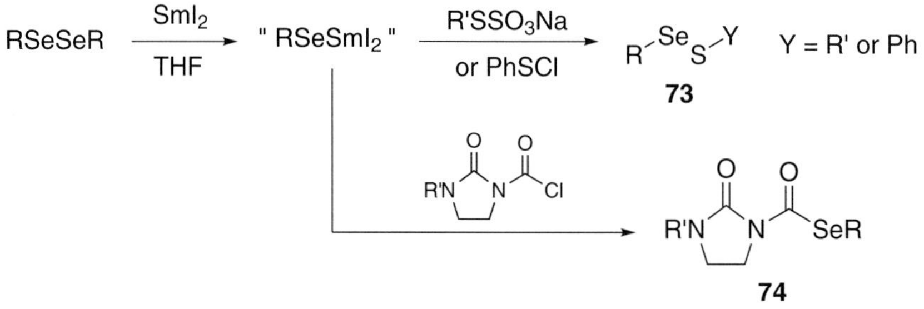

Scheme 2.42

Lanthanide benzeneselenolates (Ln[SePh]$_3$; Ln = Ce, Pr, Nd, Sm), which can be prepared by the reaction of PhSeSePh with the lanthanide and mercury in THF, reacted with SeO$_2$ to form octanuclear oxoselenido clusters with the general formula of (THF)$_8$Ln$_8$O$_2$Se$_2$(SePh)$_{16}$ **75** (Scheme 2.43) [129]. Similarly, Nd$_{12}$ [130], Pr$_{28}$ [131], and Nd$_{28}$ [131] clusters were synthesized and characterized by X-ray analysis. Such clusters exhibited unique near-IR emission properties [129–131]. Low-coordinate lanthanide structures were characterized in the solid state for LnCl(SeAr)$_2$ (Ln = Nd, Pr) **76**, which were stabilized by

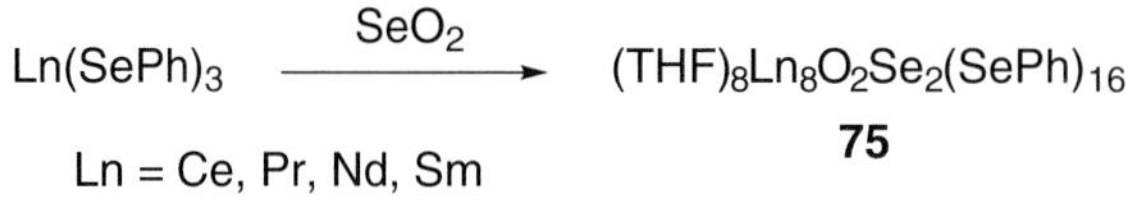

Ln(SePh)$_3$ $\xrightarrow{SeO_2}$ (THF)$_8$Ln$_8$O$_2$Se$_2$(SePh)$_{16}$

Ln = Ce, Pr, Nd, Sm

75

Scheme 2.43

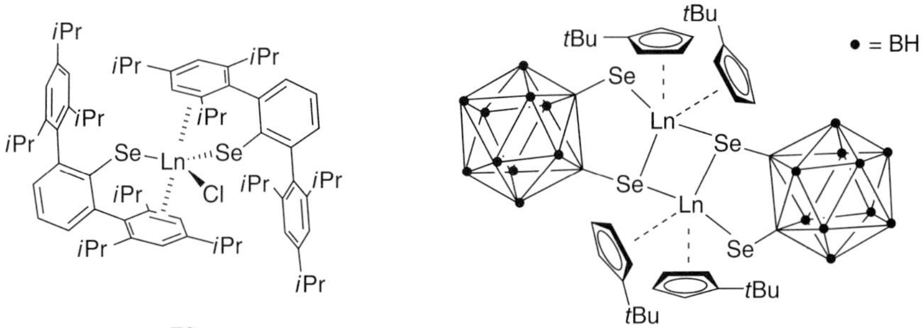

Figure 2.6 Molecular structures for lanthanide selenolates **76** and **77**.

π-coordination from two aromatic rings in both faces of the planar trigonal metal center [132]. Dimeric lanthanide carborane-1,2-diselenolate complexes ([C$_5$H$_4$*t*Bu]$_2$Ln[Se$_2$C$_2$B$_{10}$H$_{10}$])$_2$ (Ln = Nd, Gd, Dy, Er, Yb) **77** have also been prepared (Figure 2.6) [133].

Actinide selenolate (C$_5$Me$_5$)$_2$U(SePh)$_2$ **78** was prepared by the reaction of uranium hydride complex ([C$_5$Me$_5$]$_2$UH)$_2$ with PhSeSePh via four-electron reduction (Scheme 2.44) [134].

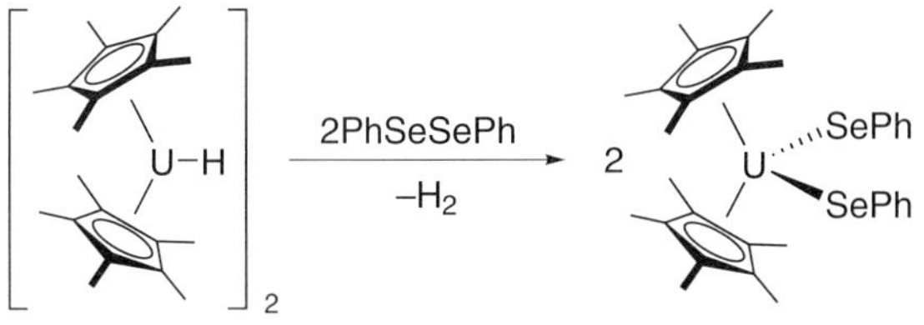

Scheme 2.44

2.4.7
Selenolates of Group 4 Elements (Ti, Zr, and Hf)

Titanium selenolate complexes with the general formula of Cp$_2$Ti(SeR)$_2$ **79** or Cp$_2$TiCl(SeR) **80** were synthesized through the reduction of Cp$_2$TiCl$_2$ with Mg and the subsequent treatment with RSeSeR depending on the stoichiometry of the reactants (Scheme 2.45) [135]. In the alternative method, group 4 metal selenolates Cp$_2$M(SeR)$_2$ (M = Ti, Zr, Hf; R = Me, *t*Bu) **81** were prepared by the reaction of Cp$_2$MCl$_2$ with a frozen THF solution of the mixture of Se and RLi (Scheme 2.46) [136]. These alkylselenolate complexes were used for the chemical vapor deposition of ZrSe$_2$ and HfSe$_2$ thin films.

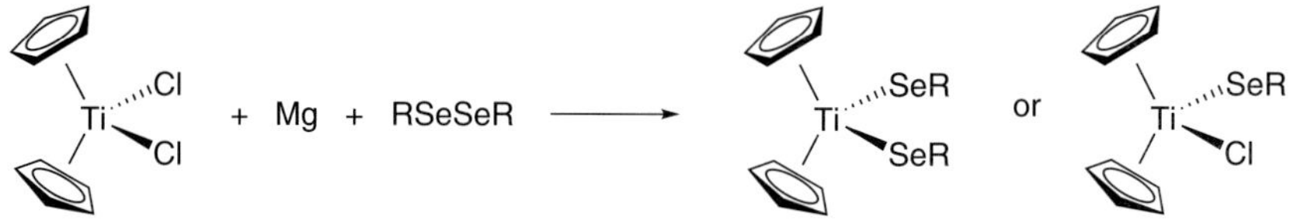

2 Nucleophilic Selenium

Scheme 2.45

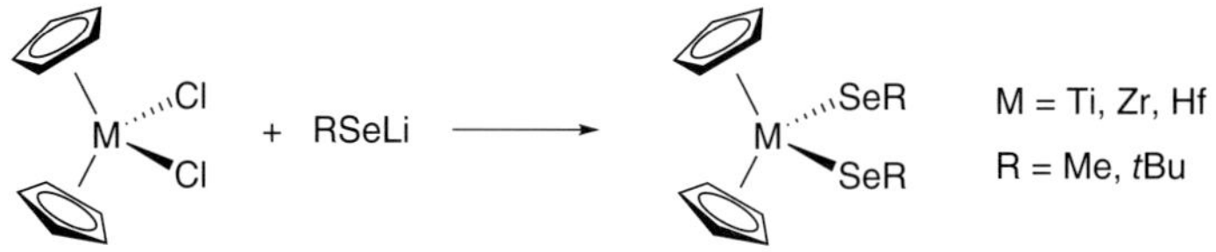

Scheme 2.46

2.4.8
Selenolates of Group 5 Elements (V, Nb, and Ta)

Vanadium selenolate Cp$_2$V(SetBu)$_2$ **82** was prepared by the reaction of Cp$_2$VCl$_2$ with two molecular equivalents of tBuSeLi in THF (Scheme 2.47) [137]. This moisture-sensitive selenolate complex was tested for the deposition of vanadium selenide (VSe$_2$), but it was not sufficiently volatile. Niobium and tantalum metallocene complexes **83** and **84** (Figure 2.7) with a diselenide or selenide ligand,

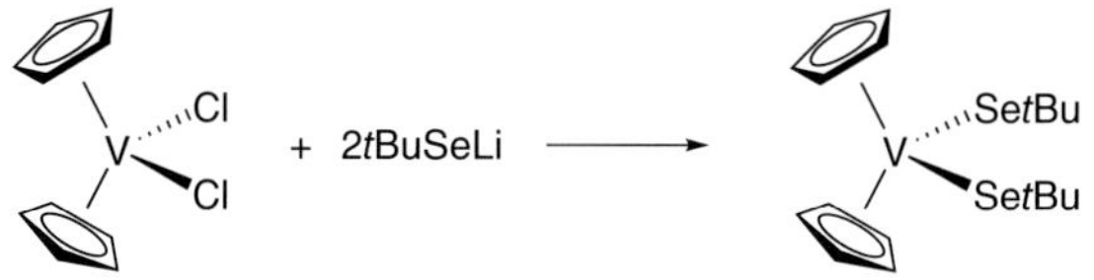

Scheme 2.47

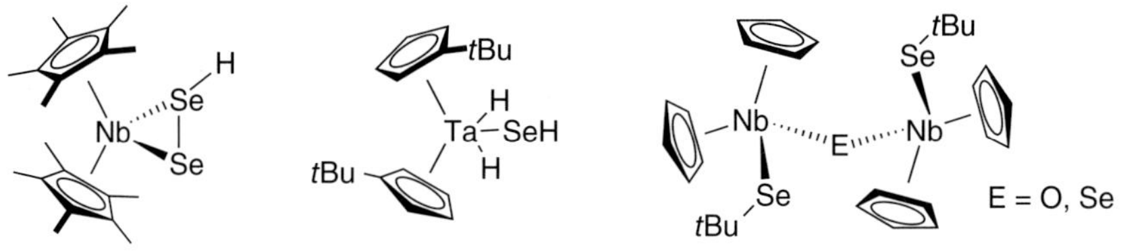

Figure 2.7 Examples of niobium and tantalum selenolates.

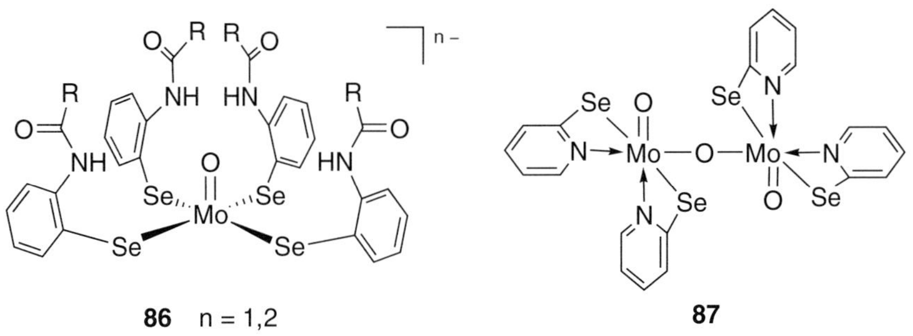

2.4 Organic Nucleophilic Selenium Reagents | 75

Figure 2.8 Molecular structures for molybdenum selenolates **86** and **87**.

respectively, were synthesized by the reaction of $(C_5Me_5)_2NbBH_4$ or $(C_5H_4tBu)_2TaH_3$ with elemental selenium [138]. Niobium selenolates $(Cp_2[tBuSe]Nb)_2E$ (E = O or Se) **85** were also prepared and characterized by X-ray analysis (Figure 2.7) [139].

2.4.9
Selenolates of Group 6 Elements (Mo and W)

Molybdenum areneselenolates $(Mo^{IV}O[SeAr]_4)^{2-}$ and $(Mo^VO[SeAr]_4)^-$ **86** were synthesized and characterized by NMR and X-ray analysis (Figure 2.8) [140]. The MoOSe$_4$ core possessed a typical square-pyramidal geometry. The conformational and molecular assembly behaviors were studied [141]. The reaction of MoO_2Cl_2(DME) with pyridine-2-selenol and *n*-butyllithium in THF resulted in the oxo-bridged molybdenum selenolate complex $(Mo_2O_3[SePy]_4)$ **87**, in which each molybdenum center has a distorted octahedral geometry [142]. The complex exhibited a catalytic activity for the oxygen-atom transfer from DMSO to PPh$_3$. A trinuclear Mo$_3$S$_7$ cluster with a diselenolate ligand **88**, which was prepared from zinc selenolate complex **89** and $(nBu_4N)_2(Mo_3S_7Br_6)$ (Scheme 2.48), was useful for the synthesis of magnetic single-component molecular conductors [143].

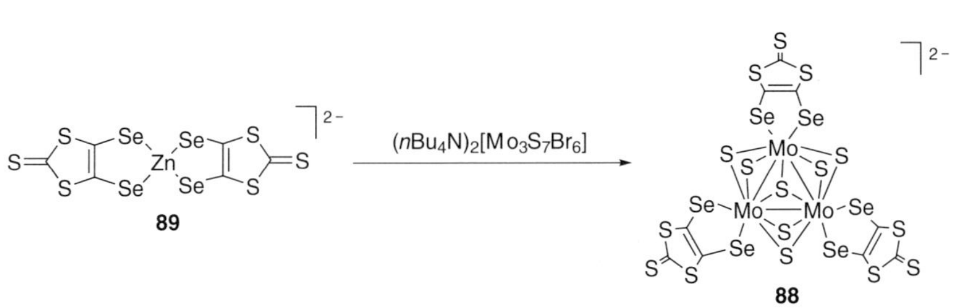

Scheme 2.48

The reaction of $(C_5Me_5)WCl_4$ with *t*BuSeLi in THF in the presence of *t*BuNC afforded W(II) selenolate complex $(C_5Me_5)W(SetBu)(CNtBu)_3$ **90** (Scheme 2.49), the molecular structure of which was clearly characterized by X-ray analysis [144].

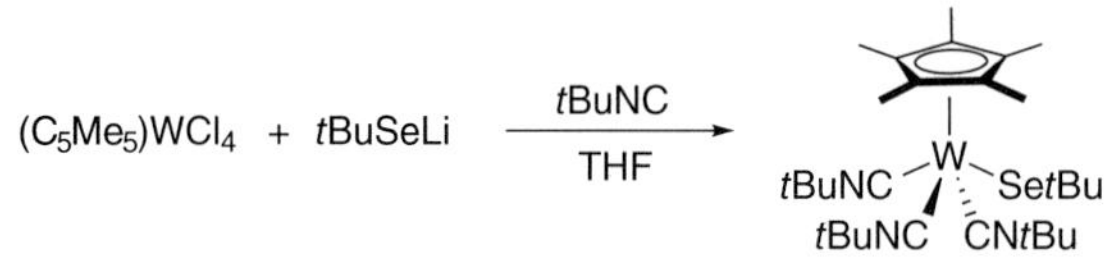

Scheme 2.49

$(C_5Me_5)WCl_4$ + tBuSeLi $\xrightarrow[\text{THF}]{t\text{BuNC}}$ **90**

Bis(dithiolene)tungsten selenolate complexes **91** were also synthesized and characterized to mimic the active site of molybdenum- or tungsten-containing enzymes, that is, formate dehydrogenases [145]. To synthesize these selenolate complexes, 2-adamantyl-, 1-adamantyl-, and *tert*-butyl-selenotrimethylsilanes (RSeSiMe$_3$) **92** were utilized (Scheme 2.50). Other types of molybdenum and tungsten selenolate complexes were also reported [146–148].

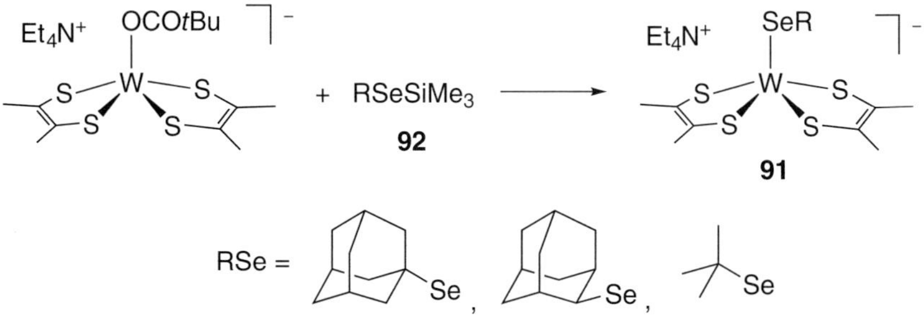

Scheme 2.50

2.4.10
Selenolates of Group 7 Elements (Mn and Re)

The reaction of Mn(OAc)$_2$ with 2 molar equivalents of PhSeSiMe$_3$ afforded polymeric manganese selenolate complex Mn(SePh)$_2$ **93**, which has tetravalent manganese(II) ions bridged by the two selenolate ligands forming one-dimensional chains [149]. The manganese complex reacted with 2,2′-bipyridine or 1,10-phenanthroline to yield the monomeric octahedral complex **94** (Scheme 2.51)

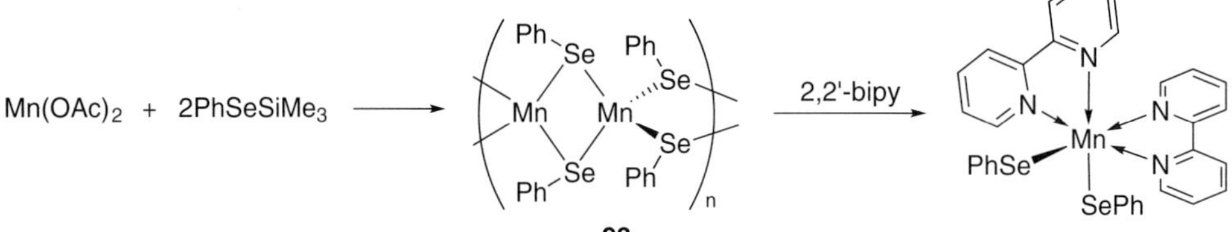

Scheme 2.51

2.4 Organic Nucleophilic Selenium Reagents

[149]. The reaction of Mn(η^5-C$_5$H$_7$)(CO)$_3$ with PhSeH yielded mononuclear MnL$_n$(SePh) **95**, dinuclear Mn$_2$L$_n$(μ-CO)(μ-SePh)$_2$ **96**, and heterocubane (Mn[CO]$_3$[μ$_3$-SePh])$_4$ **97** manganese complexes depending on the presence or absence of other ligands (L) (Scheme 2.52) [150]. The selenolate ligands of monomeric manganese complexes cis-(Mn[CO]$_4$[SeR]$_2$)$^-$ **98** exhibited nucleophilic reactivity toward nickel(II) and palladium(II) compounds to give trinuclear selenolate complexes L$_n$Mn(μ-SeR)$_2$Ni(μ-SePh)$_2$MnL$_n$ **99** containing a distorted square planar Ni(Se)$_4$ core and a dinuclear palladium complex **100** with a Pd$_2$(μ-SePh)$_2$ core, respectively (Scheme 2.53) [151].

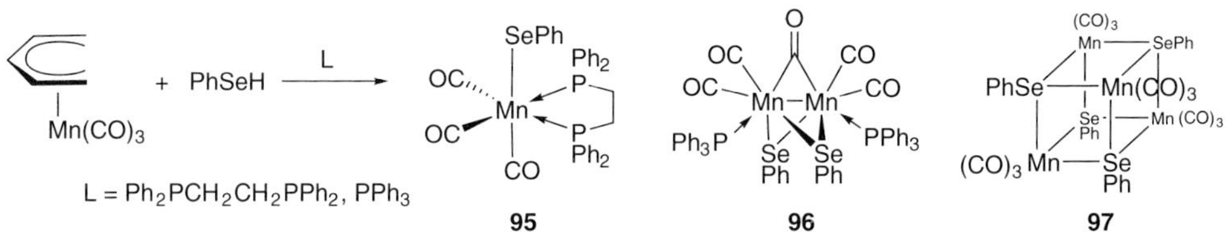

Scheme 2.52

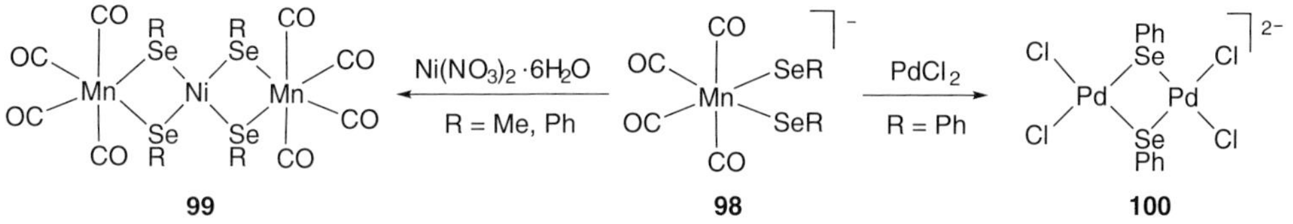

Scheme 2.53

Square-pyramidal rhenium selenolate complex (ReS[S$_2$C$_7$H$_{10}$][SeH][SH])$^-$ **101** (Figure 2.9) was produced by the addition of H$_2$Se to (ReS$_2$[S$_2$C$_7$H$_{10}$])$^-$ [152]. A similar pentacoordinate rhenium complex ReO(S$_3$C$_6$H$_{12}$)(SePh) **102** was synthesized and subjected to optical resolution [153]. The obtained enantiomers were characterized by VCD spectroscopy.

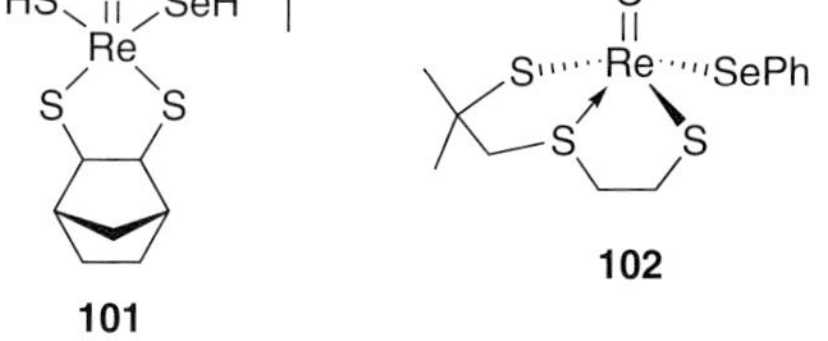

Figure 2.9 Examples of rhenium selenolates.

2.4.11
Selenolates of Group 8 Elements (Fe, Ru, and Os)

Various types of iron selenolate complexes have been synthesized and characterized. Most well studied are dinuclear iron complexes with one μ-selenolate and another μ-Y ligand to form a butterfly-shape Fe_2SeY cluster. One common starting material to synthesize these butterfly complexes is $([CO]_3Fe[\mu\text{-}SeR][\mu\text{-}CO]Fe[CO]_3)^-$ **103**, which was prepared *in situ* by the reaction of $Fe_3(CO)_{12}$, RSeH, and NEt_3 [154]. The μ-carbonyl ligand was exchanged with various ligands such as N-arylbenzimidoyl chloride **104** [155, 156], phosphaalkene **105** [157], diazo compounds **106** [158], isothiocyanates **107** [156], CS_2 and MeI [159], and PhSeBr [159] to yield the corresponding iron selenolate complexes with a Fe_2SeY core (Scheme 2.54). The μ-carbonyl ligand was also converted to a bridging carbine ligand by the treatment with electrophile Et_3OBF_4 [156, 160].

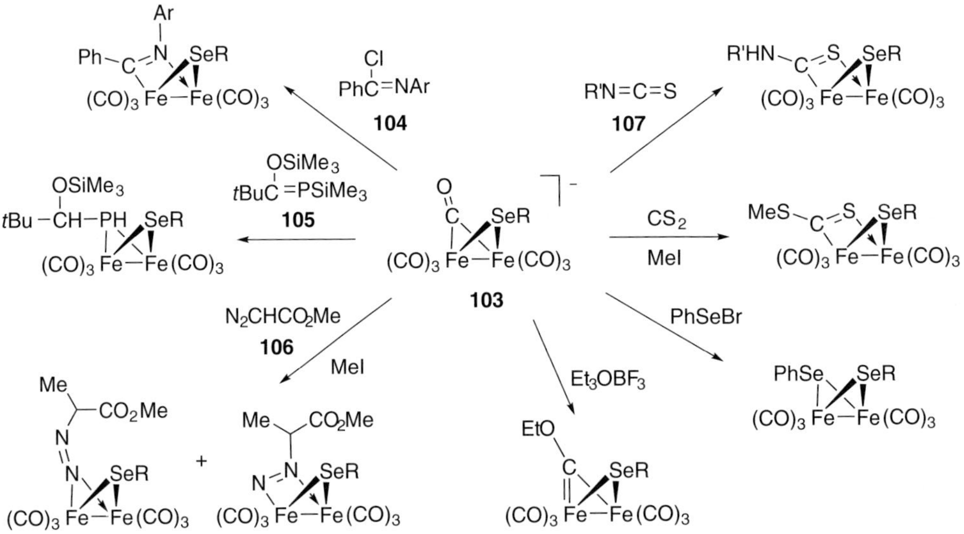

Scheme 2.54

Binuclear iron complexes with two μ-selenolate ligands were synthesized by several ways. The reaction of $M_2Fe(CO)_4$ (M = Na or K) with elemental selenium led to the formation of $M_2([CO]_6Fe_2[\mu\text{-}Se]_2)$ **108**, which was subsequently reacted with organo halides (RX) to yield selenolate complexes $(CO)_6Fe_2(\mu\text{-}SeR)_2$ **109** with a butterfly-shape Fe_2Se_2 core (Scheme 2.55) [161]. Disilyl diselenide $tBu_3SiSe\text{-}SeSitBu_3$ **110** [162] was reacted with two equivalents of $Fe(CO)_5$ under photochemical conditions to yield selenolate complex $(CO)_6Fe_2(\mu\text{-}SeSitBu_3)_2$ **111** [163]. This complex was reduced electrochemically or by sodium or potassium to generate nucleophilic selenolate $tBu_3Si\text{-}Se^-$ (Scheme 2.56). A similar butterfly-shape diiron diselenolate complex $(CO)_6Fe_2(\mu\text{-}Se_2C_3H_5CH_3)$ **112** was synthesized as a model of

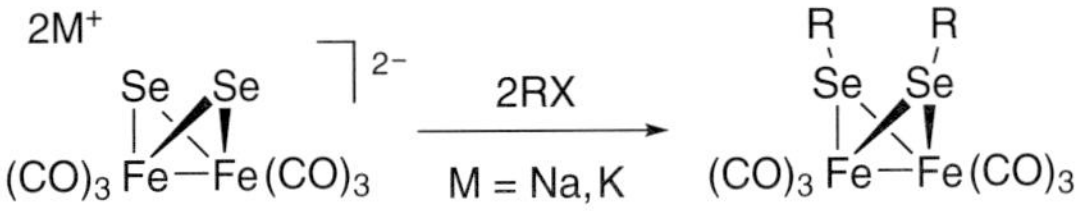

2.4 Organic Nucleophilic Selenium Reagents

Scheme 2.55

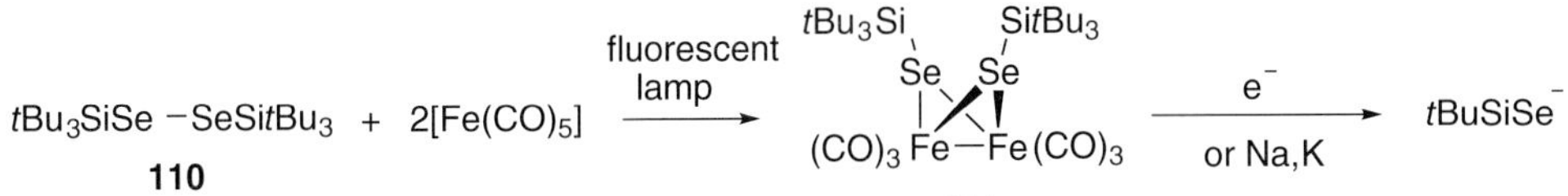

$[\text{(CO)}_3\text{Fe}-\text{Fe(CO)}_3(\mu\text{-Se})_2]^{2-} \; 2M^+$ **108** (M = Na, K) $\xrightarrow{2RX}$ (CO)$_3$Fe–Fe(CO)$_3$ with Se–R, Se–R bridges **109**

Scheme 2.56

$t\text{Bu}_3\text{SiSe}-\text{SeSi}t\text{Bu}_3$ **110** + 2[Fe(CO)$_5$] $\xrightarrow{\text{fluorescent lamp}}$ (CO)$_3$Fe–Fe(CO)$_3$ with Se–SitBu$_3$ bridges **111** $\xrightarrow{e^-\text{ or Na, K}}$ tBuSiSe$^-$

the active site of (FeFe)-hydrogenases, and the catalytic activity in the electrochemical reduction of acetic acid to molecular hydrogen was investigated by exchanging the one CO ligand with PPh$_3$ (Scheme 2.57) [164].

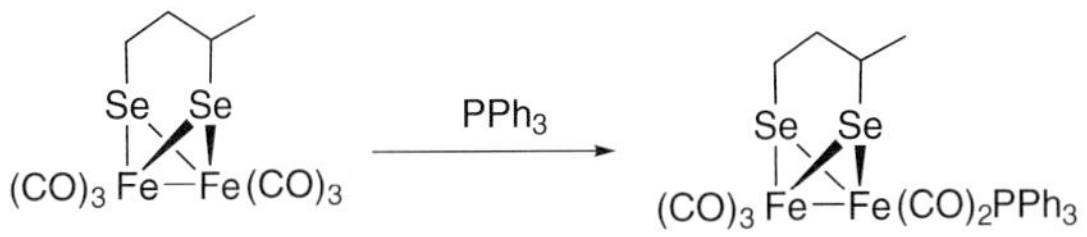

Scheme 2.57

112 $\xrightarrow{\text{PPh}_3}$ (CO)$_3$Fe–Fe(CO)$_2$PPh$_3$

On the other hand, the reaction of iron(II) bis(trimethylsilyl)amide Fe$_2$(N[SiMe$_3$]$_2$)$_4$ with sterically bulky aromatic selenols ArSeH yielded three-coordinate selenolate iron complexes **113** with a square planar Fe$_2$Se$_2$ cluster core (Figure 2.10) [165]. A similar selenolate iron complex (NBnMe$_3$)$_2$(Fe$_2$[SeiPr]$_6$) **114** was prepared by the reaction of FeCl$_2$ with iPrSeNa [166]. Complex **114** reacted with FeCl$_2$ and CuCl to afford heterometallic selenolate complexes, such as (NBnMe$_3$)$_2$(Fe$_2$Cu$_4$[SeiPr]$_8$Cl$_2$)

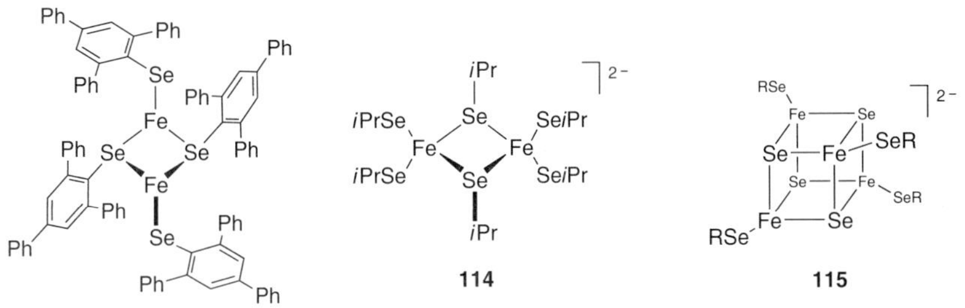

Figure 2.10 Examples of iron(II) selenolate complexes.

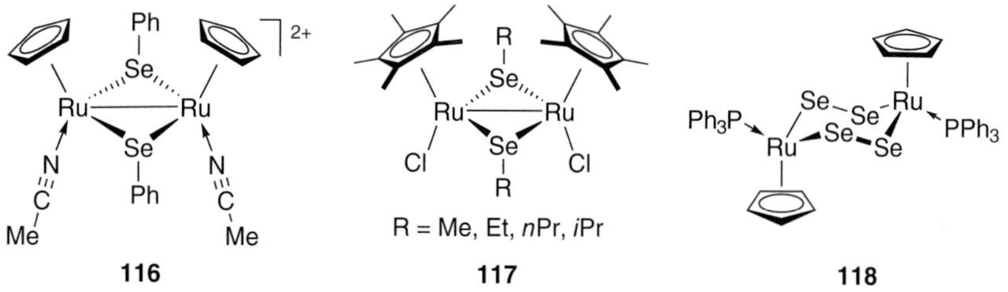

80 | *2 Nucleophilic Selenium*

Figure 2.11 Examples of ruthenium(III) selenolate complexes.

[167]. More complex iron–selenium clusters, such as cubic Fe$_4$Se$_4$ cluster **115** with organic selenolate ligands, were prepared as a model of the active site of iron–sulfur enzymes that catalyze biologically important redox reactions, such as photosynthesis [168].

Ruthenium(III) selenolate complexes **116** and **117** (Figure 2.11) with a puckered Ru$_2$Se$_2$ cluster core were prepared by the reaction of a ruthenium(II) complex with diselenides [169, 170]. Complex **117** exhibited a catalytic activity in the reactions of various nucleophiles (NuH), such as ketones, with propargylic alcohol (Scheme 2.58) [170, 171]. Other types of ruthenium selenolate complexes were also reported. For example, (RuCp[PPh$_3$][μ-Se$_2$])$_2$ **118** has a six-membered Ru$_2$Se$_4$ ring in a chair conformation [172].

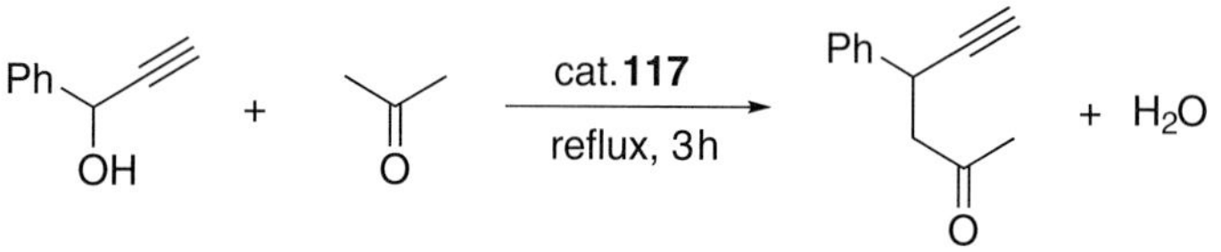

Scheme 2.58

A series of methyl-substituted 8-quinolineselenolates **119** of ruthenium, osmium, rhodium, and iridium were prepared (Scheme 2.59), and their cytotoxicity was investigated [173]. The osmium selenolate complex was found to be most toxic among them.

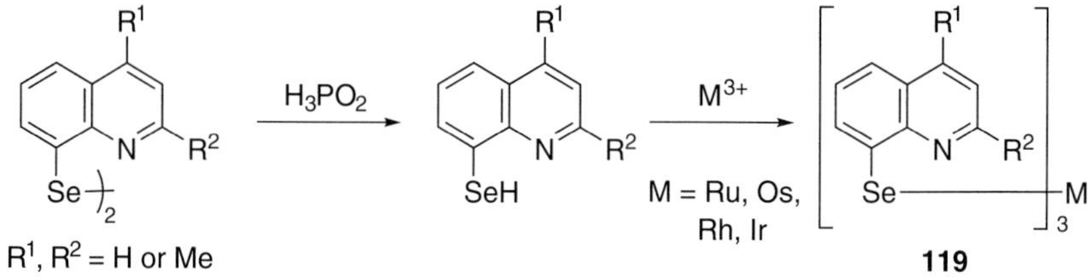

Scheme 2.59

2.4.12
Selenolates of Group 9 Elements (Co, Rh, and Ir)

Mononuclear cobalt(III) selenolate complexes $(Co[SePy]_{3-n}[en]_n)^{n+}$ (Py = 2-pyridyl, en = 1,2-ethanediamine, n = 0, 1, 2) **120** (n = 2) were synthesized by the reaction of $Co(ClO_4)_2 \cdot 6H_2O$ and diselenide PySeSePy in the presence of en (Scheme 2.60) [174]. The SePy ligand was oxidized with peracetic acid to $PySeO_2$ and $PySeO_3$ ligands, which bonded to the cobalt center with a N,O-coordination mode (**176**). Similar cobalt complexes, having 3- or 5-substituted pyridine-2-selenolato ligands, were also synthesized in electrochemical reactions [175].

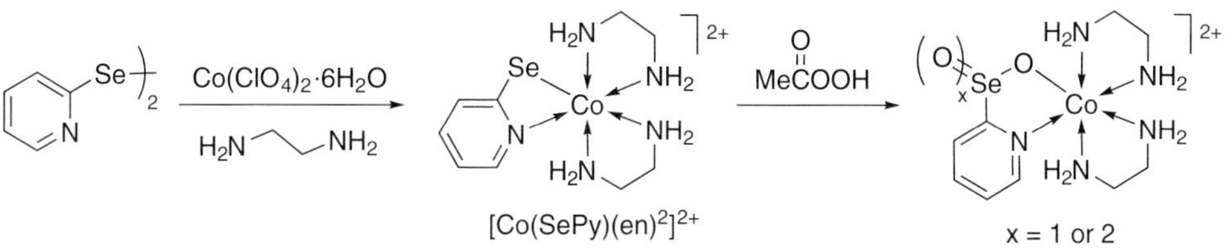

Scheme 2.60

The reaction of $CpCoI_2(CO)$ with poly(o-diselenobenzene) **121** afforded (phenylenediselenolato)cobalt complex dimer $(CpCo[Se_2C_6H_4])_2$ **122**, which exists in an equilibrium with the monomer **123** in solution (Scheme 2.61) [176]. The monomeric form reacted with molybdenum (or tungsten) carbonyl complex **124** to give heteronuclear complex **125** with bridging phenylenediselenolato ligands [177].

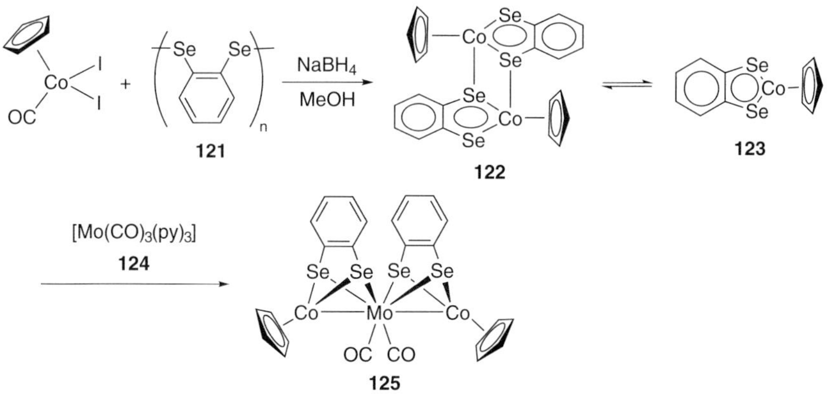

Scheme 2.61

Dilithium carborane-1,2-diselenolate $Li_2Se_2C_2B_{10}H_{10}$ **126**, which was prepared by the insertion of elemental selenium into Li–C bonds of lithium o-carborane [133, 178], reacted with cyclopentadienyl half-sandwich cobalt complexes **127** and **128** to yield monomeric and dimeric cobalt selenolate complexes **129** and **130**, respectively (Scheme 2.62) [179, 180].

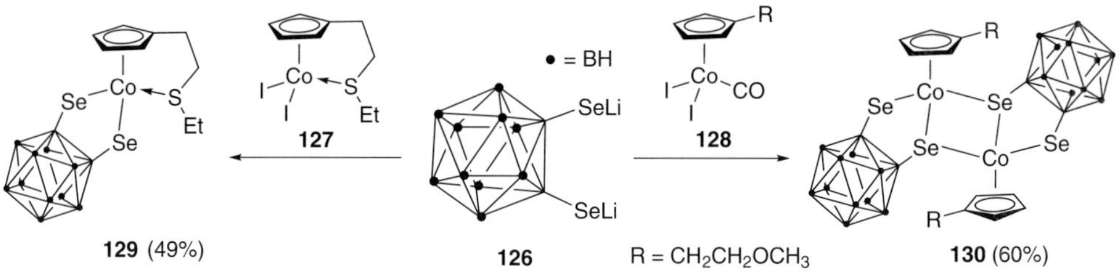

Scheme 2.62

The reaction of dimeric complex $Co_2(NO)_4(\mu\text{-}Cl)_2$ **131** with tBuSeLi produced unstable binuclear cobalt complex $Co_2(NO)_4(\mu\text{-}SetBu)_2$ **132** with a Co_2Se_2 core (Scheme 2.63) [181]. On the other hand, a reactive "*tripod*Co^0" species, which was prepared by the reaction of $CoCl_2$ with KC_8 in the presence of *tripod* (*tripod* = $CH_3C[CH_2PPh_2]_3$), was reacted with PhSeSePh to afford monomeric cobalt selenolate complex (*tripod*Co[SePh]) **133** or dimetallaselenocumulenic compound (*tripod*Co=Se=Co*tripod*) **134** depending on the reaction conditions (Scheme 2.64) [182].

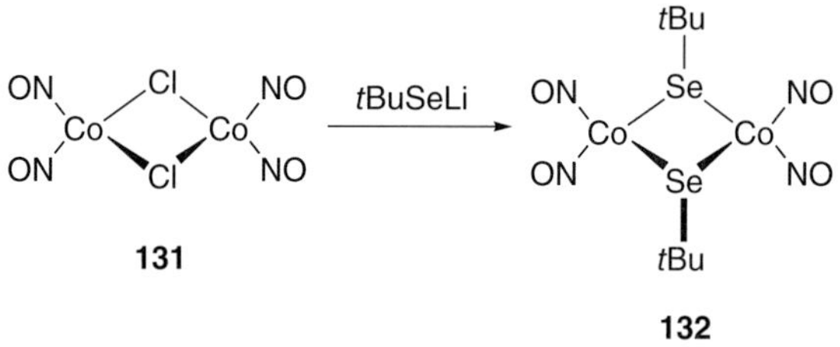

Scheme 2.63

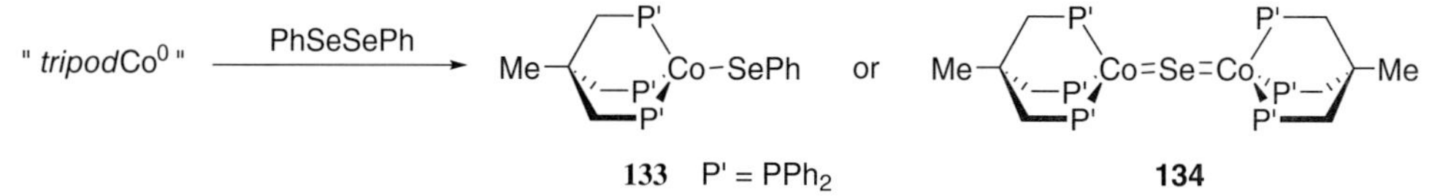

Scheme 2.64

Rhodium and iridium carborane-1,2-diselenolate complexes ($[C_5Me_5]$M$[Se_2C_2B_{10}H_{10}]$; M = Rh and Ir) **135** were synthesized by the reaction of ($[C_5Me_5]$MCl$_2$)$_2$ and dilithium carborane-1,2-diselenolate **126** (Scheme 2.65) [178, 183]. The 16 electron complexes reacted with dimethyl acetylene dicarboxylate **136** to give M,Se-adducts **137** [184].

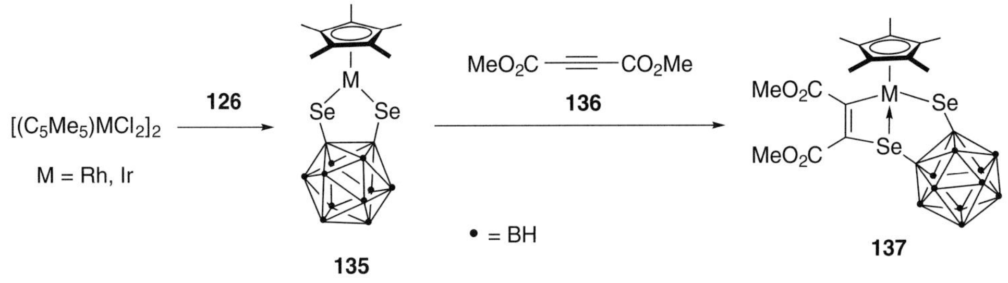

2.4 Organic Nucleophilic Selenium Reagents | 83

Scheme 2.65

Mononuclear rhodium selenolate complex $(C_5Me_5)Rh(PMe_3)(SePh)_2$ **138** was synthesized by the reaction of $(C_5Me_5)Rh(PMe_3)Cl_2$ **139** with PhSeLi (Scheme 2.66) [185]. A similar monomeric complex $Tp^{Me2}Rh(SePh)_2(MeCN)$ (Tp^{Me2} = trispyrazolylborato) **140** was prepared and applied as a catalyst for the hydrogenation of styrene or *N*-benzylideneaniline (Scheme 2.67) [186]. Binuclear rhodium and iridium selenolate complexes **141** and **142** with an M_2Se_2 (M = Rh or Ir) core were prepared by using a bidentate 1,1′-ferrocene diselenolate ligand $(Fe[C_5H_4Se]_2)^{2-}$ (Figure 2.12) [187]. Rhodium-catalyzed reductive coupling of diselenides with alkyl halides, using hydrogen as a reducing agent, was reported (Scheme 2.3) [14].

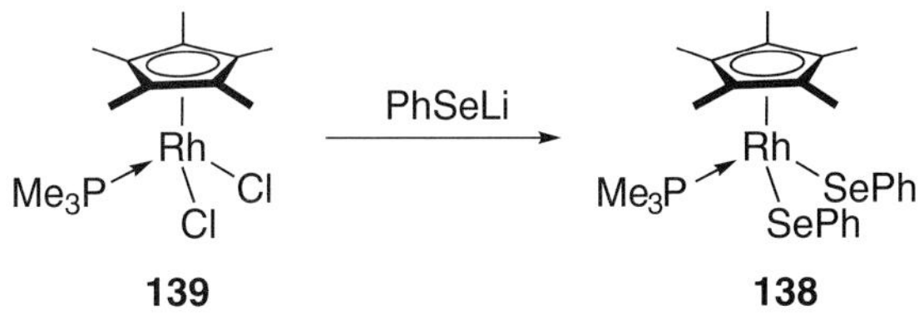

Scheme 2.66

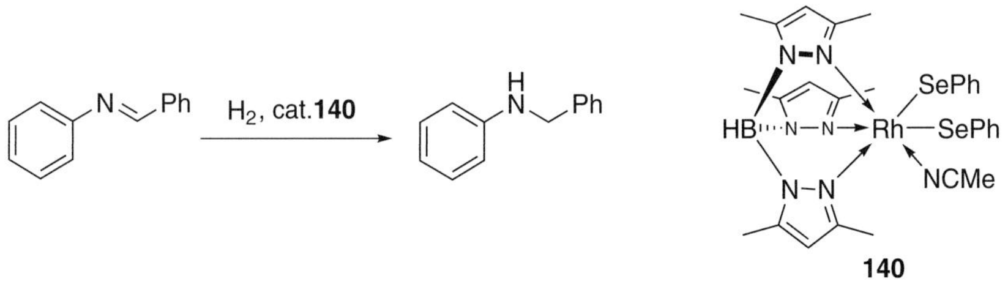

Scheme 2.67

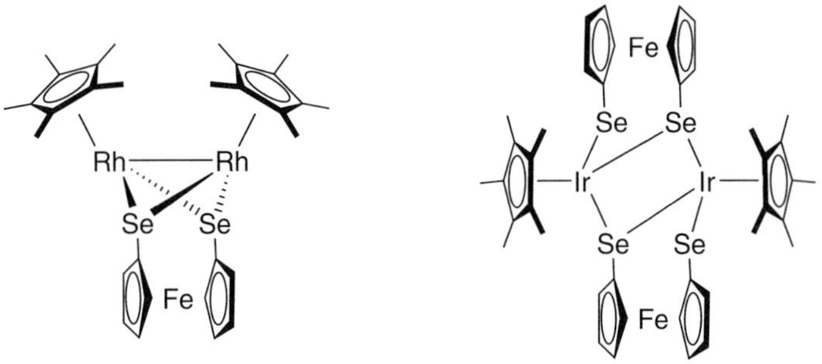

Figure 2.12 Molecular structures for binuclear rhodium and iridium selenolate complexes **141** and **142**.

2.4.13
Selenolates of Group 10 Elements (Ni, Pd, and Pt)

By a similar reaction to that shown in Scheme 2.64, nickel(0) complex (*tripod*$_4$Ni$_3$) (*tripod* = CH$_3$C[CH$_2$PPh$_2$]$_3$) reacted with PhSeSePh to afford tetrahedral 17-electron nickel(I) selenolate complex (*tripod*Ni[SePh]), which can be oxidized electrochemically to cationic nickel(II) complex (*tripod*Ni[SePh])$^+$ [182].

Various monomeric nickel(II) selenolate complexes with a square planar geometry were synthesized. (Ni[CO][SePh]$_3$)$^-$ **143** [188] was reduced with BH$_4^-$ under CO atmosphere to afford nickel(0) anion complex (Ni[CO]$_3$[SePh])$^-$ **144**, which was then converted back to **143** through the oxidative addition of PhSeSePh and PhSeH (Scheme 2.68) [189]. The PhSe$^-$/PhS$^-$ ligand exchange reaction was also reported for **144** [189]. Nickel dithiolene–diselenolene complex (Bu$_4$N)$_2$(Ni[C$_2$N$_2$S$_{2.2}$Se$_{0.8}$]$_2$) **145** was obtained by the reaction of NiCl$_2$·6H$_2$O with nucleophilic dianion **146**, which was obtained from 3,4-dichloro-1,2,5-thiadiazole by the treatment with gray selenium and NaBH$_4$ in EtOH (Scheme 2.69) [190]. Imidazol-2-ylidene complexes

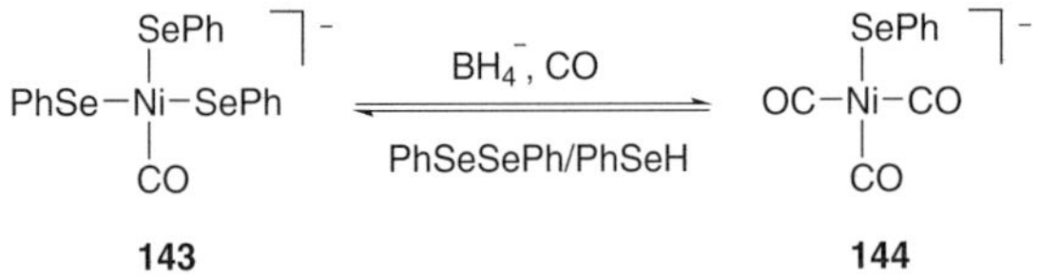

Scheme 2.68

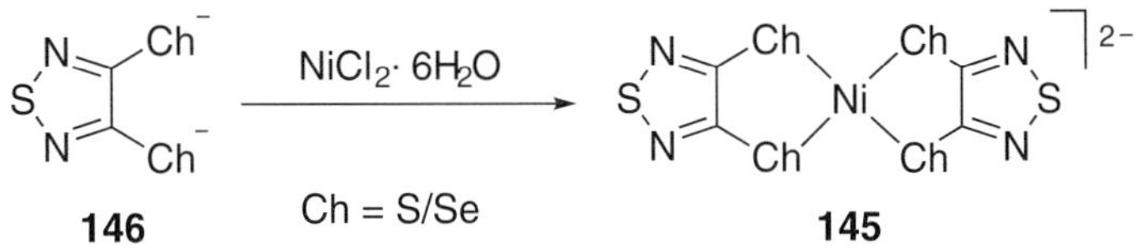

Scheme 2.69

Ni(iPr₂Im)₂(SeR)(C₆F₅) **147** were synthesized by the reaction of Ni(iPr₂Im)₂(F)(C₆F₅) **148** with RSeSiMe₃ (R = Ph, iPr) (Scheme 2.70) [191]. Distorted square planar tetracoordinate complexes [Ni(SeAr){P(o-C₆H₄S)₂(o-C₆H₄SH)}]⁻ with an intramolecular Ni···H–S interaction were also synthesized by a PPh₃/ArSe⁻ ligand exchange reaction [192].

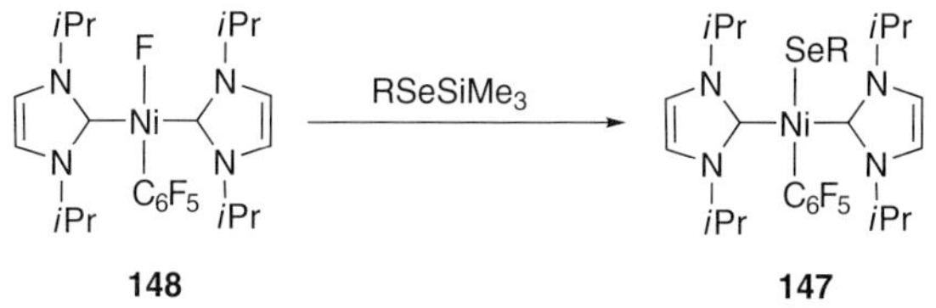

Scheme 2.70

Nickel(III) selenolate complexes were also reported. Neutral radical complexes CpNi(ddds) (ddds = 5,6-dihydro-1,4-dithiin-2,3-diselenolate) **149** and CpNi(bds) (bds = 1,2-benzenediselenolate) **150** were synthesized by the reaction of Cp₂Ni with the corresponding nickel diselenolene complexes (Ni[ddds]₂)₂ and Ni(bds)₂, respectively (Figure 2.13) [193]. Pentacoordinate trigonal bipyramidal complex [Ni(SePh){P(C₆H₃-3-SiMe₃-2-S)₃}]⁻ **151** was obtained by the reaction of nickel(II) complex [Ni(SePh){P(C₆H₃-3-SiMe₃-2-S)₂(C₆H₃-3-SiMe₃-2-SH)}]⁻ with oxygen [194]. The selenolate ligand of **151** was reductively eliminated by the treatment with carbon monoxide (Scheme 2.71) [194]. A mixed-valent Ni(II)Ni(III) dinuclear

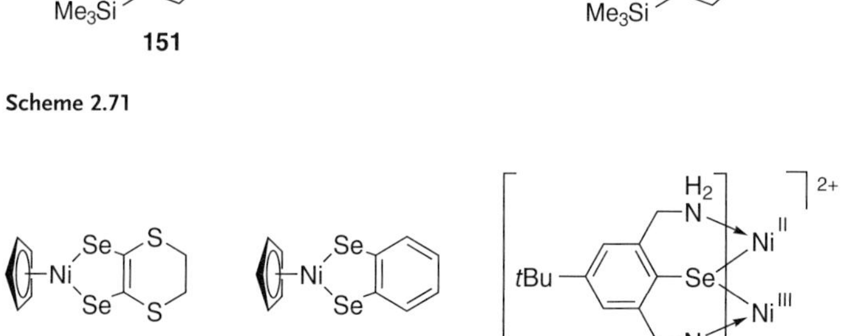

Scheme 2.71

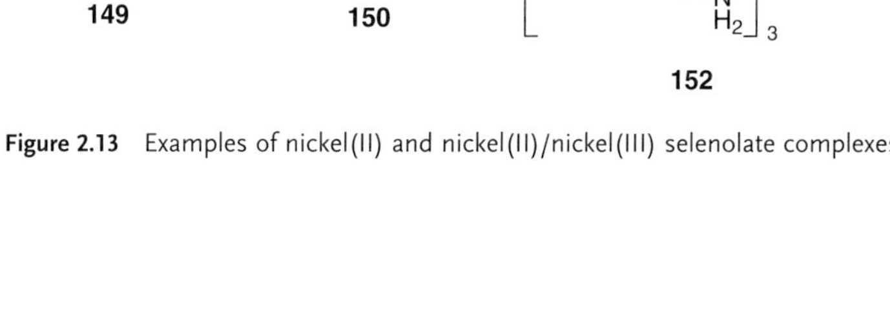

Figure 2.13 Examples of nickel(II) and nickel(II)/nickel(III) selenolate complexes.

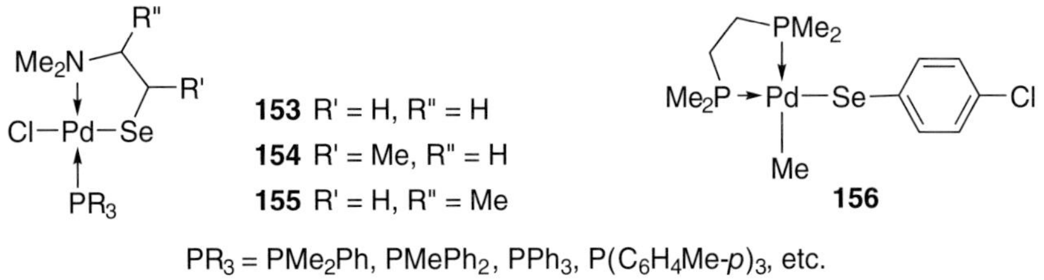

2 Nucleophilic Selenium

153 R' = H, R" = H
154 R' = Me, R" = H
155 R' = H, R" = Me

156

PR₃ = PMe$_2$Ph, PMePh$_2$, PPh$_3$, P(C$_6$H$_4$Me-*p*)$_3$, etc.

Figure 2.14 Examples of mononuclear palladium(II) selenolate complexes.

157a M = Pd
157b M = Pt

158

159

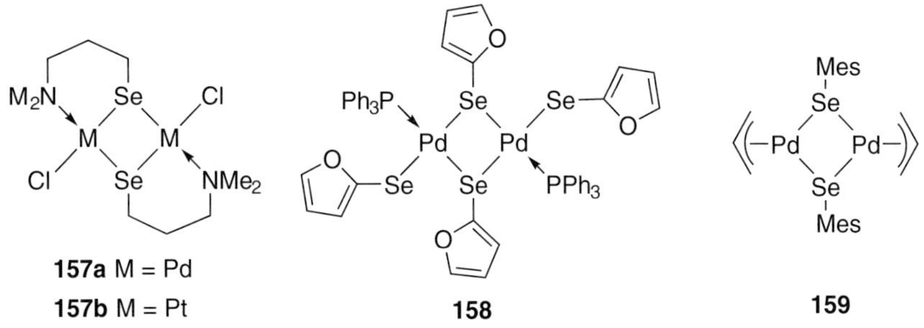

Figure 2.15 Examples of dinuclear palladium(II) selenolate complexes.

complex **152** with three bridging selenolate ligands was also synthesized and characterized (Figure 2.13) [195]. Some of the nickel(II) and nickel(III) complexes described above were designed to mimic the nickel active site of (NiFe) hydrogenases.

Mononuclear palladium selenolate complexes were obtained by ligand exchange or by oxidative addition reaction of palladium complexes to diselenides. Thus, square planar tetracoordinate complexes **153–155** [196, 197] and **156** [198] were synthesized (Figure 2.14).

Various types of dinuclear palladium(II) selenolate complexes with a Pd$_2$Se$_2$ core were also synthesized. Nucleophilic substitution of a chlorine ligand in palladium(II) complexes with a selenolate ligand yielded dinuclear palladium complexes, such as (Cl$_2$Pd[μ-SePh]$_2$PdCl$_2$)$^{2-}$ **100** [151], (PdCl[SeCH$_2$CH$_2$CH$_2$NMe$_2$])$_2$ **157a** [199], (Pd[SeC$_4$H$_3$O]$_2$[PPh$_3$])$_2$ **158** [200], and (Pd[μ-SeAr][η^3-allyl])$_2$ **159** [201] (Figure 2.15). Similarly, a palladium complex with one selenolate bridge ([PMePh$_2$][Cl]Pd[μ-Cl][μ-SeCH$_2$CH$_2$OH]Pd[Cl][PMePh$_2$]) was also synthesized [202]. In the preparation of these complexes, manganese complex *cis*-(Mn[CO]$_4$[SeR]$_2$)$^-$ **98** (Scheme 2.53) [151] and polymeric lead selenolates (Pb[SeR]$_2$)$_n$ [201] can be applied as a selenium nucleophile in addition to common lithium and sodium selenolates. An unusual dinuclear palladium(I) complex [{Pd(PPh$_3$)}$_2$(μ-SeTrip)$_2$] **160** was obtained by the reaction of 9-triptyceneselenol **161** with Pd(PPh$_3$)$_4$ (Scheme 2.72) [203].

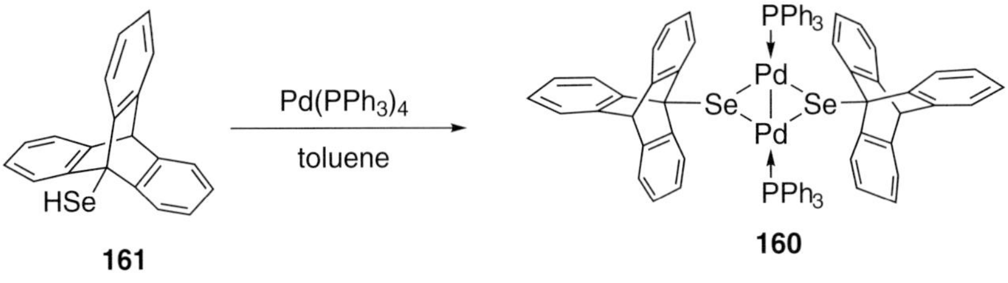

2.4 Organic Nucleophilic Selenium Reagents

Scheme 2.72

Polynuclear palladium selenolate complexes **162** and **163** with a Pd_3Se_4 or Pd_6Se_6 core (Figure 2.16) were obtained in small amounts by the reaction of $(PdCl_2[PPh]_3)$ and PhSeNa [204]. Hexanuclear palladium(II) selenolate complexes, such as $(Pd[SeCH_2CH_2CH_2NMe_2]_2)_6$ [205] and $(Pd[SeCH_2CH_2CH_2OH]_2)_6$ [202], were also synthesized by the reaction of Na_2PdCl_4 and the corresponding sodium selenolates.

Palladium-catalyzed coupling reactions between aryl halides and selenium nucleophiles have been developed recently (see Scheme 2.4) [15, 16].

Mononuclear platinum(II) monoselenolate and diselenolate complexes with a square planar geometry were usually obtained by the nucleophilic ligand exchange reaction of platinum(II) chloride complexes with various selenolating reagents. The reaction of $(PtCl[\mu\text{-}Cl][PR_3])_2$ **164** with $NaSeCH_2CH_2NMe_2$ afforded a series of monoselenolate complexes $(PtCl[SeCH_2CH_2NMe_2][PR_3])$ **165** with the Se and Cl coordinating atoms in *trans* positions (Figure 2.17) [206, 207]. The complexes

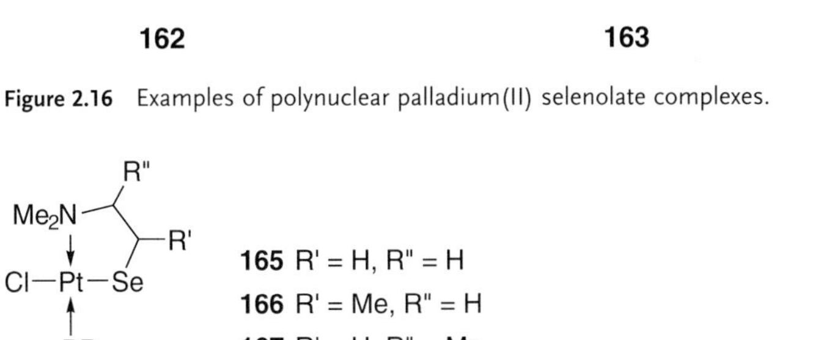

Figure 2.16 Examples of polynuclear palladium(II) selenolate complexes.

165 R' = H, R" = H
166 R' = Me, R" = H
167 R' = H, R" = Me

$PR_3 = PnBu_3, PnPr_3, PEt_3, PMe_2Ph$, etc.

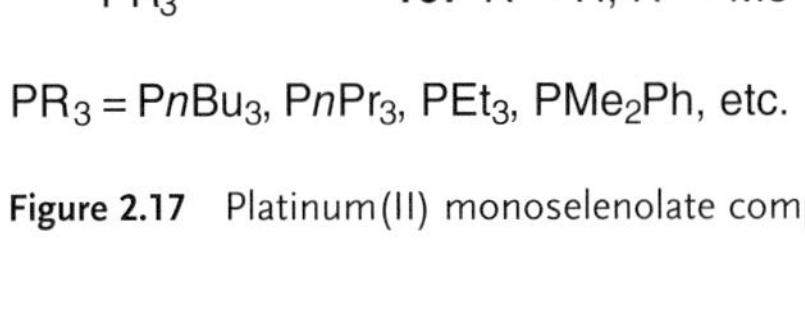

Figure 2.17 Platinum(II) monoselenolate complexes.

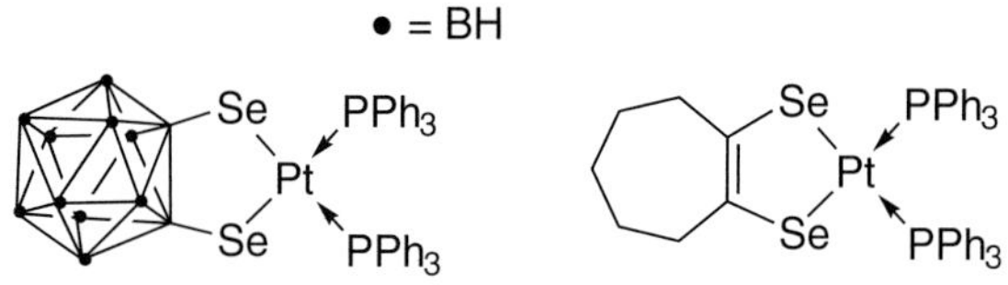

Figure 2.18 Examples of fused diselenolene platinum(II) complexes.

showed a weak absorption at about 405 nm, which was assigned to charge transfer transitions from the selenolato center to the phosphine coligand. The complexes were further converted to polynuclear platinum complexes by the reaction with **164** [206]. Similarly, mononuclear platinum(II) selenolate complexes PtCl(SeCH[CH$_3$]CH$_2$NMe$_2$)(PR$_3$) **166** and PtCl(SeCHCH$_2$[CH$_3$]NMe$_2$)(PR$_3$) **167** were synthesized (Figure 2.17) [197].

On the other hand, diselenolate complexes Pt(SeAr)$_2$(PPh$_3$)$_2$ **168** were prepared by the reaction of cis-(PtCl$_2$[PPh$_3$]$_2$) with ArSeNa or ArSeLi (Scheme 2.73) [208]. The obtained cis-platinum complex isomerized to the trans complex **169** upon prolonged standing in solution. Similarly, cis-Pt(Se–C≡C–n-pentyl)$_2$(PPh$_3$)$_2$ [209] and cis-Pt(SeCF$_3$)$_2$(PPh$_3$)$_2$ [210] were synthesized, and the cis–trans isomerization was reported. Fused diselenolene platinum(II) complexes, such as **170** and **171** (Figure 2.18), were also prepared [211, 212], and their luminescence properties were investigated.

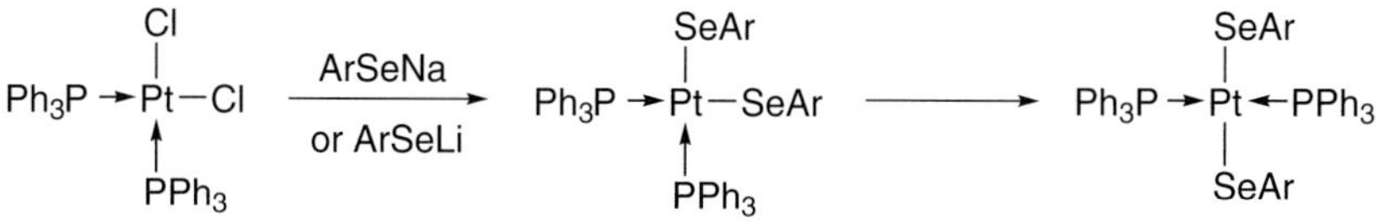

Scheme 2.73

Hydrido selenolato platinum(II) complex PtH(SeTrip)(PPh$_3$)$_2$ (Trip = 9-triptycyl) **172** was prepared by the reaction of TripSeH **161** and (Pt[PPh$_3$]$_2$[η^2-C$_2$H$_4$]) [213]. The hydrido complex was converted to selenaplatinacyclic compound **173** by the treatment with HBF$_4$ (Scheme 2.74). The reaction of **172** with acetylene derivatives gave hydroselenation syn adducts **174** preferentially [214]. Similar hydrido platinum(II) complexes with a different bulky substituent on the Se atom were also synthesized recently [215, 216].

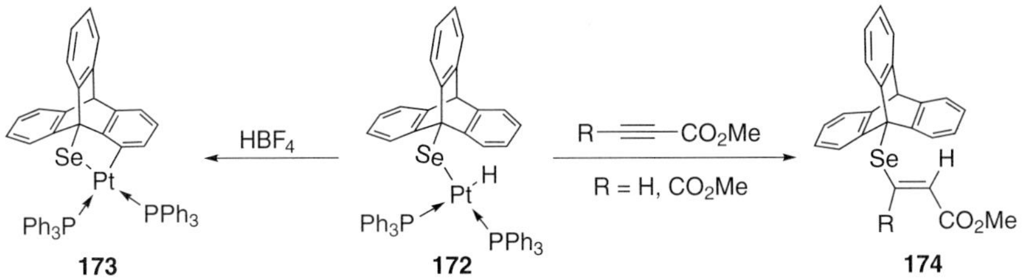

Scheme 2.74

Octahedral hexacoordinate platinum(IV) selenolate complex (PtMe$_2$[SePh][PR$_3$] [bu$_2$bpy])(ClO$_4$) (bu$_2$bpy = 4,4′-di-*tert*-butyl-2,2′-bipyridine) **175** was prepared by the reaction of platinum(II) complex PtMe$_2$(bu$_2$bpy) with mercury selenolate cluster (Hg$_4$[SePh]$_6$[PR$_3$]$_4$)(ClO$_4$)$_2$ **176** (Scheme 2.75) [217]. The addition of PhSe$^-$ and PR$_3$ ligands occurred with *trans* stereochemistry. Similarly, Pt(Tol)$_2$(SeC$_6$H$_4$-*p*-Cl)$_2$(bpy) (bpy = 2,2′-bipyridine) **177** was synthesized by the oxidative addition of diselenide to the corresponding Pt(II) complex **178** (Scheme 2.76) [198]. This reaction was suggested to attain the equilibrium between the platinum(II) and platinum(IV) complexes.

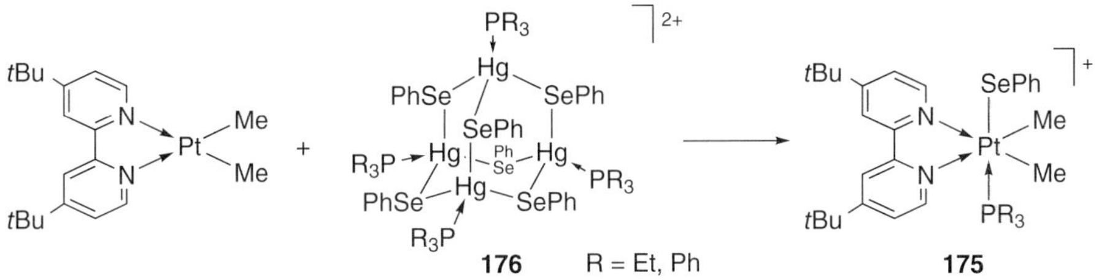

Scheme 2.75

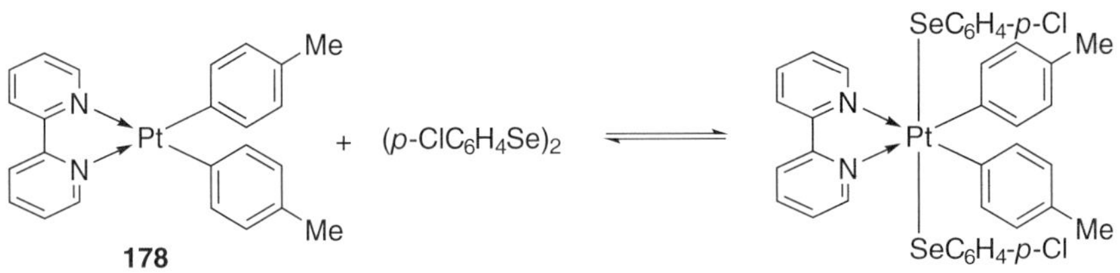

Scheme 2.76

Ferrocenylselenolate bridging dinuclear platinum(II) complex **179** was synthesized by the reaction of *trans*-PtCl$_2$(P*n*Bu$_3$)$_2$ with 1,1′-bis(trimethylsilylseleno) ferrocene **180** (Scheme 2.77) [218]. Similarly, the dinuclear complex (PtCl[SeCH$_2$CH$_2$CH$_2$NMe$_2$])$_2$ **157b** with a Pt$_2$Se$_2$ core was synthesized by the reaction of K$_2$PtCl$_4$ with an equimolar amount of NaSeCH$_2$CH$_2$CH$_2$NMe$_2$ [219].

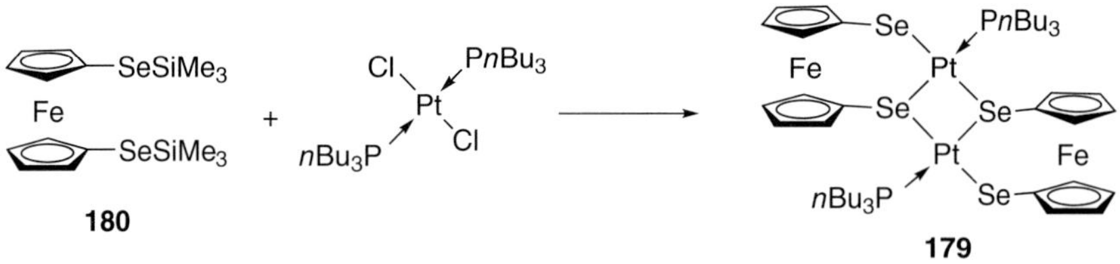

90 | 2 Nucleophilic Selenium

Scheme 2.77

2.4.14
Selenolates of Group 11 Elements (Cu, Ag, and Au)

Monomeric copper(I) and silver(I) complexes with a trimethylsilylselenolate ligand (RR′$_2$P)$_3$MSeSiMe$_3$ **13** (M=Cu) and **14** (M=Ag) were synthesized by the reaction of Se(SiMe$_3$)$_2$ with metal acetates (MOAc) (Scheme 2.13) [60]. The metal complex has a tetrahedral coordination structure with a terminal selenolate ligand. Protonolysis of **13** with EtOH yielded a terminal selenol complex (RR′$_2$P)$_3$CuSeH [60]. On the other hand, similar reactions of ferrocenoyl reagent FcC(O)SeSiMe$_3$ **181** with CuOAc or AgOAc in the presence of PPh$_3$ afforded copper(I) selenolate clusters [Cu$_2$(μ-SeC(O)Fc)$_2$(PPh$_3$)$_3$] **182** and [Cu$_4$(μ-SeC(O)Fc)$_4$(PPh$_3$)$_4$] **183** or silver(I) selenolate cluster [Ag$_4$(μ-SeC(O)Fc)$_4$(PPh$_3$)$_4$] **184**, respectively (Scheme 2.78) [220].

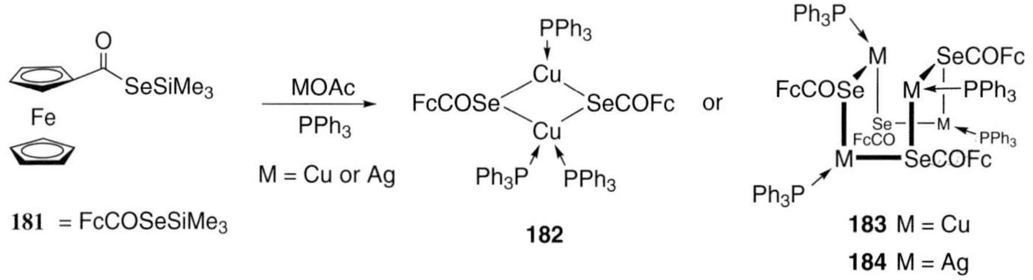

Scheme 2.78

A dinuclear copper(I) μ-selenolate complex (Cu$_2$[μ-SePh][μ-dppm]$_2$)BF$_4$ (dppm = Ph$_2$PCH$_2$PPh$_2$) **185** with a triangular Cu$_2$Se core was prepared by the reaction of (Cu$_2$[μ-dppm]$_2$[MeCN]$_2$)(BF$_4$)$_2$ with PhSeNa (Figure 2.19), and the luminescence behavior was investigated [221]. Similarly, hexanuclear copper(I) selenolate complexes (Cu$_6$[μ$_3$-SePh]$_4$[μ-dppm]$_4$)(BF$_4$)$_2$ and [Cu$_6$(μ$_3$-SePh)$_4${μ-(Ph$_2$P)$_2$NH}$_4$](BF$_4$)$_2$ were synthesized [222].

More complex copper(I) or silver(I) selenolates were synthesized by several methods. Electrochemical reactions of Cu or Ag with di(2-pyridyl) diselenide afforded the compounds with a general formula of (M[SePy]) (M = Cu, Ag) [223].

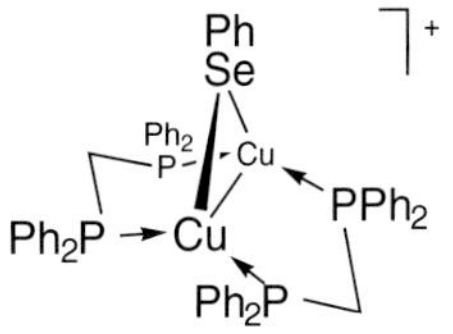

185

Figure 2.19 Molecular structure for dinuclear copper(I) μ-selenolate complex **185**.

The copper complex adopts a tetrameric form, whereas the silver complex would be hexameric. (NMe$_4$)$_2$(Cu$_4$[SePh]$_6$) was prepared by the reaction of CuCl with PhSeH [224]. All selenolate ligands can be substituted by the reaction with CS$_2$ in CH$_3$CN/CH$_3$OH or DMF/CH$_3$OH [225]. A neutral octanuclear silver cluster complex (Ag$_8$[2,4,6-*i*Pr$_3$C$_6$H$_2$Se]$_8$) (see Figure 2.1) was synthesized by the reaction of AgNO$_3$ with 2,4,6-triisopropylbenzeneselenol, and the nanosecond optical limiting property was studied [18, 226].

The cross-coupling reaction of aryl iodides and PhSeH to produce ArSePh was achieved by using CuI/2,9-dimethyl-1,10-phenanthroline in the presence of *t*BuONa or K$_2$CO$_3$ as base (Scheme 2.79) [17].

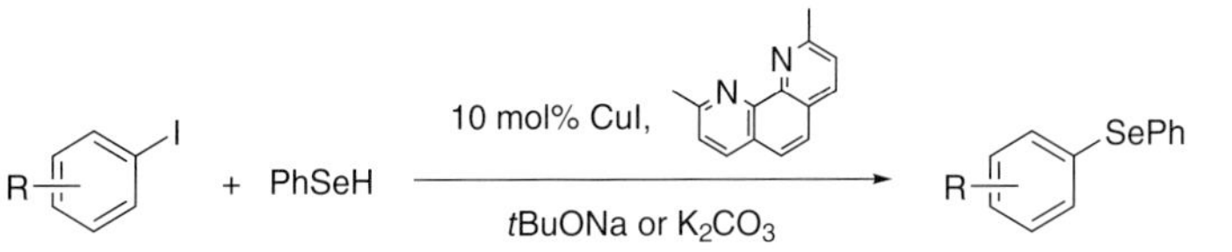

Scheme 2.79

Gold(I) selenolate complexes are usually obtained in a monomeric form. The reaction of carboraneselenols or carboraneselenolates **186** with AuCl(PPh$_3$) yielded the corresponding gold(I) selenolates **187** (Scheme 2.80) [227, 228]. The complexes obtained have Au⋯Au interactions in the solid state to make the associate pairs. When similar reaction conditions were applied to *p*-C$_6$H$_4$(SeSiMe$_3$)$_2$, the phenylene-1,4-diselenolate bridging gold(I) complex **188** was obtained (Scheme 2.81) [229]. Similarly, various gold(I) selenolate complexes were synthesized, such as (Au$_2$[SePh]$_2$[μ-dppf]) (dppf = 1,1′-bis(diphenylphosphine)ferrocene) **189** [230] and [Au{SeC$_6$H$_4$(CH$_2$NMe$_2$)-2}(PR$_3$)] **190** [231, 232] (Figure 2.20). Complex **190** not

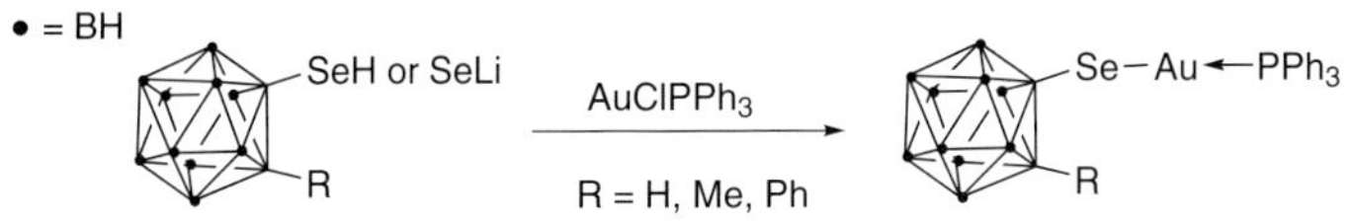

Scheme 2.80

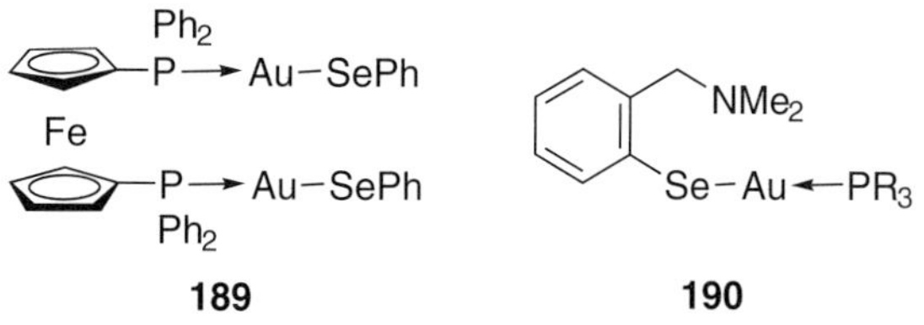

Figure 2.20 Molecular structures for gold(I) selenolate complexes **189** and **190**.

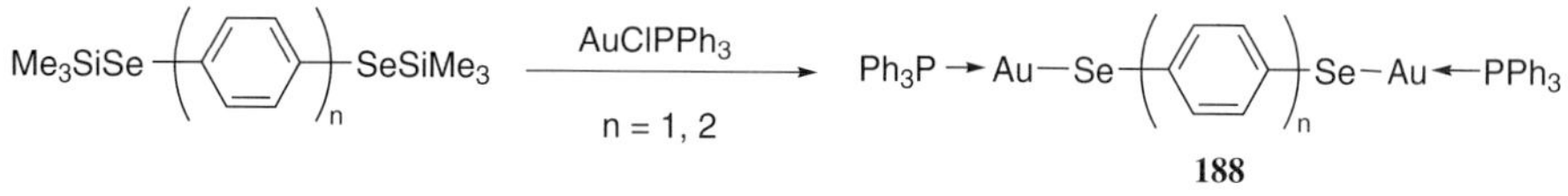

Scheme 2.81

only possessed interesting photoluminescence properties [231] but can also be utilized as a synthetic model for the inhibition of glutathione peroxidase by gold compounds [232]. On the other hand, self-assembled monolayers of n-dodecaneselenol was made on the Au(111) surface by immersing the gold electrode in the selenol solution [233].

2.4.15
Selenolates of Group 12 Elements (Zn, Cd, and Hg)

The reaction of PhSeSePh with zinc produced polymeric zinc selenolate $Zn(SePh)_2$ [125]. Similarly, various zinc selenolates $Zn(SeR)_2$, which are useful for the synthesis of selenoethers, selenol esters, and carboxylic acids from electron-deficient alkenes, acid chlorides or anhydrides, and lactones or esters, respectively, were prepared from the diselenide (RSeSeR) by the reaction with $Zn/ZrCl_4$ [87], $Zn/AlCl_3$ [88–90], or $Zn/RuCl_3$ [91] systems (see Scheme 2.28). Ring opening of unprotected aziridines by zinc selenolates in a biphasic system was also reported recently (Scheme 2.82) [234]. A monomeric zinc selenolate stabilized by a bidentate tmeda ligand ([tmeda]Zn[SePh]$_2$) **191**, which can be utilized for the preparation of ZnSe quantum dots by heating, was synthesized by the reaction of $Zn(SePh)_2$ with tmeda (Scheme 2.83) [235]. The zinc center has a distorted tetrahedral geometry. A similar monomeric complex ([tmeda]Zn[SeSiMe$_3$]$_2$) **15** was synthesized (Scheme 2.14) and was utilized for the preparation of a nanocluster, ([tmeda]$_5$Zn$_5$Cd$_{11}$Se$_{13}$[SePh]$_6$[thf]$_2$), which exhibits "band-edge" luminescence at room temperature [61].

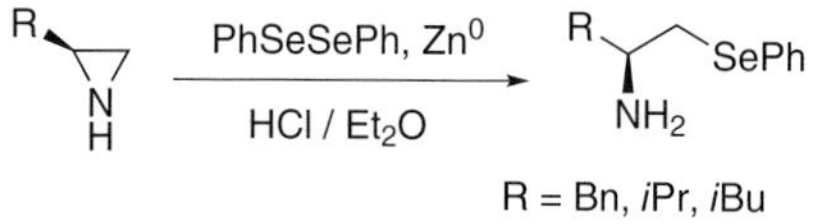

Scheme 2.82

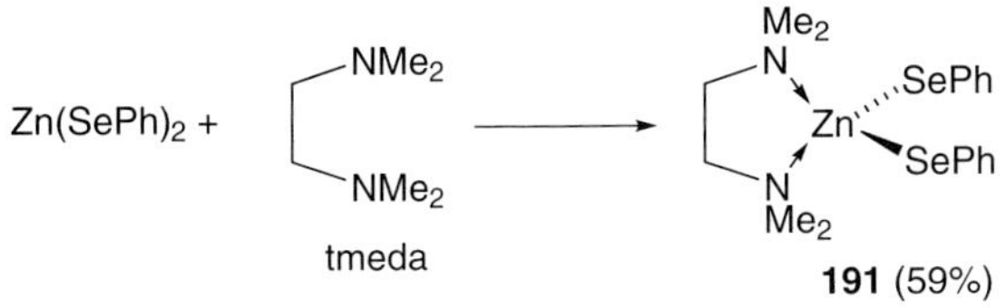

Scheme 2.83

The reaction of phenylselenenyl halides PhSeX (X = Cl, Br) with metallic zinc in THF produced new, air-stable zinc selenolate compounds PhSeZnX, which smoothly reacted with epoxides and other electrophiles in water suspension at room temperature (Scheme 2.84) [236].

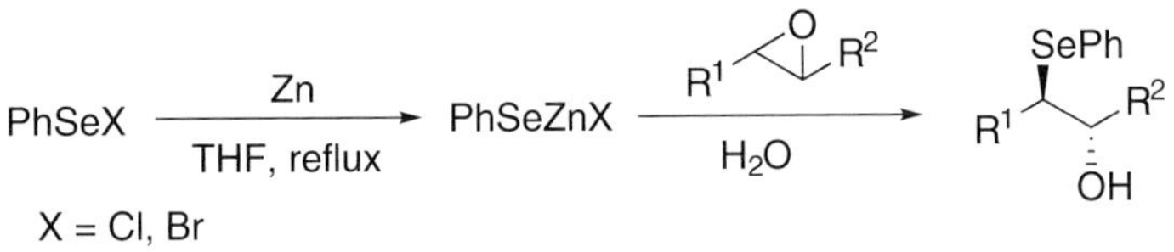

Scheme 2.84

A monomeric cadmium selenolate (Cd[SeCH$_2$CH$_2$NMe$_2$]$_2$) **192** was obtained by the reaction of Cd(OAc)$_2$ and NaSeCH$_2$CH$_2$NMe$_2$, and the complex was further converted to trinuclear selenolate cluster (Cd$_3$[OAc]$_2$[SeCH$_2$CH$_2$NMe$_2$]$_4$) **193** by the treatment with Cd(OAc)$_2$ (Scheme 2.85) [21]. Pyrolysis of these complexes gave CdSe nanoparticles. A dinuclear cadmium selenolate complex with a square planar Cd$_2$Se$_2$ core, ([phen]$_2$Cd[μ-SePh])$_2$(PF$_6$)$_2$ **194**, was prepared by the reaction of Cd(OAc)$_2$ with PhSeSePh and NaBH$_4$ in the presence of phenanthroline (phen) (Scheme 2.86) [237]. The cadmium center adopts a distorted octahedral geometry, and the complex exhibits interesting photophysical and electrochemical properties [237].

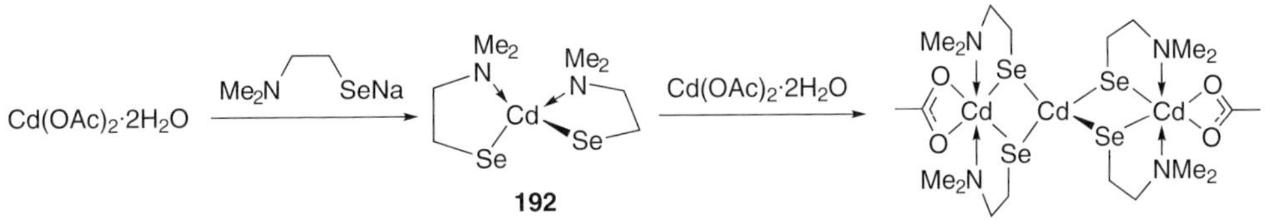

Scheme 2.85

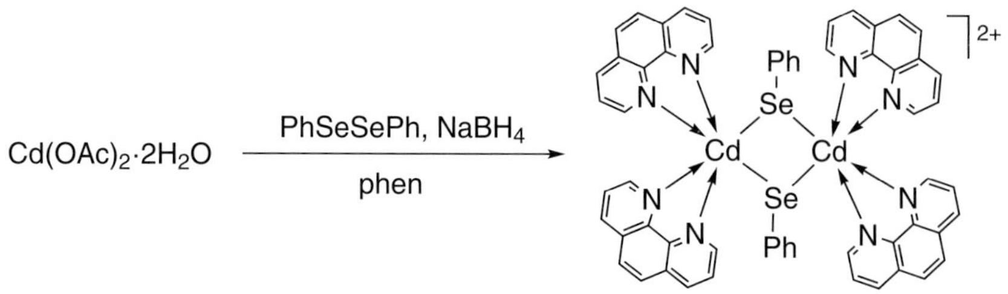

2 Nucleophilic Selenium

Cd(OAc)$_2$·2H$_2$O $\xrightarrow[\text{phen}]{\text{PhSeSePh, NaBH}_4}$ **194**

Scheme 2.86

The reaction of PhSeSePh with mercury in dioxane or refluxing xylene yielded bis(benzeneselenolato)mercury(II) (Hg[SePh]$_2$) [238, 239], which reacted with acid chlorides in the presence of a catalytic amount of nBu$_4$NBr to give selenol esters RCOSePh (Scheme 2.87) [238]. X-ray crystallography showed that (Hg[SePh]$_2$) exists as a polymeric assembly in the solid state [239]. The reaction of (Hg[SePh]$_2$) with PhSe$^-$ in the presence of nBu$_4$NBr afforded a monomeric tricoordinating mercury complex (nBu$_4$N)(Hg[SePh]$_3$) **195**, which has a trigonal planar mercury atom (Scheme 2.88) [239]. A monomeric mercury(II) complex **196** with two selenolate ligands having a coordinating nitrogen atom was prepared by the reaction of the corresponding diselenide **197** with mercury (Scheme 2.89) [240]. Polynuclear mercury selenolate complexes (Hg$_4$[SePh]$_6$[PR$_3$]$_4$[ClO$_4$]$_2$) [217], (Hg$_{10}$Te$_4$[SePh]$_{12}$[PPh nPr$_2$]$_4$) [241], and (Hg$_{34}$Te$_{16}$[SePh]$_{36}$[PPhnPr$_2$]$_4$) [241] were also synthesized, and their structures were determined by X-ray analysis.

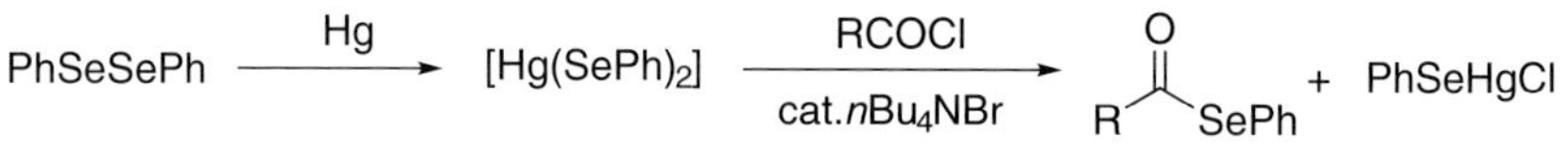

PhSeSePh $\xrightarrow{\text{Hg}}$ [Hg(SePh)$_2$] $\xrightarrow[\text{cat.}n\text{Bu}_4\text{NBr}]{\text{RCOCl}}$ R−C(=O)−SePh + PhSeHgCl

Scheme 2.87

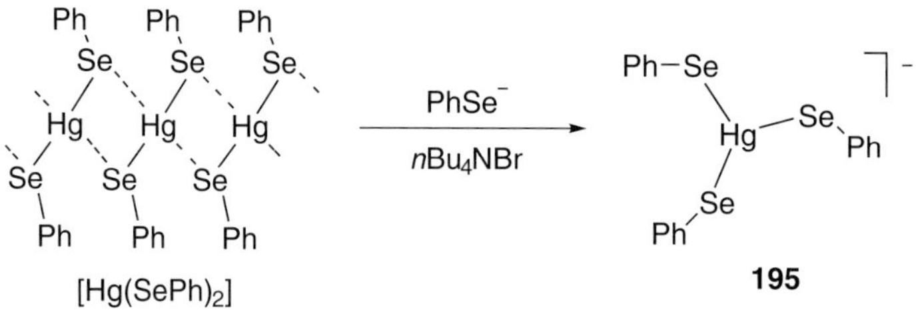

[Hg(SePh)$_2$] $\xrightarrow[n\text{Bu}_4\text{NBr}]{\text{PhSe}^-}$ **195**

Scheme 2.88

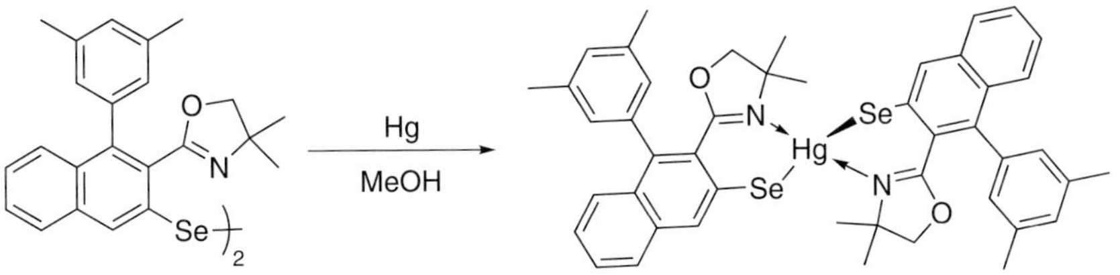

Scheme 2.89

2.4.16
Selenolates of Group 13 Elements (B, Al, Ga, and In)

Various boron selenolates B(SeR)$_3$ [242, 243] and aluminum selenolates Me$_2$AlSeMe [244], iBu$_2$AlSePh [245, 246], and Se(AlMe$_2$)$_2$ [247, 248] have been reported. Their preparation and reactions were summarized in several publications [7, 8, 33, 34, 36]. These highly reactive reagents are useful as efficient selenium nucleophiles for the preparation of various selenium compounds. For example, iBu$_2$AlSeR (R = Ph, Bn) was utilized for the synthesis of selenol esters **198** [245] and selenoformates **199** [249] (Scheme 2.90).

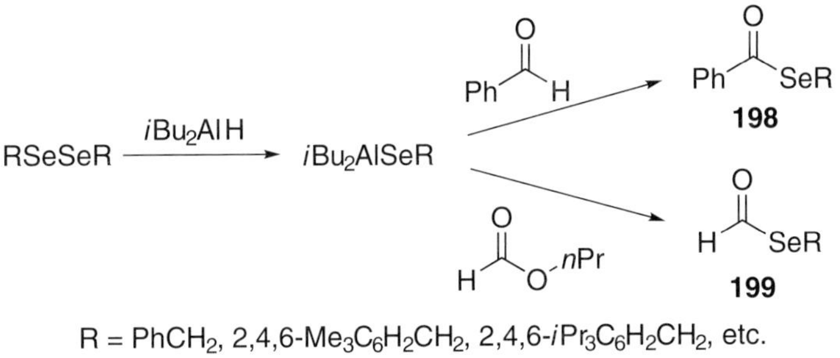

R = PhCH$_2$, 2,4,6-Me$_3$C$_6$H$_2$CH$_2$, 2,4,6-iPr$_3$C$_6$H$_2$CH$_2$, etc.

Scheme 2.90

A tetracoordinated borane compound with a selenolate ligand ArSeBPh$_2$ (Ar = quinoline-8-) **200** was prepared by the reaction of quinoline-8-selenol with BPh$_3$ (Scheme 2.91), and the structure was characterized by X-ray analysis [250].

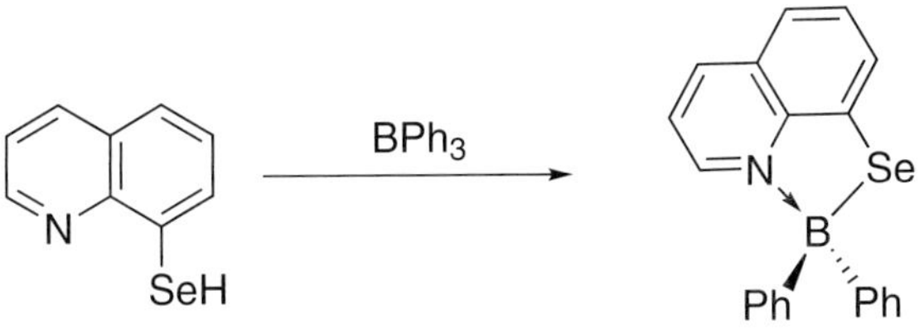

Scheme 2.91

Synthesis of gallium(III) selenolates was reviewed previously [34]. Recently, an anionic homoleptic tetrahedral gallium selenolate $(Ga[SePh]_4)^-$ **201** was synthesized by the reaction of $(Et_4N)(GaCl_4)$ with four equivalents of PhSeNa prepared from PhSeSePh and $NaBH_4$ in methanol (Scheme 2.92) [251]. Similar anionic indium selenolates, $(In[SePh]_4)^-$ and $(In[SePh]_3[SeH])^-$, are also known [252].

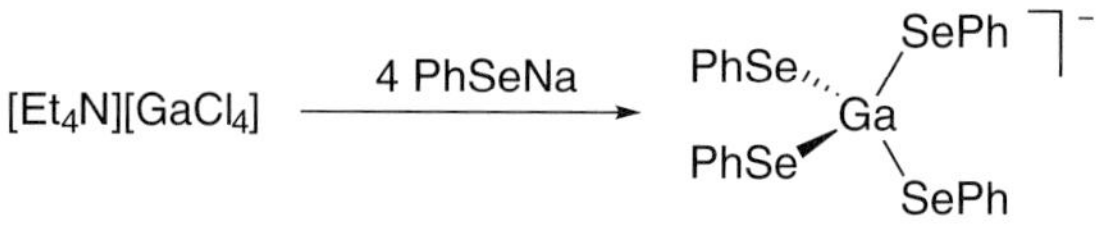

Scheme 2.92

Indium(III) monoselenolate $(EtSe)InI_2$ was prepared by the reaction of EtSeH with $iPrInI_2$ [253]. Similarly, dialkylindium(III) selenolates $(Me_2In[SeAr])$ were synthesized by the reaction of $InMe_3$ with the corresponding areneselenols ArSeH. Depending on the steric demand of the aromatic ring, the indium complexes transform from monomeric to polymeric structures (Figure 2.21) [254]. On the other hand, indium(III) diselenolates $(RSe)_2InX$ (X = I or Br) **202** were obtained by the reactions of the corresponding diselenides with indium(I) iodide [92] and indium(I) bromide [93] (see Scheme 2.28). The reagents obtained added to alkynes to give the Markovnikov adducts **203** stereoselectively [92, 93] or reacted with aziridines to give selenocysteine derivatives **204** [255] (Scheme 2.93). $In(SePh)_3$ was used for the preparation of $As(PhSe)_3$ as a nucleophilic selenolate-ligand donor (Scheme 2.94) [256].

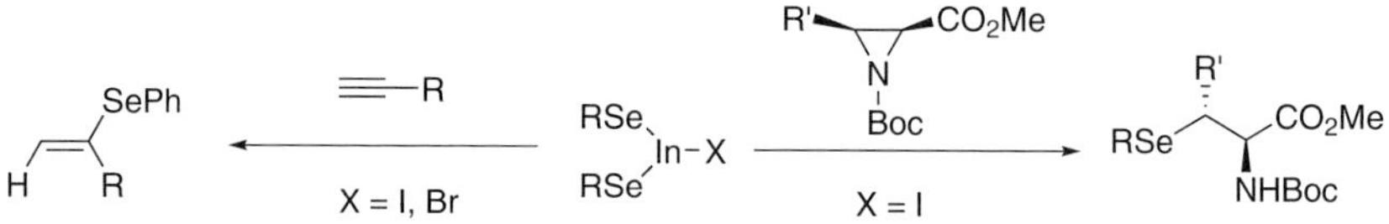

Scheme 2.93

Scheme 2.94

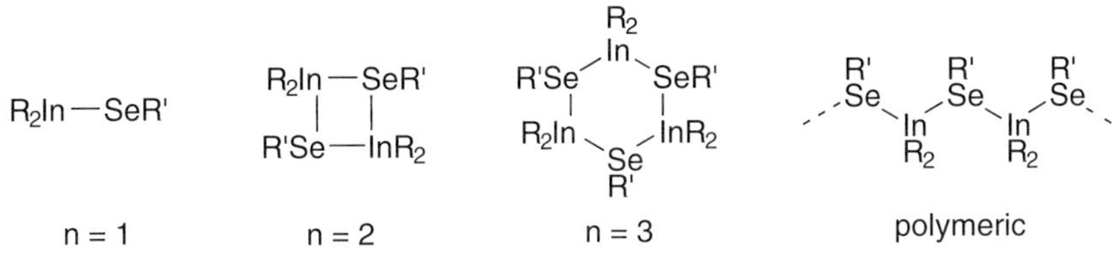

Figure 2.21 Monomeric to polymeric structures for indium(III) monoselenolates.

2.4.17
Selenolates of Group 14 Elements (Si, Ge, Sn, and Pb)

Synthesis and reactions of $Se(SiMe_3)_2$ [59], $PhSeSiMe_3$ [257, 258], and $MeSeSiMe_3$ [259, 260] were previously reviewed [32, 34]. These reagents are widely used as selenium nucleophiles not only for the introduction of a selenium moiety into organic compounds but also for the preparation of various metal selenolate complexes. The reactions of $Se(SiMe_3)_2$ are described earlier in Section 2.3.1.

$PhSeSiMe_3$ reacted as a selenol equivalent with alkenes to give Markovnikov adducts (Scheme 2.95) [261]. The reaction with $Mn(OAc)_2$ afforded polymeric manganese selenolate complex $Mn(SePh)_2$ **93** (Scheme 2.51) [149]. Similarly, $RSeSiMe_3$ (R = Ph, iPr) was used for the preparation of imidazol-2-ylidene nickel complexes $Ni(iPr_2Im)_2(SeR)(C_6F_5)$ **147** via F^-/RSe^- ligand exchange (Scheme 2.70) [191]. $PhSeSiMe_3$ was also employed for the preparation of complex CuSeIn [98] and CuSeFe [100] clusters with terminal and bridging $PhSe^-$ ligands.

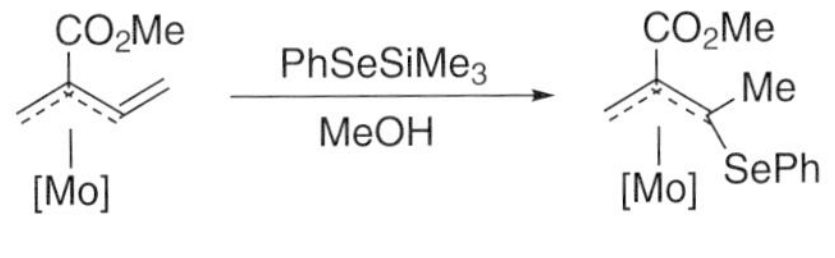

Scheme 2.95

The reaction of potassium silanide $KSiMe(SiMe_3)_2$ with elemental selenium yielded potassium silanylselenolate $KSeSiMe(SiMe_3)_2$, which was converted to selenol $HSeSiMe(SiMe_3)_2$ by treating with anhydrous acetic acid [62]. Similarly, the reaction of sodium silanides $NaSiRtBu_2$ (R = Ph, tBu) with elemental selenium afforded sodium silanylselenolates $NaSeSiRtBu_2$, which were oxidized with oxygen to give diselenides $tBu_2RSi-SeSe-SiRtBu_2$, protonated to selenols $HSeSiRtBu_2$, or crystallized to the dimeric (with a square planar Na_2Se_2 core) or tetrameric (with a cubic Na_4Se_4 core) cluster **205** or **206** depending on the crystallization conditions (Scheme 2.96) [162, 163].

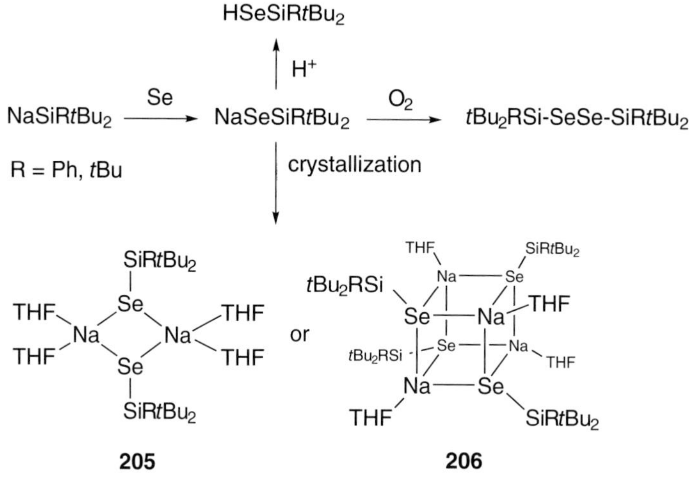

Scheme 2.96

Pentacoordinate silicon compounds with selenolate ligands were synthesized by the reaction of the pentacoodinate chlorosilicon complex **207** with PhSeH (Scheme 2.97) or by the reaction of trihydridosilane **208** with benzene-1,2-diselenol (Scheme 2.98) [262]. The coordination at the silicon varied from trigonal bipyramidal to distorted square pyramidal depending on the selenolate ligand.

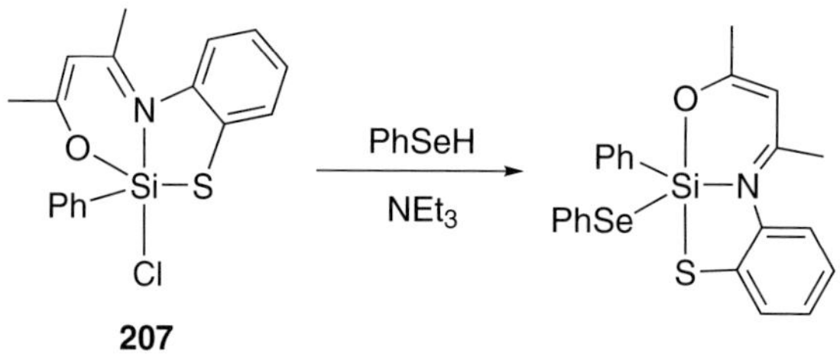

Scheme 2.97

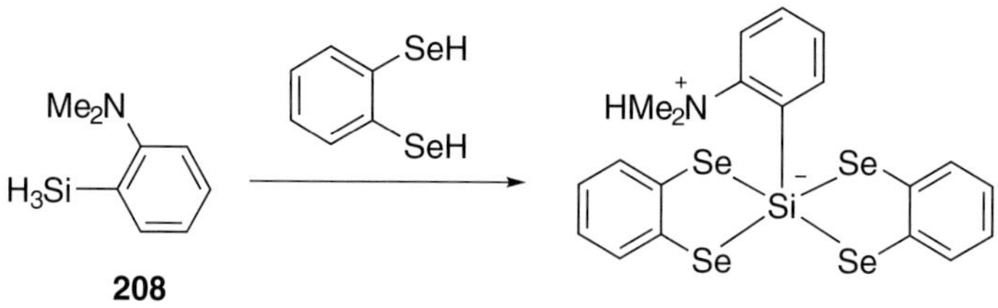

Scheme 2.98

2.4 Organic Nucleophilic Selenium Reagents

Homoleptic germanium selenolate Ge(PhSe)₄ **209** was synthesized by the reaction of GeCl₄ with PhSeLi [263] or by the reaction of activated germanium, which was prepared from GeCl₂·dioxane and LiBEt₃H, with PhSeSePh (Scheme 2.99) [264].

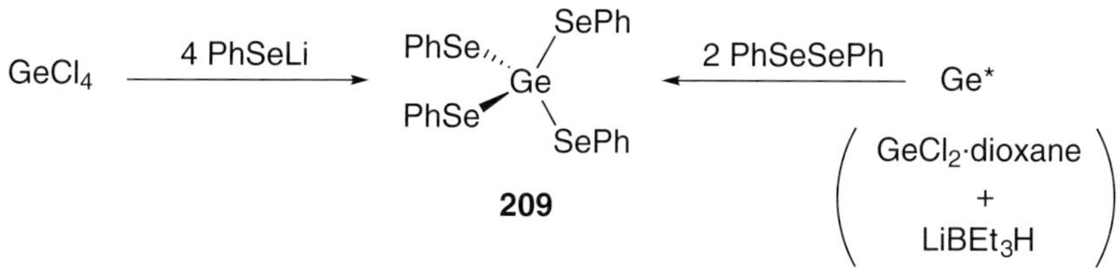

$$GeCl_4 \xrightarrow{4\ PhSeLi} \underset{\mathbf{209}}{PhSe\diagdown_{Ge}\diagup^{SePh}_{SePh}} \xleftarrow{2\ PhSeSePh} Ge^* \quad \begin{pmatrix} GeCl_2 \cdot dioxane \\ + \\ LiBEt_3H \end{pmatrix}$$

Scheme 2.99

Phenyl tributylstannyl selenide (PhSeSnBu₃) was synthesized by the reaction of Bu₃SnCl with PhSeNa [265]. The selenide was applied as a useful selenolate reagent in the synthesis of selenol esters from acyl chlorides and in the synthesis of α-phenylseleno carbonyl compounds from α-bromocarbonyl compounds catalyzed by a palladium complex (Scheme 2.100) [265]. Ferrocene-bridged bis(selenolatostannyl) compound (ArSe)₃Sn–Fc–Sn(SeAr)₃ **210** (Figure 2.22) was synthesized by the reaction of the corresponding stannyl trichloride with ArSeSeAr and NaBH₄, and the electrochemical properties were investigated [266]. Diselenastannacycle **211** was obtained by the reaction of Ph₂SnCl₂ with 1,2-diselenolato-1,2-dicarba-*closo*-docecaborane dianion ($Se_2C_2B_{10}H_{10}^{2-}$) [267]. On the other hand, the reaction of RSeLi (R = CH[3,5-Me₂Pz]₂; Pz = pyrazol-1-yl) with Ph₂SnCl₂ afforded pentacoordinate tin complex RSeSnPh₂Cl **212** with a selenolate ligand [268]. Metallacycle **213** with a Sn–Se–Sn linkage was obtained by the reaction of (PtMe₂[bu₂bpy]) with (Me₂SnSe)₃, and the ring expansion and contraction reactions were studied [269].

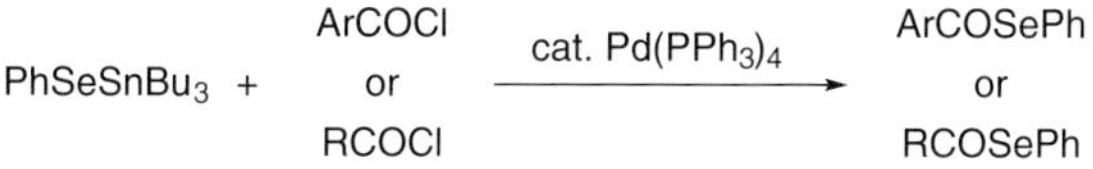

$$PhSeSnBu_3 + \begin{array}{c} ArCOCl \\ or \\ RCOCl \end{array} \xrightarrow{cat.\ Pd(PPh_3)_4} \begin{array}{c} ArCOSePh \\ or \\ RCOSePh \end{array}$$

Scheme 2.100

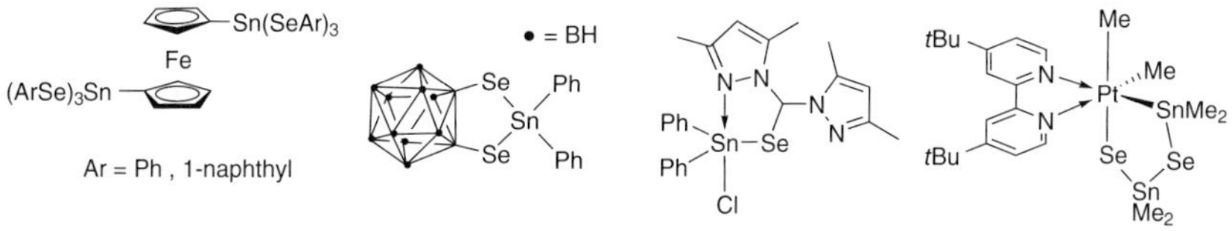

Figure 2.22 Various types of stannyl selenides.

Divalent tin and lead selenolates M(SePh)$_2$ (M = Sn, Pb) were synthesized by the acid–base reaction between M(OAc)$_2$ and PhSeH or by the ligand exchange reaction between M(SPh)$_2$ and PhSeH (Scheme 2.101) [270]. The selenolates were poorly soluble in MeOH, but when reacted with PhSe$^-$, they became soluble with the formation of (M[SePh]$_3$)$^-$ anions [270]. Pb(SeR)$_2$ (R = –CH$_2$CH$_2$NMe$_2$, Mes) was used as a nucleophilic selenolate reagent to synthesize palladium(II) selenolate complexes 159 and 214 (Scheme 2.102) [201].

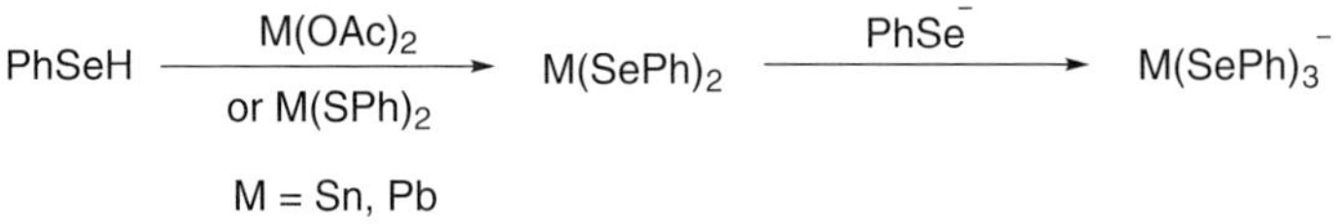

$$\text{PhSeH} \xrightarrow{\substack{\text{M(OAc)}_2 \\ \text{or M(SPh)}_2}} \text{M(SePh)}_2 \xrightarrow{\text{PhSe}^-} \text{M(SePh)}_3^-$$

M = Sn, Pb

Scheme 2.101

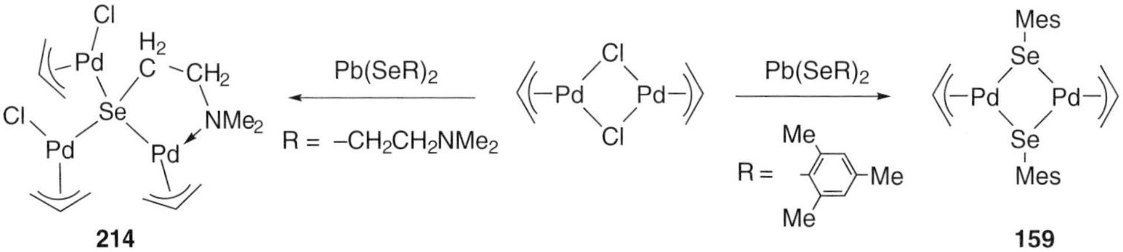

Scheme 2.102

2.4.18
Selenolates of Group 15 Elements (P, As, Sb, and Bi)

Recently, inorganic phosphorus–selenium compounds, such as P$_4$Se$_{10}$ [75–77] and W.R. [78, 80–84], have been frequently used as a nucleophilic selenium reagent (see Section 2.3.2). Various types of phosphorus compounds with an organic selenolate substituent have also been synthesized and characterized. For example, tri(alkyl- or aryl-seleno)phosphines (RSe)$_3$P were obtained by the reaction of white phosphorus with organic diselenides in the presence of a base [271] or by the reaction of PCl$_3$ with PhSeNa [272] (Scheme 2.103).

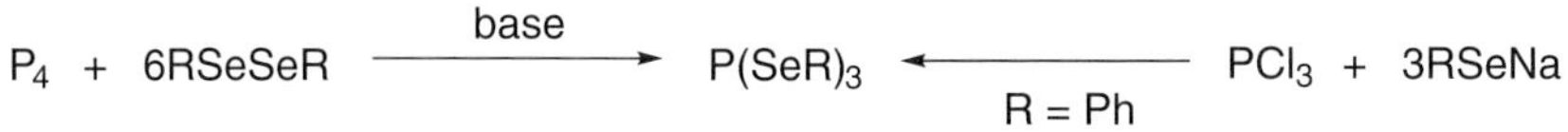

$$\text{P}_4 + 6\text{RSeSeR} \xrightarrow{\text{base}} \text{P(SeR)}_3 \xleftarrow{\text{R = Ph}} \text{PCl}_3 + 3\text{RSeNa}$$

Scheme 2.103

Arsenic selenolates were synthesized by several methods [34]. RSeAsMe$_2$ was synthesized by the reaction of Me$_2$AsNEt$_2$ with selenol RSeH [273] or by the reaction of Me$_2$AsCl with RSeNa [274]. As(PhSe)$_3$ was obtained from In(SePh)$_3$ by the ligand exchange reaction with As(SiMe$_3$)$_3$ (Scheme 2.94) [256]. Ligand exchange reactions of thiophenolates M(SPh)$_3$ (M = As, Sb, Bi) with PhSeH were also used for the preparation of M(SePh)$_3$ (Scheme 2.104) [270].

$$M(SPh)_3 + 3PhSeH \xrightarrow{M = As, Sb, Bi} M(SePh)_3$$

Scheme 2.104

Antimony selenolates Sb(SeAr)$_3$ (Ar = 2,4,6-R$_3$C$_6$H$_2$; R = Me, *i*Pr, *t*Bu) **215** (Figure 2.23) were synthesized from SbCl$_3$ by halide exchange with ArSeH, and their thermolytic decomposition to Sb$_2$Se$_3$ was investigated [275]. Preparation of metal selenide nanostructures and thin films was also investigated by using similar antimony and bismuth selenolates M(SePy)$_3$ (M = Sb, Bi; Py = 2-pyridyl) **216** as a precursor [276]. Antimony quinoline-8-selenolate **217** was also synthesized and characterized by X-ray analysis [277].

Molecular structures of monomeric bismuth selenolates Bi(SeAr)$_3$, such as **218** and **219** (Figure 2.24), were reported for complexes with selenolate ligands having a coordinating nitrogen atom [278–280]. Polynuclear bismuth selenolates clusters (Bi$_4$[SePh]$_{13}$)$^-$ and (Bi$_6$[SePh]$_{16}$Br$_2$), which have both terminal and bridging PhSe$^-$ ligands, were obtained by the reaction of BiBr$_3$ with PhSeSiMe$_3$ in the presence

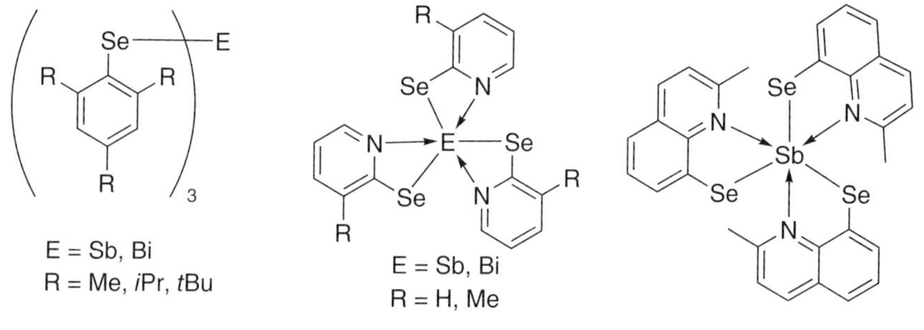

Figure 2.23 Examples of antimony and bismuth selenolates.

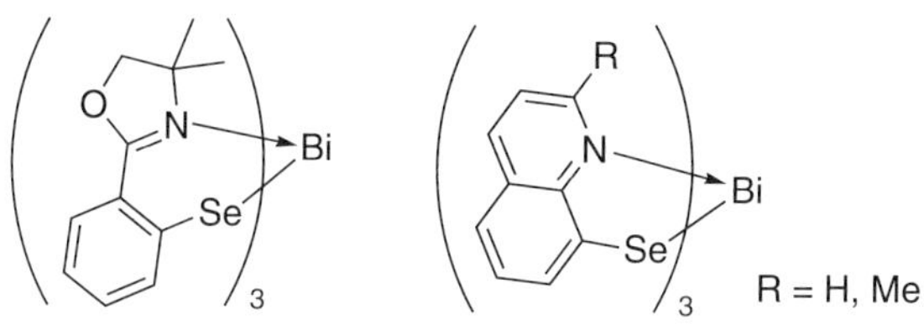

Figure 2.24 Molecular structures for monomeric bismuth selenolates **218** and **219**.

of tertiary phosphines [281]. Bismuth complexes with selenoether ligands (BiX$_3$L) (X = Cl, Br; L = MeSe(CH$_2$)$_3$SeMe, MeC(CH$_2$SeMe)$_3$) were also synthesized by the reaction of BiX$_3$ with L [282]. The complexes formed polymeric structures by using the bidentate or tridentate selenium ligand in the solid state.

References

1 Sharpless, K.B. and Lauer, R.F. (1973) *J. Am. Chem. Soc.*, **95**, 2697.
2 Anderson, J.W., Barker, G.K., Drake, J.E., and Rodger, M. (1973) *J. Chem. Soc., Dalton Trans.*, 1716.
3 Reich, H.J., Renga, J.M., and Reich, I.L. (1975) *J. Am. Chem. Soc.*, **97**, 5434.
4 Patai, S. and Rappoport, Z. (eds) (1986) *The Chemistry of Organic Selenium and Tellurium Compounds*, vol. 1, John Wiley & Sons, Ltd, Chichester.
5 Paulmier, C. (1986) *Selenium Reagents and Intermediates in Organic Synthesis*, Pergamon Press, Oxford.
6 Nicolaou, K.C. and Petasis, N.A. (1984) *Selenium in Natural Products Synthesis*, CIS, Philadelphia.
7 Iwaoka, M. and Tomoda, S. (2000) *Topics in Current Chemistry*, vol. 208 (ed. T. Wirth), Springer, Berlin, pp. 55–80.
8 Engman, L. and Gupta, V. (1999) *Organoselenium Chemistry, a Practical Approach* (ed. T.G. Back), Oxford University Press, New York, pp. 67–91.
9 Tomoda, S. and Iwaoka, M. (1988) *J. Chem. Soc., Chem. Commun.*, 1283.
10 Nishibayashi, Y., Singh, J.D., Fukuzawa, S., and Uemura, S. (1995) *J. Chem. Soc., Perkin Trans. 1*, 2871.
11 Wirth, T. (1999) *Tetrahedron*, **55**, 1.
12 Sun, J., Yang, M., Yuan, F., Jia, X., Yang, X., Pan, Y., and Zhu, C. (2009) *Adv. Synth. Catal.*, **351**, 920.
13 Hatfield, D.L., Yoo, M.-H., Carlson, B.A., and Gladyshev, V.N. (2009) *Biochim. Biophys. Acta*, **1790**, 1541.
14 Ajiki, K., Hirano, M., and Tanaka, K. (2005) *Org. Lett.*, **7**, 4193.
15 Gao, G.-Y., Colvin, A.J., Chen, Y., and Zhang, X.P. (2004) *J. Org. Chem.*, **69**, 8886.
16 Nishiyama, Y., Tokunaga, K., Kawamatsu, H., and Sonoda, N. (2002) *Tetrahedron Lett.*, **43**, 1507.

17 Gujadhur, R.K. and Venkataraman, D. (2002) *Tetrahedron Lett.*, **44**, 81.
18 Tang, K., Jin, X., Yan, H., Xie, X., Liu, C., and Gong, Q. (2001) *J. Chem. Soc., Dalton Trans.*, 1374.
19 Kumbhare, L.B., Jain, V.K., Phadnis, P.P., and Nethaji, M. (2007) *J. Organomet. Chem.*, **692**, 1546.
20 Kedarnath, G., Dey, S., Jain, V.K., and Dey, G.K. (2006) *J. Nanosci. Nanotechnol.*, **6**, 1031.
21 Kedarnath, G., Dey, S., Jain, V.K., Dey, G.K., and Varghese, B. (2006) *Polyhedron*, **25**, 2383.
22 Kedarnath, G., Kumbhare, L.B., Jain, V.K., Phadnis, P.P., and Nethaji, M. (2006) *Dalton Trans.*, 2714.
23 Ie, Y., Hirose, T., Yao, A., Yamada, T., Takagi, N., Kawai, M., and Aso, Y. (2009) *Phys. Chem. Chem. Phys.*, **11**, 4949.
24 Blum, T. (2003) *Berichte des Forschungszentrums Jülich*, **4044**, 1–128.
25 Blum, T., Ermert, J., and Coenen, H.H. (2001) *J. Labelled Cpd. Radiopharm.*, **44**, 587.
26 Ermert, J., Blum, T., Hamacher, K., and Coenen, H.H. (2001) *Radiochim. Acta*, **89**, 863.
27 Gregus, Z., Gyurasics, Á., and Csanaky, I. (2000) *Toxicol. Sci.*, **57**, 22.
28 Jackson-Rosario, S.E., and Self, W.T. (2010) *Metallomics*, **2**, 112.
29 Hondal, R.J., Nilsson, B.L., and Raines, R.T. (2001) *J. Am. Chem. Soc.*, **123**, 5140.
30 Gieselman, M.D., Xie, L., and van der Donk, W.A. (2001) *Org. Lett.*, **3**, 1331.
31 Quaderer, R., Sewing, A., and Hilvert, D. (2001) *Helv. Chim. Acta*, **84**, 1197.
32 Ricci, A. and Comes-Franchini, M. (2002) *Sci. Synth.*, **4**, 427.
33 Ooi, T. and Maruoka, K. (2004) *Sci. Synth.*, **7**, 215.

References

34 Comasseto, J.V. and Guarezemini, A.S. (2007) *Sci. Synth.*, **39**, 947.
35 Polo, A. and Real, J. (2007) *Sci. Synth.*, **39**, 961.
36 Kambe, N. (2007) *Sci. Synth.*, **39**, 1059.
37 Polo, A. and Real, J. (2007) *Sci. Synth.*, **39**, 1063.
38 Kurz, J.L. and Harris, J.C. (1970) *J. Org. Chem.*, **35**, 3086.
39 Paulmier, C. (1986) *Selenium Reagents and Intermediates in Organic Synthesis*, Pergamon Press, Oxford, pp. 25–57.
40 Lide, D.R. (ed.-in-chief) (2005) *CRC Handbook of Chemistry and Physics*, 86th edn, CRC Press, Taylor & Francis, Boca Raton, FL.
41 Guillemin, J.-C., Bouayad, A., and Vijaykumar, D. (2000) *Chem. Commun.*, 1163.
42 Riague, E.H. and Guillemin, J.-C. (2002) *Organometallics*, **21**, 68.
43 Guillemin, J.-C., Bajor, G., Riague, E.H., Khater, B., and Veszprémi, T. (2007) *Organometallics*, **26**, 2507.
44 Bajor, G., Veszprémi, T., Riague, E.H., and Guillemin, J.-C. (2004) *Chem. Eur. J.*, **10**, 3649.
45 Møllendal, H., Konovalov, A., and Guillemin, J.-C. (2009) *J. Phys. Chem. A*, **113**, 6342.
46 Laube, J., Jäger, S., and Thöne, C. (2001) *Eur. J. Inorg. Chem.*, 1983.
47 Landry, V.K., Minoura, M., Pang, K., Buccella, D., Kelly, B.V., and Parkin, G. (2006) *J. Am. Chem. Soc.*, **128**, 12490.
48 Jayaram, P.N., Roy, G., and Mugesh, G. (2008) *J. Chem. Sci.*, **120**, 143.
49 Krief, A., Van Wemmel, T., Redon, M., Dumont, W., and Delmotte, C. (1999) *Angew. Chem. Int. Ed.*, **38**, 2245.
50 Ahrika, A., Robert, J., Anouti, M., and Paris, J. (2001) *New J. Chem.*, **25**, 741.
51 Kumar, S., Kandasamy, K., Singh, H.B., Wolmershäuser, G., and Butcher, R.J. (2004) *Organometallics*, **23**, 4199.
52 Cardey, B. and Enescu, M. (2005) *ChemPhysChem*, **6**, 1175.
53 Paldán, K., Oilunkaniemi, R., Laitinen, R.S., and Ahlgrén, M. (2005) *Main Group Chem.*, **4**, 127.
54 Chadwick, S., Englich, U., Ruhlandt-Senge, K., Watson, C., Bruce, A.E., and Bruce, M.R.M. (2000) *J. Chem. Soc., Dalton Trans.*, 2167.
55 Klayman, D.L. and Griffin, T.S. (1973) *J. Am. Chem. Soc.*, **95**, 197.
56 Krief, A., Trabelsi, M., Dumont, W., and Derock, M. (2004) *Synlett*, 1751.
57 Brandsma, L. and Wijers, H. (1963) *Receuil Trav. Chim.*, **82**, 68.
58 Sekiguchi, M., Tanaka, H., Takami, W., Ogawa, A., Ryu, I., and Sonoda, N. (1991) *Heteratom Chem.*, **2**, 427.
59 Detty, M.R. and Seidler, M.D. (1982) *J. Org. Chem.*, **47**, 1354.
60 Borecki, A. and Corrigan, J.F. (2007) *Inorg. Chem.*, **46**, 2478.
61 DeGroot, M.W., Taylor, N.J., and Corrigan, J.F. (2003) *J. Am. Chem. Soc.*, **125**, 864.
62 Lange, H. and Herzog, U. (2002) *J. Organomet. Chem.*, **660**, 36.
63 Ishihara, H., Kotetsu, M., Fukuta, Y., and Nada, F. (2001) *J. Am. Chem. Soc.*, **123**, 8408.
64 Kotetsu, M., Fukuta, Y., and Ishihara, H. (2001) *Tetrahedron Lett.*, **42**, 6333.
65 Kotetsu, M., Ishida, M., Takakura, N., and Ishihara, H. (2002) *J. Org. Chem.*, **67**, 486.
66 Kotetsu, M., Fukuta, Y., and Ishihara, H. (2002) *J. Org. Chem.*, **67**, 1008.
67 Wu, X. and Hu, L. (2007) *J. Org. Chem.*, **72**, 765.
68 Müller, A., Diemann, E., Jostes, R., and Bogge, H. (1981) *Angew. Chem. Int. Ed. Engl.*, **20**, 934.
69 O'Neal, S. and Kolis, J.W. (1988) *J. Am. Chem. Soc.*, **110**, 1971.
70 Saravanan, V., Porhiel, E., and Chandrasekaran, S. (2003) *Teterahedron Lett.*, **44**, 2257.
71 Bhat, R.G., Porhiel, E., Saravanan, V., and Chandrasekaran, S. (2003) *Teterahedron Lett.*, **44**, 5251.
72 Sridhar, P.R., Saravanan, V., and Chandrasekaran, S. (2005) *Pure Appl. Chem.*, **77**, 145.
73 Saravanan, V., Mukherjee, C., Das, S., and Chandrasekaran, S. (2004) *Tetrahedron Lett.*, **45**, 681.
74 Chiu, W.-H., Zhang, Q.-F., Williams, I.D., and Leung, W.-H. (2010) *Organometallics*, **29**, 2631.
75 Rae, I.D. and Wade, M.J. (1976) *Int. J. Sulf. Chem.*, **8**, 519.

76 Hallam, H.E. and Jones, C.M. (1969) *J. Chem. Soc. A*, 1033.
77 Milewska, M.J. and Połoński, T. (1999) *Tetrahedron Asymmetry*, **10**, 4123.
78 Gray, I.P., Bhattacharyya, P., Slawin, A.M.Z., and Woollins, J.D. (2005) *Chem. Eur. J.*, **11**, 6221.
79 Ozturk, T., Ertas, E., and Mert, O. (2007) *Chem. Rev.*, **107**, 5210.
80 Bethke, J., Karaghiosoff, K., and Wessjohan, L.A. (2003) *Ahedron Lett.*, **44**, 6911.
81 Knapp, S. and Darout, E. (2005) *Lett.*, **7**, 203.
82 Mohanakrishnan, A.K. and Amaladass, P. (2005) *Tetrahedron Lett.*, **46**, 7201.
83 Makiyama, A., Komatsu, I., Iwaoka, M., and Yatagai, M. (2011) *Phosph. Sulf. Silic.*, **186**, 125.
84 Shi, W., Shafaei-Fallah, M., Zhang, L., Anson, C.E., Matern, E., and Rothenberger, A. (2007) *Chem. Eur. J.*, **13**, 598.
85 González-Calera, S., Eisler, D.J., Morey, J.V., McPartlin, M., Singh, S., and Wright, D.S. (2008) *Angew. Chem. Int. Ed.*, **47**, 1111.
86 Singh, D., Deobald, A.M., Camargo, L.R.S., Tabarelli, G., Rodrigues, O.E.D., and Braga, A.L. (2010) *Org. Lett.*, **12**, 3288–3291.
87 Zhang, S. and Tian, F. (2001) *J. Chem. Res. (S)*, 198.
88 Movassagh, B. and Mirshojaei, F. (2003) *Monatsh. Chem.*, **134**, 831.
89 Movassagh, B., Shamsipoor, M., and Joshaghani, M. (2004) *J. Chem. Res.*, 148.
90 Nazari, M. and Movassagh, B. (2009) *Tetrahedron Lett.*, **50**, 438.
91 Movassagh, B. and Tatar, A. (2007) *Synlett*, 1954.
92 do Rego Barros, O.S., Lang, E.S., de Oliveirra, C.A.F., Peppe, C., and Zeni, G. (2002) *Tetrahedron Lett.*, **43**, 7921.
93 Peppe, C., Lang, E.S., Ledesma, G.N., de Castro, L.B., do Rego Barros, O.S., and de Azevedo Mello, P. (2005) *Synlett*, 3091.
94 Li, X., Zhang, S., Wang, Y., and Zhang, Y. (2002) *J. Chin. Chem. Soc.*, **49**, 1111.

95 Konno, T., Yoshimura, T., Masuyama, G., and Hirotsu, M. (2002) *Bull. Chem. Soc. Jpn.*, **75**, 2185.
96 Yuan, M., Dirmyer, M., Badding, J., Sen, A., Dahlberg, M., and Schiffer, P. (2007) *Inorg. Chem.*, **46**, 7238.
97 Cai, S. and Jin, G.-X. (2007) *Organometallics*, **26**, 5442.
98 Ahlrichs, R., Crawford, N.R.M., Eichhöfer, A., Fenske, D., Hampe, O., Kappes, M.M., and Olkowska-Oetzel, J. (2006) *Eur. J. Inorg. Chem.*, 345.
99 Sun, C., Westover, R.D., Margulieux, K.R., Zakharov, L.N., Holland, A.W., and Pak, J.J. (2010) *Inorg. Chem.*, **49**, 4756.
100 Eichhöfer, A., Olkowska-Oetzel, J., Fenske, D., Fink, K., Mereacre, V., Powell, A.K., and Buth, G. (2009) *Inorg. Chem.*, **48**, 8977.
101 Murai, T., Hayakawa, S., and Kato, S. (2001) *J. Org. Chem.*, **66**, 8101.
102 Tani, K., Murai, T., and Kato, S. (2002) *J. Am. Chem. Soc.*, **124**, 5960.
103 Dyachenko, V.D. (2004) *Russ. J. Gen. Chem.*, **74**, 1137.
104 Kamimura, A., Mitsudera, H., Asano, S., Kidera, S., and Kakehi, A. (1999) *J. Org. Chem.*, **64**, 6353.
105 Clegg, W., Davies, R.P., Snaith, R., and Wheatley, A.E.H. (2001) *Eur. J. Inorg. Chem.*, 1411.
106 Pietschnig, R., Merz, K., and Schäfer, S. (2005) *Heteroatom Chem.*, **16**, 169.
107 Rodionova, L.S., Filanovskii, B.K., and Petrov, M.L. (2001) *Russ. J. Gen. Chem.*, **71**, 85.
108 Niyomura, O., Kato, S., and Murai, T. (1999) *Inorg. Chem.*, **38**, 507.
109 Ju-Nam, Y., Allen, D.W., Gardiner, P.H.E., Light, M.E., Hursthouse, M.B., and Bricklebank, N. (2007) *J. Organomet. Chem.*, **692**, 5065.
110 Chin, J., Tak, J., Hahn, D., Yang, I., Ko, J., Ham, J., and Kang, H. (2009) *Bull. Korean Chem. Soc.*, **30**, 496.
111 Ganesh, V. and Chandrasekaran, S. (2009) *Synthesis*, 3267.
112 Murai, T., Suzuki, A., and Kato, S. (2001) *J. Chem. Soc., Perkin Trans. 1*, 2711.
113 Pietschnig, R., Schäfer, S., and Merz, K. (2003) *Org. Lett.*, **5**, 1867.

114 Okamoto, T., Kudoh, K., Wakamiya, A., and Yamaguchi, S. (2005) *Org. Lett.*, **7**, 5301.

115 Braga, A.L., Alves, E.F., Silveira, C.C., and De Andrade, L.H. (1999) *Tetrahedron Lett.*, **41**, 161.

116 Lenardão, E.J., Silva, M.S., Sachini, M., Lara, R.G., Jacob, R.G., and Perin, G. (2009) *ARKIVOC*, 221.

117 Kawai, Y., Ando, H., Ozeki, H., Koketsu, M., and Ishihara, H. (2005) *Org. Lett.*, **7**, 4653.

118 Kamimura, A., Mitsudera, H., Omata, Y., Matsuura, K., Shirai, M., and Kakehi, A. (2002) *Tetrahedron*, **58**, 9817.

119 Mitsudera, H., Kakehi, A., and Kamimura, A. (1999) *Tetrahedron Lett.*, **40**, 7389.

120 Englich, U. and Ruhlandt-Senge, K. (2001) *Z. Anorg. Allg. Chem.*, **627**, 851.

121 Ruhlandt-Senge, K. and Englich, U. (2000) *Chem. Eur. J.*, **6**, 4063.

122 Lamsabhi, A.M., Mó, O., Yáñez, M., and Boyd, R.J. (2008) *J. Chem. Theory Comput.*, **4**, 1002.

123 Wang, L., and Zhang, Y. (1999) *Heteroatom Chem.*, **10**, 203.

124 Hillier, A.C., Liu, S.Y., Sella, A., and Elsegood, M.R.J. (2000) *J. Alloys Compounds*, **303–304**, 83.

125 Freedman, D., Kornienko, A., Emge, T.J., and Brennan, J.G. (2000) *Inorg. Chem.*, **39**, 2168.

126 Fukuzawa, S., Niimoto, Y., Fujinami, T., and Sakai, S. (1990) *Heteroatom Chem.*, **2**, 491.

127 Guo, H. and Zhang, Y. (2000) *J. Chem. Res. (S)*, 374.

128 Su, W., Gao, N., and Zhang, Y. (2002) *J. Chem. Res. (S)*, 168.

129 Banerjee, S., Huebner, L., Romanelli, M.D., Kumar, G.A., Riman, R.E., Emge, T.J., and Brennan, J.G. (2005) *J. Am. Chem. Soc.*, **127**, 15900.

130 Banerjee, S., Kumar, G.A., Riman, R.E., Emge, T.J., and Brennan, J.G. (2007) *J. Am. Chem. Soc.*, **129**, 5926.

131 Romanelli, M., Kumar, G.A., Emge, T.J., Riman, R.E., and Brennan, J.G. (2008) *Angew. Chem. Int. Ed.*, **47**, 6049.

132 Hauber, S.-O. and Niemeyer, M. (2007) *Chem. Commun.*, 275.

133 Yu, X.-Y., Jin, G.-X., Hu, N.-H., and Weng, L.-H. (2002) *Organometallics*, **21**, 5540.

134 Evans, W.J., Miller, K.A., Kozimor, S.A., Ziller, J.W., DiPasquale, A.G., and Rheingold, A.L. (2007) *Organometallics*, **26**, 3568.

135 Song, L.-C., Han, C., Hu, Q.-M., and Zhang, Z.-P. (2004) *Inorg. Chim. Acta*, **357**, 2199.

136 Hector, A.L., Levason, W., Reid, G., Reid, S.D., and Webster, M. (2008) *Chem. Mater.*, **20**, 5100.

137 Hector, A.L., Jura, M., Levason, W., Reid, S.D., and Reid, G. (2009) *New J. Chem.*, **33**, 641.

138 Brunner, H., Kubicki, M.M., Leblanc, J.-C., Meier, W., Moïse, C., Sadorge, A., Stubenhofer, B., Wachter, J., and Wanninger, R. (1999) *Eur. J. Inorg. Chem.*, 843.

139 Hector, A.L., Jura, M., Levason, W., Reid, G., Reid, S.D., and Webster, M. (2008) *Acta Cryst.*, **C64**, m321.

140 Okamura, T., Taniuchi, K., Lee, K., Yamamoto, H., Ueyama, N., and Nakamura, A. (2006) *Inorg. Chem.*, **45**, 9374.

141 Okamura, T., Taniuchi, K., Ueyama, N., and Nakamura, A. (1999) *Polym. J.*, **31**, 651.

142 Ma, X., Schulzke, C., Yang, Z., Ringe, A., and Magull, J. (2007) *Polyhedron*, **26**, 5497.

143 Llusar, R., Triguero, S., Polo, V., Vicent, C., Gómez-García, C.J., Jeannin, O., and Fourmigué, M. (2008) *Inorg. Chem.*, **47**, 9400.

144 Kawaguchi, H. and Tatsumi, K. (2000) *Chem. Commun.*, 1299.

145 Groysman, S. and Holm, R.H. (2007) *Inorg. Chem.*, **46**, 4090.

146 Song, L.-C., Shi, Y.-C., and Zhu, W.-F. (1999) *Polyhedron*, **18**, 2163.

147 Westerhoff, O., Martens, J., and Thöne, C. (1999) *Z. Anorg. Allg. Chem.*, **625**, 1823.

148 Jin, G.-X. and Herberhold, M. (2001) *Trans. Metal Chem.*, **26**, 496.

149 Eichhöfer, A., Wood, P.T., Viswanath, R., and Mole, R.A. (2007) *Eur. J. Inorg. Chem.*, 4794.

150 Reyes-Lezama, M., Höpfl, H., and Zúñiga-Villarreal, N. (2010) *Organometallics*, **29**, 1537.

151 Liaw, W.-F., Chou, S.-Y., Jung, S.-J., Lee, G.-H., and Peng, S.-M. (1999) *Inorg. Chim. Acta*, **286**, 155.

152 Dopke, J.A., Wilson, S.R., and Rauchfuss, T.B. (2000) *Inorg. Chem.*, **39**, 5014.

153 De Montigny, F., Guy, L., Pilet, G., Vanthuyne, N., Roussel, C., Lombardi, R., Freedman, T.B., Nafie, L.A., and Crassous, J. (2009) *Chem. Commun.*, 4841.

154 Seyferth, D., Womack, G.B., and Dewan, J.C. (1985) *Organometallics*, **4**, 398.

155 Yan, C.-G., Sun, J., and Sun, J. (1999) *J. Organomet. Chem.*, **585**, 63.

156 Song, L.-C., Lu, G.-L., Hu, Q.-M., Fan, H.-T., Chen, J., Sun, J., and Huang, X.-Y. (2001) *J. Organomet. Chem.*, **627**, 255.

157 Song, L.-C., Lu, G.-L., Hu, Q.-M., and Sun, J. (1999) *Organometallics*, **18**, 2700.

158 Wang, Z.-X., Miao, S.-B., and Zhang, Z.-Y. (2000) *J. Organomet. Chem.*, **604**, 214.

159 Song, L.-C., Zeng, G.-H., Mei, S.-Z., Lou, S.-X., and Hu, Q.-M. (2006) *Organometallics*, **25**, 3468.

160 Song, L.-C., Sun, Y., Hu, Q.-M., and Liu, Y. (2003) *J. Organomet. Chem.*, **676**, 80.

161 Jäger, S., Jones, P.G., Laube, J., and Thöne, C. (1999) *Z. Anorg. Allg. Chem.*, **625**, 352.

162 Kückmann, T.I., Hermsen, M., Bolte, M., Wagner, M., and Lerner, H.-W. (2005) *Inorg. Chem.*, **44**, 3449.

163 Kückmann, T., Schödel, F., Sänger, I., Bolte, M., Wagner, M., and Lerner, H.-W. (2010) *Eur. J. Inorg. Chem.*, 468.

164 Harb, M.K., Windhager, J., Daraosheh, A., Görls, H., Lockett, L.T., Okumura, N., Evans, D.H., Glass, R.S., Lichtenberger, D.L., El-khateeb, M., and Weigand, W. (2009) *Eur. J. Inorg. Chem.*, 3414.

165 Hauptmann, R., Kliß, R., and Henkel, G. (1999) *Angew. Chem., Int. Ed.*, **38**, 377.

166 Hauptmann, R., Lackmann, J., Chen, C., and Henkel, G. (1999) *Acta Cryst.*, **C55**, 1084.

167 Lackmann, J., Hauptmann, R., Weißgräber, S., and Henkel, G. (1999) *Chem. Commun.*, 1995.

168 Nakamoto, M., Fukaishi, K., Tagata, T., Kambayashi, H., and Tanaka, K. (1999) *Bull. Chem. Soc. Jpn.*, **72**, 407.

169 Becker, E., Mereiter, K., Schmid, R., and Kirchner, K. (2004) *Organometallics*, **23**, 2876.

170 Nishibayashi, Y., Imajima, H., Onodera, G., Hidai, M., and Uemura, S. (2004) *Organometallics*, **23**, 26.

171 Nishibayashi, Y., Imajima, H., Onodera, G., Inada, Y., Hidai, M., and Uemura, S. (2004) *Organometallics*, **23**, 5100.

172 Dibrov, S.M., Deng, B., Ellis, D.E., and Ibers, J.A. (2005) *Inorg. Chem.*, **44**, 3441.

173 Lukevics, E., Zaruma, D., Ashaks, J., Shestakova, I., Domracheva, I., Bridane, V., and Yashchenko, E. (2009) *Chem. Heterocycl. Comp.*, **45**, 182.

174 Kita, M., Tamai, H., Ueta, F., Fuyuhiro, A., Yamanari, K., Nakajima, K., Kojima, M., Murata, K., and Yamashita, S. (2001) *Inorg. Chim. Acta*, **314**, 139.

175 Valenzuela, E., Sousa-Pedrares, A., Durán-Carril, M.L., García-Vázquez, J.A., Romero, J., and Sousa, A. (2007) *Z. Anorg. Allg. Chem.*, **633**, 1853.

176 Habe, S., Yamada, T., Nankawa, T., Mizutani, J., Murata, M., and Nishihara, H. (2003) *Inorg. Chem.*, **42**, 1952.

177 Murata, M., Habe, S., Araki, S., Namiki, K., Yamada, T., Nakagawa, N., Nankawa, T., Nihei, M., Mizutani, J., Kurihara, M., and Nishihara, H. (2006) *Inorg. Chem.*, **45**, 1108.

178 Herberhold, M., Jin, G.-X., Yan, H., Milius, W., and Wrackmeyer, B. (1999) *Eur. J. Inorg. Chem.*, 873.

179 Hou, X.-F., Wang, X.-C., Wang, J.-Q., and Jin, G.-X. (2004) *J. Organomet. Chem.*, **689**, 2228.

180 Hou, X.-F., Liu, S., Wang, H., Chen, Y.-Q., and Jin, G.-X. (2006) *Dalton Trans.*, 5231.

181 Bitterwolf, T.E. and Pal, P. (2006) *Inorg. Chim. Acta*, **359**, 1501.

182 Mautz, J. and Huttner, G. (2008) *Eur. J. Inorg. Chem.*, 1423.

183 Herberhold, M., Jin, G.-X., Yan, H., Milius, W., and Wrackmeyer, B. (1999) *J. Organomet. Chem.*, **587**, 252.
184 Herberhold, M., Yan, H., Milius, W., and Wrackmeyer, B. (2000) *Z. Anorg. Allg. Chem.*, **626**, 1627.
185 Herberhold, M., Daniel, T., Daschner, D., Milius, W., and Wrackmeyer, B. (1999) *J. Organomet. Chem.*, **585**, 234.
186 Seino, H., Misumi, Y., Hojo, Y., and Mizobe, Y. (2010) *Dalton Trans.*, (**39**), 3072.
187 Herberhold, M., Jin, G.-X., and Rheingold, A.L. (2002) *Z. Anorg. Allg. Chem.*, **628**, 1985.
188 Liaw, W.-F., Horng, Y.-C., Ou, D.-S., Ching, C.-Y., Lee, G.-H., and Peng, S.-M. (1997) *J. Am. Chem. Soc.*, **119**, 9299.
189 Liaw, W.-F., Chen, C.-H., Lee, C.-M., Lee, G.-H., and Peng, S.-M. (2001) *J. Chem. Soc., Dalton Trans.*, 138.
190 Deplano, P., Marchio, L., Mercuri, M.L., Pilia, L., Serpe, A., and Trogu, E.F. (2003) *Polyhedron*, **22**, 2175.
191 Schaub, T., Backes, M., and Radius, U. (2008) *Eur. J. Inorg. Chem.*, 2680.
192 Chen, C.-H., Lee, G.-H., and Liaw, W.-F. (2006) *Org. Chem.*, **45**, 2307.
193 Nomura, M., Cauchy, T., Geoffroy, M., Adkine, P., and Fourmigue, M. (2006) *Inorg. Chem.*, **45**, 8194.
194 Lee, C.-M., Chuang, Y.-L., Chiang, C.-Y., Lee, G.-H., and Liaw, W.-F. (2006) *Inorg. Chem.*, **45**, 10895.
195 Kersting, B. and Siebert, D. (1999) *Eur. J. Inorg. Chem.*, 189.
196 Dey, S., Jain, V.K., Chaudhury, S., Knoedler, A., Lissner, F., and Kaim, W. (2001) *J. Chem. Soc., Dalton Trans.*, 723.
197 Dey, S., Kumbhare, L.B., Jain, V.K., Schurr, T., Kaim, W., Klein, A., and Belaj, F. (2004) *Eur. J. Inorg. Chem.*, 4510.
198 Canty, A.J., Denney, M.C., Patel, J., Sun, H., Skelton, B.W., and White, A.H. (2004) *J. Organomet. Chem.*, **689**, 672.
199 Dey, S., Jain, V.K., Knoedler, A., and Kaim, W. (2003) *Inorg. Chim. Acta*, **349**, 104.
200 Wagner, A., Hannu-Kuure, M.S., Oilunkaniemi, R., and Laitinen, R.S. (2005) *Acta Cryst.*, **E61**, m2198.
201 Ghavale, N., Dey, S., Wadawale, A., and Jain, V.K. (2008) *Organometallics*, **27**, 3297.
202 Kumbhare, L.B., Wadawale, A.P., Jain, V.K., Kolay, S., and Nethaji, M. (2009) *J. Organomet. Chem.*, **694**, 3892.
203 Nakata, N., Uchiumi, R., Yoshino, T., Ikeda, T., Kamon, H., and Ishii, A. (2009) *Organometallics*, **28**, 1981.
204 Hannu-Kuure, M.S., Palda'n, K., Oilunkaniemi, R., Laitinen, R.S., and Ahlgrén, M. (2003) *J. Organomet. Chem.*, **687**, 538.
205 Dey, S., Jain, V.K., Klein, A., and Kaim, W. (2004) *Inorg. Chem. Commun.*, **7**, 601.
206 Day, S., Jain, V.K., Knoedler, A., Kaim, W., and Zalis, S. (2001) *Eur. J. Inorg. Chem.*, 2965.
207 Dey, S., Jain, V.K., Knödler, A., Klein, A., Kaim, W., and Záliš, S. (2002) *Inorg. Chem.*, **41**, 2864.
208 Hannu-Kuure, M.S., Komulainen, J., Oilunkaniemi, R., Laitinen, R.S., Suontamo, R., and Ahlgrén, M. (2003) *J. Organomet. Chem.*, **666**, 111.
209 Schäfer, S., Moser, C., Tirrée, J.J., Nieger, M., and Pietschnig, R. (2005) *Inorg. Chem.*, **44**, 2798.
210 Kirij, N.V., Tyrra, W., Pantenburg, I., Naumann, D., Scherer, H., Naumann, D., and Yagupolskii, Y.L. (2006) *J. Organomet. Chem.*, **691**, 2679.
211 Wrackmeyer, B., García Hernández, Z., Kempe, R., and Herberhold, M. (2007) *Eur. J. Inorg. Chem.*, 239.
212 Morley, C.P., Webster, C.A., Douglas, P., Rofe, K., and Di Vaira, M. (2010) *Dalton Trans.*, (**39**), 3177.
213 Ishii, A., Nakata, N., Uchiumi, R., and Murakumi, K. (2008) *Angew. Chem., Int. Ed.*, **47**, 2661.
214 Ishii, A., Kamon, H., Murakami, K., and Nakata, N. (2010) *Eur. J. Org. Chem.*, 1653.
215 Ishii, A., Yamaguchi, Y., and Nakata, N. (2010) *Dalton Trans.*, (**39**), 6181.
216 Nakata, N., Yamaguchi, Y., and Ishii, A. (2010) *J. Organomet. Chem.*, **695**, 970.
217 Janzen, M.C., Jennings, M.C., and Puddephatt, R.J. (2002) *Can. J. Chem.*, **80**, 41.
218 Brown, M.J. and Corrigan, J.F. (2004) *J. Organomet. Chem.*, **689**, 2872.

219 Dey, S., Jain, V.K., and Butcher, R.J. (2007) *Inorg. Chim. Acta*, **360**, 2653.
220 MacDonald, D.G. and Corrigan, J.F. (2008) *Dalton Trans.*, 5048.
221 Yam, V.W.-W., Lam, C.-H., and Cheung, K.-K. (2001) *Chem. Commun.*, 545.
222 Yam, V.W.-W., Lam, C.-H., Fung, W.K.-M., and Cheung, K.-K. (2001) *Inorg. Chem.*, **40**, 3435.
223 Rodriguez, A., Romero, J., García-Vázquez, J.A., Durán, M.L., Sousa-Pedrares, A., Sousa, A., and Zubieta, J. (1999) *Inorg. Chim. Acta*, **284**, 133.
224 Jin, X., Tank, K., Long, Y., and Tang, Y. (1999) *Acta Cryst.*, **C55**, 1799.
225 Tang, K., Jin, X., Long, Y., Cui, P., and Tang, Y. (2000) *J. Chem. Res. (S)*, 452.
226 Liu, C., Wang, X., Gong, Q., Tang, K., Jin, X., Yan, H., and Cui, P. (2001) *Adv. Mater.*, **13**, 1687.
227 Canales, S., Crespo, O., Gimeno, M.C., Jones, P.G., Laguna, A., and Romero, P. (2003) *Dalton Trans.*, 4525.
228 Laromaine, A., Teixidor, F., Kivekäs, R., Sillanpää, R., Arca, M., Lippolis, V., Crespo, E., and Viñas, C. (2006) *Dalton Trans.*, 5240.
229 Taher, D., Taylor, N.J., and Corrigan, J.F. (2009) *Can. J. Chem.*, **87**, 380.
230 Canales, S., Crespo, O., Gimeno, M.C., Jones, P.G., and Laguna, A. (2004) *Inorg. Chem.*, **43**, 7234.
231 Crespo, O., Gimeno, M.C., Laguna, A., Kulcsar, M., and Silvestru, C. (2009) *Inorg. Chem.*, **48**, 4134.
232 Bhabak, K.P. and Mugesh, G. (2009) *Inorg. Chem.*, **48**, 2449.
233 Protsailo, L.V., Fawcett, W.R., Russell, D., and Meyer, R.L. (2002) *Langmuir*, **18**, 9342.
234 Braga, A.L., Schwab, R.S., Alberto, E.E., Salman, S.M., Vargas, J., and Azeredo, J.B. (2009) *Tetrahedron Lett.*, **50**, 2309.
235 Jun, Y.-W., Koo, J.-E., and Cheon, J. (2000) *Chem. Commun.*, 1243.
236 Santi, C., Santoro, S., Battistelli, B., Testaferri, L., and Tiecco, M. (2008) *Eur. J. Org. Chem.*, 5387.
237 Yam, V.W.-W., Pui, Y.-L., and Cheung, K.-K. (1999) *New J. Chem.*, **23**, 1163.
238 Silveira, C.C., Braga, A.L., and Larghi, E.L. (1999) *Organometallics*, **18**, 5183.
239 Lang, E.S., Dias, M.M., Abram, U., and Vázquez-López, E.M. (2000) *Z. Anorg. Allg. Chem.*, **626**, 784.
240 Kandasamy, K., Singh, H.B., and Wolmershäuser, G. (2005) *Inorg. Chim. Acta*, **358**, 207.
241 Eichhöfer, A. and Deglmann, P. (2004) *Eur. J. Inorg. Chem.*, 349.
242 Schmidt, M. and Block, H.D. (1970) *J. Organomet. Chem.*, **25**, 17.
243 Clive, D.L.J. and Menchen, S.M. (1979) *J. Org. Chem.*, **44**, 4279.
244 Kozikowski, A.P. and Ames, A. (1985) *Tetrahedron*, **41**, 4821.
245 Inoue, T., Takeda, T., Kambe, N., Ogawa, A., Ryu, I., and Sonoda, N. (1994) *J. Org. Chem.*, **59**, 5824.
246 Maruoka, K., Miyazaki, T., Ando, M., Matsumura, Y., Sakane, S., Hattori, K., and Yamamoto, H. (1983) *J. Am. Chem. Soc.*, **105**, 2831.
247 Segi, M., Koyama, T., Nakajima, T., Suga, S., Murai, S., and Sonoda, N. (1989) *Tetrahedron Lett.*, **30**, 2095.
248 Segi, M., Takahashi, T., Ichinose, H., Li, M.G., and Nakajima, T. (1992) *Tetrahedron Lett.*, **33**, 7865.
249 Maeda, H., Tanabe, T., Hotta, K., and Mizuno, K. (2005) *Tetrahedron Lett.*, **46**, 2015.
250 Tokoro, Y., Nagai, A., Kokado, K., and Chujo, Y. (2009) *Macromolecules*, **42**, 2988.
251 Han, Y.-G., Xu, C., Duan, T., Zhang, Q.-F., and Leung, W.-H. (2009) *J. Mol. Struct.*, **936**, 15.
252 Smith, D.M. and Ibers, J.A. (1998) *Polyhedron*, **17**, 2105.
253 Hoffmann, G.G. and Faist, R. (1990) *J. Organomet. Chem.*, **391**, 1.
254 Briand, G.G., Decken, A., and Hamilton, N.S. (2010) *Dalton Trans.*, (**39**), 3833.
255 Braga, A.L., Schneider, P.H., Paixão, M.W., Deobald, A.M., Peppe, C., and Bottega, D.P. (2006) *J. Org. Chem.*, **71**, 4305.
256 Baldwin, R.A., Rahbarnoohi, H., Jones, L.J., III, McPhail, A.T., Wells, R.L., White, P.S., Rheingold, A.L., and Yap, G.P.A. (1996) *Heteroatom Chem.*, **7**, 409.
257 Liotta, D., Paty, P.B., Johnston, J., and Zima, G. (1978) *Tetrahedron Lett.*, 5091.

258 Miyoshi, N., Ishii, H., Kondo, K., Murai, S., and Sonoda, N. (1979) *Synthesis*, 300.
259 Schmidt, M., Kiewert, E., Lux, H., and Sametschek, C. (1986) *Phosphorus Sulfur Relat. Elem.*, **26**, 163.
260 Clarembeau, M., Cravador, A., Dumont, W., Hevesi, L., Kreif, A., Lucchetti, J., and Van Ende, D. (1985) *Tetrahedron*, **41**, 4793.
261 Hodson, A.G.W., Thind, R.K., and McPartlin, M. (2002) *J. Organomet. Chem.*, **664**, 277.
262 Theis, B., Metz, S., Burschka, C., Bertermann, R., Maisch, S., and Tacke, R. (2009) *Chem. Eur. J.*, **15**, 7329.
263 Gysling, H.J. and Luss, H.R. (1989) *Organometallics*, **8**, 363.
264 Schlecht, S. and Friese, K. (2003) *Eur. J. Inorg. Chem.*, 1411.
265 Nishiyama, Y., Kawamatsu, H., Funato, S., Tokunaga, K., and Sonoda, N. (2003) *J. Org. Chem.*, **68**, 3599.
266 Nayek, H.P., Hilt, G., and Dehnen, S. (2009) *Eur. J. Inorg. Chem.*, 4205.
267 Wrackmeyer, B., Hernández, Z.G., Kempe, R., and Herberhold, M. (2007) *Appl. Organomet. Chem.*, **21**, 108.
268 Tan, R.-Y., Song, H.-B., and Tang, L.-F. (2006) *J. Organomet. Chem.*, **691**, 5964.
269 Janzen, M.C., Jenkins, H.A., Jennings, M.C., Rendina, L.M., and Puddephatt, R.J. (2002) *Organometallics*, **21**, 1257.
270 Arsenault, J.J.I. and Dean, P.A.W. (1983) *Can. J. Chem.*, **61**, 1516.
271 Maier, L. (1976) *Helv. Chim. Acta*, **59**, 252.
272 Keder, N.L., Shibao, R.K., and Eckert, H. (1992) *Acta Cryst.*, **C48**, 1670.
273 Sagan, L.S., Zingaro, R.A., and Irgolic, K.J. (1972) *J. Organomet. Chem.*, **39**, 301.
274 Banks, C.H., Daniel, J.R., and Zingaro, R.A. (1979) *J. Med. Chem.*, **22**, 572.
275 Bochmann, M., Song, X., Hursthouse, M.B., and Karaulov, A. (1995) *J. Chem. Soc., Dalton Trans.*, 1649.
276 Sharma, R.K., Kedarnath, G., Jain, V.K., Wadawale, A., Nalliath, M., Pillai, C.G.S., and Vishwanadh, B. (2010) *Dalton Trans.*, (**39**), 8779.
277 Silina, E., Belyakov, S., Ashaks, J., Pecha, L., and Zaruma, D. (2007) *Acta Cryst.*, **C63**, m62.
278 Mugesh, G., Singh, H.B., and Butcher, R.J. (1999) *J. Chem. Res. (S)*, 416.
279 Silina, E., Ashaks, J., Belyakov, S., Tokmakov, A., Pech, L., and Zaruma, D. (2007) *Chem. Heterocycl. Comp.*, **43**, 1582.
280 Silina, E., Belyakov, S., Ashaks, J., Tokmakov, A., Pech, L., and Zaruma, D. (2009) *Chem. Heterocycl. Comp.*, **45**, 860.
281 DeGroot, M.W., and Corrigan, J.F. (2000) *J. Chem. Soc., Dalton Trans.*, 1235.
282 Barton, A.J., Genge, A.R.J., Levason, W., and Reid, G. (2000) *J. Chem. Soc., Dalton Trans.*, 859.

3
Selenium Compounds in Radical Reactions

W. Russell Bowman

Selenium compounds are used in a wide variety of synthetic radical reactions as radical precursors, for incorporation of selenium into target products and for a number of other important applications such as radical trapping. Many precursors are suitable for use in common radical chemistry protocols, for example, reactions facilitated by use of tributyltin hydride (Bu_3SnH).

The polarizability (softness) of the selenium atom and weakness of the C–Se and Se–H bonds facilitate many of the radical applications. Selenium compounds are generally available, easy to synthesize and not overly costly. While selenium is an essential element for good health, excess amounts can cause health problems. Therefore, general care and safe work needs to be undertaken in the laboratory. Certain selenium compounds, for example, benzeneselenol (PhSeH), smell foul, and careful containment is required.

The chapter is divided into several sections based on the reaction types. The various methods of synthesis of selenium radical precursors are referred to under synthetic protocols. Most of this chemistry is covered in detail in other chapters, and in particular, in Chapters 1 and 2.

The radical chemistry of selenium has been reviewed [1]. This chapter concentrates on newer chemistry or areas not covered in detail in the earlier reviews.

3.1
Homolytic Substitution at Selenium to Generate Radical Precursors

For clarity, the section only discusses bimolecular homolytic substitution S_H2 reactions in which the PhSe moiety is abstracted by the radical reagent, thereby yielding radicals which are used in the synthesis. The radicals $Bu_3Sn\bullet$ and $(TMS)_3Si\bullet$ generated from the common radical reagents tributyltin hydride (Bu_3SnH) and tris(trimethylsilyl)silane ($[TMS]_3SiH$), respectively, readily undergo S_H2 substitutions at Se centers. The typical S_H2 substitution on Se using tributyltin radicals generated from Bu_3SnH in a typical synthetic reaction is shown in Scheme 3.1. The PhSe moiety is transferred from the "radical" group to tributyltin.

Organoselenium Chemistry: Synthesis and Reactions, First Edition. Edited by Thomas Wirth.
© 2012 Wiley-VCH Verlag GmbH & Co. KGaA. Published 2012 by Wiley-VCH Verlag GmbH & Co. KGaA.

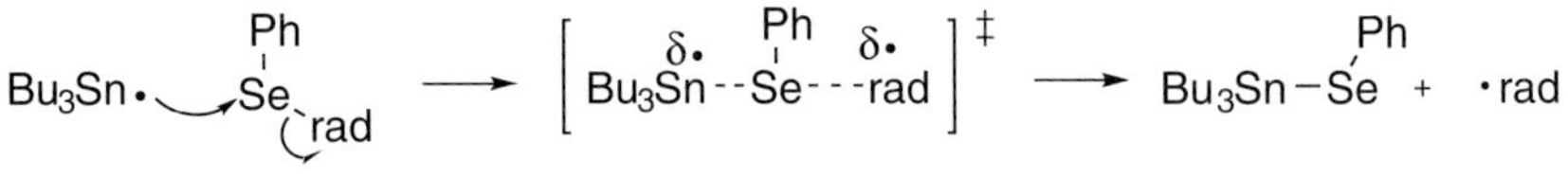

$$Bu_3Sn\bullet \underset{\text{rad}}{\overset{\text{Ph}}{\underset{|}{Se}}} \longrightarrow \left[Bu_3\overset{\delta\bullet}{Sn}\text{--}\overset{\overset{\text{Ph}}{|}}{Se}\text{---}\overset{\delta\bullet}{\text{rad}} \right]^{\ddagger} \longrightarrow Bu_3Sn\text{--}\overset{\text{Ph}}{\underset{|}{Se}} + \bullet\text{rad}$$

Scheme 3.1 S_H2 substitution with $Bu_3Sn\bullet$ abstraction of PhSe.

In planning new applications of radical selenium chemistry, phenylselenides could be used as precursors for any group of radicals in which the Se-radical bond can be broken in S_H2 reactions with common radical reagents. The radicals that can be generated include alkyl, acyl, imidoyl, alkenyl, and oxycarbonyl.

The large bivalent selenium atom is not subject to much steric hindrance and S_H2 reactions proceed without difficulty with tertiary-C leaving groups or tertiary C-centered radicals as the attacking species.

3.1.1
Bimolecular S_H2 Reactions: Synthetic Considerations

Selenide groups, normally PhSe, can replace halides, commonly bromine or iodine, in radical-generating reactions with tributyltin hydride (Bu_3SnH) and tris(trimethylsilyl)silane ($[TMS]_3SiH$) for synthetic purposes.

Phenylselenyl groups are abstracted by primary alkyl ($RCH_2\bullet$), $Bu_3Sn\bullet$, and $Bu_3Ge\bullet$ radicals at rates comparable with bromine, but several orders of magnitude smaller than iodine. For example, abstraction of PhSe by primary alkyl radicals from $PhSeCH_2CO_2Et$, $BrCH_2CO_2Et$, and ICH_2CO_2Et react at rates of 1.0×10^5, 0.7×10^5, and $2.6 \times 10^7 \, M^{-1}s^{-1}$, respectively, at 50 °C [2]. The abstraction rate of PhSe by $Bu_3Sn\bullet$ and $Bu_3Ge\bullet$ radicals is at least an order of magnitude faster than that by primary alkyl radicals [3]. The fastest useful radical abstraction of PhSe is from the diselenide (PhSeSePh) and therefore the source of choice for transferring phenylselenyl groups [2]. Some useful rates of S_H2 abstractions are shown in Table 3.1.

Evidence from substituent effects indicates that PhSe transfer reactions proceed via an S_H2 transition state **1** rather than a stepwise process via a reactive

Table 3.1 Reaction constants for the reactions: $A\bullet + PhSe\text{--}X \rightarrow A\text{--}SePh + X\bullet$.

A•	X	Temperature (°C)	Rate constant ($M^{-1}s^{-1}$)
$Bu_3Sn\bullet$	CH_2OBu	25	5.8×10^6
$Bu_3Ge\bullet$	CH_2OBu	25	2.3×10^7
$Bu_3Sn\bullet$	CH_2CO_2Et	25	1.2×10^8
$Bu_3Ge\bullet$	CH_2CO_2Et	25	9.2×10^8
$RCH_2\bullet$	CH_2CO_2Et	50	1.0×10^5
$RCH_2\bullet$	CMe_2CO_2Et	50	2.3×10^5
$RCH_2\bullet$	$CMe(CO_2Et)_2$	50	8.0×10^5
$RCH_2\bullet$	SePh	25	2.6×10^7

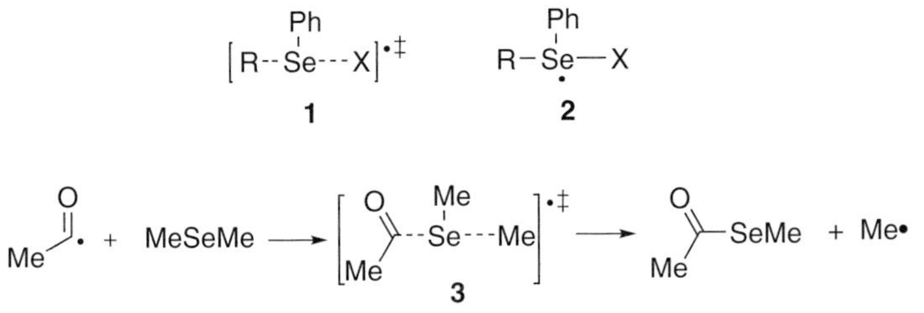

3.1 Homolytic Substitution at Selenium to Generate Radical Precursors

[Scheme 3.2 depicts transition states 1, 2, and 3 for homolytic substitution at selenium]

Scheme 3.2

intermediate **2** that subsequently fragments (Scheme 3.2) [2]. It is generally agreed that S_H2 substitution on S and Se proceeds via the T-shaped transition state **1** which has low steric hindrance. These conclusions are supported by *ab initio* theoretical calculations for attack on Se by alkyl, silyl, germyl, and stannyl radicals [4]. Similar conclusions from *ab initio* and DFT calculations predict that homolytic substitution at S, Se, and Te by acetyl radicals also proceeds via a transition state **3** (Scheme 3.2) [5].

Abstraction of PhSe from acyl-selenides (R-CO-SePh) is faster than abstraction from alkyl phenylselenides and also proceeds by an S_H2 mechanism. The rate of abstraction from (TMS)$_3$Si• and Bu$_3$Sn• has been measured as $2 \times 10^8 \, M^{-1} s^{-1}$ at 80 °C [6].

Bromides and iodides are very good groups for facilitating S_H2 reactions but do not survive many synthetic transformations and hence need to be introduced shortly before the radical reaction. This common use of halides suffers from a number of disadvantages, for example, elimination reactions in base, nucleophilic substitution, and homolytic cleavage under heat or light. The huge advantage of selenide radical precursors, normally containing PhSe groups, in facilitating S_H2 substitutions in place of halides is their stability to most reaction conditions. In contrast, the phenyl sulfide group is lesser useful due to the C–S bond strength and lower polarizability and is not easily abstracted by common radical reagents. However, sulfenamides (RS-NR$_2$) [7], which have a much weaker N–S bond are used for radical reactions to generate aminyl radicals. In contrast, the N–Se bond of selenamides is too weak and not commonly used to generate aminyl radicals [7].

The PhSe group or related selenides can be introduced early on in a synthetic sequence, survive other transformations, and are finally removed in a radical reaction. These selenyl groups are often referred to as "radical protective groups," that is, protect the potential radical center until "deprotected" with a radical reagent. Phenyl selenide groups are stable to most common synthetic reactions, for example, hydrolysis, common nucleophiles, reducing conditions (LAH, DIBAL, NaBH$_4$), acid-catalyzed imine formation, S_NAr, NaNH$_2$/NH$_3$, metathesis, and Wittig reactions. The group is even stable to Swern reaction conditions in which oxidation and elimination could be expected [8, 9].

3 Selenium Compounds in Radical Reactions

Selenium chemistry is well developed and there are a plethora of protocols for introducing PhSe and related groups. Commonly, the anion (PhSe⁻) is used as a nucleophile to displace halides in S_N2 reactions. The PhSe⁻ anion is most commonly generated *in situ*, or shortly before reaction, by $NaBH_4$ reduction of the corresponding diselenide. This avoids forming the anion from benzeneselenol that smells terrible. Conversely, the use of phenylselenyl halides facilitates the introduction of the group as an electrophile. These synthetic protocols are not discussed separately in the chapter but are shown in various synthetic sequences. The procedures are normally reported in the papers using the radical chemistry and other chapters in this book.

Another major advantage of PhSe groups is that they are not attacked by nucleophilic N centers present in many target molecules or building blocks containing amine [8, 9] or imine [8, 9] groups. An example of the use of the PhSe group in radical reactions involving amines and imines is shown in Scheme 3.3 [8]. The PhSe group was inserted by S_N2 substitution on the alkyl bromide. The PhSe group is compatible with $NaBH_4$, DIBAL, amines, imines, and acid catalysis. Building blocks with $n = 1, 2,$ and 3 are easily prepared.

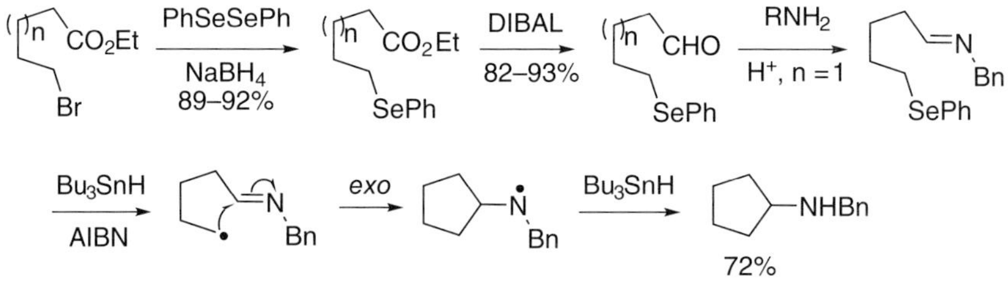

Scheme 3.3

A further synthetic advantage of PhSe groups over halides is that they are not attacked by nucleophilic N centers of basic heteroarenes present in many biologically active target molecules. Examples include imidazoles [10, 11], benzimidazoles [10–12], pyrazoles [11, 13], and azines [14]. Selenide building blocks can be easily introduced onto diazoles or other NH-heteroareanes by simple alkylation with the N anion as the nucleophile. This protocol provides a quick entry into a large number of heteroarene targets that could not easily be prepared by nonradical routes. The use of the PhSe "protective group" facilitates this alkylation protocol. An example is shown of part of the synthetic studies of promising anticancer compounds, mimics of the important anticancer alkaloid mitomycin [15] (Scheme 3.4). Alkylation of **4** yields the disubstituted precursor **5**. The alkylation of the tautomer of **4** is not shown. The use of selenide-building blocks facilitates the synthesis of five-, six-, and seven-membered analogues **6**. As observed in cyclizations onto diazoles, the six-membered ring cyclization gives the best yield. Some monocyclization and reduction were also reported. The radical cyclization onto the aromatic rings is formally an aromatic homolytic substitution.

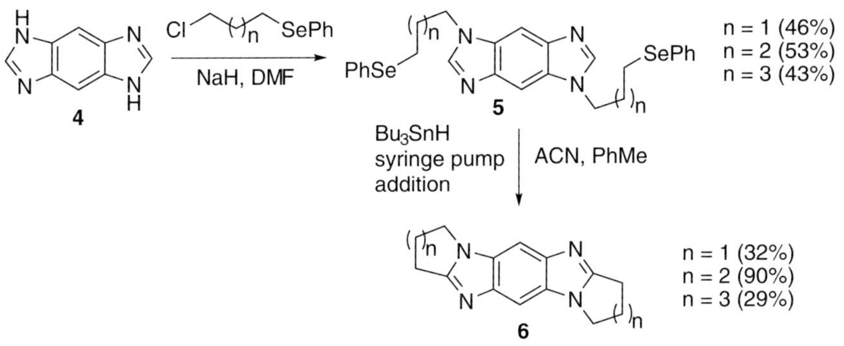

Scheme 3.4

3.1.1.1 Radical Reagents

While Bu$_3$SnH and tris(trimethylsilyl)silane are the most commonly used reagents in reactions using selenide precursors, most other radical reagents used for radical abstraction of halides have also proved useful, for example, tributylgermanium hydride [16] and phosphorus reagents such as ethylpiperidine hypophosphite and diphosphine oxide. An example of the use of diphosphine oxide is shown for the synthesis of the alkaloid horsfiline (Scheme 3.5) [17]. In this example, a tertiary center is displaced showing the low steric requirement of substitutions on Se. The PhSe group survives the use of Me$_3$Al in the amide synthesis and alkylation conditions in the NH protection. The radical cyclization onto the aromatic ring is another example of aromatic homolytic substitution. AIBN and related azo compounds are most commonly used as initiators but the use of triethylborane (Et$_3$B) has been finding increasing use with selenide precursors and also has the advantage of being able to be used at room temperature or below and in the presence of oxygen.

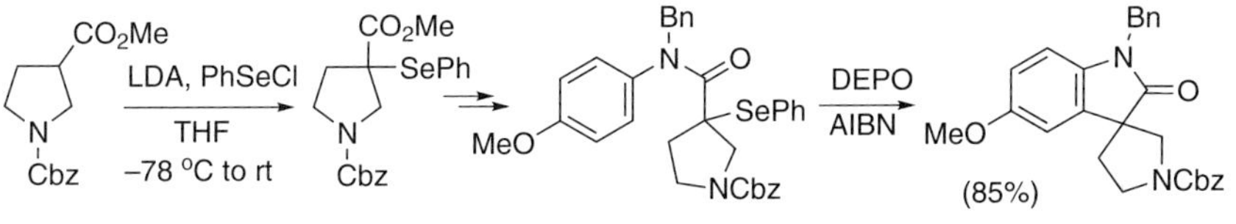

Scheme 3.5

3.1.2
Alkyl Radicals from Selenide Precursors

The use of alkyl selenide precursors is a common procedure and the majority of precursors are on ordinary alkyl chains –(CH$_2$)$_n$–SePh. Selenyl groups in alpha positions to ethers, including the anomeric position of carbohydrates [18], and amines [19] have also been used. The advantage of selenides is again exemplified because the corresponding halogen derivatives are not commonly stable. Phenyl selenides in alpha positions to electron withdrawing groups have been widely used, for example, amides [17], nitriles [20], esters, and ketones.

An interesting example of the use of alkyl selenides is shown in Scheme 3.6 that facilitates alkyl cyclization onto the alkene of 2-alkylaziridines [21]. The PhSe group survives the use of NaNH$_2$ in the preparation of the methylene aziridine precursor **7**. The intermediate radical **8** undergoes 5-*exo* cyclization to yield **9** that, in turn, undergoes ring opening to yield the product **10**. A second example shows the use of alkyl selenide precursors **11/12** to generate radical intermediates **13** and **14** that are able to equilibrate to yield the required radical **14** for cyclization as shown in Scheme 3.7 [22]. The protocol provides a novel route to cyclic imines, for example, **15**. Note that the PhSe group survives imine formation in the synthesis of the precursor. The selenide precursor has the advantage over the corresponding halogeno ketones which would be unstable and very lachrymatory.

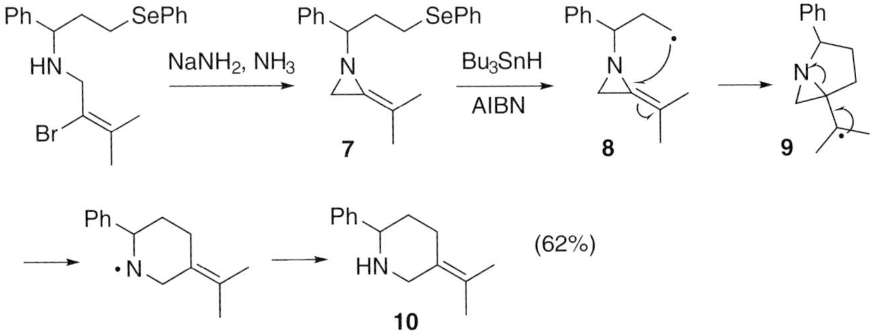

Scheme 3.6

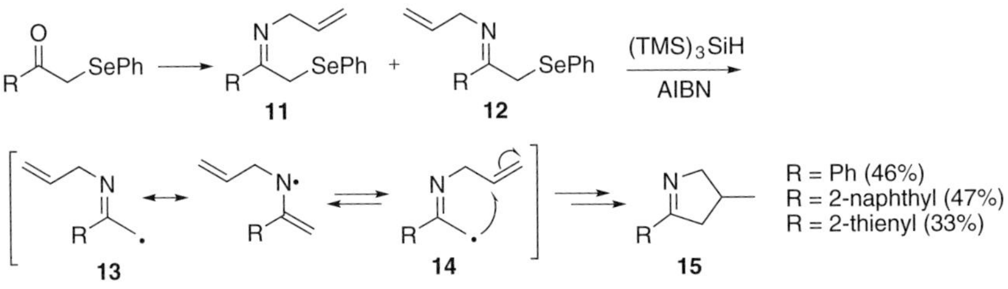

Scheme 3.7

Alkyl selenide precursors have been widely used to generate alkyl radicals for cyclization onto heteroarenes by aromatic homolytic substitution, for example, imidazoles [10, 11], benzimidazoles [10–12], pyrazoles [11, 13], and azines [14]. Aromatic homolytic substitution, including many examples with selenyl precursors, has been recently reviewed [23]. Examples are shown in Schemes 3.4, 3.27, and 3.28 for solid-phase reactions.

3.1 Homolytic Substitution at Selenium to Generate Radical Precursors | 117

Alkyl selenide precursors have been used to cyclize onto aldehydes [24] and nitriles [25]. A simple substrate-controlled asymmetric synthesis of enantio pure tetrahydrofuranols uses alkyl selenide precursors for cyclization onto aldehydes as shown in Scheme 3.8 [24]. The precursor **16** was prepared by epoxide opening and DIBAL reduction; the selenide was unchanged by the reaction conditions. Intermediate radical **17** cyclizes onto the aldehyde but not with much stereoselectivity to yield two diastereoisomers. The diastereoisomers **18** and **19** were converted to the products **20** and **21** that were easily separated to give pure diastereomers. The other two possible isomers were prepared by using opposite enantiomers of the starting material.

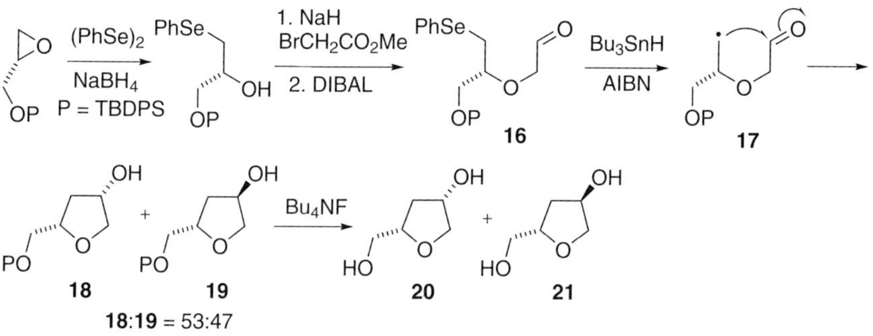

Scheme 3.8

Alkyl radicals **22**, generated from alkyl selenide precursors, cyclize onto α,β-unsaturated esters to yield 3′-substituted 2′,3′-dideoxyribonucleosides with high diastereoselectivity (Scheme 3.9) [26]. The nucleosides were formed in a ratio of **23**:**24** = 98:2. Other examples of alkyl selenide cyclization onto α,β-unsaturated esters have been reported [27, 28]. Alkyl selenide precursors have also been used for generating alkyl radicals **25** to add to carbon monoxide to form intermediate acyl radicals **26** for cyclization onto α,β-unsaturated esters for the synthesis of tetrahydrofuran-3-ones, for example, **27** and **28** (Scheme 3.10) [27]. The most favored TS for the Beckwith model is represented by **26** that favors the *cis* product

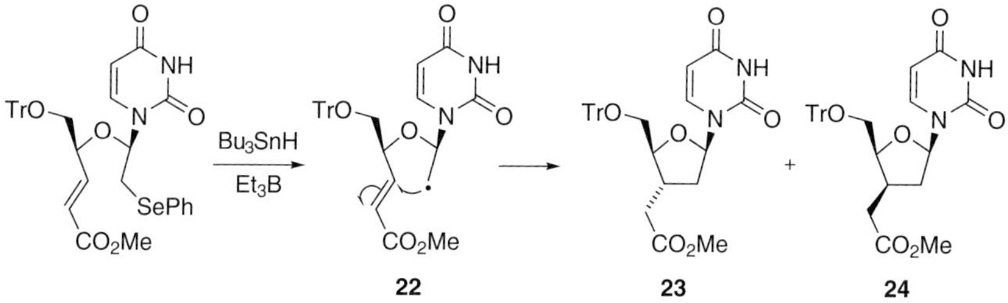

Scheme 3.9

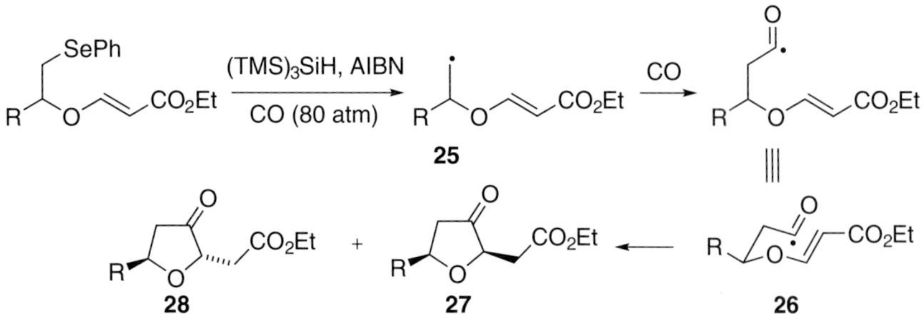

Scheme 3.10

27. The yields and *cis* : *trans* ratios for some of the reactions are R = Ph (64%, 8/1), R = Bn (59%, 9/1), and R = Bu (77%, 7/1). The protocol was also successfully used for a wide range of starting materials in which oxygen was replaced by NH to yield pyrrolidin-3-ones in good yields. The PhSe groups were inserted by ring opening with PhSe⁻ anions of epoxides (see Scheme 3.8) and aziridines for the synthesis of tetrahydrofuran-3-ones and pyrrolidin-3-ones, respectively.

The use of chiral selenium-containing compounds for enantioselective reactions which leave the Se attached has been developed. Wirth has developed the use of chiral selenyl compounds to impart the stereoselectivity. The general protocol has been reviewed [29] and discussed in earlier chapters. In these stereoselective reactions, the chiral arylselenyl moiety commonly needs to be removed and one of the common methods is radical reduction. Another general protocol uses chirality in the molecule to direct the stereoselectivity of the selenyl addition. In the example shown in Scheme 3.11, the hydroxyl group in the optically pure starting material directs the diastereoselectivity [30]. Two diastereoisomers **29** and **30** were formed in reasonable yield, separated, and the PhSe group removed by radical reduction to yield the chirally pure isomers **31** and **32**. All the possible isomers were prepared by using suitable starting materials.

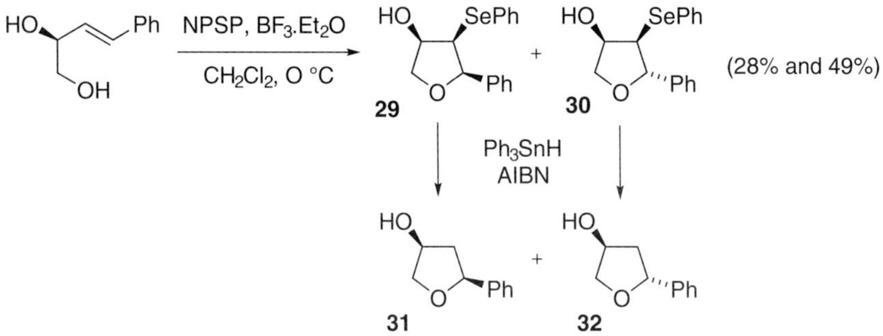

Scheme 3.11

The use of phenyl vinyl selenide as the dienophile in highly regio- and stereoselective inverse electron-demand Diels–Alder reactions has provided a route to highly substituted bicyclo[2.2.2]octenones containing a PhSe group [31, 32]. The removal of the phenylselenyl group can be carried out with Bu$_3$SnH to yield radical intermediates that can either be reduced or rearranged [31, 32]. The synthesis of vinyl selenides has been reviewed in detail [33]. Scheme 3.12 shows an example of these Diels Alder reactions of addition of vinyl selenide to **33** to yield the bicyclo[2.2.2]octenone **34** [32]. The phenylselenyl intermediate **34** was treated with Bu$_3$SnH to yield the intermediate radical **35** that cyclizes onto the ketone followed by a ring opening to yield the bicyclo[3.2.1]octenones **36** and **37**. A range of other Diels–Alder reactions using both phenyl vinyl selenide and sulfide were also reported [32]. When phenyl vinyl sulfide was used, different radical reactions were obtained because the Bu$_3$Sn• radicals react with the carbonyl group rather than abstracting the phenyl sulfide group. The abstraction of the PhSe by Bu$_3$Sn• radicals was faster than the addition of Bu$_3$Sn• to the carbonyl group.

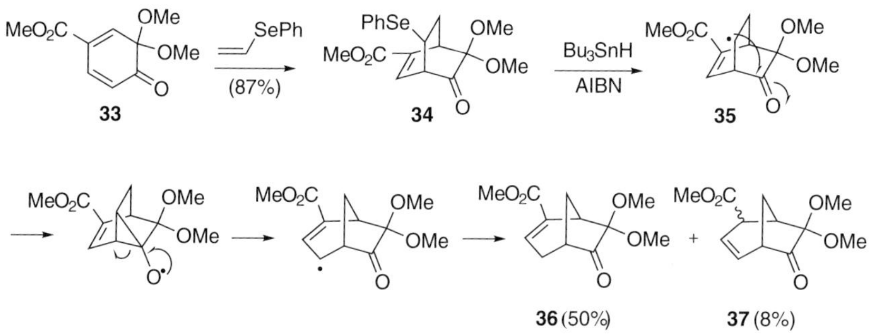

Scheme 3.12

3.1.3
Acyl Radicals from Acyl Selenide Precursors

Acyl radicals have been reviewed in detail. The chapter contains considerable detail of the synthesis and use of acyl-selenides in radical reactions [34]. The PhSe moiety is easily abstracted from acyl selenides by Bu$_3$Sn• and (TMS)$_3$Si• radicals to yield acyl radicals. This facile abstraction has led to acyl-selenides being the precursors of choice for generating acyl radicals in synthetic reactions. Unlike acyl halides, they are relatively stable and survive careful chromatographic purification. However, synthetic reactions need to be carefully planned to avoid homolysis with loss of CO from the intermediate acyl radicals prior to further reaction. The protocol also provides a high yielding route for the replacement of carboxylic acids with hydrogen when there are no competing radical reactions. With aryl-acyl selenides, there is no problem of homolysis of the acyl radicals. Although Bu$_3$SnH and (TMS)$_3$SiH are most commonly used to abstract the PhSe group, other reagents such as Bu_3GeH may also be used to generate acyl radicals from acylselenides [16].

Acyl selenides are easily prepared from the corresponding carboxylic acids in high yield with mild reagents that do not interfere with most other functional groups and excellent detail is given in reviews and the papers referred to in this section [34]. PhSeH has a most unpleasant smell and protocols avoiding its use are preferred. The first of the common methods involves reaction of the anion of benzeneselenol ($PhSe^-$) with the respective acid chloride. The anion is prepared *in situ* by sodium borohydride reduction of diphenyl diselenide in methanol. The most common synthetic route avoids the use of acyl halides and uses diphenyl diselenide or N-(phenylselenyl)phthalimide directly with the carboxylic acid, facilitated by the use of tributylphosphine.

The rate of abstraction of the PhSe moiety from acyl selenides by $Bu_3Sn\bullet$ and $(TMS)_3Si\bullet$ radicals is several orders of magnitude faster than the rate of abstraction from alkyl selenides. The rate of abstraction by $(TMS)_3Si\bullet$ and $Bu_3Sn\bullet$ has been measured as $2 \times 10^8 \, M^{-1} s^{-1}$ at 80 °C [6]. Fortunately, the rate of CO scission from primary alkyl acyl radicals is relatively slow ($2 \times 10^2 \, M^{-1} s^{-1}$ at 80 °C) and is much slower than the rate of CO addition to primary alkyl radicals to give alkyl acyl radicals ($6.3 \times 10^5 \, M^{-1} s^{-1}$ at 80 °C [34]. Most reactions using acyl selenide precursors have fast reactions for the resulting acyl radical and therefore CO loss is not a major problem. However, if the radical reaction is slow, then CO loss becomes a serious synthetic problem. Reactions can be carried out under a high pressure of CO to avoid these problems but is not a procedure easily available in the laboratory.

One protocol for avoiding the problem of CO loss is to use the differing rates of CO loss and addition to advantage. The cyclization of acyl radicals onto pyrrole is a relatively slow reaction and decarbonylation gave a plethora of products [11, 35]. However, when the reactions were repeated under an atmosphere of CO, only traces of decarbonylation products were observed with acyl cyclization predominant (Scheme 3.13). The study shows the use of building blocks in selenide chemistry (see Section 3.2). The acyl selenide precursors **38** were prepared in

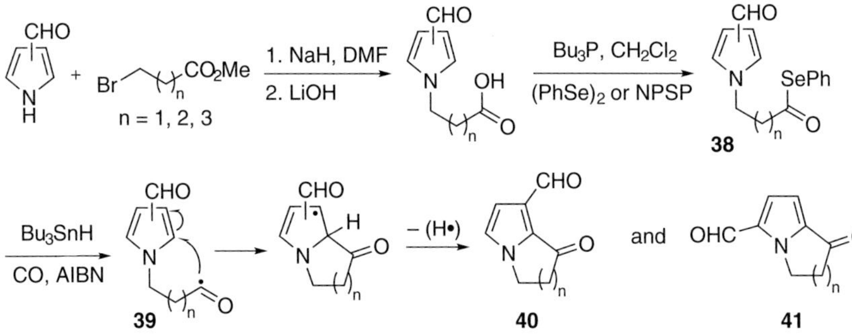

Scheme 3.13

35–65% yield from the 2- and 3-pyrrolecarbaldehyde. Cyclization of the intermediate acyl radicals **39** under an atmosphere of CO gave reasonable yields of [1,2-*a*]-fused pyrroles **40** and **41** with five-, six-, and seven-membered ketone rings attached.

When aryl-acyl selenides are used as precursors, there is no problem with CO loss because of the much stronger aryl–CO bond. An excellent example of this advantage has been the development of useful methodology for the synthesis of indole alkaloids using indole-2- and 3-acyl radical intermediates by Bennasar [36–40]. The first protocol was used to synthesize indole quinones such as calothrixin B **45**, a potent antimalarial alkaloid (Scheme 3.14) [36]. Related studies are also reported for the synthesis of the anticancer alkaloid ellipticine [37]. In this protocol, the rate of cyclization onto the quinoline is slow and decarbonylation would otherwise result. Indole-2-carboxylic acids are easily prepared and in this synthesis converted to the acyl selenide **42** using PhSeCl and Bu$_3$P. The resulting acyl radical **43** cyclizes in high yield by aromatic homolytic substitution. The cyclized product **44** exists as its stable aromatic phenolic tautomer. Oxidation of the labile benzylic methylene and deprotection gives calothrixin B **45** in 90% yield.

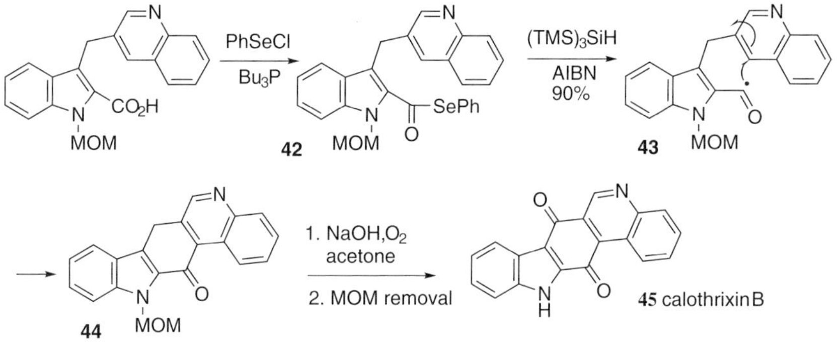

Scheme 3.14

In a second protocol, 2-indolylacyl radicals are cyclized onto a dihydropiperidine ring for the synthesis of the alkaloid dasycarpidone **49** and its epimer epidasycarpidone **50** (Scheme 3.15) [38]. Related studies for the synthesis of similar alkaloid skeletons have also been reported by Bennasar [39]. In this protocol, the acyl selenide precursor **46** was converted to the 2-indolylacyl radical **47** that undergoes 6-*endo* cyclization to yield the intermediate radical **48**. The 6-*endo* cyclization is favored over the 5-*exo* cyclization because of the ethyl substituent creating steric hindrance. The two epimers were prepared in 41% overall yield (**49** : **50** = 1 : 2.2) from the carboxylic acid used for the precursor.

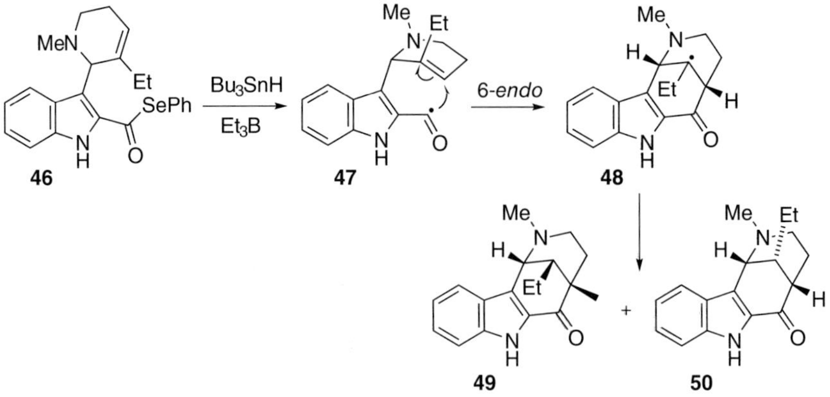

Scheme 3.15

Alkyl-acyl selenides have been extensively used by Pattenden for a range of impressive syntheses including many natural products [41–46]. Pattenden has reviewed radical-mediated macrocyclizations and transannular cyclizations in synthesis that includes many examples with acyl selenide precursors [42]. Examples of domino cyclizations using acyl selenides are shown in Section 3.4, Schemes 3.30 and 3.31 [43, 44]. Pattenden's studies also include cyclopropyl and α,β-unsaturated acyl-selenide precursors [45, 46]. Domino cyclizations involving α-ketenyl radicals from α,β-unsaturated acyl selenide precursors have been used for the synthesis of the triquinanane, pentalenene **56** (Scheme 3.16) [45]. The selenide precursor **51** was converted to the α,β-unsaturated acyl radical **52** which is in equilibrium with the α-ketenyl radical **53**, which in turn is able to undergo 5-*exo* cyclization to **54**. The final cyclization in the domino sequence gave the tricyclic ketone **55**. Alkylation, reduction, and dehydration yield the product terpene, pentalenene **56**. The sequence illustrates the interesting ambident properties and synthetic potential of radicals from α,β-unsaturated acyl-selenide precursors.

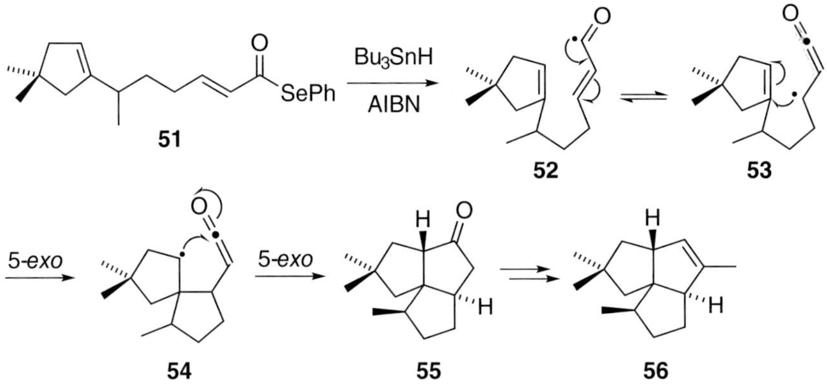

Scheme 3.16

3.1.4
Imidoyl Radicals from Imidoyl Selenides

Imidoyl selenides **58** have been successfully used as precursors for imidoyl radicals and provide a parallel to the chemistry of acyl selenides for acyl radicals [47–50]. This protocol has the major synthetic advantage that two relatively complex molecules can be joined by amide formation and then converted to the selenide precursor, thereby setting up functionalities for radical cyclizations. The use of imidoyl radicals in heteroarene synthesis is advantageous in that one unsaturated bond is introduced with the imidoyl radical.

PhSe⁻ anions are normally generated by $NaBH_4$ reduction in methanol. However, methanol would react with the imidoyl chloride intermediate **57**. Therefore, the imidoyl selenide precursors were prepared with PhSe⁻ generated from the diselenide using K-selectride in unreactive solvent (Scheme 3.17) [48–50]. Although the preparation of the precursors is essentially quantitative, there is some hydrolysis during chromatographic purification leading to good to moderate yields. The imidoyl selenides conveniently have the correct (Z)-stereochemistry required for cyclization so that isomerization of the imidoyl radical does not need to take place prior to cyclization [49].

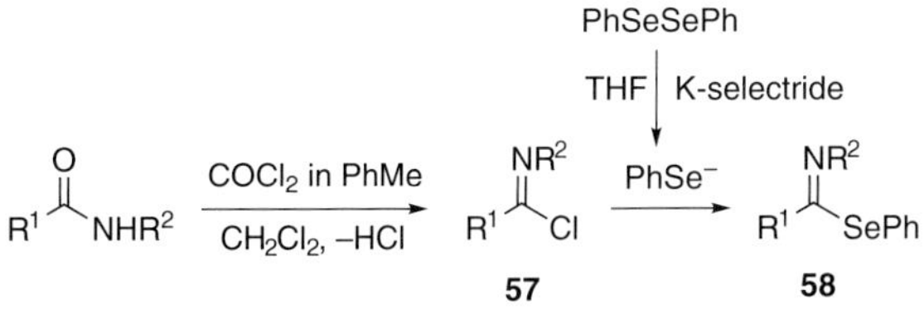

Scheme 3.17

The synthesis of indoles and quinolines from imidoyl selenides **59** is shown in Scheme 3.18 and Table 3.2. The PhSe group is easily abstracted by $Bu_3Sn\cdot$ radicals to yield imidoyl radicals **60** that undergo 5-*exo* or 6-*endo* cyclization depending on

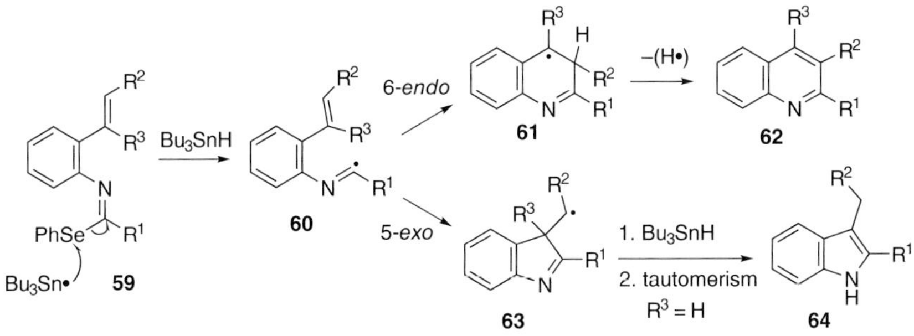

Scheme 3.18

Table 3.2 Synthesis of quinolines and indoles by imidoyl radical cyclization.

R^1	R^2	R^3	Product (% yield)
Me	H	Me	Quinoline (80%)
p-tolyl	Me	Et	Quinoline (76%)
p-tolyl	CO_2Et	H	Indole (95%)
p-tolyl	Pr	H	Indole (66%)
Me	CO_2Et	H	Indole (98%)

the substituents. When the alkene α-substituent R^3 was not H, the cyclization proceeded selectively via the 6-*endo* mode. The cyclized radical **61** aromatizes by loss of hydrogen to yield quinolines **62** in good yield. The preferred and expected 5-*exo* cyclization takes place when an alkene α-substituent is not causing steric hindrance to yield **63**. The cyclized products rapidly tautomerize the respective indoles **64**. The synthesis of ellipticine using this protocol is shown in Section 3.4 on domino reactions (Scheme 3.29) [49, 50].

Se-Phenyl benzoselenohydroximates **65** have been prepared by a similar route using reaction between chloro-oximes and PhSe⁻ anions (Scheme 3.19) [51]. The seleno-oximes were converted into suitable precursors to generate alkyl, aminyl, and alkoxyl radicals. The initially formed imidoyl radical is set up to dissociate to generate the required radicals and does not otherwise react as an imidoyl radical. An example of their use for the generation of aminyl radicals is shown in Scheme 3.19. The seleno-precursor yields an imidoyl radical under Bu_3SnH radical conditions that rapidly breaks down to the aminyl radical **66** which undergoes a cascade reaction to the pyrrolidine **67**.

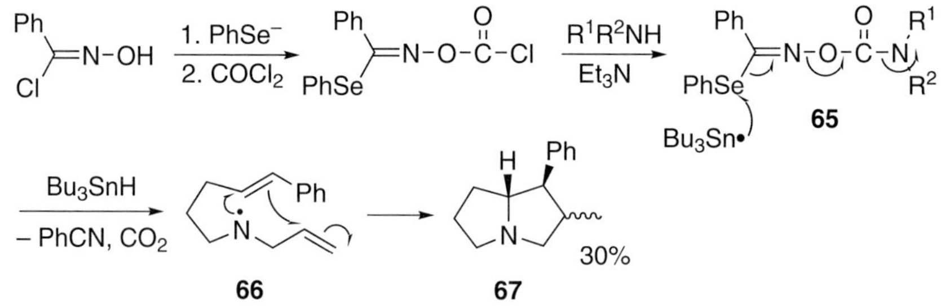

Scheme 3.19

3.1.5
Other Radicals from Selenide Precursors

Phenyl arylselenides (Ar-SePh) have not been used as precursors. The bond strengths would be similar and lead to a mixture of aryl and phenyl radicals which would not be of synthetic use. Likewise, alkenyl selenides have not been extensively used for generating alkenyl radicals, possibly because alkenyl halides are stable and commonly available as radical precursors. The synthesis of vinyl selenides has been reviewed in detail and provides many routes for preparing radical precursors [32]. A vinyl selenide has been used in the synthesis of a C-nucleoside as shown in Scheme 3.20 [52]. The abstraction is favored because the bond is weakened by the ketone and α-imide. The intermediate alkenyl radical undergoes a 1,6-H translocation followed by cyclization to yield the C-nucleoside as two epimers at the C-5 position of the carbohydrate ring.

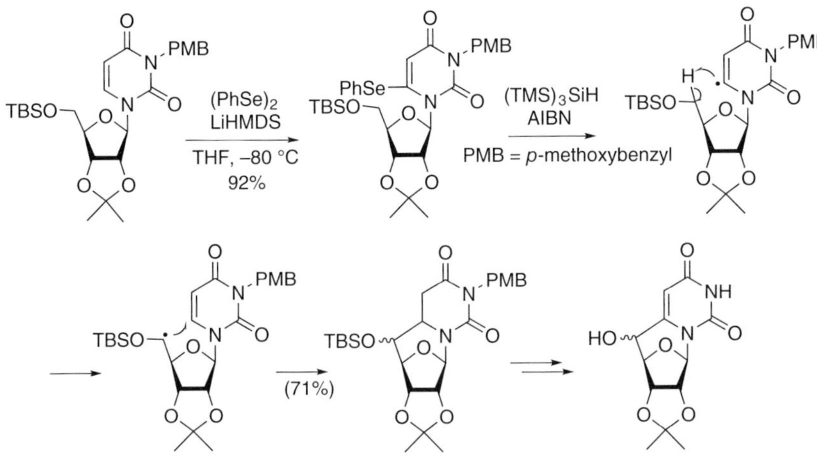

Scheme 3.20

Oxycarbonyl radicals **69** can be generated from (phenylselenyl)formates **68** [53–55]. Depending on the reaction conditions, the intermediate oxycarbonyl radical can cyclize, abstract a hydrogen from the radical reagent, or lose carbon dioxide to yield an alkyl radical (Scheme 3.21). Carbamoyl radicals **71** can also be generated from Se-phenylselenocarbamates **70** and used in cyclization reactions to prepare cyclic amides [56]. The use of the cyclization of oxycarbonyl radicals for the synthesis of a γ-lactone ring as part of the synthesis of juruenolide C **76** is shown in Scheme 3.22 [53]. The (phenylselenyl)formate precursor **73** is prepared by standard procedures from the carboxylic acid **72**. In this domino reaction, the cyclization of the oxycarbonyl radical **74** results in an alkenyl radical which abstracts hydrogen for the silyl substituent, which in turn cyclizes onto the alkene to yield a key intermediate **75** in the synthesis.

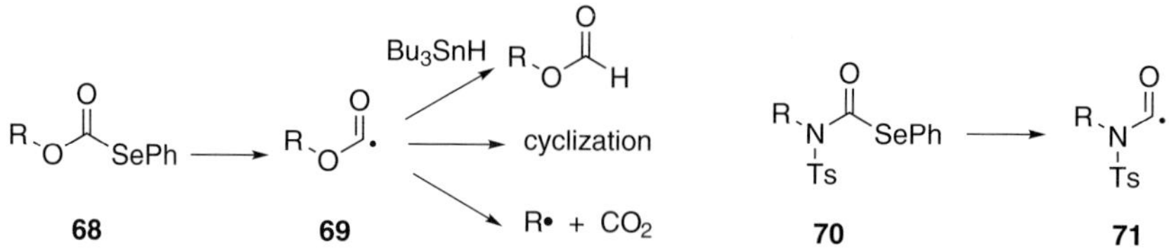

Scheme 3.21

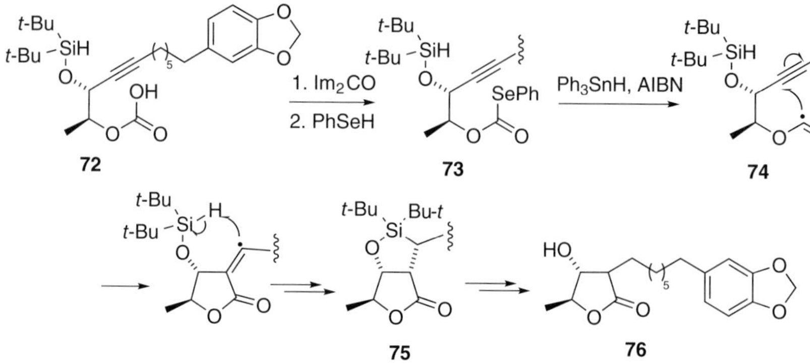

Scheme 3.22

Trimethylsilyl radicals have been generated using Bu_3SnH from phenyl trimethylsilyl selenide precursors and used in hydrosilylation of carbonyl compounds (Scheme 3.23) [56]. The silyl radicals were successfully added to a range of aryl and alkyl aldehydes in high yields and in more moderate yields to ketones. The procedure is an interesting view of polarizability of the atoms. The more polarizable (softer) Se prefers the polarizable Sn, whereas the less polarizable (harder) O and Si are favored.

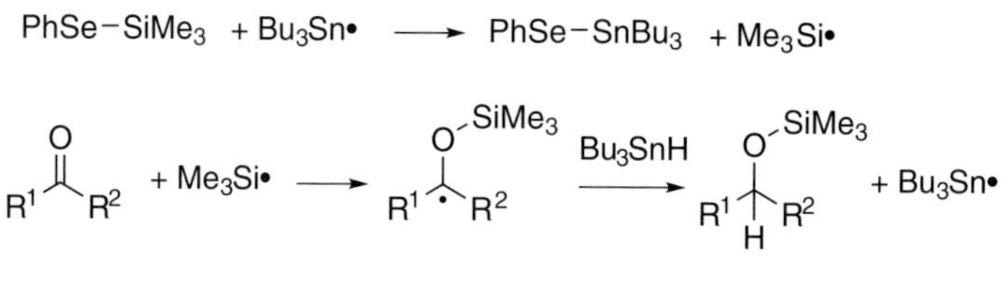

Scheme 3.23

3.2
Selenide Building Blocks

The use of PhSe-containing building blocks is an important aid to radical synthesis. The compounds are stable to most reactions conditions, can be added at a suitable point in the synthesis, act as a "radical protective group," and provide useful radical precursors for standard radical-generating conditions. Most important is to facilitate diversity by synthesizing different lengths of the same building block for five-, six-, and seven-membered ring cyclizations. Examples are shown in Schemes 3.3 [8, 9] and 3.4 [15]. Phenyl selenide building blocks have been successfully used for cyclization onto diazoles [10–13, 15]. The cyclization of acyl radicals onto pyrroles used acyl selenide building blocks (Scheme 3.13) [11, 35].

The most common simple building blocks include $PhSeCH_2(CH_2)_nX$ (X = halogen) which can be added by alkylation with a nucleophile, for example, C- or N-centered anions. Examples of the use with diazoles are shown in Scheme 3.4. Other useful building blocks include $PhSeCH_2(CH_2)_nX$ [10–15]; X = CO_2R [8, 9, 57], X = CHO [8, 9, 57, 58], and X = CN and NH_2 [8, 9, 59]. If further diversity is required, two different building blocks can be put together. The synthesis of 5,6-, 6,5-, 5,5-, and 6,6-spirocyclic amines have been synthesized in moderate yield (25–40%) using two building blocks for cyclization onto imines (Scheme 3.24) [9]. $MgBr_2$ was needed to increase the electrophilicity of the imine to speed up both steps in the domino cyclization. Similarly, the phenyl group is required to make the second radical cyclization more favorable.

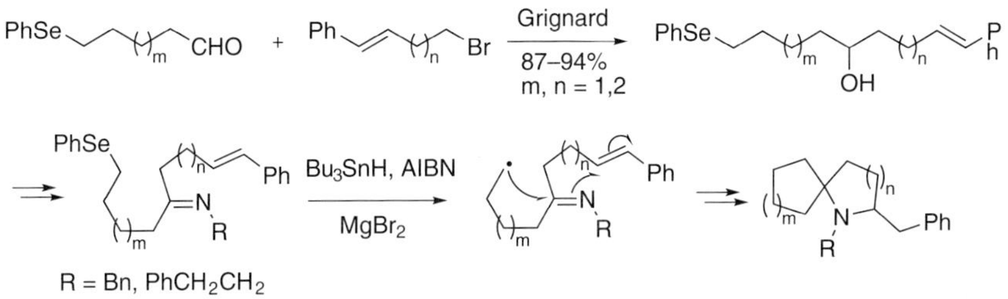

Scheme 3.24

An interesting application of building blocks was used in conjunction with metathesis with the Grubbs catalyst for the synthesis of bicyclic compounds as illustrated in Scheme 3.25 [60]. Standard lithium or Grignard reagents were used

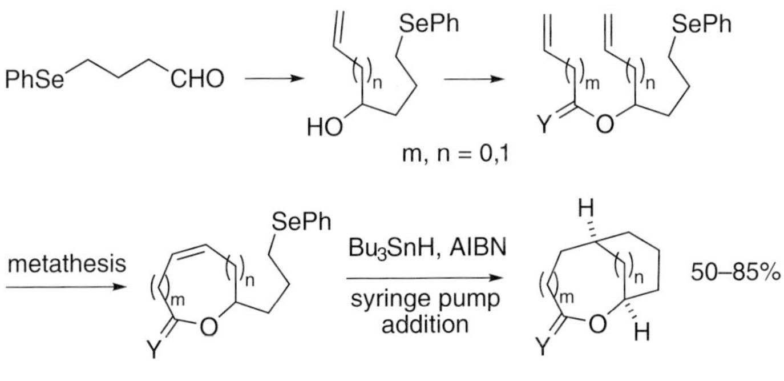

Scheme 3.25

to add to the 4-(phenylselenyl)butyraldehyde building block, followed by alkylation or acylation. Radical cyclization was achieved in good yield via 5- or 6-*exo* mode. The stability of the PhSe group is again exemplified by the compatibility with metathesis.

3.3
Solid-Phase Synthesis

With the advance of solid-phase synthesis and combinatorial chemistry, several procedures involving radical selenium chemistry have been published. The use of selenide-based solid phase has been reviewed [61]. The synthesis of libraries of heterocyclic compounds has proved attractive using selenyl resins. Solid phase selenyl bromide has been used in a range of synthetic transformations as an electrophilic reagent [62, 63]. After the solid-phase reaction, the solid phase selenyl moiety can be removed by Bu_3SnH reduction to act as a traceless linker or to generate radicals for further cyclization. An example showing the use of the solid phase selenyl bromide in a cyclization followed by radical formation in the release step and cyclization is shown in Scheme 3.26 [62]. The combinatorial value of the protocol was demonstrated by the synthesis of a wide range of indoles and tetracycles containing an indole ring.

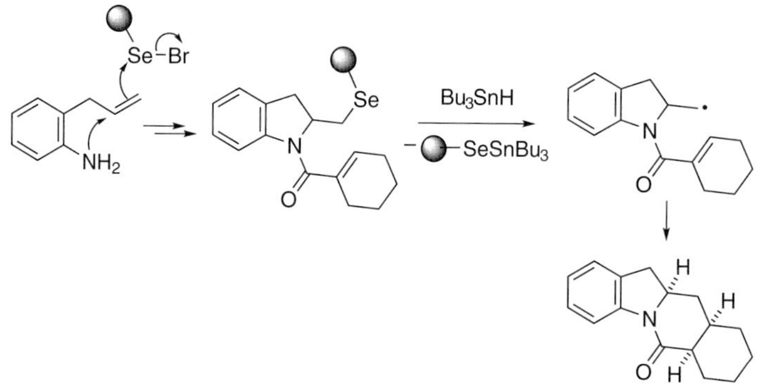

Scheme 3.26

Solid phase arylselenyl leaving groups in radical cyclizations have also been used. The solid phase allows the use of building blocks thereby facilitating diversity. In the example shown in Scheme 3.27, the protocol allows for different building blocks depending on the length of alkyl chain or azole introduced [11]. The arylselenyl bromide **77** generated on Quadragel® was reacted with a 4-C unit to give **78** for attaching azole rings, but other chain lengths and azoles can be used. An example is shown for the synthesis of a six-membered ring analogue **80** of the pyrazole natural product withasomnine by aromatic homolytic substitution onto

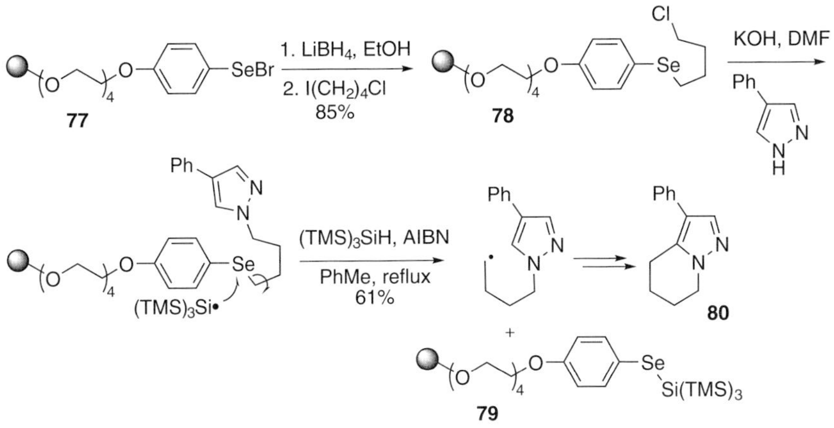

Scheme 3.27

a pyrazole ring. In the radical reaction, the silyl radical reagent ends up attached to the arylselenyl leaving group attached to the Quadragel® solid phase resin **79** and is removed from the reaction, thereby facilitating purification.

Solid phase chiral selenyl bromides have been used to carry out stereoselective additions of solid phase arylselenyl groups to alkenes [64]. The selenyl group can be removed by Bu$_3$SnH radical reduction thereby acting as a traceless linker or as a precursor for further radical reactions. This protocol is discussed in detail in Section 3.2. The solid phase arylselenyl bromide protocol has been used for the preparation of tetrahydrofuran-3-ones and pyrrolidin-3-ones [27].

Abstraction of PhSe has also been used on Wang solid phase resin to generate alkyl radicals that undergo aromatic homolytic substitution with the solid phase as the leaving group in an aromatic homolytic substitution (Scheme 3.28) [12]. This protocol results in the cyclized benzimidazole being cleaved from the resin during the reaction. The tricyclic benzimidazole was easily removed from reagents by acid wash to yield pure product in a 60% yield.

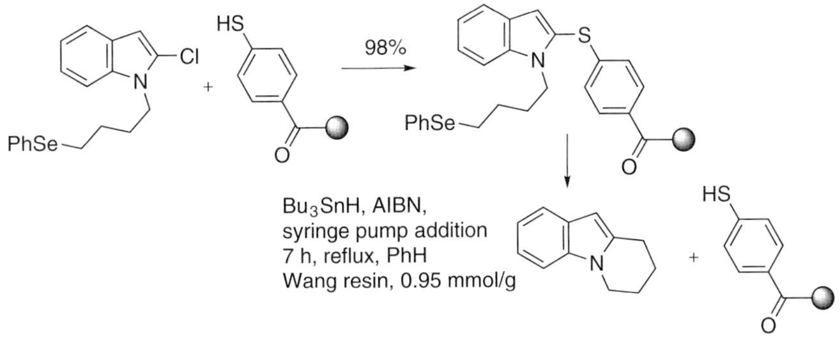

Scheme 3.28

Seleno-sulfonation has also been carried out on solid phase to yield alkenyl and alkynyl sulfones (Section 3.6.2, Scheme 3.40).

3.4
Selenide Precursors in Radical Domino Reactions

The stable PhSe group has found application in setting up domino radical reactions. A complex array of rings in domino cyclizations can be introduced into complex molecules using abstraction of the PhSe group as a radical trigger to start the domino reactions. The term "domino" is used in this chapter although the term "cascade" is also commonly used and means the same, that is, two or more steps in one reaction. Domino reactions, including radical reactions, are extensively reviewed in a recent book [65].

The anticancer alkaloid ellipticine **84** has been synthesized in a short five-step route in 19% overall yield using a domino radical reaction from the imidoyl selenide precursor **81** (Scheme 3.29) [49]. In this protocol, a complex heteroarene can be "zipped up" from a relatively simple selenide precursors. The imidoyl radical **82** undergoes 5-*exo* cyclization to yield an intermediate alkenyl radical **83** that in turn undergoes 5-*exo* cyclization (followed by neophyl rearrangement) or 6-*endo* cyclization to the π-radical intermediate **86**. H-abstraction yielded the cyclized product **85** that undergoes rapid tautomerism to ellipticine **84**. A range of analogues was also prepared using this protocol from imidoyl selenides [49, 50].

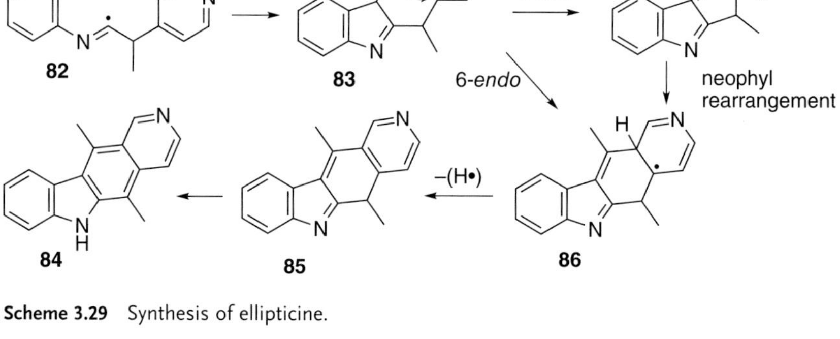

Scheme 3.29 Synthesis of ellipticine.

3.4 Selenide Precursors in Radical Domino Reactions

The most impressive radical domino reactions from selenide precursors are the acyl radical cyclizations by Pattenden for the synthesis of steroids and polycyclic terpenes [41–46]. The acyl radicals are generated from acyl selenides and largely undergo 6-*endo* cyclizations. In these cyclizations, the substituents are set up to favor 6-*endo* cyclization. However, many apparently 6-*endo* acyl radical cyclizations are 5-*exo* cyclizations followed by rearrangement to the 6-*endo* product.

The first example of this impressive acyl selenide protocol is shown in Scheme 3.30 [44]. This example shows the domino synthesis of C-nor-D-homosteroids **87** in which the acyl radical generated from the selenide **88** undergoes a macrocyclization followed by cyclopropyl ring opening and finally a 6-*exo trig* cyclization. The cyclization gives a 1:1 mixture of epimers at C10 in 25% yield as well as a cyclized product (25%) resulting from reducing the intermediate **89**. The second example shows the biomimetic "zipping up" of a long terpene-like acyl selenide precursor **90** in seven 6-*endo trig* cyclizations to yield a novel all-*trans*, *anti*-heptacycle **91** in 20% yield (Scheme 3.31) [43].

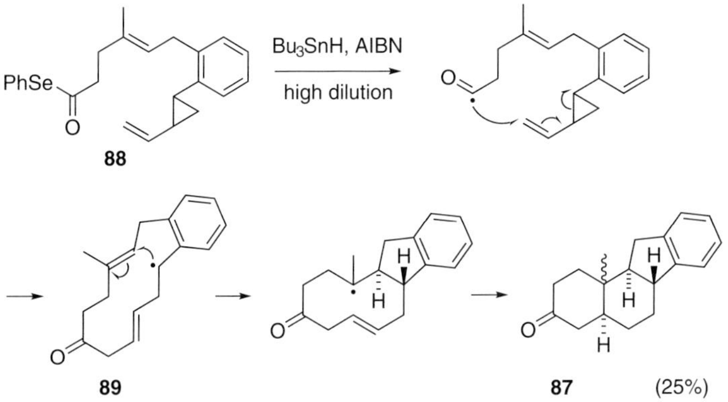

Scheme 3.30

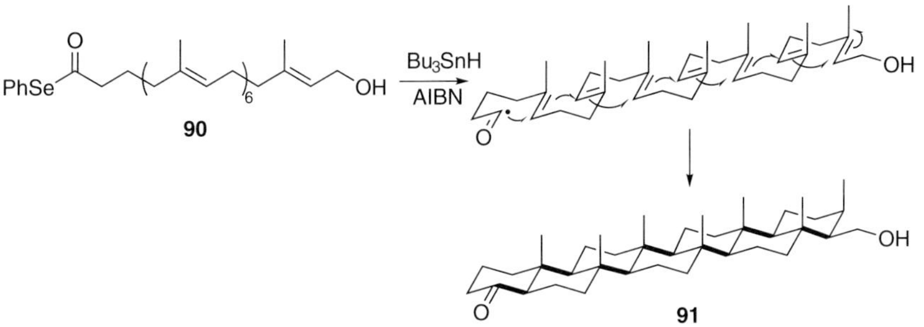

Scheme 3.31

3.5
Homolytic Substitution at Selenium for the Synthesis of Se-Containing Products

3.5.1
Intermolecular S$_H$2 onto Se

Diselenides have been used for trapping radicals to yield selenides. In these reactions, the product radical undergoes bimolecular S$_H$2 attack on selenium and displaces a selenyl radical. In older examples, the intermediate radicals were generated by the breakdown of thiohydroxamate esters (Barton esters) [66]. More recently, tosylhydroxylamines were generated *in situ* and decomposed to yield amidyl radicals (Scheme 3.32) [67]. The cyclized lactam radicals were trapped by S$_H$2 substitution on PhSeSePh with PhSe• as the leaving group.

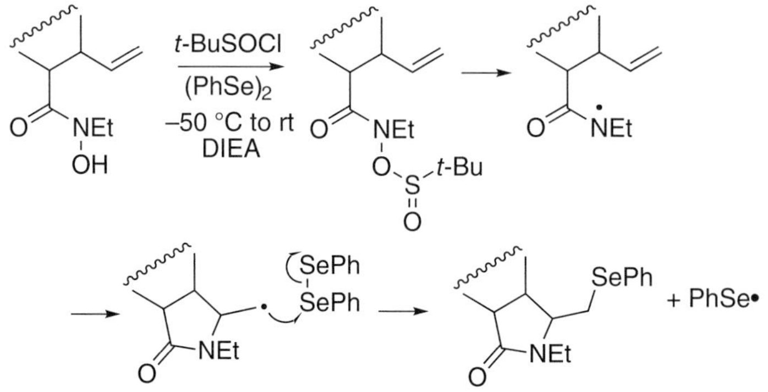

Scheme 3.32

3.5.2
Intramolecular S$_H$2: Cyclization onto Se

Intramolecular S$_H$2 on selenium that retains the Se atom in a ring has become important because current synthetic targets include Se-containing antibiotics, carbohydrates, antioxidants, and anti-inflammatory agents. These selenide analogues have improved radical quenching properties. The research groups of Schiesser and Engman have made great strides in using intramolecular S$_H$2 on selenium as a synthetic methodology for Se-containing biologically active compounds. An example is α-selenotocopherol in which the oxygen atom of α-tocopherol is replaced by Se. In these five- and six-membered ring cyclizations, the Se atom is retained and needs to be attached to a good leaving group, for example, benzyl. Radical substitution at higher heteroatoms including selenium has been reviewed [68].

The rates of cyclizations have been investigated. The rate of cyclization of aryl radicals **92** onto Se with benzyl radicals as the leaving group has been measured

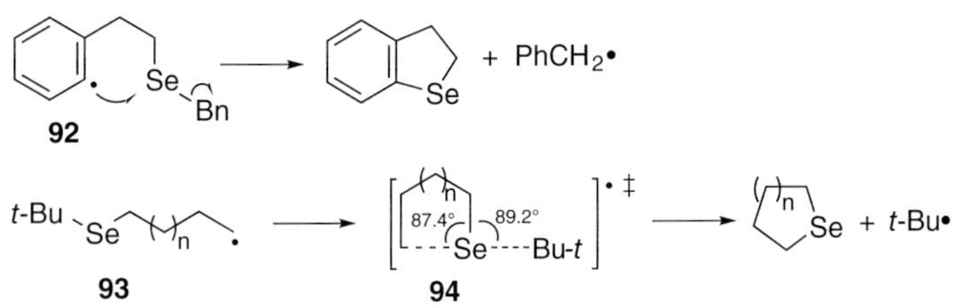

3.5 Homolytic Substitution at Selenium for the Synthesis of Se-Containing Products

Scheme 3.33

at $3 \times 10^7 \, s^{-1}$ at 80 °C (Scheme 3.33) [69]. High-level *ab initio* calculations have shown that intramolecular cyclization by alkyl radicals at Se is synthetically useful at rates of 10^4–$10^6 \, s^{-1}$ at 20 °C, largely governed by the stabilization energy of the leaving group but with some steric and polar contribution [5, 70]. Studies on the cyclization of alkyl radicals **93** show that the angle of attack and *tert*-butyl radical leaving group are approaching linearity in the transition state **94** showing favorable cyclization [71]. The modeling correctly predicted that the orders of reactivity of five- and six-membered ring cyclizations are similar but faster than that of seven-membered ring cyclization.

Six-membered ring alkyl radical cyclization onto selenium has been used for the synthesis of selenochromanes [72], selenochromanes in seleno-tocopherols and analogues [73], 2-methylselenane [74], and selenocephams and selenopenams [75]. Five-membered ring alkyl radical cyclizations have also been reported, which include the synthesis of 5-selenopentopyranose carbohydrates from (phenylseleno)formate pentose precursors [54]. An example of this five-ring cyclization is shown for the synthesis of 2,3-dihydrobenzoselenolo[2,3-*b*]pyridines **96** (Scheme 3.34) [14]. The selenide group was introduced by an S_NAr reaction on the pyridine

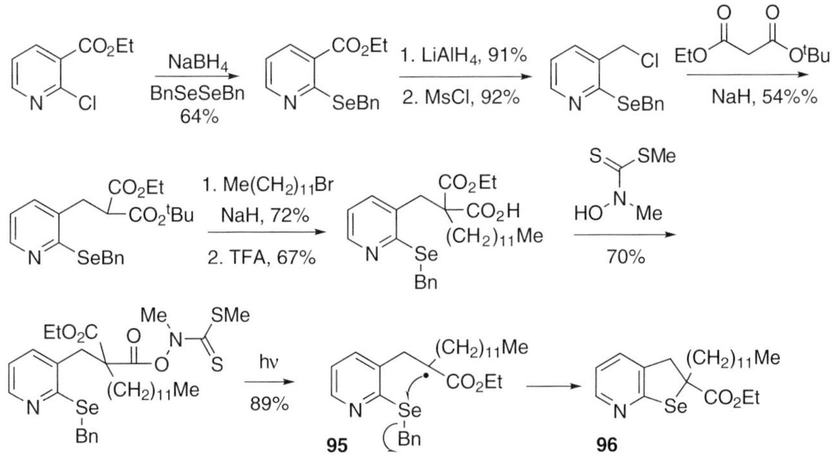

Scheme 3.34

ring. The stability of selenide moiety is again illustrated in the synthetic sequence involving several different reactions. This protocol uses the different strengths of the aryl–Se and benzyl–Se bonds to direct the cyclization with benzyl radicals acting as a good leaving group. In the key intramolecular step, the tertiary radical **95** displaces the benzyl radical leaving group. The cyclization also illustrates the low steric requirement of radical reactions on selenium.

Vinyl radicals have been used in cyclization onto Se to synthesize selenoheterocycles [76–78] and aromatic selenophenes [77] and benzoselenophenes [78]. In the first example, the vinyl radical **97** cyclizes to yield a dihydroselenophene that is converted to the selenophene **98** (Scheme 3.35) [77]. The selenophene **98** is a building block for the synthesis of seleno analogues of the antihypertensive drugs milfasartan and eprosartan in which Se replaces S. Note that the benzylselenyl group is introduced by ring opening of an expoxide.

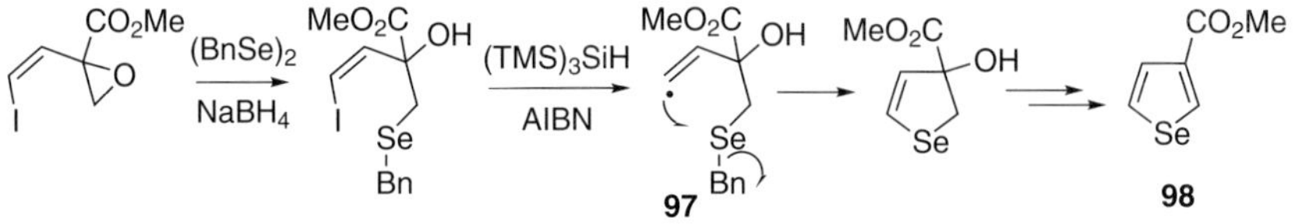

Scheme 3.35

In the second example, the benzoselenophenes **100** were synthesized as building blocks for the preparation of analogues of milfasartan and eprosartan. The benzoselenophenes were prepared by a domino radical protocol including aryl radical addition onto alkynes followed by intramolecular S_H2 on Se of the intermediate vinyl radicals **99** to yield the benzoselenophenes **100** (Scheme 3.36) [78]. The methylselenyl group was introduced by an S_NAr substitution.

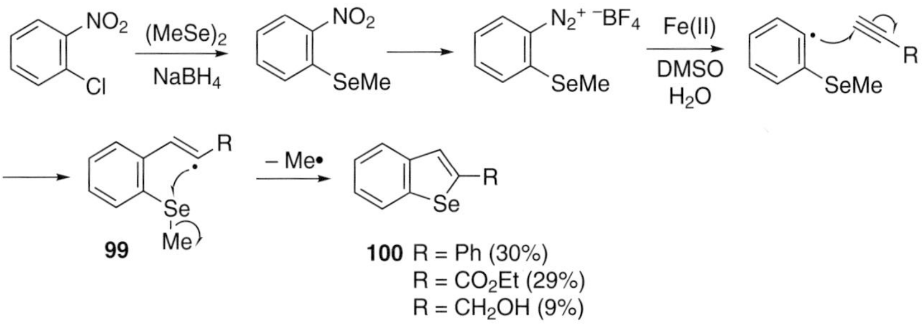

Scheme 3.36

3.6
Seleno Group Transfer onto Alkenes and Alkynes

The radical addition of X–Y to alkenes and alkynes is referred to as the Kharasch reaction, named after the discoverer. Phenylselenyl transfer has been successfully developed for a range of precursors with weak PhSe–X bonds (Scheme 3.37). These transfers include seleno-selenation, seleno-sulfonation, and seleno-alkylation. Azido-selenation is more complex and requires the use of azide with an oxidant and is not covered here. Reviews provide more details of the mechanistic parameters of Kharasch reactions [1b, 79].

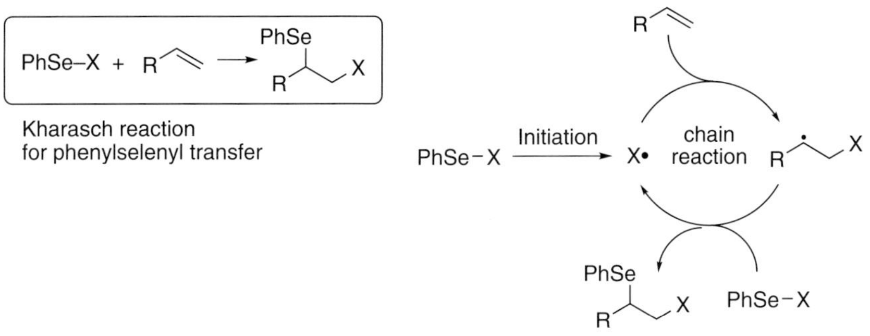

Scheme 3.37

The additions are synthetically attractive because of atom efficiency and the addition of two functionalities. These selenation reactions proceed by a chain reaction and require initiation (Scheme 3.37). Only a trace amount of X• needs to be generated to set the chain reaction underway. Light irradiation, heat, or chemical initiators can be used to initiate the reaction. The key step is an S_H2 abstraction of PhSe from PhSe–X by the intermediate alkyl radical. The resulting radical (X•) needs to be sufficiently stable but also reactive enough to add to the alkene (or alkyne).

3.6.1
Seleno-Selenation

Alkyl radicals react rapidly with PhSeSePh to yield PhSe• radicals in the key step. The rate has been measured at $2.6 \times 10^7 \, M^{-1} s^{-1}$ at 25 °C [2]. While not commonly used synthetically, there are some interesting examples reported in the literature. Reaction between PhSeSePh and methylenecylopropane gave addition and ring opening with the addition of both PhSe groups (Scheme 3.38) [80]. Cu(II) acetate and irradiation were used. Different solvents gave a number of other products.

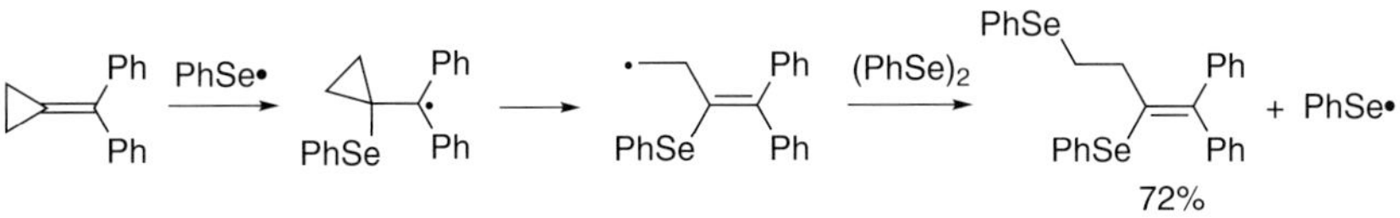

Scheme 3.38

Another example reports the ring opening of allenyl analogue [81]. Benzeneselenyl halides can also be used [82].

3.6.2
Seleno-Sulfonation

Selenide group transfers from sulfonyl selenides have proved useful due to the weak S–Se bond and are widely used in synthesis. The use of sulfonyl radicals in synthesis has been reviewed [83]. One of the common protocols is to use *p*-TolSO$_2$-SePh to facilitate stereoselective 5-*exo-trig* cyclizations of 1,5-dienes. An example is shown in Scheme 3.39 for the synthesis of 3-silylhepta-1,6-dienes [84]. The *cis* regioselectivity of up to 9:1 increased as the temperature was lowered with selective addition of the sulfonyl radical on the alkene adjacent to the silyl group. The intermediate radical **101** adopts the Beckwith TS with the large silyl group in the equatorial position to give *cis* cyclization to yield the intermediate radical **102**, which abstracts PhSe from PhSeTs to complete the chain reaction to yield cyclized addition product **103**. Other examples of this addition–cyclization protocol include dienes [85, 86], enynes and diynes, and alkynes [83]. An interesting example is the regio- and stereoselective TsSePh-mediated cyclization of bis(allenes) to yield the *trans*-fused five-membered rings containing vinyl sulfones and vinyl selenide substituents [87]. Tosyl bromide and iodide have also been used in this general protocol.

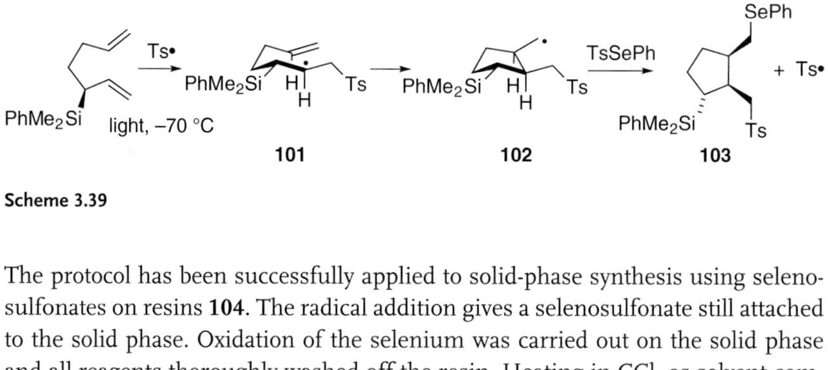

Scheme 3.39

The protocol has been successfully applied to solid-phase synthesis using selenosulfonates on resins **104**. The radical addition gives a selenosulfonate still attached to the solid phase. Oxidation of the selenium was carried out on the solid phase and all reagents thoroughly washed off the resin. Heating in CCl$_4$ as solvent completed the elimination to give alkenes or alkynes. The procedure with solid phase has been used for alkenyl sulfones starting with addition onto alkenes [88] and the equivalent addition to alkynes to give alkynyl sulfones [89]. The procedure was also used for addition–cyclization reactions as shown in Scheme 3.40 [90]. The initial Ts• radical addition was followed by cyclization and abstraction of PhSe to complete the chain reaction to give the solid phase addition–cyclization–addition product **105**. The oxidation, wash, and heat protocol gave the alkenyl sulfone **106**.

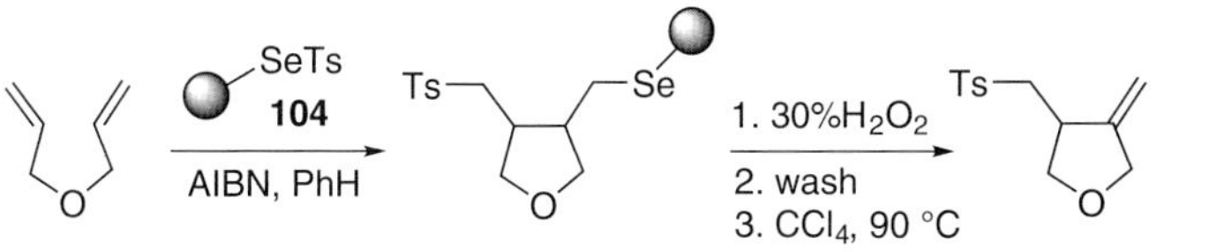

3.6 Seleno Group Transfer onto Alkenes and Alkynes

Scheme 3.40

3.6.3
Seleno-Alkylation

Seleno-alkylation with 2-(phenylselenyl)-propanedioates and -malonitriles onto alkenes and alkynes provides a useful synthetic procedure [2, 91, 92]. The procedure has been coupled with carbonylation to yield acyl selenides [93]. In this procedure, the alkyl radical adds onto the alkene and the resulting radical adds CO to give an acyl radical before abstraction of PhSe takes place to complete the chain reaction. Stereoselective seleno-alkylation has been carried out for seleno-alkylation of chiral 2-phenylcyclohexanol and 8-phenylmenthol enol ethers [94]. The rates of PhSe abstraction by various alkyl radicals involved in seleno-alkylation have been measured (Table 3.1) to show the feasibility of the methodology [2].

Intramolecular seleno-alkylation which is coupled with radical cyclization has found particular use [95, 96]. An example of enantioselective seleno-alkylation is shown in Scheme 3.41 [95]. The seleno-alkylation takes place on the alkene in the molecule thereby facilitating radical cyclization. This is a good example of the use of Et$_3$B (5 equiv.) as the radical initiator in selenium reactions at low temperature that helps facilitate high enantioselectivity. Mg(II) coupled with a ligand **107** were used to facilitate the enantioselectivity. In this example, the selenyl precursor **108** was enantioselectively converted to the bicycle **109** in high yield.

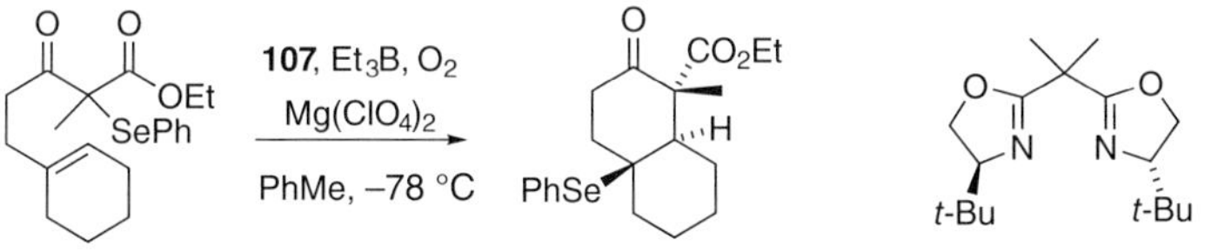

Scheme 3.41

3.7
PhSeH in Radical Reactions

3.7.1
Radical Clock Reactions

The rate of reaction of PhSeH with C-centered radicals is very fast and is used in radical clock measurements when other radical trapping species are not fast enough [97]. H-abstraction from PhSeH is an order of magnitude faster than PhSH and an order of magnitude slower than diffusion control. The rate of H-abstraction from PhSeH by primary, secondary, and tertiary C-centered radicals is the same: $k = 1.2 \times 10^9 \, M^{-1} s^{-1}$ at 20 °C [98]. Scheme 3.42 shows an example in which PhSeH has been used to measure the rate of ring opening of (*trans*-2-ethoxycarbonylcyclopropyl)methyl radicals as $k_1 = 12 \times 10^{10} \, s^{-1}$ at 20 °C [99]. The rates of H-abstraction from PhSeH (k_2 and k_3) are known and the ring-opening rate k_1 can be calculated by measuring the relative amounts of products.

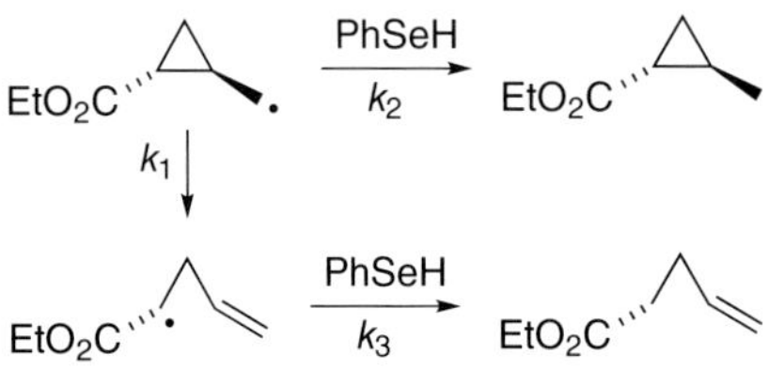

Scheme 3.42

3.7.2
Problem of Unwanted Trapping of Intermediate Radicals

The fast rate of trapping of intermediate radicals in the presence of PhSeH also causes problems. Traces of PhSeH or PhSeSePh left in the purification of phenylselenyl precursors leads to premature trapping of radicals prior to the intended reactions of these intermediate radicals. PhSeSePh is rapidly reduced *in situ* by Bu_3SnH to yield PhSeH. Slower reacting reagents such as Bu_3SnH allow intermediates to undergo the required reactions. Considerable care needs to be taken experimentally to ensure purity of radical precursors.

There are two mechanistic proposals for the exact mechanism of the interaction of PhSeH and $Bu_3Sn\cdot$ and related radicals. Scheme 3.43 shows the polarity reversal catalysis (PRC) mechanism proposed by Roberts and provides a general understanding [100]. The nucleophilic alkyl radical reacts rapidly with the nucleophilic source of hydrogen (PhSeH) as shown in the transition state (TS) in Eq. (3.1). The electrophilic radical PhSe· reacts rapidly with the nucleophilic source of hydrogen (Bu_3SnH) (Eq. (3.2)). However, the nucleophilic alkyl radical reacts slower with the nucleophilic source of hydrogen (Bu_3SnH) via a higher energy TS (Eq. (3.3)).

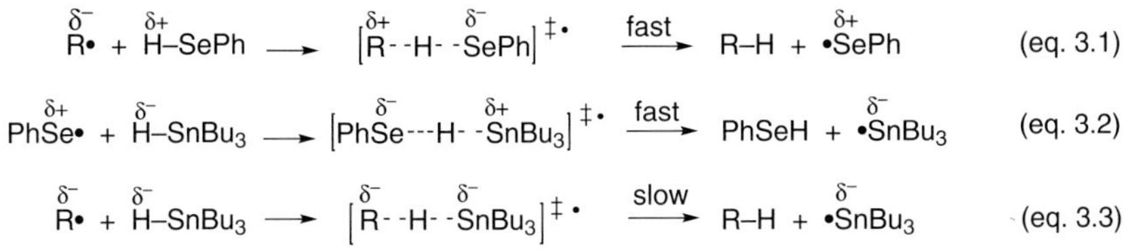

3.7 PhSeH in Radical Reactions

$$R\bullet + H-SePh \longrightarrow [R\cdots H\cdots SePh]^{\ddagger\bullet} \xrightarrow{fast} R-H + \bullet SePh \quad (eq.\ 3.1)$$

$$PhSe\bullet + H-SnBu_3 \longrightarrow [PhSe\cdots H\cdots SnBu_3]^{\ddagger\bullet} \xrightarrow{fast} PhSeH + \bullet SnBu_3 \quad (eq.\ 3.2)$$

$$R\bullet + H-SnBu_3 \longrightarrow [R\cdots H\cdots SnBu_3]^{\ddagger\bullet} \xrightarrow{slow} R-H + \bullet SnBu_3 \quad (eq.\ 3.3)$$

Scheme 3.43

Two very fast reactions replace a much slower reaction thus also speeding up the overall reaction.

3.7.3
Catalysis of Stannane-Mediated Reactions

The above "problem" (Scheme 3.43) has been exploited by Crich to provide a clever new protocol for a variety of radical reactions in which reactive intermediate radicals are trapped to yield targets which cannot otherwise be synthesized. The protocol has been recently reviewed [101].

First, the protocol has been used to trap unstable radical intermediates to provide evidence for mechanistic proposals. The protocol has been used to provide clear evidence for the intermediacy of spirodienyl radicals in cyclizations proceeding via aromatic homolytic substitution (Scheme 3.44) [102]. A recent review provides an up-to-date view on homolytic aromatic substitution [23]. When the cyclization was carried out under normal Bu$_3$SnH and AIBN conditions, <5% of the spirodienyl product **113** was obtained and mainly the phenanthridone **114** was formed. This

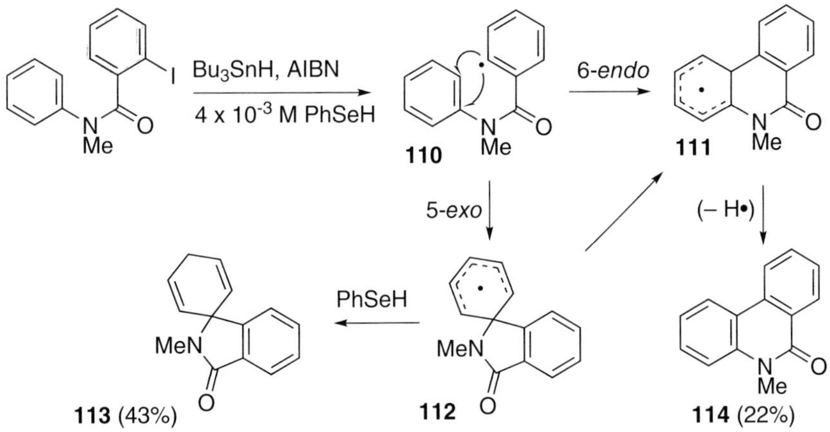

Scheme 3.44

product suggests that intermediate **110** cyclized by 6-*endo* cyclization to **111** followed by slow H-abstraction to complete the nonchain homolytic aromatic substitution. However, 5-*exo* cyclization to the spirodienyl radical **112** followed by a relatively slow neophyl rearrangement to **111** would yield the same product. Addition of PhSeH speeded up the reaction by creating a chain process and trapped the spirodienyl radical to yield **113** in 43% yield, clearly proving that 5-*exo* cyclization followed by rearrangement was the predominant mechanism.

Second, the addition of PhSeH has been used to trap unstable radical intermediates to provide new synthetic protocols [101]. A recent example is shown in Scheme 3.45 for the addition of benzene to *o*-substituted aryl halides to yield aryl-cyclohexadiene products. The addition is shown in Scheme 3.45 in which the cyclohexadienyl intermediate which would otherwise lose hydrogen in a slower nonchain aromatic homolytic substitution is trapped as the product. The conversion to a fast chain reaction also gives higher yields. The protocol has also been used to add to heteroarenes, for example, pyridines and pyrroles [103, 104] and thiophenes and furans [104, 105].

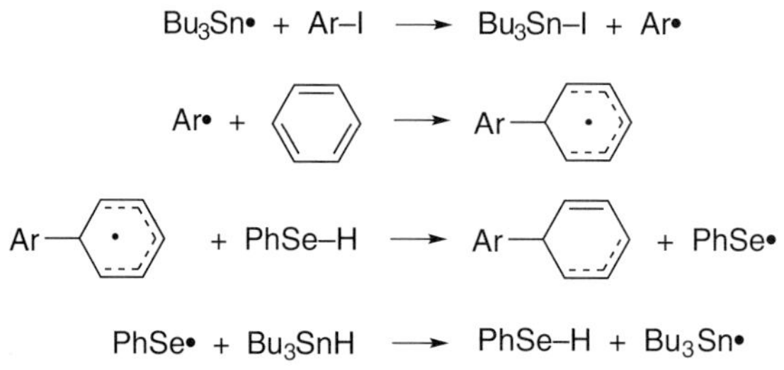

$$Bu_3Sn\bullet + Ar\text{-}I \longrightarrow Bu_3Sn\text{-}I + Ar\bullet$$

$$Ar\bullet + \text{benzene} \longrightarrow Ar\text{-cyclohexadienyl}$$

$$Ar\text{-cyclohexadienyl} + PhSe\text{-}H \longrightarrow Ar\text{-cyclohexadiene} + PhSe\bullet$$

$$PhSe\bullet + Bu_3SnH \longrightarrow PhSe\text{-}H + Bu_3Sn\bullet$$

Scheme 3.45

These products provide excellent substrates for further cyclization by nonradical routes to important natural products (Scheme 3.46) [101]. The synthesis of carbazomycin B, a compound with important biological activity, is shown in Scheme 3.47 to illustrate both the radical protocol and the use of the product in further cyclization [106]. The example illustrates that arenes with multisubstitution provide good radical substrates for this protocol of radical arylation of benzene. Other natural products have been synthesized using this protocol starting with suitably substituted *o*-iodo-phenols and -carboxylic acids [107].

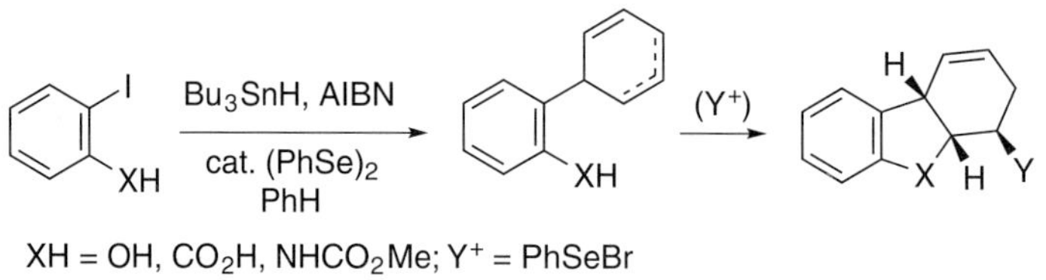

XH = OH, CO$_2$H, NHCO$_2$Me; Y$^+$ = PhSeBr

Scheme 3.46

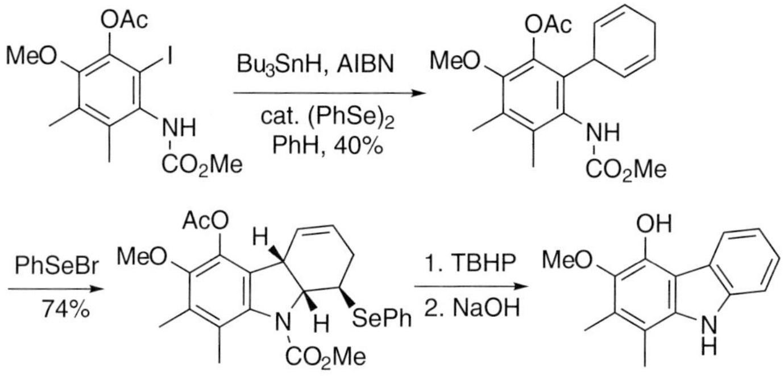

Scheme 3.47

PRC has also been used to trap out an unwanted radical by-product that hinders the required chain reaction. Cyclization onto O-trityl oximes provides a useful synthetic route to cyclized oximes [108]. Initially the formation of the stable and nucleophilic trityl radicals hindered the reaction because of its slow rate of reaction with Bu_3SnH to complete the chain reaction. PhSeSePh, which is rapidly reduced by Bu_3SnH to PhSeH, was added to facilitate PRC and fast trapping of the trityl radical to speed up the chain reaction. An example of this procedure is shown in Scheme 3.48. The nucleophilic trityl radicals react rapidly with the electrophilic source of hydrogen PhSeH to provide a fast chain reaction and higher yields.

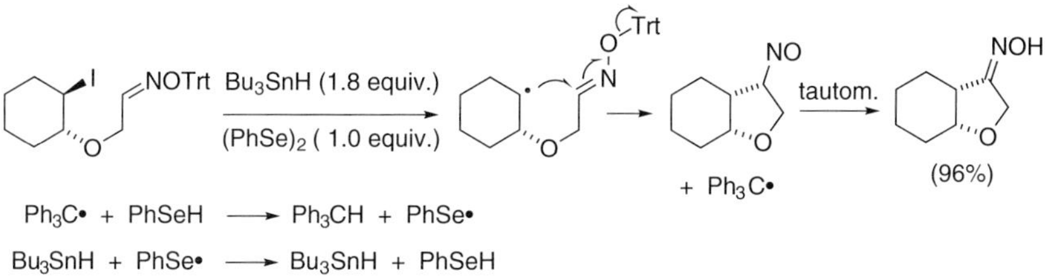

Scheme 3.48

3.8
Selenium Radical Anions, $S_{RN}1$ Substitutions

Aryl selenide anions undergo single electron transfer (SET) to suitable electron acceptors yielding PhSe• radicals and radical anions. PhSe⁻ anions are also able to add to radicals to form radical anions in which the unpaired electron resides in the low-lying C–Se σ* MO. These processes are observed in $S_{RN}1$ (substitution radical-nucleophilic, unimolecular) reactions (Scheme 3.49). The area has been thoroughly reviewed [109]. The initial step is normally light-induced SET in which an unstable halogeno-arene radical anion is formed as well of the arylselenyl

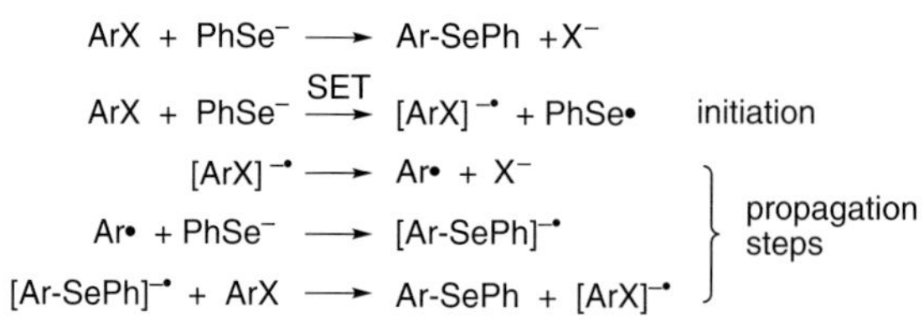

```
ArX + PhSe⁻  ⟶  Ar-SePh + X⁻
ArX + PhSe⁻  —SET→  [ArX]⁻• + PhSe•        initiation
[ArX]⁻•  ⟶  Ar• + X⁻                ⎫
Ar• + PhSe⁻  ⟶  [Ar-SePh]⁻•         ⎬ propagation steps
[Ar-SePh]⁻• + ArX  ⟶  Ar-SePh + [ArX]⁻•  ⎭
```

Scheme 3.49

radical. The propagation steps involve dissociation of the initially formed radical anions to aryl radicals which form selenyl radical anions by the addition of the selenide anions to the aryl radicals.

A range of substituted aryl halides, 1-iodoadamantane, and heteroaryl halides have been substituted with aryl selenide anions, normally PhSe⁻, in good yields (Scheme 3.49) [109–111]. An example of the use in synthesis is shown in Scheme 3.50 for the preparation of 2-, 3-, and 4-(phenylselenyl)benzonitriles [112].

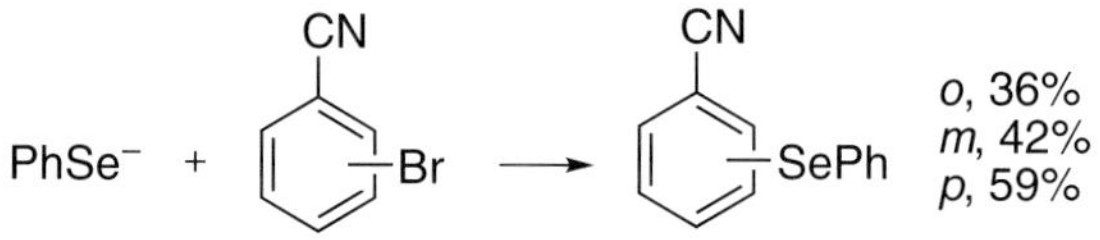

PhSe⁻ + [Ar(CN)Br] ⟶ [Ar(CN)SePh]   o, 36%
                                      m, 42%
                                      p, 59%

Scheme 3.50

A synthetic problem can arise when the intermediate selenyl radical anion ([ArSePh]⁻•) does not only undergo SET to complete the chain reaction but also dissociates in the opposite direction to yield ArSe⁻ and phenyl (Ph•) radicals [113]. This leads to scrambling of products with ArSeAr and PhSePh also being formed. This can be largely avoided by using an excess of the aryl halide thereby increasing the rate of k_{SET}(ArX). This depends on the relative energies of π and C-Se σ MOs. For instance, reactions between 1-naphthyl, 2-quinolyl, and 4-biphenyl with PhSe⁻ only yield the required product as shown in Scheme 3.51 [114]. This indicates that the π* MO is of lower energy than the σ* MO, that is, SET is favored over dissociation. Dissociation of the intermediate radical anion can be used to advantage

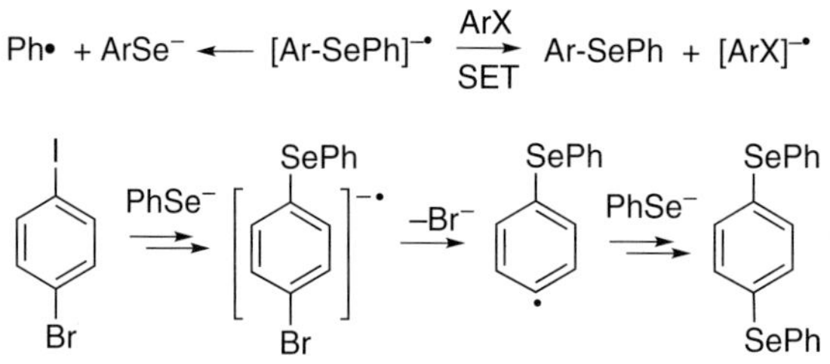

Ph• + ArSe⁻ ⟵ [Ar-SePh]⁻• —ArX/SET→ Ar-SePh + [ArX]⁻•

[4-I-C₆H₄-Br] —PhSe⁻→ [4-SePh-C₆H₄-Br]⁻• —-Br⁻→ [4-SePh-C₆H₄•] —PhSe⁻→ [1,4-(SePh)₂C₆H₄]

Scheme 3.51

with dihalogeno-arenes to synthesize di-(phenylseleno)benzenes (Scheme 3.51) [112]. The rate of dissociation in this case is much faster than SET and only the disubstituted product is obtained.

References

1. (a) Freudendahl, D.M., Shahzad, S.A., and Wirth, T. (2009) *Eur. J. Org. Chem.*, 1649–1664; (b) Renaud, P. (2000) *Topics in Current Chemistry*, vol. 208 (ed. T. Wirth), Springer, Berlin, pp. 81–112.
2. Curran, D.P., Martin-Esker, A.A., Ko, S.-B., and Newcomb, M. (1993) *J. Org. Chem.*, 58, 4691–4695.
3. (a) Beckwith, A.L.J. and Pigou, P.E. (1986) *Aust. J. Chem.*, 39, 1151–1155; (b) Beckwith, A.L.J. and Pigou, P.E. (1986) *Aust. J. Chem.*, 39, 77–87.
4. Schiesser, C.H. and Smart, B.A. (1995) *Tetrahedron*, 51, 6051–6060.
5. Horvat, S.M. and Schiesser, C.H. (2010) *New J. Chem.*, 34, 1692–1699.
6. Chatgilialoglu, C., Ferreri, C., Lucarini, M., Pedrielli, P., and Pedulli, G.F. (1995) *Organometallics*, 14, 2672–2676.
7. Bowman, W.R., Clark, D.N., and Marmon, R.J. (1994) *Tetrahedron*, 50, 1275–1294.
8. (a) Bowman, W.R., Stephenson, P.T., Terrett, N.K., and Young, A.R. (1994) *Tetrahedron Lett.*, 35, 6369–6372; (b) Bowman, W.R., Stephenson, P.T., Terrett, N.K., and Young, A.R. (1995) *Tetrahedron Lett.*, 51, 7959–7980.
9. (a) Bowman, W.R., Stephenson, P.T., and Young, A.R. (1995) *Tetrahedron Lett.*, 36, 5623–5626; (b) (1996) *Tetrahedron*, 52, 11445–11462.
10. (a) Aldabbagh, F. and Bowman, W.R. (1997) *Tetrahedron Lett.*, 38, 3793–3794; (b) Aldabbagh, F., Bowman, W.R., and Mann, E. (1997) *Tetrahedron Lett.*, 38, 7937–7940; (c) Aldabbagh, F., Bowman, W.R. (1999) *Tetrahedron*, 55, 4109–4122; (d) Aldabbagh, F., Bowman, W.R., Mann, E., and Slawin, A.M.Z. (1999) *Tetrahedron*, 55, 8111–8128.
11. Allin, S.M., Barton, W.R.S., Bowman, W.R., Bridge, E., Elsegood, M.R.J., McInally, T., and McKee, V. (2008) *Tetrahedron*, 64, 7745–7758.
12. Allin, S.M., Bowman, W.R., Karim, R., and Rahman, S.S. (2006) *Tetrahedron*, 62, 4306–4316.
13. Allin, S.M., Barton, W.R.S., Bowman, W.R., and McInally, T. (2002) *Tetrahedron Lett.*, 43, 4191–4193.
14. Fenner, T., White, K.M., and Schiesser, C.H. (2006) *Org. Biomol. Chem.*, 4, 466–474.
15. Fagan, V., Bonham, S., Carty, M.B., and Aldabbagh, F. (2010) *Org. Biomol. Chem.*, 8, 3149–3156.
16. Bowman, W.R., Krintel, S.L., and Schilling, M.B. (2004) *Org. Biomol. Chem.*, 2, 585–592.
17. Murphy, J.A., Tripoli, R., Khan, T.A., and Mali, U.W. (2005) *Org. Lett.*, 7, 3287–3289.
18. (a) Clive, D.L.J. and Zhang, J. (1987) *Chem. Commun.*, 549–550; (b) De Mesmaeker, A., Hoffman, P., Ernst, B., Hag, P., and Winkler, T. (1989) *Tetrahedron Lett.*, 30, 6311–6314.
19. Stojanovic, A. and Renaud, P. (1998) *Helv. Chim. Acta*, 81, 353–373.
20. Clive, D.L.J. and Angom, A.G. (1985) *J. Chem. Soc. Chem. Commun.*, 941–942.
21. Prévost, N. and Shipman, M. (2002) *Tetrahedron*, 58, 7165–7175.
22. Srivastava, P. and Engman, L. (2010) *Tetrahedron Lett.*, 51, 1149–1151.
23. Bowman, W.R. and Storey, J.M.B. (2007) *Chem. Soc. Rev.*, 36, 1803–1822.
24. Tiecco, M., Testaferri, L., Marini, F., Sternativo, S., Santi, C., Bagnoli, L., and Temperini, A. (2007) *Tetrahedron*, 63, 5482–5489.
25. Bowman, W.R., Bridge, C.F., and Brookes, P. (2000) *Tetrahedron Lett.*, 41, 8989–8995.
26. Kumamoto, H., Ogamino, J., Tanak, H., Suzuki, H., Haraguchi, K., Miyasaka, T.,

Yokomatsu, T., and Shibuya, S. (2001) *Tetrahedron*, **57**, 3331–3341.
27 Berlin, S., Ericcson, C., and Engman, L. (2003) *J. Org. Chem.*, **68**, 8386–8398.
28 Berlin, S., Ericsson, C., and Engman, L. (2002) *Org. Lett.*, **4**, 3–6.
29 (a) Browne, D.M. and Wirth, T. (2006) *Curr. Org. Chem.*, **10**, 1893–1903; (b) Mukherjee, A.J., Zade, S.S., Singh, H.B., and Sunoj, R.B. (2010) *Chem. Rev.*, **110**, 4357–4416.
30 Tiecco, M., Testaferri, L., Bagnoli, L., Terlizzi, R., Temperini, A., Marini, F., Santi, C., and Scarponi, C. (2004) *Tetrahedron Assymmetry*, **15**, 1949–1955.
31 Markó, I.E., Warriner, S.L., and Augustyns, B. (2000) *Org. Lett.*, **2**, 3123–3125.
32 Gao, S.Y., Chittimalla, S.K., Chuang, G.J., and Liao, C.-C. (2009) *J. Org. Chem.*, **74**, 1632–1639.
33 Perin, G., Lenardâo, E.J., Jacob, R.G., and Panatieri, R.G. (2009) *Chem. Rev.*, **109**, 1277–1301.
34 Chatgilialoglu, C., Crich, D., Komatsu, M., and Ryu, I. (1999) *Chem. Rev.*, **99**, 1991–2069. Yet, L. (1999) *Tetrahedron*, **55**, 9349–9403.
35 Allin, S.M., Barton, W.R.S., Bowman, W.R., and McInally, T. (2001) *Tetrahedron Lett.*, **42**, 7887–7890.
36 Bennasar, M.L., Roca, T., and Ferrando, F. (2006) *Org. Lett.*, **8**, 561–564.
37 Bennasar, M.L., Roca, T., and Ferrando, F. (2005) *J. Org. Chem.*, **70**, 9077–9080.
38 Bennasar, M.L., Roca, T., and Garcia-Diaz, D. (2008) *J. Org. Chem.*, **73**, 9033–9039.
39 (a) Bennasar, M.L., Roca, T., and Garcia-Diaz, D. (2008) *Synlett*, 1487–1490; (b) Bennasar, M.L., Roca, T., Garcia-Diaz, D. (2007) *J. Org. Chem.*, **72**, 4562–4565.
40 (a) Bennasar, M.L., Roca, T., Griero, R., Bassa, M., and Bosch, J. (2001) *J. Org. Chem.*, **66**, 7547–7551; (b) Bennasar, M.L., Roca, T., Griero, R., Bassa, M., and Bosch, J. (2002) *J. Org. Chem.*, **67**, 6268–6271.
41 Leading references: (a) Handa, S., Pattenden, G., Li, W.-S. (1999) *Chem. Commun.*, 311–312; (b) Hitchcock, S.A., Houldsworth, S.J., Pattenden, G., Pryde, D.C., Thompson, N.M., and Blake, A J. (1998) *J. Chem. Soc., Perkin Trans. 1*, 3181–3206; (c) Chen, I., Gill, G.B., Pattenden, G., and Simonian, H. (1996) *J. Chem. Soc., Perkin Trans. 1*, 31–43.
42 Handa, S. and Pattenden, G. (1997) *Contemp. Org. Synth.*, **4**, 196–215.
43 Handa, S. and Pattenden, G. (1999) *J. Chem. Soc. Perkin Trans. 1*, 843–845.
44 Pattenden, G., Stoker, D.A., and Winne, J.M. (2009) *Tetrahedron*, **65**, 5767–5775.
45 De Boeck, B., Harrington-Frost, N.M., and Pattenden, G. (2005) *Org. Biomol. Chem.*, **3**, 340–347.
46 (a) De Boeck, B., Herbert, N.M.A., Harrington-Frost, N.M., and Pattenden, G. (2005) *Org. Biomol. Chem.*, **3**, 328–339; (b) Hayes, C.J., Herbert, N.M.A., Harrington-Frost, N.M., and Pattenden, G. (2005) *Org. Biomol. Chem.*, **3**, 316–327.
47 (a) Bachi, M.D. and Denenmark, D. (1989) *J. Am. Chem. Soc.*, **111**, 1886–1888; (b) Bachi, M.D., Balanov, A., Bar-Ner, N., Bosch, E., Denenmark, D., and Mizhiritskii, M. (1993) *Pure Appl. Chem.*, **65**, 595–601.
48 Bowman, W.R., Fletcher, A.J., Lovell, P.J., and Pedersen, J.M. (2004) *Synlett*, 1905–1908.
49 Pedersen, J.M., Bowman, W.R., Elsegood, M.R.J., Fletcher, A.J., and Lovell, P.J. (2005) *J. Org. Chem.*, **70**, 10615–10618.
50 Bowman, W.R., Fletcher, A.J., Pedersen, J.M., Lovell, P.J., Elsegood, M.R.J., Hernández López, E., McKee, V., and Potts, G.B.S. (2007) *Tetrahedron*, **63**, 191–203.
51 Kim, S. and Lee, T.A. (1997) *Synlett*, 950–952.
52 Yoshimura, Y., Yamazaki, Y., Katsunori, W., Satoh, S., and Takahata, H. (2007) *Synlett*, 111–114.
53 Clive, D.L.J. and Ardelean, E.S. (2001) *J. Org. Chem.*, **66**, 4841–4844.
54 Lucas, M.A., Nguyen, O.T.K., Schiesser, C.H., and Zheng, S.-L. (2000) *Tetrahedron*, **56**, 3995–4000.
55 (a) Singh, A.K., Bakshi, R.K., and Corey, E.J. (1987) *J. Am. Chem. Soc.*, **109**,

6187–6189; (b) Bachi, M.D. and Bosch, E. (1992) *J. Org. Chem.*, **57**, 4696–4705.
56 Nishiyama, Y., Kajimoto, H., Kotani, K., and Sonoda, N. (2001) *Org. Lett.*, **3**, 3087–3089.
57 Clive, D.L.J. and Bergstra, R.J. (1990) *J. Org. Chem.*, **55**, 1786–1792.
58 Della, E.W. and Knill, A.M. (1996) *J. Org. Chem.*, **61**, 7529–7533.
59 Middleton, D.M., Simpkins, N.S., and Terrett, N.K. (1989) *Tetrahedron Lett.*, **30**, 3865–3868.
60 Clive, D.L.J. and Cheng, H. (2001) *Chem. Commun.*, 605–606.
61 McAllister, L.A., McCormick, R.A., and Procter, D.J. (2005) *Tetrahedron*, **61**, 1527–11576.
62 Nicolaou, K.C., Roecker, J., Pfefferkorn, J.A., and Cao, G.-Q. (2000) *J. Am. Chem. Soc.*, **122**, 2966–2967.
63 (a) Nicolaou, K.C., Pfefferkorn, J.A., Cao, G.-Q., Kim, S., and Kessabu, J. (1999) *Org. Lett.*, **1**, 807–810; (b) Ruhland, T., Andersen, K., and Pedersen, H. (1998) *J. Org. Chem.*, **63**, 9204–9211.
64 Uehlin, L. and Wirth, T. (2001) *Org. Lett.*, **3**, 2931–2933.
65 Tietze, L.F., Brasche, G., and Gericke, K.M. (2006) *Domino Reactions in Organic Synthesis*, Wiley-VCH Verlag GmbH, Weinheim.
66 (a) Damm, W., Hoffman, U., Macko, L., Neuburger, M., Zehnder, M., and Giese, B. (1994) *Tetrahedron*, **50**, 702–7048; (b) Srojanovic, A., Renaud, P., and Schenk, K. (1998) *Helv. Chim. Acta.*, **81**, 268–284; (c) Barton, D.H.R., Ozbalik, N., and Schmitt, M. (1989) *Tetrahedron Lett.*, **30**, 3263–3266; (d) Barton, D.H.R., Bridon, D., Hervé, Y., Potier, P., Thierry, J., and Zard, S.Z. (1986) *Tetrahedron*, **42**, 4983–4990.
67 Artman, G.D., Waldman, J.H., and Weinreb, S.M. (2002) *Synthesis*, 2057–2063.
68 Schiesser, C.H. (2006) *Chem. Commun.*, 4055–4065.
69 Lyons, J.E., Schiesser, C.H., and Sutej, K. (1993) *J. Org. Chem.*, **58**, 5632–5638.
70 Lobachevsky, S., Schiesser, C.H., Lin, C.Y., and Coote, M.L. (2008) *J. Phys. Chem. A*, **112**, 13622–13627.
71 Benjamin, L.J., Schiesser, C.H., and Sutej, K. (1993) *Tetrahedron*, **49**, 2557–2566.
72 Staples, M.K. and Schiesser, C.H. (2010) *Chem. Commun.*, **46**, 565–567.
73 (a) Shanks, D., Amorati, R., Fumo, M.G., Pedulli, G.F., Valgimigli, L., and Engman, L. (2006) *J. Org. Chem.*, **71**, 1033–1038 ; (b) Al-Maharik, N., Engman, L., Malström, J., and Schiesser, C.H. (2001) *J. Org. Chem.*, **66**, 6286–6290.
74 Lucas, M.A. and Schiesser, C.H. (1996) *J. Org. Chem.*, **61**, 5754–5761.
75 Carland, M.A., Martin, R.L., and Schiesser, C.H. (2001) *Tetrahedron Lett.*, **48**, 4737–4739.
76 Lobachevsky, S., Schiesser, C.H., and Gupta, V. (2007) *Tetrahedron Lett.*, **48**, 9077–9079.
77 Grange, R.L., Ziogas, J., North, A.J., Angus, J.A., and Schiesser, C.H. (2008) *Bioorg. Med. Chem. Lett.*, **18**, 1241–1244.
78 Staples, M.K., Grange, R.L., Angus, J.A., Ziogas, J., Tan, N.P.H., Taylor, M.K., and Schiesser, C.H. (2010) *Org. Biomol. Chem.*, personal communication.
79 Zard, S.Z. (2003) *Radical Reactions in Organic Synthesis*, Oxford University Press, Oxford, chapter 6.
80 Yu, L. and Huang, X. (2007) *Synlett*, 1371–1374.
81 Shi, M., Lu, J.-M., and Xu, G.-C. (2005) *Tetrahedron Lett.*, **46**, 4745–4748.
82 Miao, M. and Huang, X. (2009) *J. Org. Chem.*, **74**, 5636–5639.
83 Betrand, M.P. (1994) *Org. Prep. Proc. Int.*, **26**, 257–290.
84 James, P., Schenk, K., and Landais, Y. (2006) *J. Org. Chem.*, **71**, 3630–3633.
85 Caddick, S., Shering, C.L., and Wadman, S.N. (2000) *Tetrahedron*, **56**, 465–473.
86 (a) Bertand, M.P., Gastaldi, S., and Nouguier, R. (1997) *Synlett*, 1420–1422; (b) Bertand, M.P., Gastaldi, S., and Nouguier, R. (1996) *Tetrahedron Lett.*, **37**, 1229–1232; (c) Brumwell, J.E., Simpkins, N.S., and Terrett, N.K. (1993) *Tetrahedron Lett.*, **34**, 1219–1222.
87 Kang, S.-K., Ha, Y.-H., Kim, D.-H., Lim, Y., and Jung, J. (2001) *Chem. Commun.*, 1306–1307.

88 Qian, H. and Huang, X. (2001) *Synlett*, 1913–1916.
89 Qian, H. and Huang, X. (2002) *Tetrahedron Lett.*, **43**, 1059–1061.
90 Qian, H. and Huang, X. (2003) *J. Comb. Chem.*, **5**, 569–576.
91 Byers, J.H. and Lane, G.C. (1993) *J. Org. Chem.*, **58**, 3355–3360.
92 Renaud, P. and Vionnet, J.-P. (1993) *J. Org. Chem.*, **61**, 5895–5896.
93 Ryu, I., Muraoka, H., Kambe, N., Komatsu, M., and Sonoda, N. (1996) *J. Org. Chem.*, **61**, 6396–6403.
94 Curran, D.P., Geib, S.J., and Kuo, K.H. (1994) *Tetrahedron Lett.*, **35**, 6235–6238.
95 Yang, D., Zheng, B.-F., Gao, Q., Gu, S., and Zhu, N.-Y. (2006) *Angew. Chem. Int. Ed.*, **45**, 255–258.
96 Yang, D., Gao, Q., and Lee, O.-Y. (2002) *Org. Lett.*, **4**, 1239–1241.
97 Newcomb, M. (2001) *Radicals in Organic Synthesis*, vol. 1 (eds P. Renaud and M.P. Sibi), Wiley-VCH Verlag GmbH, Weinheim, pp. 317–336.
98 (a) Newcomb, M., Choi, S.-Y., and Horner, J.H. (1999) *J. Org. Chem.*, **64**, 1225–1231; (b) Newcomb, M., Varick, T.R., Ha, C., Manek, M.B., and Yue, X. (1992) *J. Am. Chem. Soc.*, **114**, 8158–8163.
99 Newcomb, M. (1993) *Tetrahedron*, **49**, 1151–1176.
100 Roberts, B.P. (1999) *Chem. Soc. Rev.*, **28**, 25–36.
101 Crich, D., Grant, D., Krishnamurthy, V., and Patel, M. (2007) *Acc. Chem. Res.*, **40**, 453–463.
102 Crich, D. and Hwang, J.-T. (1998) *J. Org. Chem.*, **63**, 2765–2770.
103 Crich, D. and Patel, M. (2006) *Tetrahedron*, **62**, 7824–7837.
104 Crich, D. and Patel, M. (2004) *Heterocycles*, **64**, 499–564.
105 Crich, D. and Patel, M. (2005) *Org. Lett.*, **7**, 3625–3626.
106 Crich, D. and Rumthao, S. (2004) *Tetrahedron*, **60**, 1513–1516.
107 (a) Crich, D. and Krishnamurthy, V. (2006) *Tetrahedron*, **62**, 6830–6840; (b) Crich, D. and Grant, D. (2005) *J. Org. Chem.*, **70**, 2384–2386.
108 Clive, D.L.J., Pham, M.P., and Subedi, R. (2007) *J. Am. Chem. Soc.*, **129**, 2713–2717.
109 Rossi, R.A., Pierini, A.B., and Penénory, A.B. (2003) *Chem. Rev.*, **103**, 71–167.
110 Rossi, R.A. and Penénory, A.B. (1981) *J. Org. Chem.*, **46**, 4580–4582.
111 Pierini, A.B. and Rossi, R.A. (1979) *J. Org. Chem.*, **44**, 4667–4673.
112 Palacios, S.M., Alonso, R.A., and Rossi, R.A. (1985) *Tetrahedron*, **41**, 4147–4156.
113 Penénory, A.B. and Rossi, R.A. (1990) *J. Phys. Org. Chem.*, **3**, 266–272.
114 Degrand, C. (1987) *J. Org. Chem.*, **52**, 1421–1424.

4
Selenium-Stabilized Carbanions
João V. Comasseto, Alcindo A. Dos Santos, and Edison P. Wendler

4.1
Introduction

The first attempts to prepare selenium-stabilized carbanions were reported by Gilman [1, 2]. However, only the selenium–carbon bond cleavage could be detected when diphenyl selenide (**1**) [1] and phenyl(methyl) selenide (**2**) [2] were reacted with *n*-BuLi in diethyl ether (Scheme 4.1).

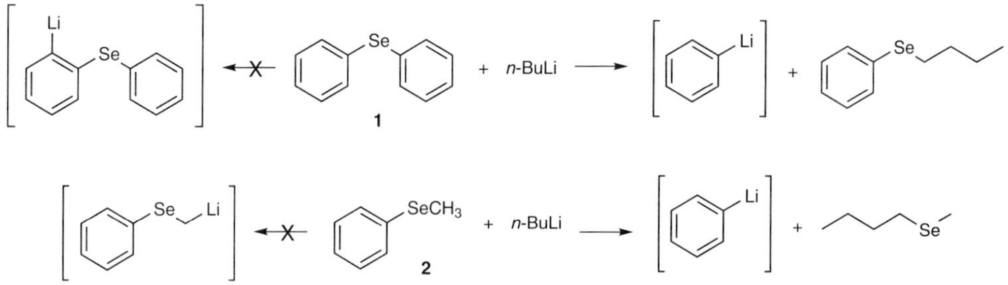

Scheme 4.1 First attempts to prepare selenium-stabilized carbanions.

Even though this initial attempt of preparing selenium-stabilized carbanions was unsuccessful, this pioneering work showed that organoselenide **3** can lead to an organolithium compound **5** when treated with another organolithium compound **4**, provided that **5** is more stable than **4** (Scheme 4.2).

R−Se−R¹ + R²−Li ⟶ [R−Li] + R²-Se-R¹
   3   4   5

Scheme 4.2 Selenium–lithium exchange.

Organoselenium Chemistry: Synthesis and Reactions, First Edition. Edited by Thomas Wirth.
© 2012 Wiley-VCH Verlag GmbH & Co. KGaA. Published 2012 by Wiley-VCH Verlag GmbH & Co. KGaA.

In 1969 Seebach and Peleties succeeded in preparing the α-lithium selenide **7**, by using lithium dialkylamines instead of *n*-BuLi to promote the deprotonation of the selenoacetal **6** [3]. When *n*-BuLi was employed, the selenium–lithium exchange reaction on the selenoacetal **6** took place generating the selenium-stabilized carbanion **8** (Scheme 4.3) [3–5].

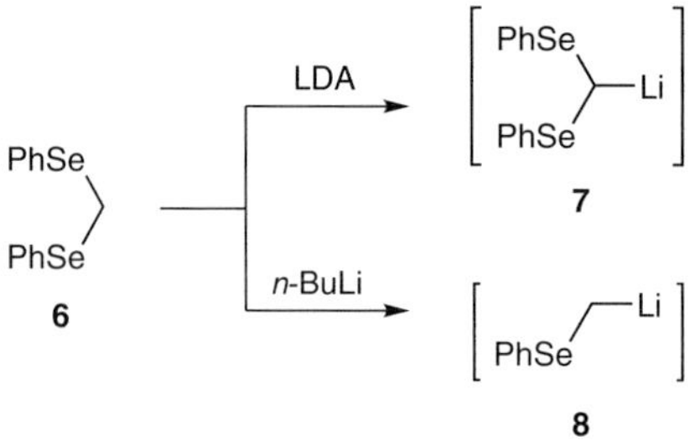

Scheme 4.3 Selenium-stabilized carbanions by deprotonation or selenium–lithium exchange.

The reaction shown in Scheme 4.3 has widely been employed in subsequent years to prepare selenium-stabilized carbanions. A third general approach to prepare selenium-stabilized carbanions consists in the addition of nucleophiles to unsaturated selenides **9** and **10**, leading to sp^3 (**11**) and sp^2 (**12**) hybridized selenium-stabilized carbanionic species (Scheme 4.4).

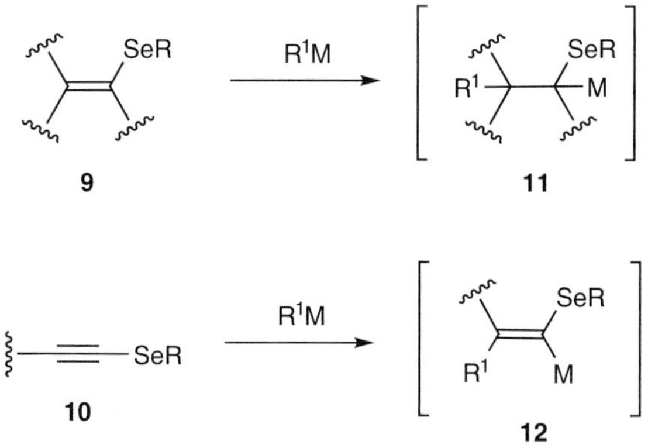

Scheme 4.4 Selenium-stabilized carbanions by conjugate addition to unsaturated selenides.

In the first part of this chapter, these three general methods will be discussed. The chapter does not intend to be an exhaustive review of the literature, since exhaustive reviews [6] and book chapters [7] were published on this subject in the past, and very few new ideas have been developed in recent years. This chapter intends to be a comprehensive account on the preparation, reactivity, and synthetic applications of selenium-stabilized carbanions. In this way, only

4.2 Preparation of Selenium-Stabilized Carbanions

4.2.1 Deprotonation of Selenides

Alkyllithiums are not suitable bases for the deprotonation of selenides. In most cases, the selenium–lithium exchange reaction prevails over the deprotonation. Entries 11, 12, 14, and 28 (Table 4.1) are some of the few examples, where the

Table 4.1 Selenium-stabilized carbanions prepared by deprotonation of the parent selenide.

$$R-Se-CHR^1(H) \xrightarrow{\text{base, solvent}} [R-Se-CR^1(M)]$$

Entry	Selenium-stabilized carbanion	Conditions	Reference
1	PhSe–CLi(Ph)	LDA, THF, −78 °C, 5 min	[12, 13]
2	PhSe–CLi(CH=CH$_2$)	LDA, THF, −78 °C, <10 min	[14]
3	PhSe–CLi(C≡C–Li)	LDA, THF or DME, −78 °C, 5 min	[15]
4	PhSe–CLi(OMe)	LiTMP[a], THF, −78 °C, 5 min	[12]
5	PhSe–CLi(SePh)	LiN(i-Bu)$_2$, THF, −78 °C, 1 h	[3, 4]
6	PhSe–CLi(TePh)	LDA, THF, −78 °C, 30 min	[16]

(Continued)

4 Selenium-Stabilized Carbanions

Table 4.1 (Continued)

Entry	Selenium-stabilized carbanion	Conditions	Reference
7	PhSe–CHLi–SiMe$_3$	LDA, THF, –78 °C, 45 min	[17]
8	PhSe–CLi(SiMe$_3$)–SePh	LDA, THF, –78 °C, 1.5 h	[18]
9	diselenetane with Li	LDA, THF, –78 °C, 3.5 h	[19]
10	PhSe–CLi(SiMe$_3$)–Ph	LiN(Et)$_2$, THF, 0 °C, 30 min	[13]
11	PhSe–CHLi–P(=O)(OEt)$_2$	n-BuLi, THF, –78 °C, 4 h	[20, 21]
12	(polystyrene)–Se–CHLi–P(=O)(OEt)$_2$	n-BuLi, THF, –60 °C, 30 min	[22]
13	(polystyrene)–Se–CHLi–Ph	LDA, THF, –60 °C, 30 min	[23]

For entries 12 and 13, polystyrene = –CH$_2$–CH(Ph)–CH$_2$–CH(Ph)–CH$_2$–CH(Ph)–

Table 4.1 (Continued)

Entry	Selenium-stabilized carbanion	Conditions	Reference
14	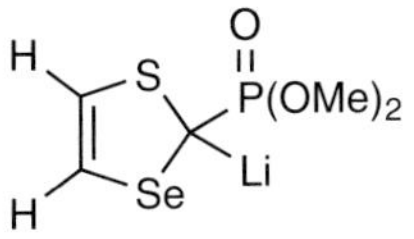 H-C(S)=C(H)-Se-C(Li)(P(O)(OMe)$_2$) (1,3-selenathiole ring with P(O)(OMe)$_2$ and Li at C2)	n-BuLi, THF, −78 °C	[24]
15	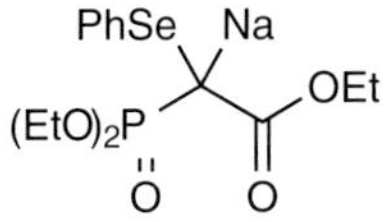 (EtO)$_2$P(O)-C(PhSe)(Na)-C(O)OEt	NaH, THF, 0 °C	[25]
16	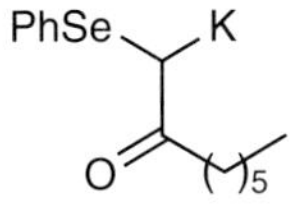 PhSe-C(K)(H)-C(O)-(CH$_2$)$_5$CH$_3$	t-BuOK, t-BuOH, 25 °C	[26]
17	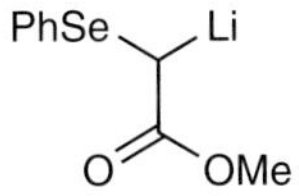 PhSe-C(Li)(CH$_3$)-C(O)OMe	LDA, THF, −78 °C to −45 °C, 30 min	[27]
18	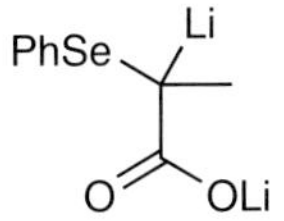 PhSe-C(Li)(CH$_3$)-C(O)OLi	LDA, THF, 0 °C (15 min) to 40 °C (30 min)	[28]
19	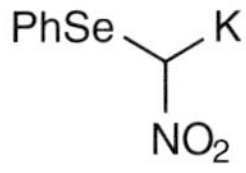 PhSe-C(K)(H)-NO$_2$	KF, solvent-free, 24 °C, 8 h	[29]
20	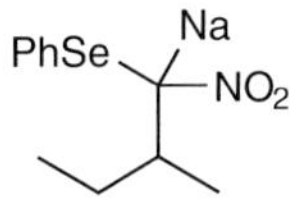 PhSe-C(Na)(NO$_2$)-CH(CH$_3$)$_2$	Na$_2$CO$_3$, benzene, reflux, ~6 h	[29]
21	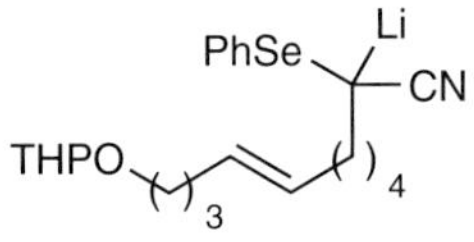 THPO-(CH$_2$)$_3$-CH=CH-(CH$_2$)$_4$-C(Li)(PhSe)-CN	LiHMDS, THF, −78 °C, 5 h	[30]
22	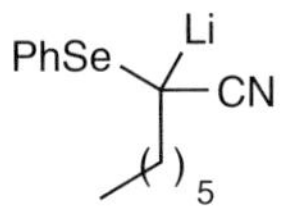 PhSe-C(Li)(CN)-(CH$_2$)$_5$-CH$_3$	LDA, THF, −78 °C	[31]

(Continued)

Table 4.1 (Continued)

Entry	Selenium-stabilized carbanion	Conditions	Reference
23	(bicyclic lactone with H, Li, SePh substituents)	LDA, THF, −78 °C	[32]
24	(PhSe)₂C(Li)(SePh)	LiN(i-Bu)₂, THF, −78 °C	[3]
25	Ph-Se(=O)-CH(Li)-CH₂-Ph	LDA, THF, −78 °C, few min	[10]
26	CH₂=C(SePh)(Li) with methyl	LDA, THF, −78 °C, 2 h	[8]
27	(CH₃)₃C=C(SePh)(K)	KDA, THF, −78 °C	[11]
28	(selenophene-fused bicycle with Li)	n-BuLi, diethyl ether, −10 °C, 1 h	[33]
29	(bis-diselenoacetal with 4 Li)	LDA (4 equiv), THF, −100 °C, 30 min	[34]
30	EtO-CH=CH-C(SPh)=C(SePh)(Li)	t-BuOK/LiTMP[a] (3 equiv), −78 °C, 10 min	[35]

a) LiTMP = lithium 2,2′,6,6′-tetramethyl piperidinyl.

deprotonation of selenides with alkyllithiums as bases were reported in the literature. The bases of choice for the deprotonation of selenides are the alkali metal dialkylamines. The most widely used base for this purpose is lithium diisopropylamide (LDA), as can be observed in Table 4.1. The presence of an extra-stabilizing group α to the selenium atom enhances the acidity of the selenide, making the

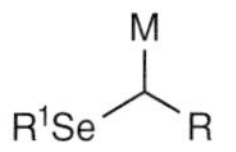

4.2 Preparation of Selenium-Stabilized Carbanions

Figure 4.1 Selenium-stabilized carbanions bearing an extra-stabilizing group.

$R = R^2S$; RSe; R^2O; $(R^2)_3Si$; $ArSO_2$; $(R^2O)_2PO$;
Ar; vinyl; alkynyl; cyano; carbonyl; carboxyl

deprotonation easier. Figure 4.1 illustrates some of the most commonly found stabilizing groups.

Another way to enhance the acidic character of the selenide is to use the *m*-(trifluoromethyl)phenylseleno group instead of the phenylseleno group [8]. However, for practical reasons, the *m*-(trifluoromethyl)phenylseleno group has been little employed. In contrast, the methylseleno group is less acidifying than the phenylseleno group [9]. For preparative purposes, the phenylseleno group is efficient enough in the majority of the cases and has been used in most studies where the deprotonation of a selenide is used to generate a selenium-stabilized carbanion. The transformation of a selenide **13** into a selenoxide or selenone **14** increases the acidity of the α-proton and allows for the deprotonation in systems where the parent selenide is not easily deprotonated (Scheme 4.5) [10].

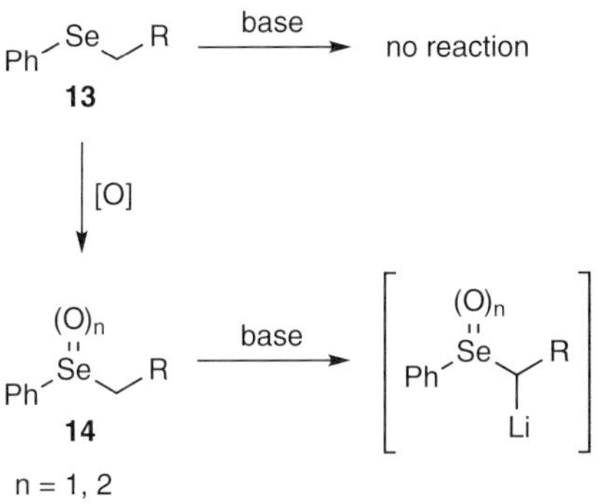

Scheme 4.5 Deprotonation of selenoxides and selenones.

This strategy is limited, however, since selenoxides and selenones are unstable, making their use unattractive for the preparation of carbanions. In view of this fact, few studies on the use of selenoxides and selenones as precursors of selenium-stabilized carbanions are available.

The presence of an alkyl group in α-position to the selenium atom lowers the acidic character of the α proton and the formation of a tertiary lithium selenide is difficult if strong electron-withdrawing groups are absent (e.g., **15**). In such cases,

4 Selenium-Stabilized Carbanions

the use of potassium as the counterion facilitates the deprotonation. For this purpose, a mixture of LDA and t-BuOK (KDA) is used as base (Scheme 4.6) [11].

The use of KDA is mandatory for the success of the deprotonation also in the case of disubstituted vinylic selenides as **16** (Scheme 4.7) [10]. When a monosubstituted vinylic selenide **17** is employed, LDA is an efficient base for the deprotonation [8].

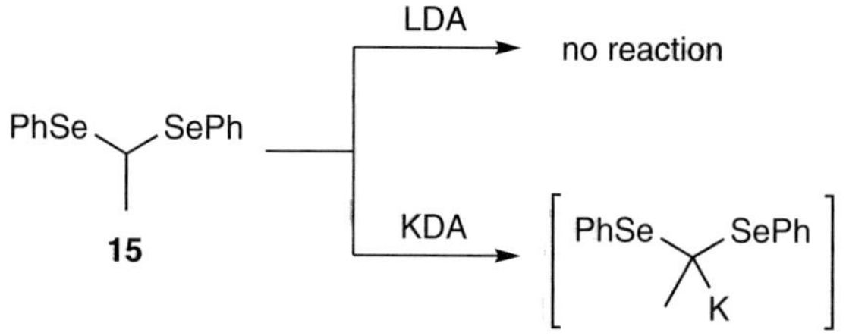

Scheme 4.6 Deprotonation of bis(phenylseleno)acetals with KDA.

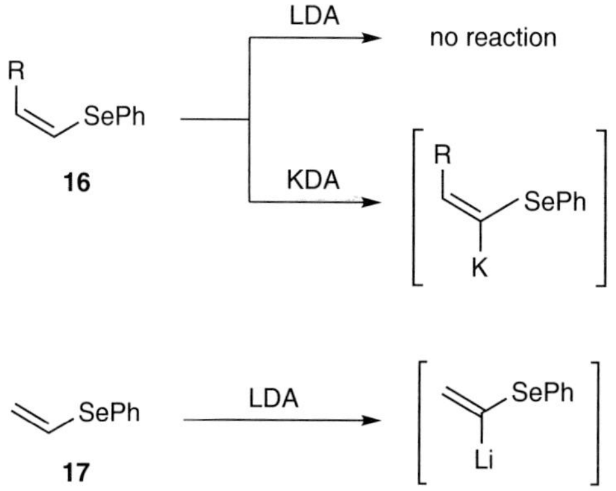

Scheme 4.7 Deprotonation of vinylic selenides.

Table 4.1 shows representative examples of selenium-stabilized carbanions prepared by deprotonation of the parent selenides. As can be observed, in most cases, the phenylseleno group is employed and the most widely used base is LDA in THF. In a few cases, as mentioned above, KDA is used as the base.

4.2.2
Element-Lithium Exchange

As mentioned in the introduction, the selenium atom of a selenide can be replaced by the lithium atom of an organolithium, when this exchange can lead to a more stable organolithium compound (Scheme 4.2). This strategy has been widely used to prepare selenium-stabilized carbanions starting from selenoacetals. Besides the selenium–lithium exchange, other elements can also be exchanged when the organoelemental compound reacts with a commercially available organolithium, leading to selenium-stabilized organolithiums. Table 4.2 lists the examples of selenium-stabilized carbanions generated by element-lithium exchange, along with the reaction conditions employed for the generation of carbanions.

Table 4.2 Selenium-stabilized carbanions by element-lithium exchange.

Entry	Selenium-stabilized carbanion	Conditions	Reference
Selenium–lithium exchange			
1	(CH$_3$)$_2$C(Li)(SePh)	n-BuLi, THF, −78 °C, 6 min	[38]
2	(CH$_3$)$_2$C(Li)(SeMe)	n-BuLi, THF, −78 °C, 30 min	[38]
3	CH$_3$C(Li)(SePh)(SePh)	LiN(i-Bu)$_2$, THF, −78 °C to −50 °C, 30 min	[4]
4	(PhSe)$_2$C(Li)(SePh)	LiN(i-Bu)$_2$, THF, −78 °C, 1 h	[3]
5	t-Bu-cyclohexyl-C(Li)(SePh)	n-BuLi, THF, −78 °C, 30 min	[38]
6	PhCH=C(Li)(SePh)	n-BuLi, THF, −78 °C, 1 h	[39]
7	CH$_3$(CH$_2$)$_9$C(Li)=C(SePh)	n-BuLi, THF, −78 °C, 30 min	[40]
Halogen–lithium exchange			
8	(CH$_3$)$_2$C=C(Li)(SeTIPP)	n-BuLi, THF, −78 °C, 20 min	[41]
9	PhCH=C(Li)(SePh)	n-BuLi, hexane, r.t., 5 min	[42]

(Continued)

4 Selenium-Stabilized Carbanions

Table 4.2 (Continued)

Entry	Selenium-stabilized carbanion	Conditions	Reference
Tellurium–lithium exchange			
10	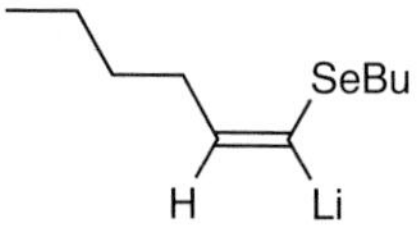 (1-SeBu, 1-Li alkene with propyl chain)	n-BuLi, THF, −78 °C	[43]
Tin–lithium exchange			
11	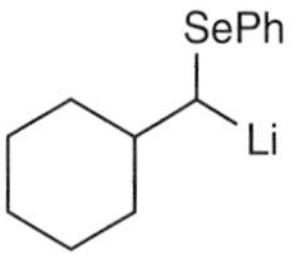 (cyclohexyl-CH(SePh)Li)	n-BuLi, THF, −78 °C, 15 min	[44]
12	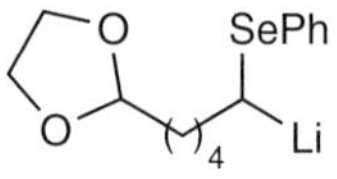 (1,3-dioxolane-CH(SePh)(CH$_2$)$_4$Li)	n-BuLi, THF, −78 °C, 15 min	[44]

TIPP = 2,4,6-triisopropylphenyl.

However, in view of the relatively easy preparation of selenoacetals [36], this class of compounds is mostly used for preparative purposes. Normally, the bis(phenylseleno)acetals (e.g., **18**) are employed, instead of the bis(methylseleno) acetals (e.g., **19**). The main reason for this preference is the very unpleasant smell of the methylbutylselenide (**20**) formed as the by-product of the selenium–lithium exchange reaction when bis(methylseleno)acetals are employed, since the exchange is similar effective in both cases (Scheme 4.8) [37, 38].

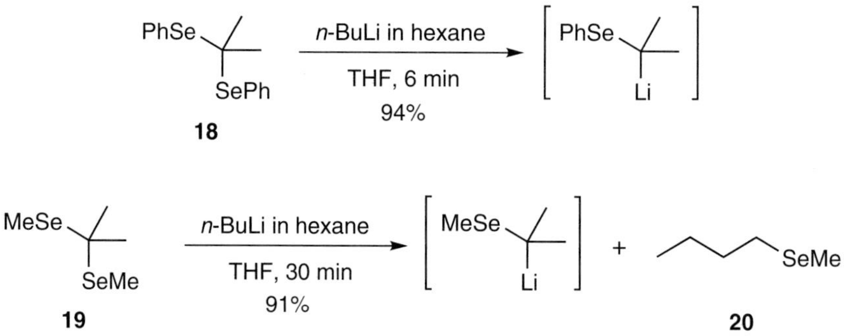

Scheme 4.8 Selenium-stabilized carbanions from the corresponding selenoacetals.

4.2 Preparation of Selenium-Stabilized Carbanions

The selenium–lithium exchange with *n*-BuLi is fast for selenoacetals derived from aldehydes (e.g., **21**) and ketones (e.g., **22**). When selenoacetals are derived from long chain or hindered ketones (e.g., **23**), the exchange is slower. In these cases, the use of *s*-BuLi instead of *n*-BuLi enhances the exchange rate (Scheme 4.9) [38].

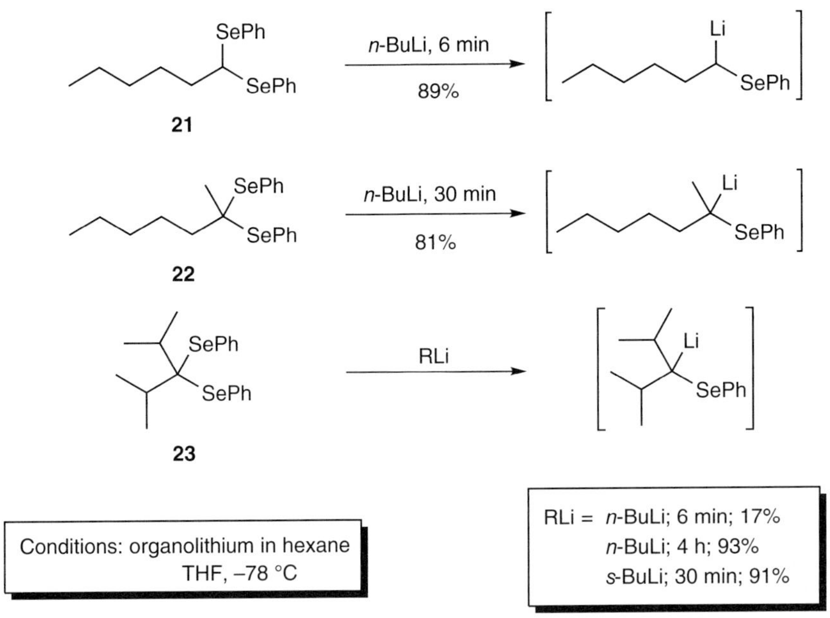

Scheme 4.9 Influence of the structure of the selenoacetal in the selenium–lithium exchange reaction.

Conditions: organolithium in hexane THF, –78 °C

RLi = *n*-BuLi; 6 min; 17%
n-BuLi; 4 h; 93%
s-BuLi; 30 min; 91%

The nature of the solvent is important in the selenium–lithium exchange reaction. The exchange reaction using *n*-BuLi in diethyl ether (Condition B) is slower than that in THF (Condition A). The reaction is more selective in diethyl ether. Only the selenoacetals derived from aromatic aldehydes (e.g., **24**) and bis(phenylseleno) methane **6** are completely cleaved by *n*-BuLi in diethyl ether [38]. (Phenylseleno) acetals derived from aliphatic aldehydes (e.g., **25**) are only partially cleaved in this mixture of solvents at –78 °C (Scheme 4.10), but they are completely cleaved in 30 min if the temperature is raised to –50 °C. Under the latter conditions (phenylseleno)acetals derived from ketones react slowly and methylselenoacetals derived from aliphatic aldehydes do not react at all [38].

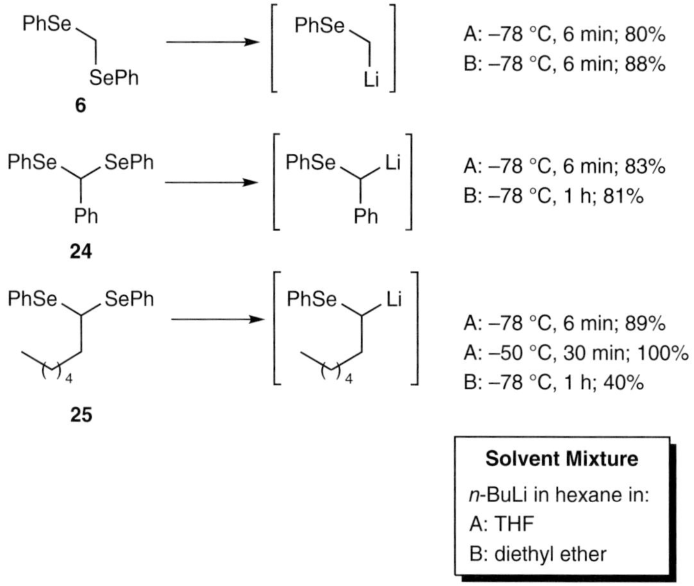

Scheme 4.10 Influence of the solvent in the cleavage of selenoacetals with *n*-BuLi at low temperatures.

4.2.3
Conjugate Addition of Organometallics to Vinyl- and Alkynylselenides

A third method to access selenium-stabilized carbanions is the conjugate addition of nucleophiles to vinyl-(**26**) and alkynylselenides (**10**). The presence of a second stabilizing group R^1 at the carbon atom in α-position of the vinylic selenide facilitates the carbanion formation (Scheme 4.11).

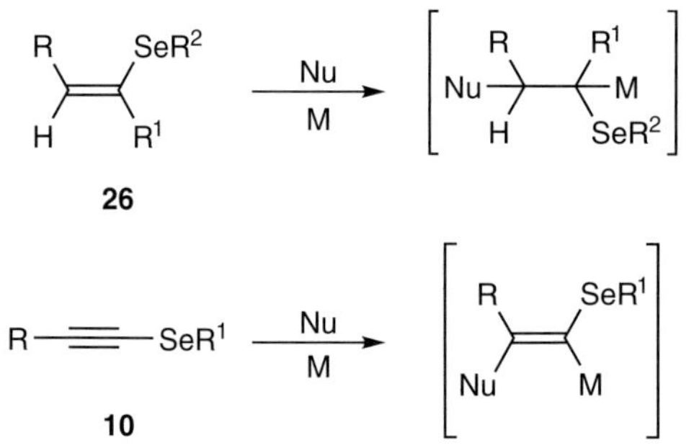

Scheme 4.11 Selenium-stabilized carbanion by conjugate addition of organometallics to vinyl- and alkynylselenides.

4.2 Preparation of Selenium-Stabilized Carbanions

Vinyl(phenyl)selenide in dimethoxymethane or diethyl ether at 0 °C reacts with n-, s- and t-alkyllithiums to give α-lithium selenides (Table 4.3, entries 1–3). MeLi, n-Bu$_2$CuLi and n-BuMgBr failed in the addition reaction [45]. However, 2-(phenylseleno)enones react with Me$_2$CuLi in diethyl ether to give the corresponding enolate (Table 4.3, entry 4) [46, 47]. Other nucleophiles are also able to perform a 1,4-addition to 2-phenylseleno enones, leading to the corresponding enolates (Table 4.3, entries 5 and 6) [48].

Table 4.3 Selenium-stabilized carbanions by alkyllithiums and organometallic addition to vinyl- (26) and alkynylselenides (10).

Entry	Selenium-stabilized carbanion	Conditions	Reference
1	(SePh, Li)	n-BuLi, dimethoxymethane, 0 °C, 1 h	[45]
2	(SePh, Li)	i-PrLi, diethyl ether, 0 °C, 1 h	[45]
3	(SePh, Li)	t-BuLi, diethyl ether, 0 °C, 1 h	[45]
4	OCuMeLi, SePh	Me$_2$CuLi, diethyl ether, −20 °C, 20 min	[46]
5	OLi, SePh, SiMe$_3$	Me$_3$SiLi, THF/HMPA, −78 °C, 5 min	[48]
6	OCu(Me)(CN)Li, SePh, SiMe$_3$	Me$_3$Si(Me)Cu(CN)Li$_2$, THF/HMPA, −78 °C, 5 min	[48]
7	Cu(CN)Li, SePh	n-BuCu(CN)Li, THF, −50 °C, 1 h	[49, 50]
8	SePh, ZrCp$_2$Cl	Cp$_2$Zr(H)Cl, THF, r.t.	[52]

(Continued)

Table 4.3 (Continued)

Entry	Selenium-stabilized carbanion	Conditions	Reference
9	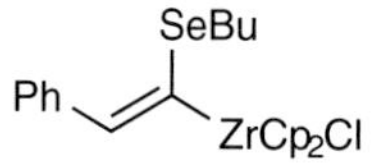 Ph–CH=C(SeBu)–ZrCp$_2$Cl	Cp$_2$Zr(H)Cl, THF, r.t.	[51]
10	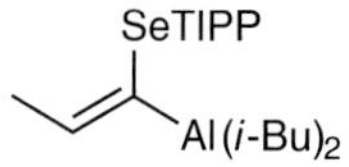 CH$_2$=C(SeTIPP)–Al(i-Bu)$_2$	DIBAL-H, hexane, 0 °C (1 h) to r.t. (3 h)	[41]
11	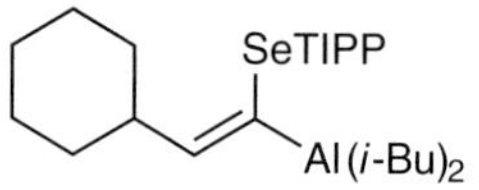 (cyclohexyl)CH$_2$–CH=C(SeTIPP)–Al(i-Bu)$_2$	DIBAL-H, hexane, 0 °C (1 h) to r.t. (3 h)	[41]
12	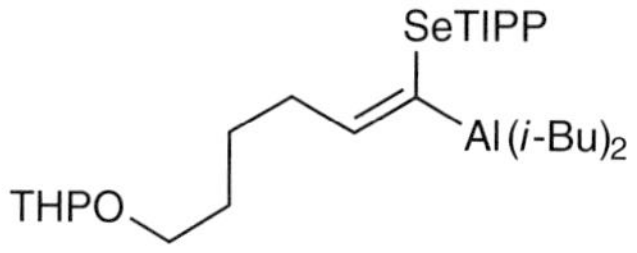 THPO–(CH$_2$)$_3$–CH=C(SeTIPP)–Al(i-Bu)$_2$	DIBAL-H, hexane, 0 °C (1 h) to r.t. (3 h)	[41]

TIPP = 2,4,6-triisopropylphenyl.

Reaction of (phenylseleno)ethyne with lithium butylcyanocuprate gives a vinylcyanocuprate (Table 4.3, entry 7) with a *trans* relationship between the phenylseleno group and the butyl group [49, 50].

In this same reaction category, we can include the hydrometallation of alkynylselenides **10**, which gives rise to an organometallic compound with the metal linked to the same carbon atom to which the organoseleno group is attached. The hydrozirconation and hydroalumination of alkynylselenides are among the transformations that can lead to this kind of nucleophilic species (Scheme 4.12).

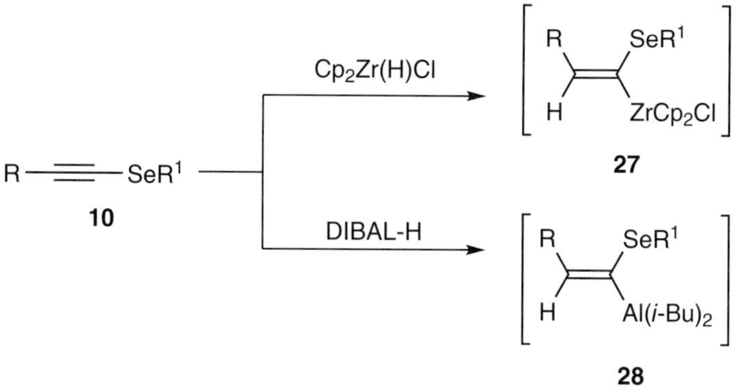

$$R\!\!\equiv\!\!SeR^1 \xrightarrow{Cp_2Zr(H)Cl} \left[\begin{array}{c} R\quad SeR^1 \\ \diagdown=\diagup \\ H\quad ZrCp_2Cl \end{array}\right] \mathbf{27}$$

$$\mathbf{10} \xrightarrow{DIBAL\text{-}H} \left[\begin{array}{c} R\quad SeR^1 \\ \diagdown=\diagup \\ H\quad Al(i\text{-}Bu)_2 \end{array}\right] \mathbf{28}$$

Scheme 4.12 Hydrozirconation and hydroalumination of alkynylselenides.

4.3 Reactivity of the Selenium-Stabilized Carbanions

The hydroalumination is stereo- and regioselective, leading to the desired organometallic compounds **28** in good yields [41, 42], especially when the 2,4,6-triisopropylphenyl group (TIPP) is employed as the R^1 group [41]. The hydrozirconation is less stereo- and regioselective and in some cases gives **27** as mixture of regio- and stereoisomers [50, 51]. It is worth mentioning that the products of the reaction of the vinylic zirconium and aluminum derivatives with iodine and organotellurenyl halides (Table 4.4, entries 20 and 21) are vinylic selenides bearing an iodine or tellurium atom linked to the same carbon atom to which selenium is attached, transforming these compounds into synthetic equivalents of selenium-stabilized vinyllithium (Table 4.3, entries 9–10). In Table 4.3 (entries 8–12) representative examples of vinylzirconium and vinylaluminum compounds prepared by hydrometallation of selenoalkynes **10** are given.

4.3 Reactivity of the Selenium-Stabilized Carbanions with Electrophiles and Synthetic Transformations of the Products

The selenium-stabilized carbanions react with electrophiles in a similar way as the analogous nonselenylated carbanionic species, leading to carbon–carbon and carbon–heteroatom bond formation. The selenium atom in these nucleophiles serves to direct the carbanion formation to a specific site, due to the ability of selenium to stabilize negative charges. A second function of the organoseleno group is to promote functional group transformations and/or new carbon–carbon bond formation after the reaction of the selenium-stabilized carbanion with electrophiles. A good example of these principles is the reaction sequence shown in Scheme 4.13 [41].

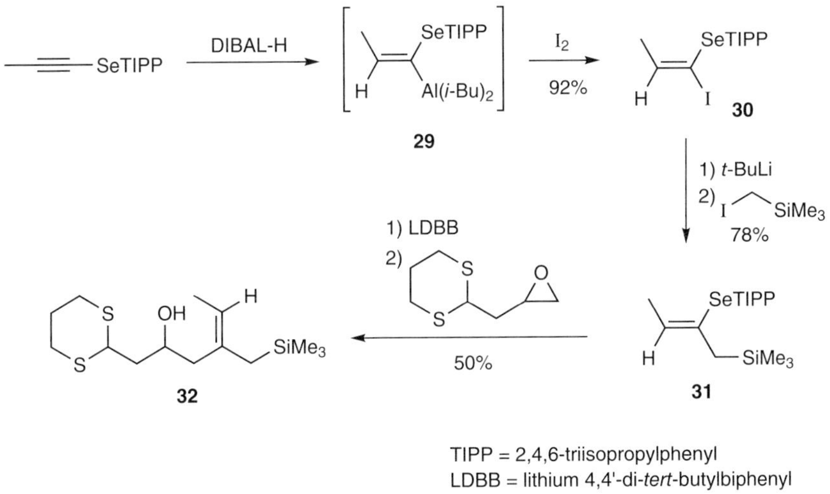

TIPP = 2,4,6-triisopropylphenyl
LDBB = lithium 4,4'-di-*tert*-butylbiphenyl

Scheme 4.13 Reaction sequence involving selenium-based transformations.

162 | *4 Selenium-Stabilized Carbanions*

Table 4.4 Reaction of selenium-stabilized carbanions with electrophiles.

Entry	Reaction	Reference
1	PhSe-C(SePh)(Me)-K + n-BuI →(THF, −78 °C) PhSe-C(SePh)(Me)-n-pentyl, 95%	[11]
2	PhSe-C(SePh)(Me)-K + i-PrI →(THF, −78 °C) PhSe-C(SePh)(Me)-CH(Me)(i-Pr), 70%	[11]
3	SePh-CH(Li)-C≡C-Li + furan-dioxolane-CH₂CH₂I →(DME, −78 °C) furan-dioxolane-CH₂CH₂-CH(SePh)-C≡CH, 90%	[15]
4	SePh-C(Me)(Li)-C(=O)-OLi + cyclohexene oxide →(THF, 0 °C to 60 °C, 24 h, H⁺) bicyclic lactone with SePh, 54%	[28]
5	SePh-CH(K)-C(=O)-(CH₂)₄-CH₃ + allyl-Br →(t-BuOH, 25 °C, 30 min) SePh-CH(allyl)-C(=O)-(CH₂)₄-CH₃, 87%	[26]
6	MeO-C(=O)-CH(Li)-SePh + cyclohexenone →(THF/HMPT, −45 °C, 6 h) 3-oxocyclohexyl-CH(SePh)-C(=O)-OMe, 73%	[27]

7	(lithiated lactone with SePh) + MeI	THF/HMPA, −78 °C to 0 °C	(methylated lactone) [32]
8	SeTIPP-vinyl-Li + ICH₂SiMe₃	THF, −78 °C, 1 h, 78%	SeTIPP-vinyl-CH₂SiMe₃ [41]
9	PhSeCH₂Li + PhC(O)Ph	THF, −35 °C, 14 h, 85%	PhSe-CH₂-C(OH)Ph₂ [3]
10	PhSeCH(SePh)Li + PhC(O)Ph	THF, −20 °C, 14 h, 60%	PhSe-CH(SePh)-C(OH)Ph₂ [3]
11	(PhSe)₂CHLi + PhCHO	THF, 86%	PhSe-C(SePh)(OH)-CH(OH)Ph [3]
12	(MeSe)CMe₂Li + 2,2,6,6-tetramethylcyclohexanone	THF, −45 °C, 3 h, 77%	(SeMe/HO adduct on cyclohexane) [55]

(Continued)

Table 4.4 (Continued)

Entry	Reaction	Reference
13	dioxolane-CH(SePh)-(CH₂)₄-Li + iPrCH₂CHO → THF, −78 °C, 30 min, 83% (7:3), product: dioxolane-CH(SePh)-(CH₂)₄-CH(OH)-CH₂iPr	[44]
14	cyclopropyl(PhSe)CH-Li + p-tolyl-C(O)-CH₃ → diethyl ether, −78 °C, 81%, product: cyclopropyl(PhSe)C(OH)-C(CH₃)(p-tolyl)	[56]
15	(MeSe)₂C(Me)-Li + 2-methyl-5-isopropenyl-cyclohex-2-enone → THF/HMPT, −78 °C, 12 min, 66%	[57]
16	MeSe-CH(Li)-C(O)OMe + cyclohex-2-enone → THF, −78 °C, 57%	[27]
17	nBuSe-C(Li)=CH-nBu + DMF → THF, −78 °C to r.t., 35 min, 70%, product: nBuSe-C(CHO)=CH-nBu	[43]

164 | 4 Selenium-Stabilized Carbanions

4.3 Reactivity of the Selenium-Stabilized Carbanions | 165

18 (EtO)$_2$P(=O)CH(SePh)Na + PhCHO $\xrightarrow[\text{−78 °C to reflux}]{\text{THF}}$ Ph–CH=CH–SePh (E) + Ph–CH=CH–SePh (Z)   80% (6:94)   [20]

19 (EtO)$_2$P(=O)C(SePh)(Na)C(=O)OEt + CH$_3$(CH$_2$)$_3$CHO $\xrightarrow[\text{86% (7:3)}]{\text{THF, reflux, 3 h}}$ EtO$_2$C–C(SePh)=CH–C$_4$H$_9$ (E/Z mixture)   [25]

20 (SeTIPP)(CH$_3$)C=CH–Al(i-Bu)$_2$ + I$_2$ $\xrightarrow[\text{92%}]{\text{THF, −78 °C, 30 min}}$ (SeTIPP)(CH$_3$)C=CH–I   [41]

21 Ph(Se^{n}Bu)C=CH–ZrCp$_2$Cl + BuTeBr $\xrightarrow[\text{92%}]{\text{THF, r.t., 20 min}}$ Ph(Se^{n}Bu)C=CH–TeBu   [43]

22 (SePh)(cyclohexylidene)C=Cu(CN)Li + allyl-Br $\xrightarrow[\text{75%}]{\text{THF, −50 °C, 75 min}}$ (SePh)(cyclohexylidene)C=CH–CH$_2$–CH=CH$_2$ [49]

TIPP = 2,4,6-triisopropylphenyl.

4 Selenium-Stabilized Carbanions

In the first step diisobutyl aluminum hydride adds regio- and stereoselectively to the triple bond of an alkynylselenide to give a vinyl organometallic **29**, which reacts with iodine in the second step to give vinyliodide **30**. In the next reaction, the vinyliodide **30** gives a selenium-stabilized vinyllithium through an iodine–lithium exchange, which is subsequently alkylated with iodomethyl(trimethylsilane) to compound **31**. In the last steps the selenium atom is reductively removed to give a vinyllithium using lithium 4,4′-di-*tert*-butylbiphenyl, which reacts with an epoxide to give the final product **32**. In the sequence from **30** to **32**, the stereochemistry of the double bond is retained.

4.3.1
Reaction of Selenium-Stabilized Carbanions with Electrophiles

As mentioned before, the reactivity of the selenium-stabilized carbanions parallels that of similar nonselenylated species. As such, these carbanionic entities react with alkyl halides, epoxides, carbonyl and carboxyl compounds [53], halogenating species, as well as with others electrophiles such as silyl, sulfenyl, selenenyl, and tellurenyl halides, to give the expected products. Exhaustive reviews [6] and book chapters [7] describe the capture of the *in situ* generated selenium-stabilized carbanions with electrophiles. Scheme 4.14 shows the most common reactions,

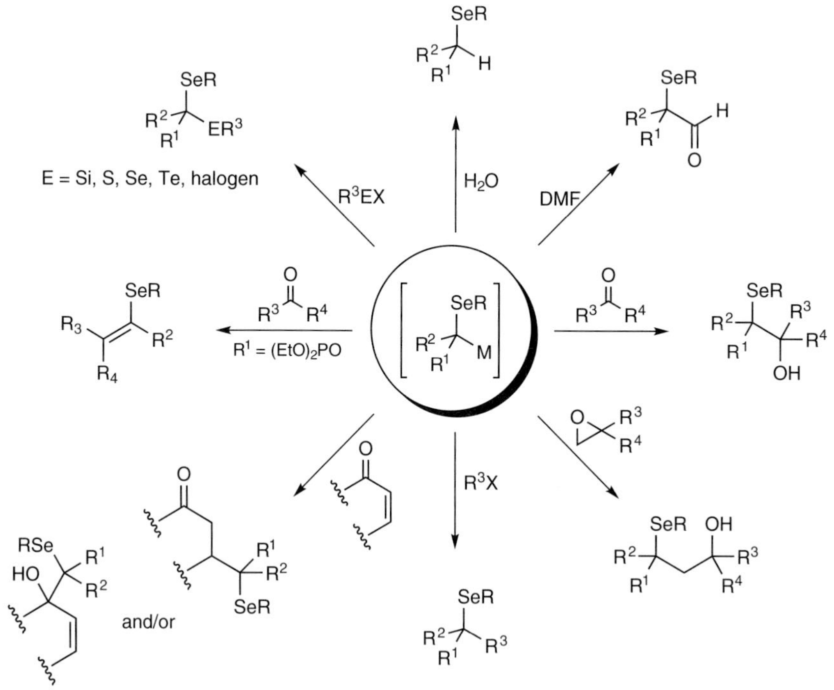

Scheme 4.14 Reactions of selenium-stabilized carbanions.

which have found use in synthesis. Table 4.4 gives selected examples of these transformations. As can be seen in Table 4.4, the reaction is fast in most cases and occurs at −78 °C in diethyl ether or in THF. Even hindered ketones react with α-selenium alkyllithiums in good yields (Table 4.4, entry 12). In the case of α,β-unsaturated carbonyl compounds, the 1,4-addition is preferred when the reaction occurs in THF/HMPA (Table 4.4, entry 15) and the 1,2-addition is predominant when the solvent is THF (Table 4.4, entry 16). When the R^1 group is a phosphonate or phosphinoxide, the product is a vinylic selenide (Table 4.4, entries 18 and 19) [54].

4.3.2
Selenium-Based Transformations on the Reaction Products of Selenium-Stabilized Carbanions with Electrophiles

The removal of the selenium atom from the reaction products of selenium-stabilized carbanions with electrophiles can be made in several ways, giving selenium-free compounds, which can be a target compound or a synthetic intermediate in total synthesis. Considering the reactions shown in Scheme 4.14 and in Table 4.4, the most synthetically useful transformations are the alkylation of 2-selenocarbonyl compounds (Table 4.4, entry 7) and the formation of β-hydroxyselenides (Table 4.4, entries 9–14).

The alkylated α-selenocarbonyl compounds are the precursors of α,β-unsaturated selenium-free carbonyl compounds, through an intermediate selenoxide (Scheme 4.15).

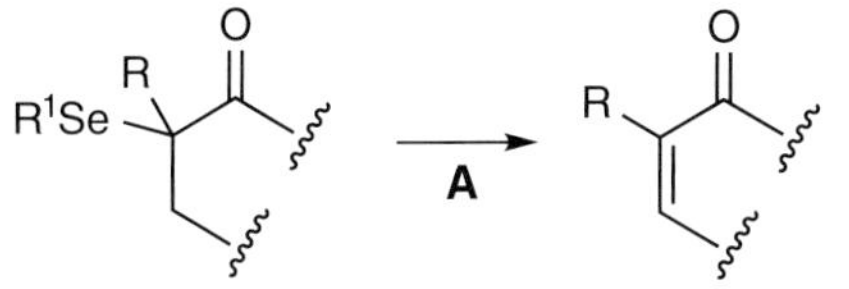

Scheme 4.15 Selenide as synthetic equivalent of an olefin.

The oxidation of a selenide, followed by a *syn* elimination of the selenoxide (Transformation **A**, Scheme 4.15) constitutes a key selenium-based synthetic transformation [6, 7] and deserves further comments here.

The β-hydroxyselenides (**33**) constitute versatile synthetic intermediates as shown in Scheme 4.16.

Transformation **B** (Scheme 4.16) constitutes a reductive removal of the selenium atom from the product. This transformation is not restricted to β-hydroxyselenides, other classes of organoselenium compounds can also be reductively deselenated [6, 7]. Transformation **C** (Scheme 4.16) shows the oxidative removal using selective oxidants, normally peroxides and peracids [6, 7], which oxidize the selenium atom leaving the hydroxy group unchanged. It must

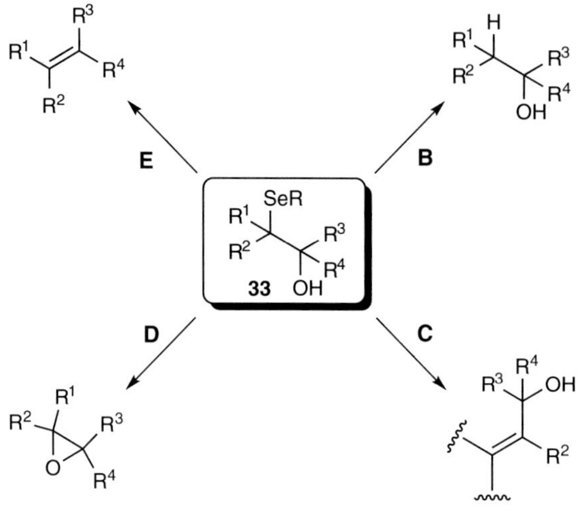

Scheme 4.16 Most important synthetic transformations of β-hydroxyselenides.

be emphasized that the selenoxide *syn* elimination gives exclusively the allylic alcohol; the carbinolic hydrogen is not eliminated [6, 7]. Transformation **D** (Scheme 4.16) consists in an epoxide forming reaction, which involves the attack of the hydroxy group to the neighboring carbon atom bearing the organoseleno group. The elimination reaction represented by transformation **E** (Scheme 4.16) can be performed by several reagents.

A particularly interesting transformation of β-hydroxyselenides is an acid catalyzed ring expansion, when the organoseleno group is attached to a carbocycle (Transformation **F**, Scheme 4.17) [55, 56].

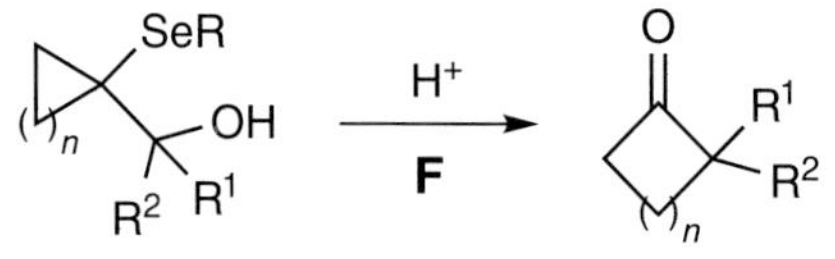

Scheme 4.17 Ring expansion of β-hydroxyselenides.

Following Schemes 4.14–4.17, Table 4.5 shows specific examples of Transformations **A** to **F**.

4.4
Stereochemical Aspects

Chiral selenium-stabilized carbanions should be valuable intermediates in organic synthesis, provided they are configurationally stable. However, several studies

Table 4.5 Selenium-based transformations on the products of the reaction between selenium-stabilized carbanions and electrophiles.

Entry	Transformation	Reference
Transformation of type **A**		
1	SePh lactone → H$_2$O$_2$ (30%), THF, AcOH, 0 °C, 30 min, 96% → methylene lactone	[58]
2	SePh iodolactone → H$_2$O$_2$ (30%), THF, AcOH, 0 °C, 30 min, 90% → methylene iodolactone	[28]
Transformation of type **B**		
3	SeMe alcohol → Bu$_3$SnH / benzene, 80 °C, 24 h, 97% → alcohol	[6c]
4	SeMe alcohol → Li / EtNH$_2$, −10 °C, 12 min, 75% → alcohol	[6c]
5	SeTIPP vinylsilane dithiane → 1) LDBB (2 equiv); 2) MeI; 3) BF$_3$·Et$_2$O; 4) epoxide, THF, −78 °C, 2 h, 50% → product	[41]

(*Continued*)

Table 4.5 (Continued)

Entry	Transformation	Reference
Transformation of type C		
6	(4-pentyl)CH(OH)CH(SeMe)(butyl) → O₃, CH₂Cl₂, Et₃N (2 equiv), −78 °C to 25 °C, 1.5 h, 96% → (E)-alkene-CH(OH)(CH(propyl)₂) with (CH₂)₄ chain	[59]
7	(4-pentyl)CH(OH)CH(SeMe)(CH₂)₅ → O₃, CH₂Cl₂, Et₃N (2 equiv), −78 °C to 25 °C, 3 h, 66% → allylic alcohol with (CH₂)₄ and (CH₂)₅	[59]
8	(6-alkyl)CH(OH)CH(SeMe)Me → t-BuOOH/Al₂O₃, THF, 55 °C, 96% → allylic alcohol with (CH₂)₆	[60]
Transformation of type D		
9	Ph-CH=CH-CH(OH)CH(SeMe)Me → 1) MeI, 25 °C, 4 h; 2) t-BuOK, DMSO, 25 °C, 1 h, 85% (2 steps) → Ph-CH=CH-epoxide-Me	[61]

4.4 Stereochemical Aspects | 171

10	(β-hydroxy tert-butylcyclohexyl SeMe)	1) CH$_3$OSO$_2$F, diethyl ether, −78 °C, 1.5 h 2) KOH (10% aq.), 0 °C, 1 h 95%	(spiro epoxide product) [55]

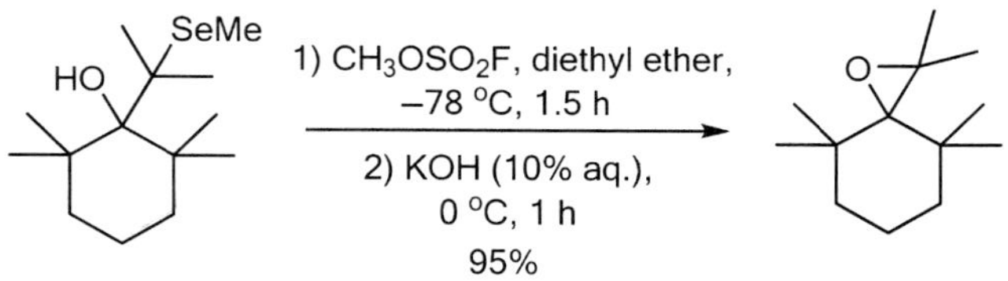

Transformation of type E

11	(4-tert-butylcyclohexanol with CH$_2$SePh)	p-TSA pentane reflux, 18 h 68%	(4-tert-butyl methylenecyclohexane) [62]
12	(cyclopropyl SeMe, OH, C$_{10}$H$_{21}$, H)	CDI (2 equiv) toluene 110 °C, 15 h 65%	(cyclopropylidene with C$_{10}$H$_{21}$, H) [63]
13	(cyclopropyl SeMe, OH, Bn, Ph)	PI$_3$ (2 equiv) Et$_3$N, CH$_2$Cl$_2$ 20 °C, 2 h 75%	(cyclopropylidene with Bn, Ph) [63]

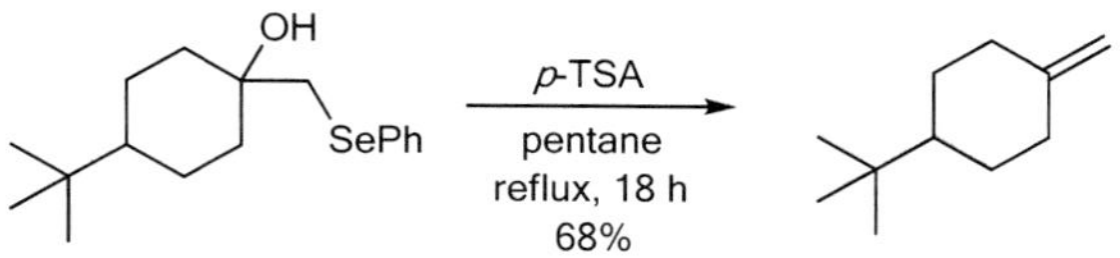

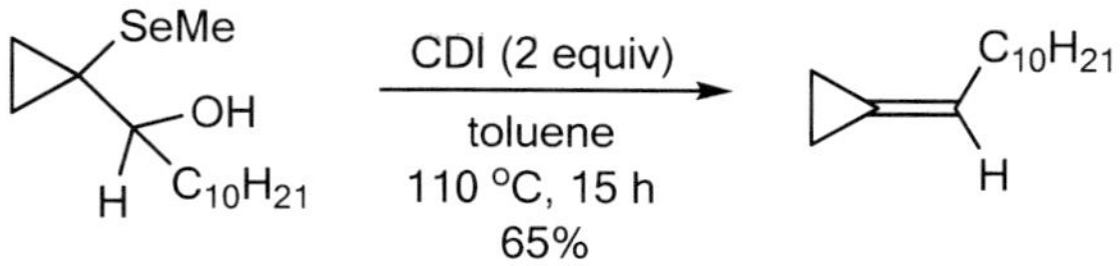

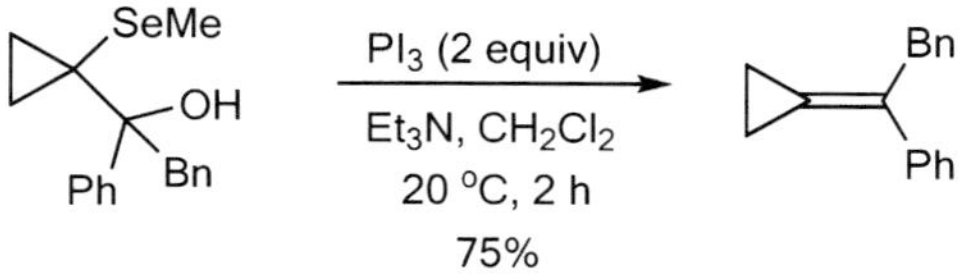

Transformation of type F

14	(cyclopropyl PhSe, HO, p-tolyl, Me)	p-TSA benzene / H$_2$O 80 °C, 12 h 80%	(cyclobutanone with Me, p-tolyl) [56]

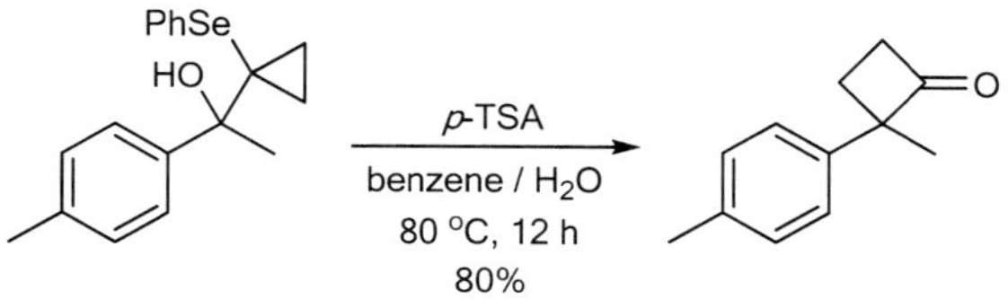

(Continued)

172 | 4 Selenium-Stabilized Carbanions

Table 4.5 (Continued)

Entry	Transformation	Reference
15	SeMe-substituted cyclobutanol with p-tolyl group → 2,2-dimethyl-3-(p-tolyl)cyclopentanone; CH$_3$OSO$_2$F, diethyl ether, 20 °C, 1 h, 82%	[55]
16	1-(PhSe)-1-(tert-butyl)cyclopentanol → 2,2-dimethylcyclohexanone; AgBF$_4$, CH$_2$Cl$_2$, 20 °C, 5 h, 75%	[64]
17	PhSe/HO-substituted cyclopentyl–cyclohexyl → spirocyclic ketone; AgBF$_4$, CH$_2$Cl$_2$, 20 °C, 5 h, 50%	[64]

LDBB = lithium 4,4′-di-*tert*-butylbiphenyl.
CDI = carbonyl diimidazole.

indicated that these organometallic species are configurationally unstable and equilibrate at low temperature [65–70].

4.4.1 Cyclic Selenium-Stabilized Carbanions

Sequential treatment of selenoacetal **34** with *n*-BuLi in THF at −78 °C, followed by capture of the selenium-stabilized carbanion with electrophiles gave a mixture of **35** and **36** in a 92 : 8 ratio in favor of the axially substituted product **35**. As the electrophiles presented very different reactivities and steric demands, it was concluded that the 92 : 8 represents the equilibrium mixture of the axial/equatorial organolithiums **37** and **38** (Scheme 4.18) [65].

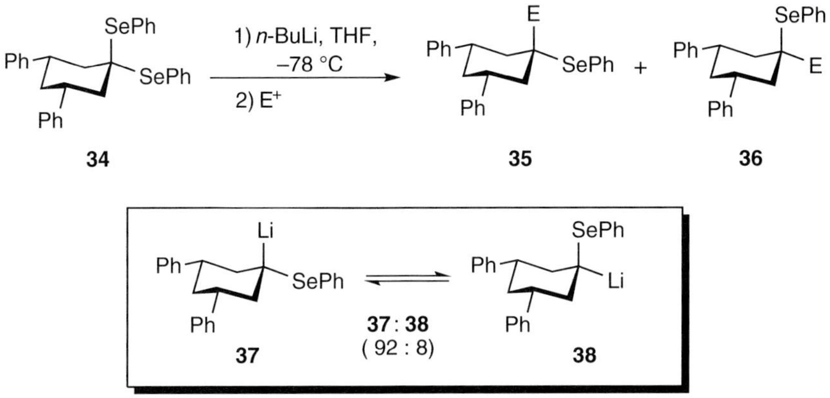

Scheme 4.18 Equilibrium mixture of axial/equatorial lithium selenophenyl-cyclohexanes.

When compound **34** was cleaved with *t*-BuLi in the presence of a large excess of Me_3SiCl, a mixture of **39** and **40** in a 98 : 2 ratio was formed, the axially substituted **39** predominating, which indicates a kinetic preference for the selenium–lithium exchange with the axial group (Scheme 4.19) [65].

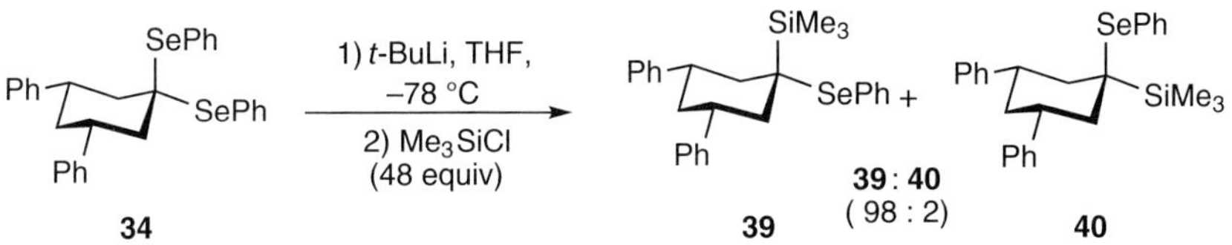

Scheme 4.19 Kinetic preference for the axial organolithium derivative.

The opposite kinetic preference was observed when the cyclic selenoacetal **41** was reacted under the same conditions (Scheme 4.20) [65].

4 Selenium-Stabilized Carbanions

Scheme 4.20 Kinetic preference for the equatorial organolithium derivative.

In this case, the axially substituted product **42** was the minor product (3%), the equatorial selenium atom being preferentially exchanged, leading to the equatorially substituted product **43** (97%).

However, the sequential addition of *t*-BuLi, followed by capture of the selenium-stabilized carbanion with chlorotrimethylsilane at different equilibration times, led to the preferential formation of the axially substituted product **42** (96%) after 60 min equilibration time, confirming that the axial organolithium **44** is thermodynamically more stable than the equatorial intermediate **45** (Scheme 4.21) [65].

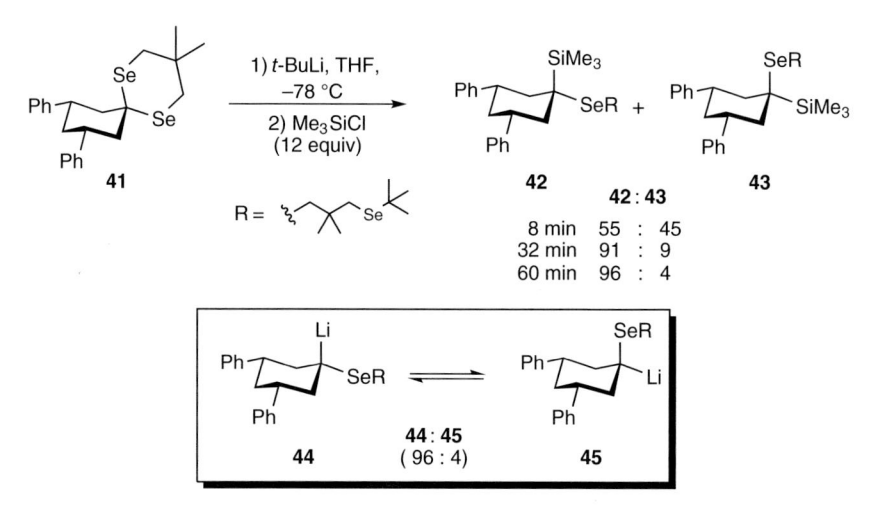

Scheme 4.21 Thermodynamic preference for the axial organolithium derivative.

Similar results were obtained when selenoacetal **46** was treated with *n*-BuLi in THF at −78 °C, followed by protonation leading to a mixture of **47** and **48**. In this case too, the equilibration favors the axial organolithium **49** over the equatorial species **50** (Scheme 4.22) [66].

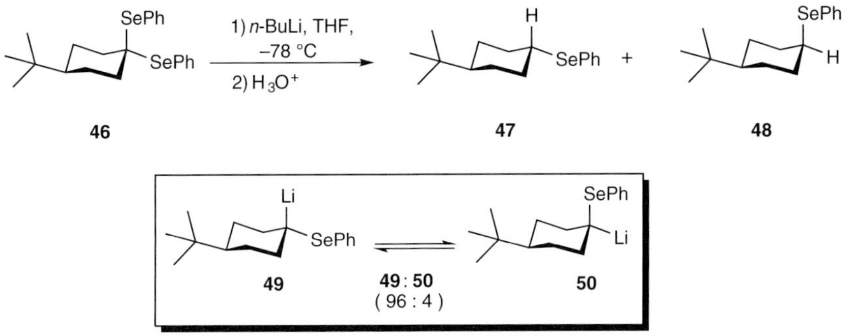

Scheme 4.22 Thermodynamic preference for the axial organolithium derivative.

A different behavior was observed for 1,3-diselenanes **51–55**. Herein, the metal atom (Li, K) prefers the equatorial position (Scheme 4.23) [19].

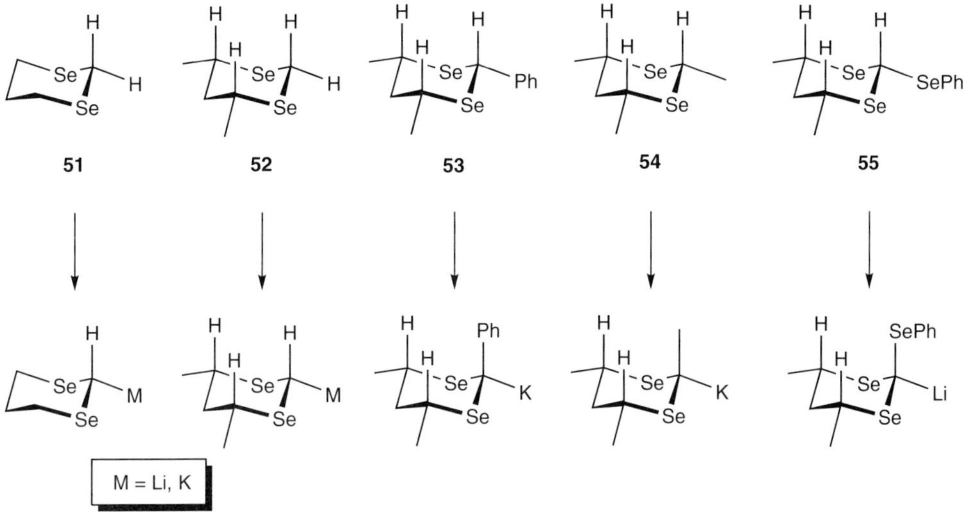

Scheme 4.23 Equatorial preference for the organometallics derived from 1,3-diselenanes by deprotonation with LDA/KDA.

Compound **51** reacts with both LDA and KDA to give the corresponding carbanions, which on reaction with electrophiles give the equatorial derivatives. The similar behavior is observed for compound **52**. Compound **53** does not react with LDA (see Section 2.1, Scheme 4.6) it requires the use of KDA for the equatorial products to be formed on reaction of the carbanion with electrophiles. The similar behavior is observed for compound **54**. Compound **55** is deprotonated with LDA giving the equatorial product. The three last examples indicate that the originally formed organometallics with the metal atom at the axial position isomerize to the equatorial isomers.

4.4.2
Acyclic Selenium-Stabilized Carbanions

Chiral acyclic selenium-stabilized lithium carbanions racemize rapidly at low temperature. However, addition of chiral ligands to the lithium carbanions generates diastereoisomeric complexes, which reacts with electrophiles with some degree of stereoselection as illustrated in Scheme 4.24 [67].

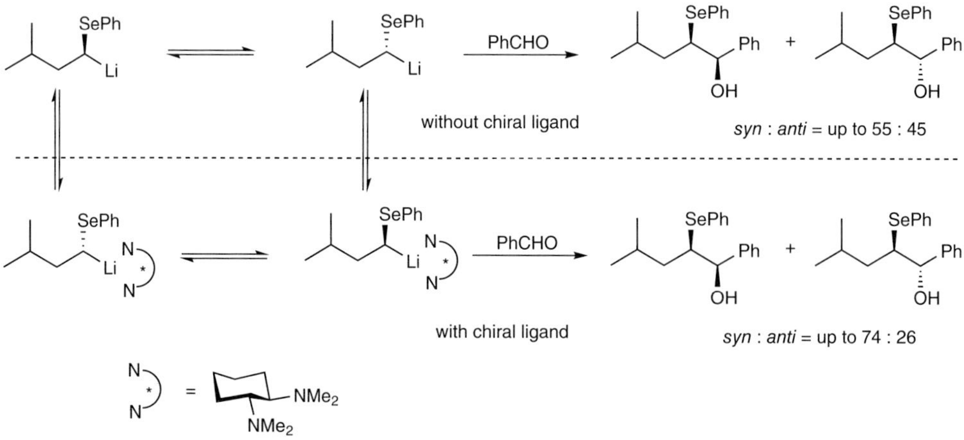

Scheme 4.24 Stereoselection of addition of a selenium-stabilized carbanion to benzaldehyde promoted by a chiral ligand.

4.5
Application of Selenium-Stabilized Carbanions in Total Synthesis

The final proof of the usefulness of a synthetic methodology is its application to the preparation of compounds with a practical appeal. The selenium-stabilized carbanions offer a wide range of possibilities toward this end, in view of the peculiar reactivity associated with organoselenium compounds. The selenium atom in these molecules allow the introduction of useful functionalities into the organic skeleton or even the formation of new carbon–carbon bonds, via reaction of

4.5 Application of Selenium-Stabilized Carbanions in Total Synthesis

selenium-stabilized carbanions with electrophiles. In the following sections, we will present some representative examples of synthetic applications of reactions involving selenium-stabilized carbanions.

4.5.1
Examples Using Alkylation Reactions of Selenium-Stabilized Carbanions

A total synthesis of (±)-Vernomenin (**56**) employed the alkylation of a selenium-stabilized carbanion to introduce a prenyl group and an unsaturation in α-position to a carbonyl group (Scheme 4.25) [71].

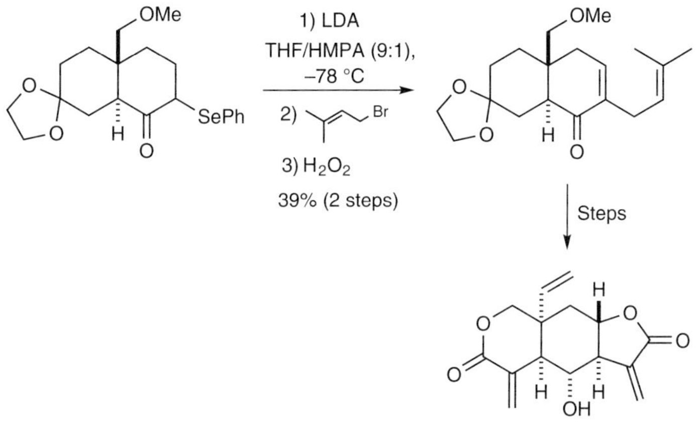

Scheme 4.25 One-pot procedure for the introduction of a prenyl group and an unsaturation in α-position to a carbonyl group in the total synthesis of (±)-Vernomenin (**56**).

The allergenic lactone (±)-Frulanolide (**57**) was prepared in few steps featuring the alkylation of the selenium-stabilized carbanion **58** with the allyl halide **59**, followed by an iodolactonization of **60**. Deiodination of **61** with DBU and oxidative removal of the phenylseleno group led to (±)-Frulanolide (**57**) as shown in Scheme 4.26 [72].

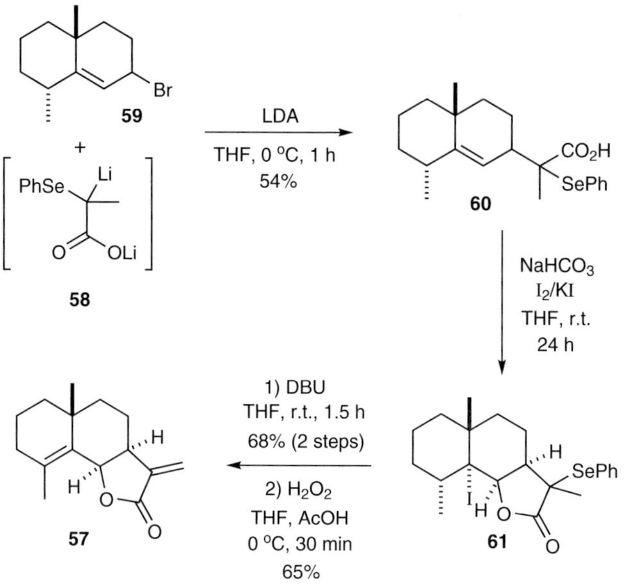

Scheme 4.26 Alkylation of a selenium-stabilized carbanion with an allyl bromide in the key step of (±)-Frulanolide (**57**) synthesis.

The antitumor agent Asperdiol (**62**) was synthesized in a convergent manner. The two fragments were coupled by an alkylation of the lithium dianion **63** with the bromide **64** (Scheme 4.27) [73].

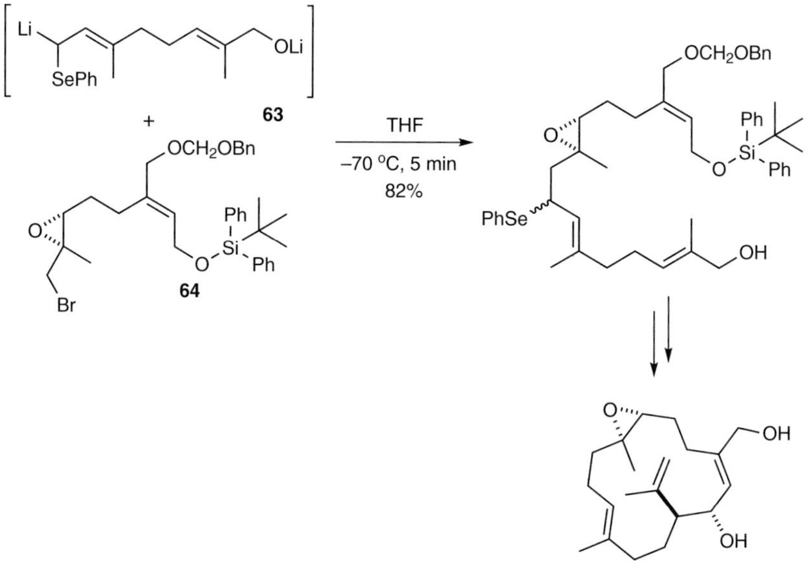

Scheme 4.27 Alkylative coupling of two fragments in the total synthesis of Asperdiol (**62**).

4.5 Application of Selenium-Stabilized Carbanions in Total Synthesis | 179

Butenolides **65–68** were prepared starting from epoxy triflates **69** and **70**, according to Scheme 4.28 [73]. The first step consists in the alkylation of the selenium-stabilized carbanion **71** by the epoxide-opening reaction of **69** and **70**. Upon work-up the hydroxy carboxylic acids cyclize to lactones **72–75**. Separation of lactone **72** from **73** and lactone **74** from **75**, followed by an oxidative elimination led to *exo-* or

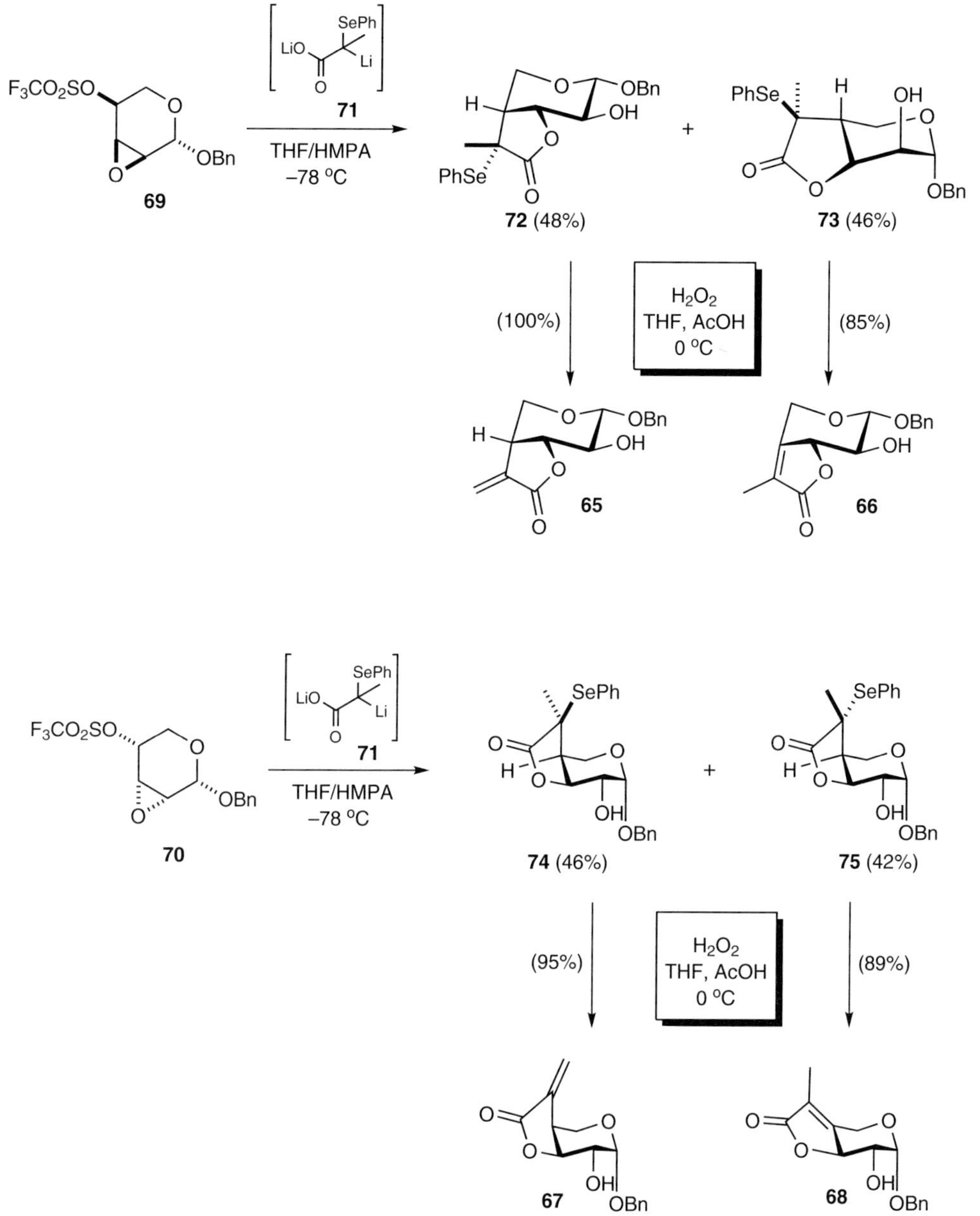

Scheme 4.28 Butenolides from alkylation of a selenium-stabilized carbanion by an epoxide opening reaction.

endo-cyclic butyrolactones **65–68**. In the first case (lactones **72** and **74**), the phenylseleno group is *trans*-oriented to H-4 and in the second case (lactones **73** and **75**) it is *cis*-oriented to H-4 (Scheme 4.28) [74].

4.5.2
Examples Using the Addition of Selenium-Stabilized Carbanions to Carbonyl Compounds

The natural product α-Cuparenone (**76**) was prepared according to Scheme 4.29 [56]. This synthetic strategy features ring expansions assisted by selenium. Use of the selenomethyl group led to better yields than the use of the phenylseleno group.

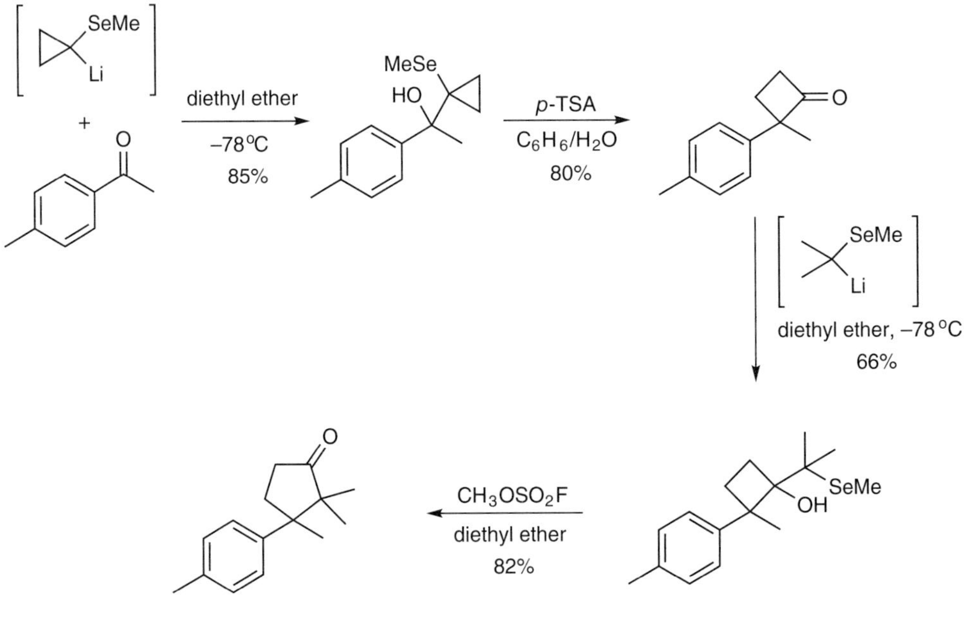

Scheme 4.29 Synthesis of a natural product through the addition of selenium-stabilized carbanions to ketones followed by ring expansion of the resulting β-hydroxyselenides.

Permethyl cyclohexane (**77**) was prepared by a sequence of a selenium-stabilized carbanion addition to ketones, followed by ring-expansion of the resulting β-hydroxy selenides (Scheme 4.30) [75].

Scheme 4.30 Synthesis of permethyl cyclohexane (**77**) through sequential addition of a selenium-stabilized carbanion to ketones followed by ring expansion.

A similar sequence was employed to prepare compound **78** (Scheme 4.31) [76, 77].

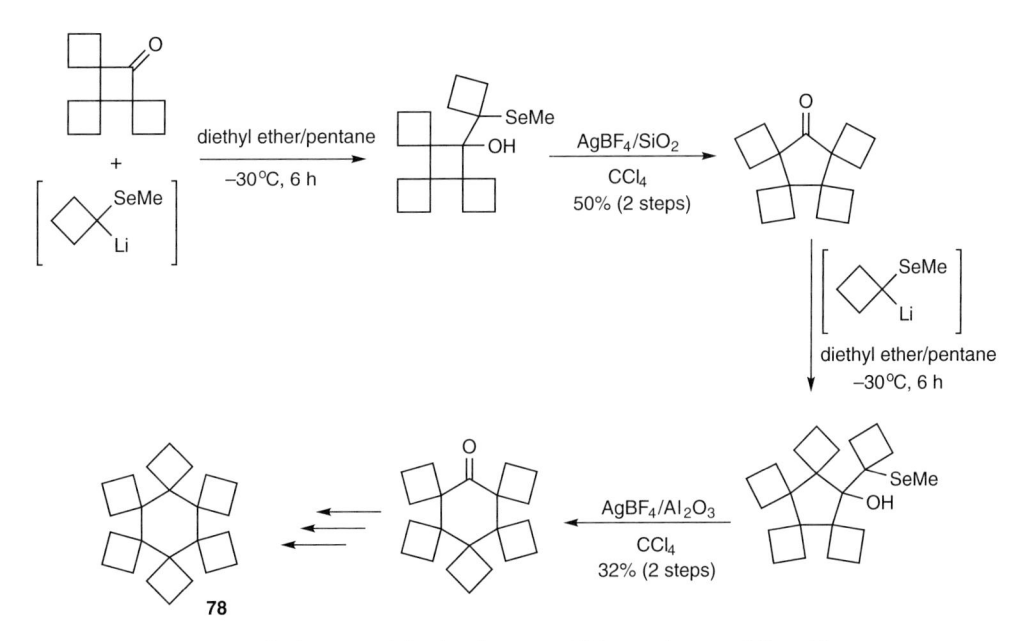

Scheme 4.31 Sequences of selenium-stabilized carbanions additions to ketones, followed by ring-expansions in the synthesis of a non-natural product.

An alkylated taxane model system **79** was prepared making use of a selenium-stabilized carbanion addition to an aldehyde to obtain the highly substituted diene **80** (Scheme 4.32) [78].

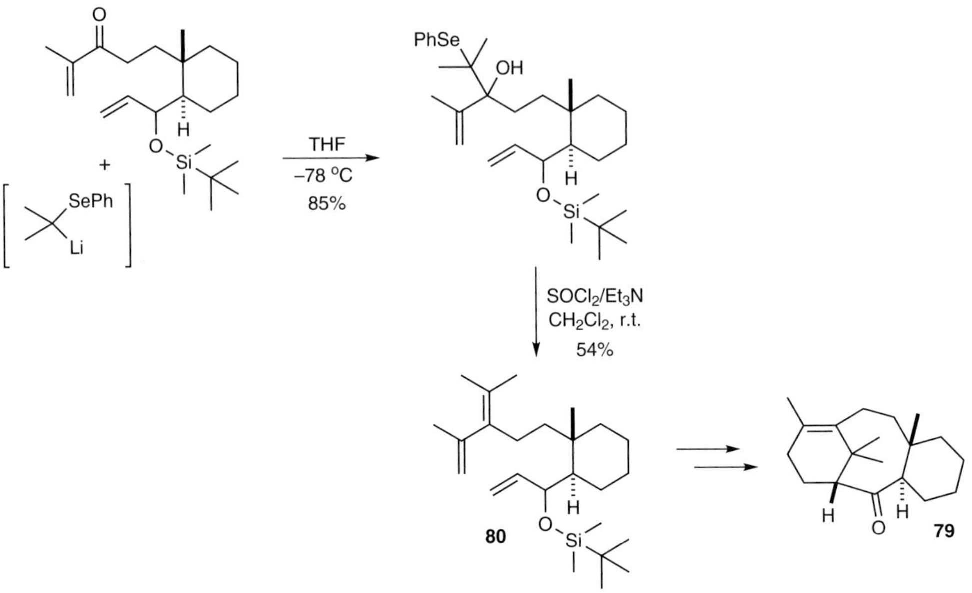

Scheme 4.32 Synthesis of a taxane model system.

A general approach to the total synthesis of *Lauraceae* lactones **81–84** was developed, making use of selenium-stabilized carbanion (Scheme 4.33) [79]. The carbanion addition and the selenoxide *syn*-elimination were performed in a one-pot procedure. After chromatographic separation, the *Z*- and *E*-methyl esters were hydrolyzed to the corresponding carboxylic acids **85–88**, which were then transformed into natural products **81–84**.

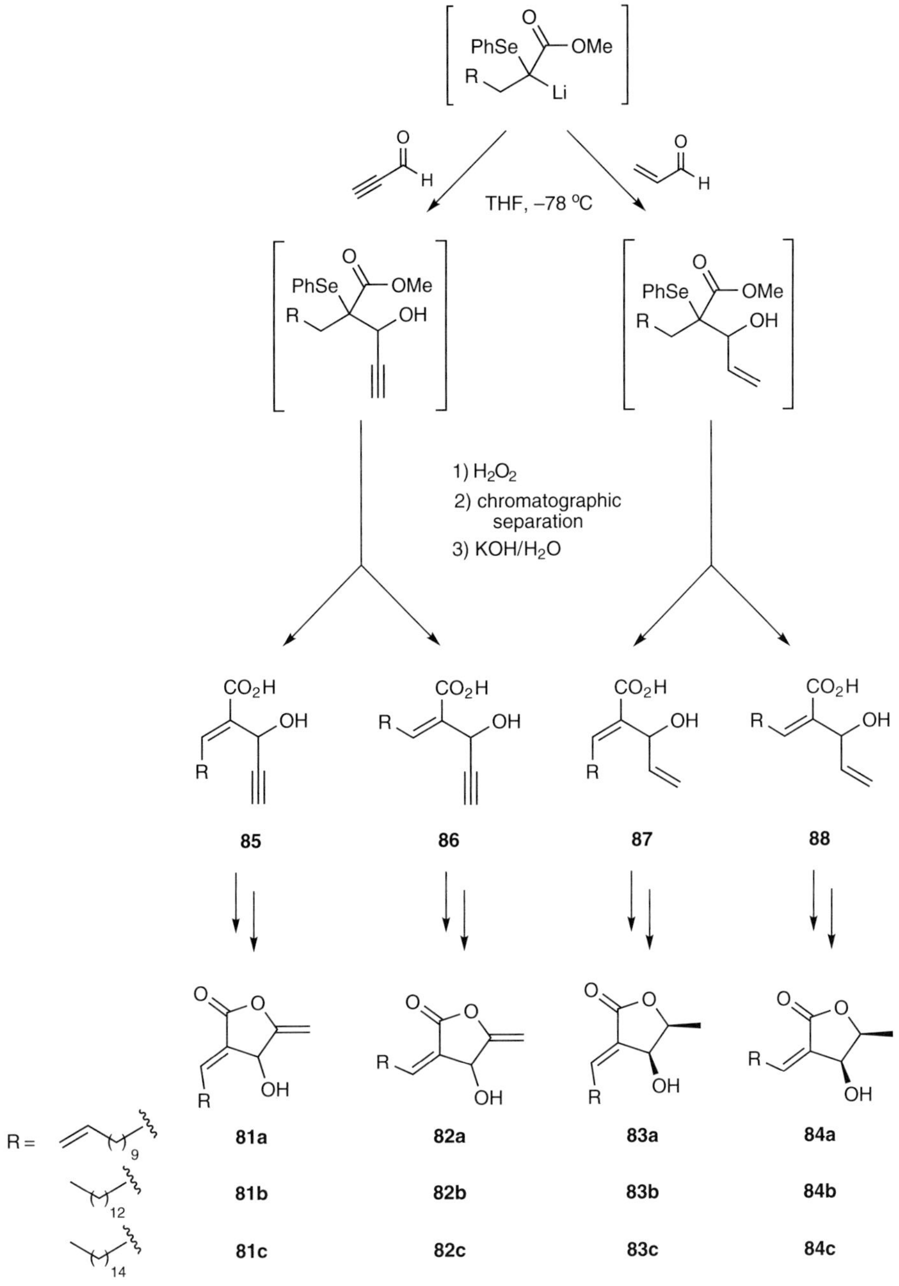

4.5 Application of Selenium-Stabilized Carbanions in Total Synthesis

Scheme 4.33 Synthesis of *Lauraceae* lactones.

184 | *4 Selenium-Stabilized Carbanions*

The sesquiterpene Khusimone (**89**) was synthesized featuring a radical cyclization in the key step. The free radical **90** was generated by homolytic cleavage of selenide **91**, which was obtained by addition of the selenium-stabilized carbanion **92** to aldehyde **93**, followed by functional transformations on the hydroxyselenide **94** (Scheme 4.34) [80].

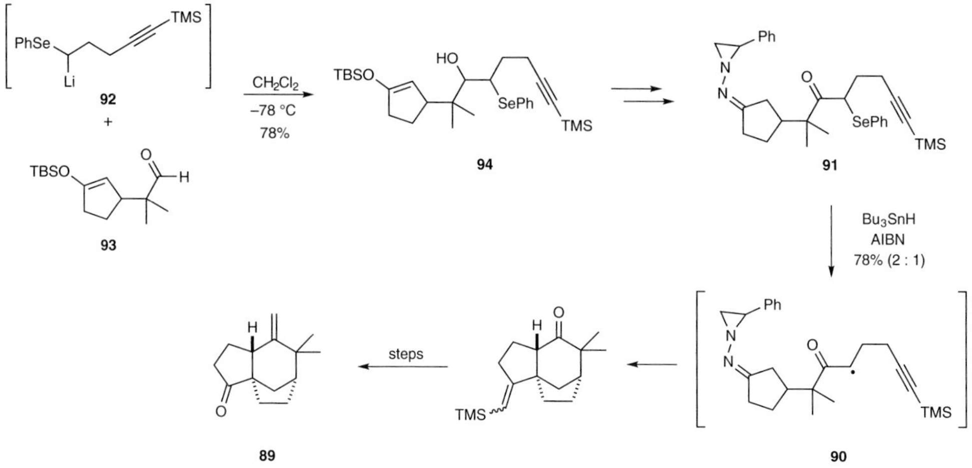

Scheme 4.34 Synthesis of Khusimone (**89**) by a radical cyclization.

The same strategy was employed for the synthesis of *d,l*-Pentalenene (**95**) (Scheme 4.35) [81].

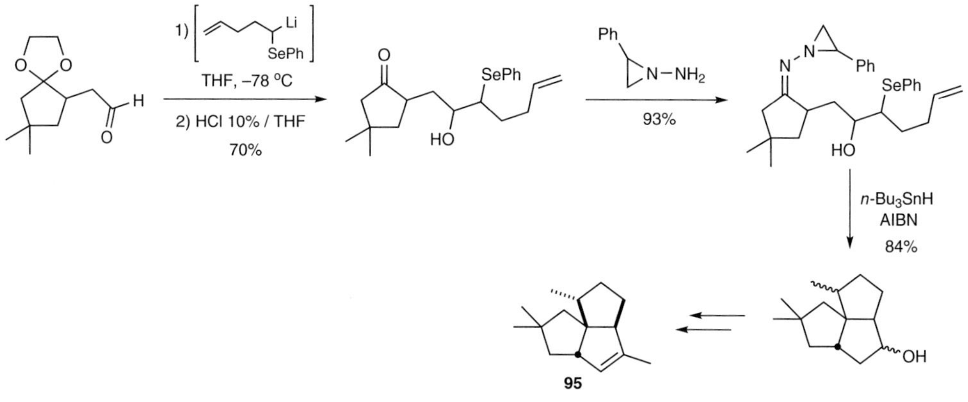

Scheme 4.35 Synthesis of *d,l*-Pentalenene (**95**) by a radical cyclization.

4.5.3
Examples Using 1,4-Addition of Selenium-Stabilized Carbanions to α,β-Unsaturated Carbonyl Compounds

An 1,4-addition of phenylselenomethyllithium in THF/HMPA to the α,β-unsaturated lactone **96** led to compound **97**. The phenylseleno group was

4.5 Application of Selenium-Stabilized Carbanions in Total Synthesis | 185

oxidatively removed after several steps to give compound **98**, which after two further steps gave (4R,5R)-Antillatoxin (**99**), an ichthyotoxin metabolite isolated from the marine cyanobacterium *Lyngbya majuscule* (Scheme 4.36) [82].

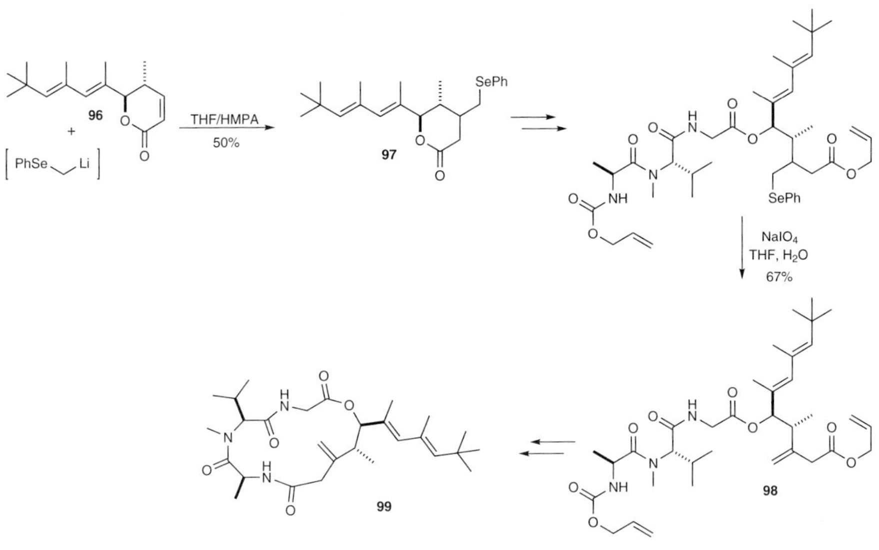

Scheme 4.36 Synthesis of (4R,5R)-Antillatoxin (**99**).

The 1,4-addition of phenylselenomethyllithium to the α,β-unsaturated lactone **100** was the key step in the synthesis of (±)-Pederin (**101**). The phenylseleno group, introduced in the first step resisted to several steps, being removed only in the penultimate step of the total synthesis (Scheme 4.37) [83].

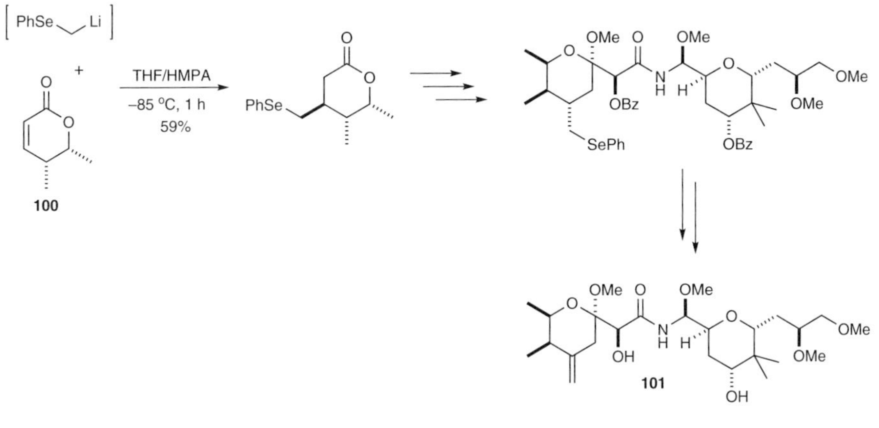

Scheme 4.37 Total synthesis of (±)-Pederin (**101**).

4 Selenium-Stabilized Carbanions

The antifungal agent (−)-Avenaciolide (**102**) was prepared in an enantioselective way, featuring a tandem 1,4-addition of the selenium-stabilized anion **103** to the enone **104**, followed by capture of the intermediate enolate by iodine. Cyclization of **105** followed by oxidative removal of the phenylseleno group gave the target compound **102** (Scheme 4.38) [84].

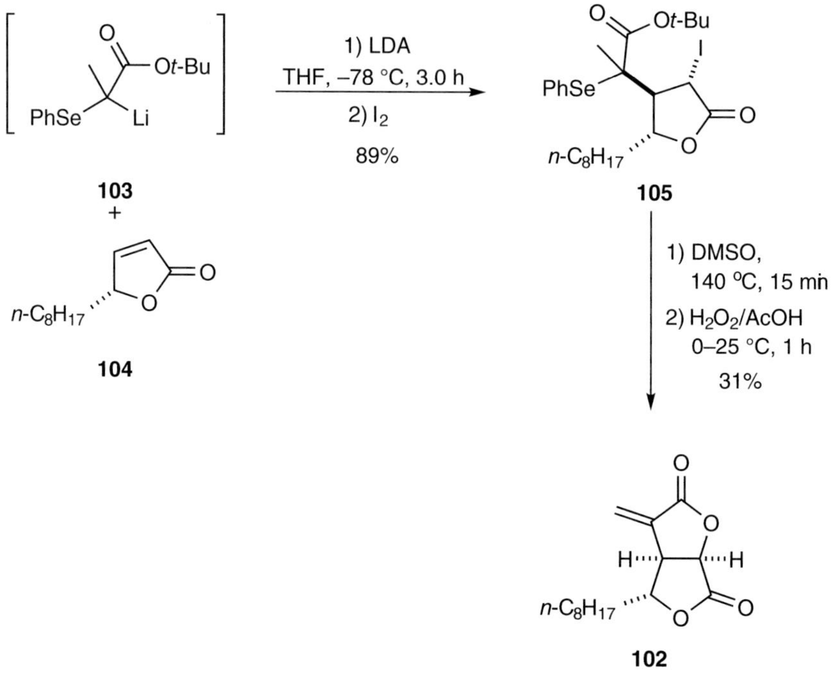

Scheme 4.38 Enantioselective synthesis of (−)-Avenaciolide (**102**).

4.6
Conclusion

Practical and high yields routine procedures for the preparation of selenium-stabilized carbanions from easily prepared precursors are now available. The reactivity of such species with electrophiles is well understood. The reaction products of these transformations constitute valuable precursors of new functionalities and carbon skeletons, which can be explored for the assembly of more complex organic compounds.

References

1 Gilman, H. and Bebb, R.L. (1939) *J. Am. Chem. Soc.*, **61**, 109–112.
2 Gilman, H. and Webb, F.J. (1949) *J. Am. Chem. Soc.*, **71**, 4062–4066.
3 Seebach, D. and Peleties, N. (1969) *Angew. Chem. Int. Ed. Engl.*, **8**, 450–451.
4 Seebach, D. and Peleties, N. (1972) *Chem. Ber.*, **105**, 511–520.
5 Seebach, D., Meyer, N., and Beck, A.K. (1977) *Liebigs Ann. Chem.*, 846–858.
6 (a) Clive, D.L.J. (1978) *Tetrahedron*, **34**, 1049–1132; (b) Reich, H.J. (1979) *Acc. Chem. Res.*, **12**, 22–30; (c) Krief, A. (1980) *Tetrahedron*, **36**, 2531–2640.
7 (a) Reich, H.J. (1978) *Oxidation in Organic Chemistry, Part C* (ed. W.S. Trahanowsky), Academic Press, New York; (b) Nicolaou, K.C. and Petasis, N.A. (1984) *Selenium in Natural Products Synthesis*, CIS, Philadelphia; (c) Paulmier, C. (1986) *Selenium Reagents and Intermediates in Organic Chemistry*, Pergamon Press, Oxford; (d) Patai, S. and Rappoport, Z. (1986 and 1987) *The Chemistry of Organic Selenium and Tellurium Compounds*, vol. 1 and 2, John Wiley & Sons, Ltd, Chichester, UK; (e) Liotta, D. (1987) *Organoselenium Chemistry*, John Wiley & Sons, Inc., New York; (f) Krief, A. and Hevesi, L. (1988) *Organoselenium Chemistry*, Springer, Berlim; (g) Gough, M.J. and Stelle, J. (1995) *Comprehensive Organic Functional Group Transformations*, vol. 4 (ed. G.W. Kirby), Pergamon Press, Oxford; (h) Chieffi, A. and Comasseto, J.V. (1999) *Organoselenium Chemistry: A Practical Approach* (ed. T.G. Back), Oxford University Press, New York; (i) Ponthieux, S. and Paulmier, C. (2000) *Organoselenium Chemistry – Modern Developments in Organic Synthesis* (ed. T. Wirth), Springer, Berlin; (j) Wirth, T. (2007) *Comprehensive Organometallic Chemistry III*, vol. 9 (eds R.H. Crabtree and D.M.P. Mingos), Elsevier, Oxford.
8 Reich, H.J., Willis, W.W., and Clark, P.D. (1981) *J. Org. Chem.*, **46**, 2775–2784.
9 Van Ende, D., Cravador, A., and Krief, A. (1979) *J. Organomet. Chem.*, **177**, 1–4.
10 Reich, H.J., Shah, S.K., and Chow, F. (1979) *J. Am. Chem. Soc.*, **101**, 6648–6656.
11 Raucher, S. and Koolpe, G.A. (1978) *J. Org. Chem.*, **43**, 3794–3796.
12 Reich, H.J., Chow, F., and Shah, S.K. (1979) *J. Am. Chem. Soc.*, **101**, 6638–6648.
13 Reich, H.J. and Shah, S.K. (1977) *J. Org. Chem.*, **42**, 1773–1776.
14 Reich, H.J. and Shah, S.K. (1977) *J. Am. Chem. Soc.*, **99**, 263–265.
15 Reich, H.J., Shah, S.K., Gold, P.M., and Olson, R.E. (1981) *J. Am. Chem. Soc.*, **103**, 3112–3120.
16 Brandt, C.A., Comasseto, J.V., Nakamura, W., and Petragnani, N. (1983) *J. Chem. Res. (S)*, 156–157.
17 Sachdev, K. and Sachdev, H.S. (1976) *Tetrahedron Lett.*, **17**, 4223–4226.
18 Gröbel, B.T. and Seebach, D. (1977) *Chem. Ber.*, **110**, 852–866.
19 Krief, A. and Defrère, L. (1996) *Tetrahedron Lett.*, **37**, 2667–2670.
20 Comasseto, J.V. and Petragnani, N. (1978) *J. Organomet. Chem.*, **152**, 295–304.
21 Coutrot, P., Grison, C., and Youssefi-Tabrizi, M. (1987) *Synthesis*, 169–170.
22 Xu, W.M., Tang, E., and Huang, X. (2004) *Synthesis*, 2094–2098.
23 Huang, X. and Xu, W. (2002) *Tetrahedon Lett.*, **43**, 5495–5497.
24 Bryce, M.R., Moore, A.J., Lorcy, D., Dhindsa, A.S., and Robert, A. (1990) *J. Chem. Soc., Chem. Commun.*, 470–472.
25 Silveira, C.C., Nunes, M.R.S., Wendling, E., and Braga, A.L. (2001) *J. Organomet. Chem.*, **623**, 131–136.
26 Takahashi, T., Nagashima, H., and Tsuji, J. (1978) *Tetrahedron Lett.*, **19**, 799–802.
27 Lucchetti, J. and Krief, A. (1978) *Tetrahedron Lett.*, **19**, 2697–2700.
28 Petragnani, N. and Ferraz, H.M.C. (1978) *Synthesis*, 476–478.
29 Sakakibara, T., Manandhar, M., and Ishido, Y. (1983) *Synthesis*, 920–923.
30 Jahn, U. and Curran, D.P. (1997) *Synlett*, 565–567.
31 Grieco, P.A. and Yokoyama, Y. (1977) *J. Am. Chem. Soc.*, **99**, 5210–5211.
32 Fuchs, S., Berl, V., and Lepoittevin, J.-P. (2007) *Eur. J. Org. Chem.*, 1145–1152.

33 Nakayama, J., Dong, H., Sawada, K., Ishii, A., and Kumakura, S. (1996) *Tetrahedron*, **52**, 471–488.
34 Cooke, G., Bryce, M.R., Petty, M.C., Ando, D.J., and Hursthouse, M.B. (1993) *Synthesis*, 465–467.
35 Yoshimatsu, M., Matsuura, Y., and Gotoh, K. (2003) *Chem. Pharm. Bull.*, **51**, 1405–1412.
36 Clarembeau, M., Cravador, A., Dumont, W., Hevesi, L., Krief, A., Lucchetti, J., and Van Ende, D. (1985) *Tetrahedron*, **41**, 4793–4812.
37 Krief, A., Dumont, W., Clarembeau, M., and Badaoui, E. (1989) *Tetrahedron*, **45**, 2023–2030.
38 Krief, A., Dumont, W., Clarembeau, M., Bernard, G., and Badaoui, E. (1989) *Tetrahedron*, **45**, 2005–2022.
39 Gröbel, B.T. and Seebach, D. (1977) *Chem. Ber.*, **110**, 867–877.
40 Denis, J.N. and Krief, A. (1982) *Tetrahedron Lett.*, **23**, 3411–3414.
41 Pérez-Balado, C. and Markó, I.E. (2006) *Tetrahedron*, **62**, 2331–2349.
42 Braga, A.L., Zeni, G., de Andrade, L.H., and Silveira, C.C. (1997) *Synlett*, 595–596.
43 Dabdoub, M.J., Begnini, M.L., Guerrero, P.G., Jr., and Baroni, A.C.M. (2000) *J. Org. Chem.*, **65**, 61–67.
44 Fernandopulle, S.C., Clive, D.L.J., and Yu, M. (2008) *J. Org. Chem.*, **73**, 6018–6021.
45 Raucher, S. and Koolpe, G.A. (1978) *J. Org. Chem.*, **43**, 4252–4253.
46 Liotta, D., Barnum, C.S., and Saindane, M. (1981) *J. Org. Chem.*, **46**, 4301–4304.
47 Zima, G., Barnum, C., and Liotta, D. (1980) *J. Org. Chem.*, **45**, 2736–2737.
48 Kusuda, S., Ueno, Y., Hagiwara, T., and Toru, T. (1993) *J. Chem. Soc., Perkin Trans. 1*, 1981–1988.
49 Braga, A.L., Reckziegel, A., Silveira, C.C., and Comasseto, J.V. (1994) *Synth. Commun.*, **24**, 1165–1170.
50 Comasseto, J.V., Menezes, P.H., Stefani, H.A., Zeni, G., and Braga, A.L. (1996) *Tetrahedron*, **52**, 9687–9702.
51 Dabdoub, M.J., Begnini, M.L., and Guerrero, P.G., Jr. (1998) *Tetrahedron*, **54**, 2371–2400.
52 Guerrero, P.G., Jr., Dabdoub, M.J., and Baroni, A.C.M. (2008) *Tetrahedron Lett.*, **49**, 3872–3876.
53 Denis, J.N., Dummont, W., and Krief, A. (1976) *Tetrahedron Lett.*, **17**, 453–456.
54 Perin, G., Lenardão, E.J., Jacob, R.G., and Panatieri, R.B. (2009) *Chem. Rev.*, **109**, 1277–1301.
55 Labar, D. and Krief, A. (1982) *J. Chem. Soc., Chem. Commun.*, 564–566.
56 Halazy, S., Zutterman, F., and Krief, A. (1982) *Tetrahedron Lett.*, **23**, 4385–4388.
57 Luchetti, J., Dumont, W., and Krief, A. (1979) *Tetrahedron Lett.*, **20**, 2695–2696.
58 Grieco, P.A. and Miyashita, M. (1974) *J. Org. Chem.*, **39**, 120–122.
59 Labar, D., Hevesi, L., Dumont, W., and Krief, A. (1978) *Tetrahedron Lett.*, **19**, 1141–1144.
60 Labar, D., Hevesi, L., Dumont, W., and Krief, A. (1978) *Tetrahedron Lett.*, **19**, 1145–1148.
61 Van Ende, D. and Krief, A. (1976) *Tetrahedron Lett.*, **17**, 457–460.
62 Rémion, J., Dumont, W., and Krief, A. (1976) *Tetrahedron Lett.*, **17**, 1385–1388.
63 Halazy, S. and Krief, A. (1979) *J. Chem Soc. Chem. Commun.*, 1136–1137.
64 Labar, D., Laboureur, J.L., and Krief, A. (1982) *Tetrahedron Lett.*, **23**, 983–986.
65 Reich, H.J. and Bowe, M.D. (1990) *J. Am. Chem. Soc.*, **112**, 8994–8995.
66 Krief, A., Edward, G., Badaoui, E., De Beys, V., and Dieden, R. (1989) *Tetrahedron Lett.*, **30**, 5635–5638.
67 Hoffmann, R.W. and Klute, W. (1996) *Chem. Eur. J.*, **2**, 694–700.
68 Ruhland, T., Dress, R., and Hoffmann, R.W. (1993) *Angew. Chem. Int. Ed. Engl.*, **32**, 1467–1468.
69 Reich, H.J. and Dykstra, R.R. (1993) *Angew. Chem. Int. Ed. Engl.*, **32**, 1469–1470.
70 Klute, W., Krüger, M., and Hoffmann, R.W. (1996) *Chem. Ber.*, **129**, 633–638.
71 Grieco, P.A., Nishizawa, M., Oguri, T., Burke, S.D., and Marinovic, N. (1977) *J. Am. Chem. Soc.*, **99**, 5773–5779.
72 Petragnani, N., Ferraz, H.M.C., and Yonashiro, M. (1985) *Synthesis*, 27–29.
73 Still, W.C. and Mobilio, D. (1983) *J. Org. Chem.*, **48**, 4785–4786.
74 Al-Abed, Y., Al-Tel, T.H., and Voelter, W. (1993) *Tetrahedron*, **49**, 9295–9306.
75 Fitjer, L., Scheuermann, H.J., and Wehle, D. (1984) *Tetrahedron Lett.*, **25**, 2329–2332.

76 Fitjer, L., Justus, K., Puder, P., Dittmer, M., Hassler, C., and Noltemeyer, M. (1991) *Angew. Chem. Int. Ed. Engl.*, **30**, 436–438.

77 Fitjer, L. and Quabeck, U. (1987) *Angew. Chem. Int. Ed. Engl.*, **26**, 1023–1025.

78 Bonnert, R.V. and Jenkins, P.R. (1989) *J. Chem. Soc., Perkin Trans. 1*, 413–418.

79 Rollinson, S.W., Amos, R.A., and Katzenellenbogen, J.A. (1981) *J. Am. Chem. Soc.*, **103**, 4114–4125.

80 Kim, S. and Cheong, J.H. (1997) *Synlett*, 947–949.

81 Kim, S., Cheong, J.H., and Yoo, J. (1998) *Synlett*, 981–982.

82 Yokokawa, F., Fujiwara, H., and Shioiri, T. (1999) *Tetrahedron Lett.*, **40**, 1915–1916.

83 Willson, T., Kocienski, P., Faller, A., and Campbell, S. (1987) *J. Chem. Soc., Chem. Commun.*, 106–108.

84 Tsuboi, S., Sakamoto, J., Sakai, T., and Utaka, M. (1989) *Chem. Lett.*, **18**, 1427–1428.

5
Selenium Compounds with Valency Higher than Two
Józef Drabowicz, Jarosław Lewkowski, and Jacek Ścianowski

5.1
Introduction

Discovery of an efficient method for the synthesis of alkenes under mild conditions, by *syn*-elimination of selenoxides [1], initiated a rapid growth of methodologies using selenoorganic compounds as useful intermediates in organic transformations. Due to specific reactivities of these compounds, it is possible to introduce or interconvert different functional groups in a chemo-, regio-, and stereoselective manner. There are many publications that present many details of selenoorganic compounds in organic synthesis [2]. Development of efficient catalytic reactions with these compounds is probably one of the most important directions in modern organic chemistry (green chemistry with selenium reagents) [3].

Herein recent synthetic applications of five groups of organoselenium compounds at higher oxidation state are presented: trivalent, tetravalent, pentavalent, hexavalent, and hypervalent derivatives. According to the classification based on formal oxidation state and number of ligands they can be devided into the following classes [2f].

Trivalent compounds:

- dicoordinated selenonium salts;
- tricoordinated derivatives (selenonium salts, seleniranium salts, and selenonium ylides).

Tetravalent compounds:

- dicoordinated derivatives (selenium dioxide and diimidoselenium compounds);
- tricoordinated derivatives (selenoxides and seleninic acids derivatives).

Pentavalent compounds (selenonium salt oxides):

- hexavalent compounds: tetracoordinated compounds (selenones, selenonic acids devatives, and selenoximines).

Organoselenium Chemistry: Synthesis and Reactions, First Edition. Edited by Thomas Wirth.
© 2012 Wiley-VCH Verlag GmbH & Co. KGaA. Published 2012 by Wiley-VCH Verlag GmbH & Co. KGaA.

5 Selenium Compounds with Valency Higher than Two

Hypervalent derivatives:

- tetravalent, tetracoordinated compounds (selenuranes);
- pentacoordinated derivatives (selenurane oxides);
- hexacoordinated derivatives (perselenuranes).

New applications of these groups of organoselenium compounds in asymmetric synthesis are also presented.

5.2
Trivalent, Dicoordinated Selenonium Salts

Dicoordinated selenonium salts can be formed in the reaction of selenocarbonyl compounds with alkylating agents. There are only few examples in which applications of these salts as synthetically useful reagents have been reported. Most recently, it has been described that 2-methylseleno-1,3-dithiolium trifluoromethanesulfonate 2, obtained from the reaction of selone 1 with methyl trifluoromethanesulfonate, was used for the synthesis of new unsymmetrical tetrathiafulvalene (TTF) derivatives (Scheme 5.1) [4]. The four acetoxyphenyl-TTFs 4a–d were synthesized via Wittig-type condensations in 15–33% yield. Deacetylation of TTFs 4a–d with hydrazine in methanol yielded p-hydroxy-TTFs as red crystals in 85–90% yield.

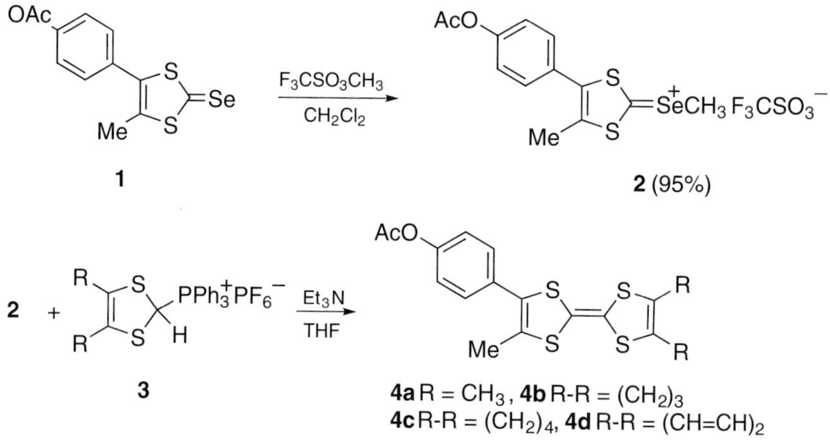

Scheme 5.1

The corresponding 1,3-diselenolium salts 6a–b were obtained via the methylation of selones 5a–b using methyl triflate. Treatment of 6a and 6b with malononitrile, using pyridine as a base, afforded molecules 7a and 7b in 53 and 65% yield, respectively, after the in situ elimination of methaneselenol. These compounds are simplest systems containing electron-donor and electron-acceptor units separated by a conjugated spacer group (D–π–A system) (Scheme 5.2) [5].

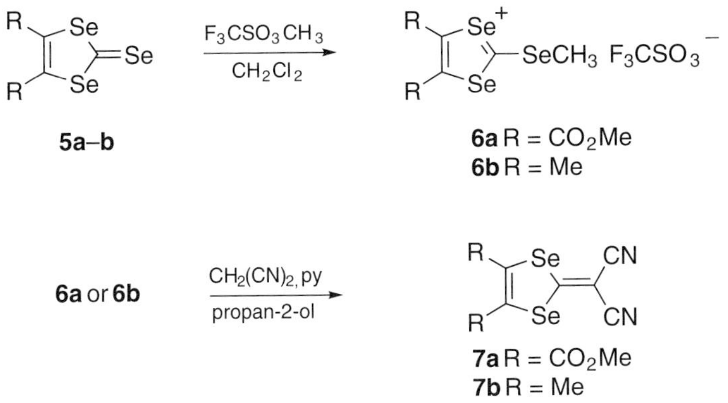

Scheme 5.2

The reaction of compound **6b** with triphenylphosphonium salts **8a–b** in the presence of triethylamine gave dithiadiselenafulvalenes **9a–b** in 52–57% yield (Scheme 5.3) [6, 7]. Reduction of **6b** with sodium cyanoborohydride yielded the unstable selenide **10** that was used for the synthesis of tetramethyltetraselenafulvalene **11** in a five-step procedure (Scheme 5.4) [8]. Starting from compound **6a** after reduction the series of 2-ylidene-1,3-diselenones **12a–e** was obtained in moderate yield (21–36%) (Scheme 5.5) [9].

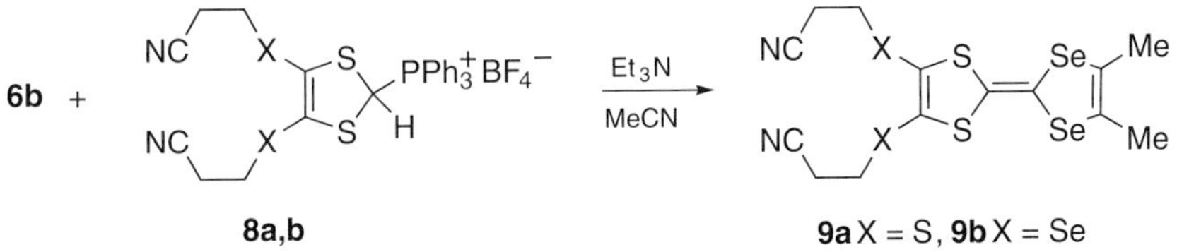

Scheme 5.3

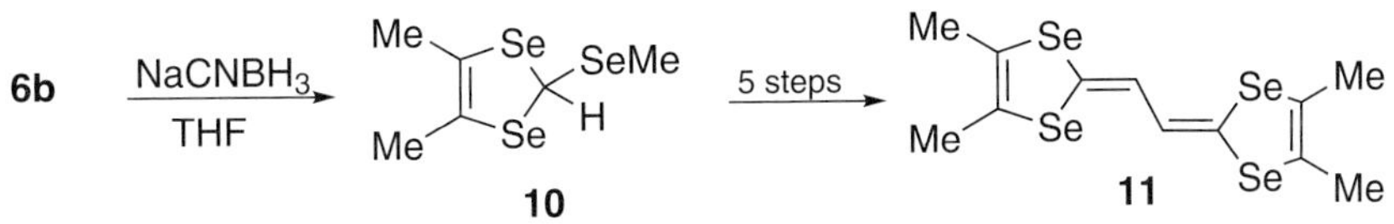

Scheme 5.4

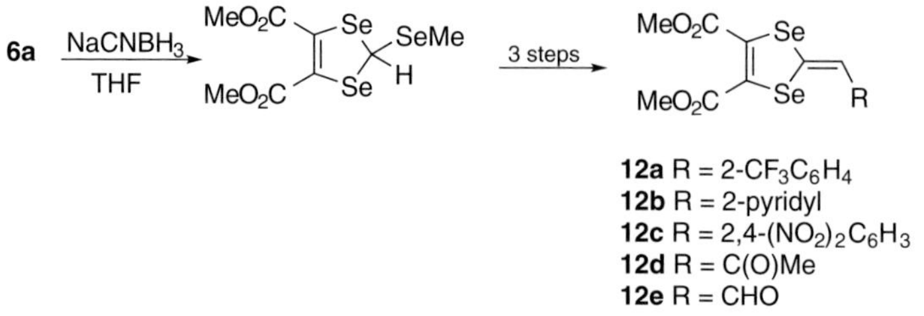

Scheme 5.5

The highly polarized carbon–selenium bond of selenourea reacts with alkyl halides to the corresponding selenouronium salts. A series of ^{75}Se-labeled salts was prepared, for example, 1,3-dicyclohexylselenourea **13** reacted with N-*tert*-butoxycarbonyl-2-amino-4-bromobutyrate **14** to give the selenouronium bromide **15** in 80% yield (Scheme 5.6) [10]. Treatment of the selenouronium salt **15** with tetrabutylammonium hydroxide (TBAH) gave selenolate **16**, which was methylated *in situ* with iodomethane to give the unsymmetrical selenide **17** in 90% yield. Using this two-step alkylating procedure benzylmethyl, benzylbutyl, benzylisopropyl selenides, and 1-phenyl-(1-methylseleno)ethane were synthesized with nearly quantitative yields.

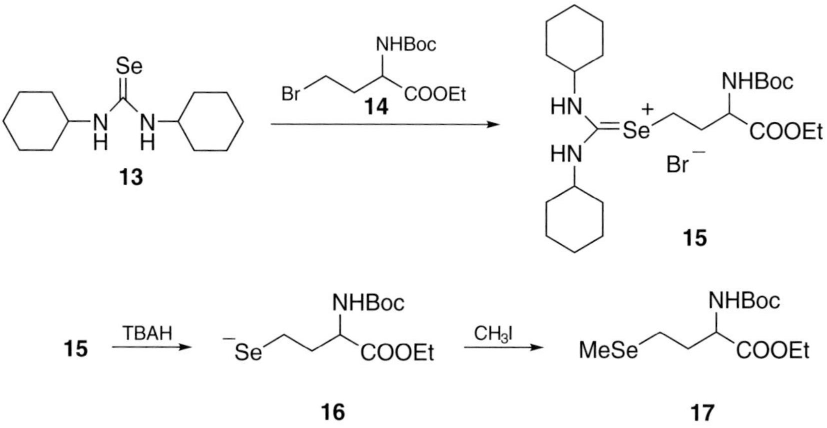

Scheme 5.6

5.3
Trivalent, Tricoordinated Derivatives

Tricoordinated derivatives are represented by three groups of selenoorganic compounds: seleniranium (episelenonium) salts **18**, five-membered or lager cyclic and acyclic selenonium salts **19**, and selenonium ylides **20**:

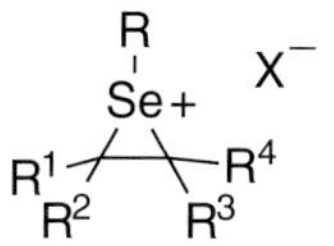

5.3 Trivalent, Tricoordinated Derivatives

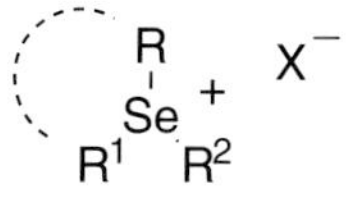

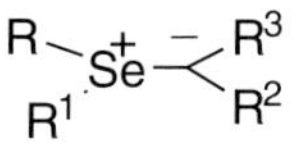

18: R-Se+ with R¹, R², R³, R⁴ substituents, X⁻ counterion

19: cyclic Se+ with R, R¹, R², X⁻ counterion

20: R-Se+ with R¹, R², R³ substituents

Seleniranium salts **18** are compounds, which are in most cases unstable and difficult to isolate. Their structures were established by low-temperature NMR spectroscopy [11], in one case the stable seleniranium salt **21** was isolated and its structure was confirmed by X-ray analysis (Scheme 5.7) [12].

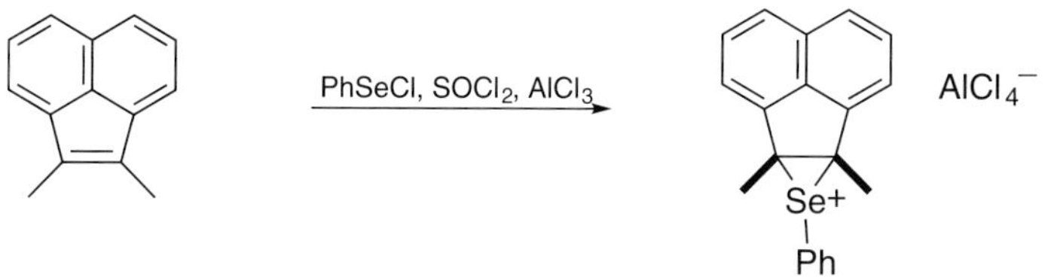

[acenaphthylene with dimethyl] →(PhSeCl, SOCl₂, AlCl₃)→ **21** (Se+–Ph bridged acenaphthene), AlCl₄⁻

Scheme 5.7

Cyclic episelenonium salts may be easily obtained from the addition of electrophilic organoselenium reagents to double bonds, further reaction with nucleophiles proceed with Markovnikov orientation. Especially interesting are chiral electrophilic selenium reagents, obtained from the optically active diselenides, which have been used for the asymmetric formation of carbon–oxygen, carbon–nitrogen, and carbon–carbon bonds, via the addition to the double bonds or by cyclization reactions [13]. Recently, these types of reactions have been widely used in organic synthesis and their detailes are discussed in Chapter 1.

In recent years, selenonium salts **19** have been found in many interesting applications in organic synthesis. The parent member of this family is trimethylselenonium hydroxide **23**, a strongly methylating agent, which was prepared by addition of silver oxide to the solution of trimethylselenonium iodode **22** in methanol–water (100:1) at room temperature (Scheme 5.8). Trimethylacetic acid, alcohols, phenol, thiophenol, aniline, imidazole, pyrazole, and xanthine were efficiently esterified, etherified, and N-methylated in 85–100% yield [14].

$Me_3Se^+I^-$ (**22**) →(Ag₂O, MeOH, H₂O)→ $Me_3Se^+OH^-$ (**23**)

Scheme 5.8

Similar properties were reported for Se-(trifluoromethyl)dibenzoselenophenium triflate (**25**), which was used as a trifluoromethylating reagent for nucleophilic substrates and was also successful used in the trifluoromethylation of aniline giving o- and p-trifluoromethylaniline (38% and 19% yield). Compound **25** can be obtained by the fluorination of selenide **24** or by cyclization of the selenoxide **26** with triflic anhydride (Scheme 5.9) [15].

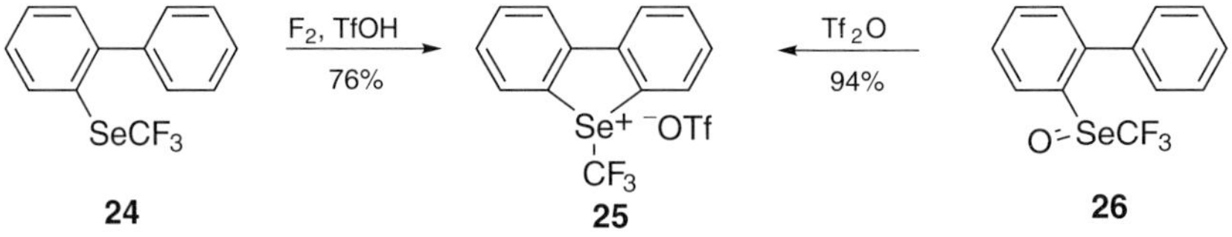

Scheme 5.9

Optically active selenonium salts were used for asymmetric alkylations of enolate anions. Chiral (+)-ethylmethylphenylselenonium perchlorate **28** was obtained by fractional recrystallization of the racemic salt **27**, then by the ion exchange with NaClO$_4$ (Scheme 5.10). When 2-methoxycarbonyl-1-indanone (**29**) was treated with the optically active salt **28** in the presence of potassium carbonate in dichloromethane, (−)-2-methylated **30** and (+)-2-ethylated **31** products were obtained in 41% and 36%, respectively. Unfortunately, small enantiomeric exceses of 2.4% and 6.2% were observed (Scheme 5.11) [16].

Scheme 5.10

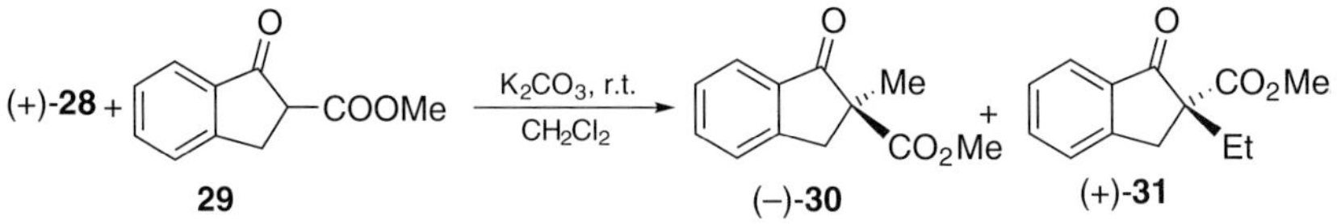

Scheme 5.11

In recent years, many papers that describe selenonium salts have been devoted to unsaturated selenonium salts. Alkynylselenonium salts such as dimethyl(phenylethynyl)selenonium tetrafluoroborate **33** were prepared by methylation of methylphenylethynyl selenide **32** with Meerwein's reagent. Diphenyl(phenylethynyl)selenonium triflate **35** was synthesized by the reaction of trimethyl(phenylethynyl)silane **34** and diphenyl selenoxide with trifluoromethanesulfonic anhydride [17]. The cyclic analogue **37** was also obtained by the reaction of **34** with dibenzoselenophene 5-oxide **36** [18].

A general methodology for the synthesis 1-alkynyl(diphenyl)selenonium salts **39a–g** is based on the heteroatom transfer reaction between 1-alkynyl(phenyl)-λ^3-iodanes **38** and diphenylselenide, which takes place selectively under mild conditions in very good yields (Scheme 5.12) [19].

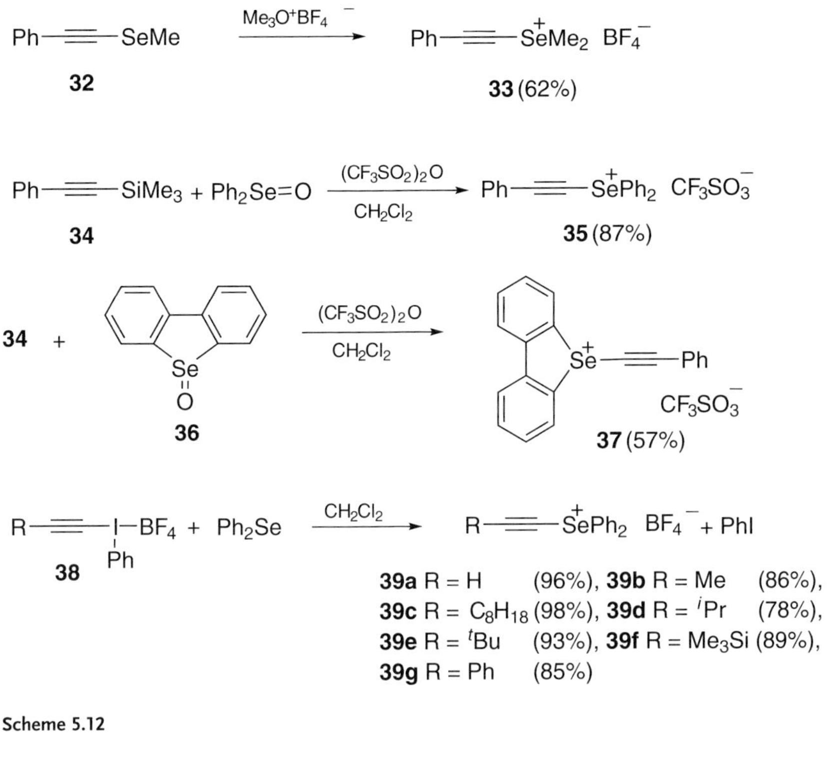

Scheme 5.12

The reaction of alkynylselenonium salts **33** and **35** with benzenesulfinic acid in isopropanol afforded (Z)-(β-phenylsulfonyl)vinylselenonium salts **40** and **41** in 76% and 72% yield, respectively [17]. When the sodium benzenesulfinate in alcohol at room temperature was used for these reactions, the corresponding (Z)-β-alkoxyvinylsulfones **42a–c** were obtained as main products (Scheme 5.13).

The same products **42a–c** were prepared from salt **37** in moderate yields (51–77%) [18].

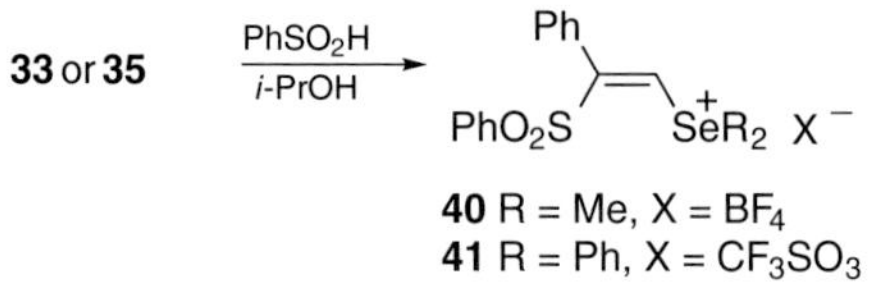

33 or 35 $\xrightarrow{\text{PhSO}_2\text{H}}{i\text{-PrOH}}$

Ph\
PhO$_2$S $\overset{+}{\text{SeR}_2}$ X$^-$

40 R = Me, X = BF$_4$
41 R = Ph, X = CF$_3$SO$_3$

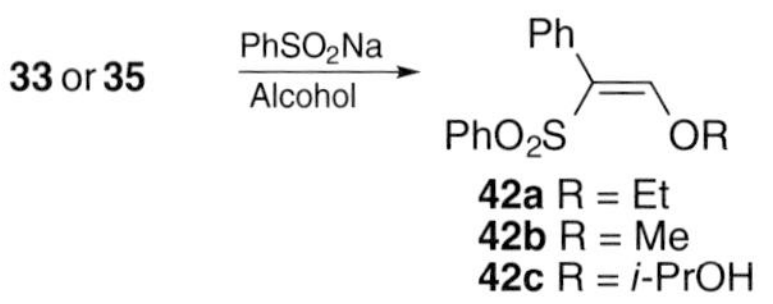

33 or 35 $\xrightarrow{\text{PhSO}_2\text{Na}}{\text{Alcohol}}$

Ph\
PhO$_2$S OR

42a R = Et
42b R = Me
42c R = i-PrOH

Salt	Alcohol	Product	Yield (%)
33	EtOH	42a	74
33	MeOH	42b	60
33	i-PrOH	42c	68
35	EtOH	42a	69
35	MeOH	42b	64
35	i-PrOH	42c	71

Scheme 5.13

It was demonstrated that the reaction of salt **35** with active methylene compounds and addition of *t*-BuOK proceed via an oxyselenurane intermediate. Benzoylacetonitrile at room temperature gave the selenurane **43**, which after refluxing in chloroform was converted to the furan derivative **44** [20]. If amides are used instead of active methylene compounds, the formation of oxazol was observed. The reaction of **35** with benzamide and NaH in THF gave 2,4-diphenyloxazol **45** (50%), diphenylselenide **46** (56%), and diphenylselenoxide **47** (40%) [20]. Treatment of **35** with thiophenol derivatives with catalytic amount of triethylamine yielded β-arylthiovinylselenonium salts **48a–e** [21]. The reaction of alkynylselenonium salt **35** with *n*-Bu$_4$NI or *n*-Bu$_4$NBr in CH$_2$Cl$_2$ proceeded with the formation of 1-halo-alkynes **49a–b**. Alkynyl chloride was not obtained from the reaction of **35** with *n*-Bu$_4$NCl, but instead 2-chloroacetophenone was isolated in 23% yield. In contrast, the reaction with *n*-Bu$_4$NF in CH$_2$Cl$_2$ gave diphenyl selenoxide (Ph$_2$SeO) and phenylacetylene in good yield. Different behavior of *n*-Bu$_4$NX in these reactions was observed, because the alkynyl moiety of **35** acted as the alkynyl cation or the acetylide ion depending upon the kind of halide [22]. When the salt **35** reacted with one equivalent of phenyllithium in THF, beside a mixture of compounds **50–54**, 1-(*o*-biphenylyl)-2-phenylethyne **55** (formed via a benzyne intermediate) was obtained (Scheme 5.14) [23].

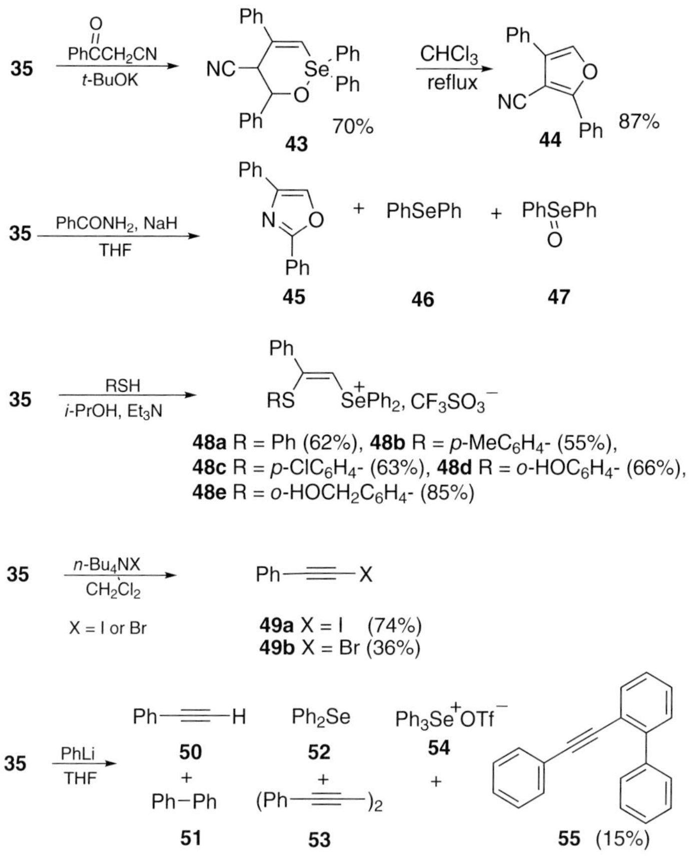

Scheme 5.14

The alkenylselenonium salts **40**, **41**, and **48** have found many interesting synthetic applications. The reactions of primary and secondary alkoxides with **40** and **41** afforded (Z)-O-alkyl enol ethers **42a–d** in good yields (Scheme 5.15). Selenonium salt **41** can react with Bu$_4$NCl, NaI, LiBr, NaBr, KBr, NaCl, RC≡CLi (R = Ph, ^{n}Bu, ^{t}Bu) to produce the corresponding (Z)-β-halogenovinyl or (Z)-β-alkynylvinyl sulfones [24].

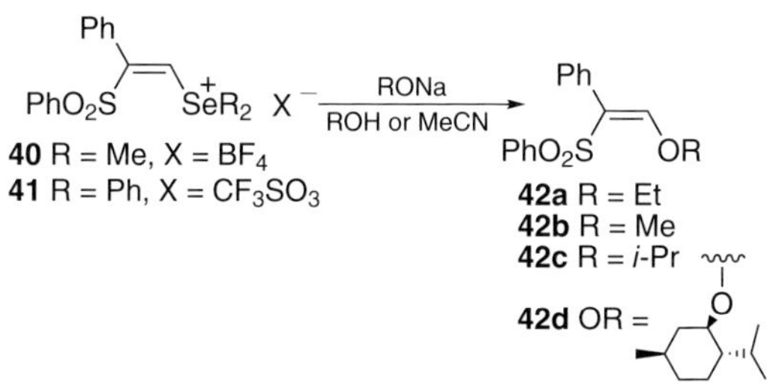

Salt	Alcohol	Product	Yield (%)
40	EtOH	42a	80
40	MeOH	42b	82
40	i-PrOH	42c	70
40	(−)-Menthol	42d	67
41	EtOH	42a	75
41	MeOH	42b	71
41	i-PrOH	42c	89
41	(−)-Menthol	42d	47

Scheme 5.15

Another utilization of alkenylselenonium salts **40** and **41** in organic synthesis is the alkenylation of enolates. The reaction of lithium enolates, which were prepared *in situ* by the reactions of lithium hexamethyldisilazide and carbonyl compounds **56** or **58**, with alkenylselenonium salts **40** and **41**, gave the ethenylated carbonyl compounds **57** and **59** in high yields. The application of **40** in the diastereoselective alkenylation of enolates gave high diastereoselectivities (up to 95% *de*) (Scheme 5.16) [25].

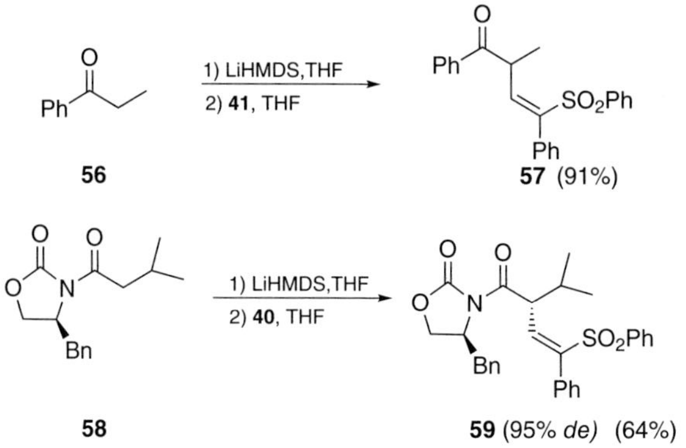

Scheme 5.16

Similar reactions were performed using β-arylthiovinylselenonium salts **48**. Salt **48a** can react with nucleophiles to give (Z)-β-arylthio-α-functionalized ethenes **60a–c** with retention of configuration. An application of the above reactions to an intramolecular cyclization reaction resulted in the synthesis of medium-ring sized heterocyclic compounds containing sulfur and oxygen atoms such as **61** and **62** (Scheme 5.17) [21].

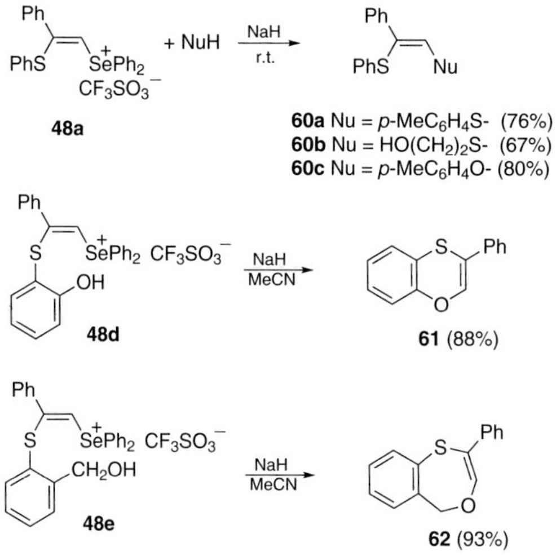

Scheme 5.17

5.3 Trivalent, Tricoordinated Derivatives

Recently, the first syntheses of allenylselenonium salts **64a–b** were reported. They were achieved by alkylation of the corresponding allenyl methyl selenides **63a–b** with methyl triflate (Scheme 5.18). The reaction of salts **64a–b** with deprotonated acetylacetone produced highly functionalized dihydrofurans **65** and furan derivative **66** [26].

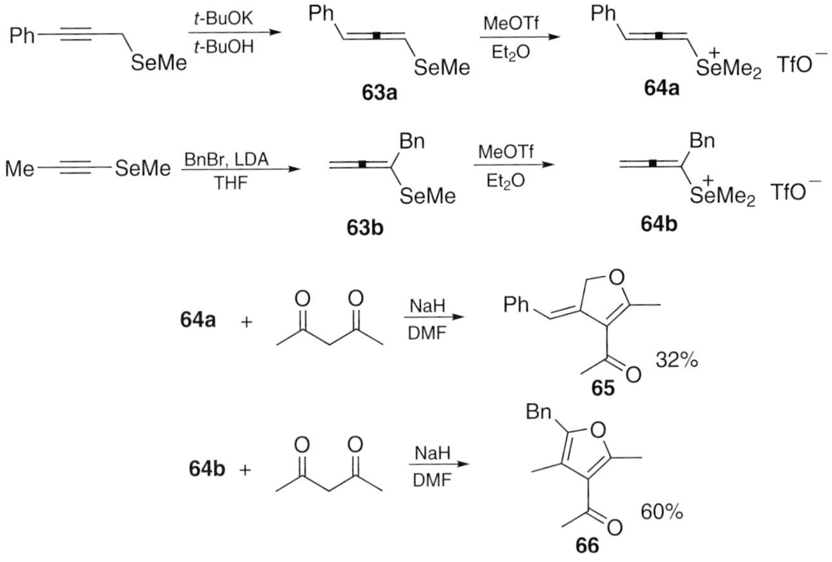

Scheme 5.18

Dimethyl(4-methylphenyl)selenonium trifluoromethanesulfonate **67** was examined in a palladium-catalyzed Mizoroki–Heck reaction. Salt **67** reacted with butyl acrylate **68** in the presence of palladium(II) catalysts and three equivalents of silver acetate to give product **69**, although the yields were low. PdCl$_2$, and palladium catalysts possessing phosphine, and bipyridine ligands were tested. The best result was obtained with the bipyridine ligand in DMF at 150 °C leading to product **69** in 38% yield (Scheme 5.19) [27].

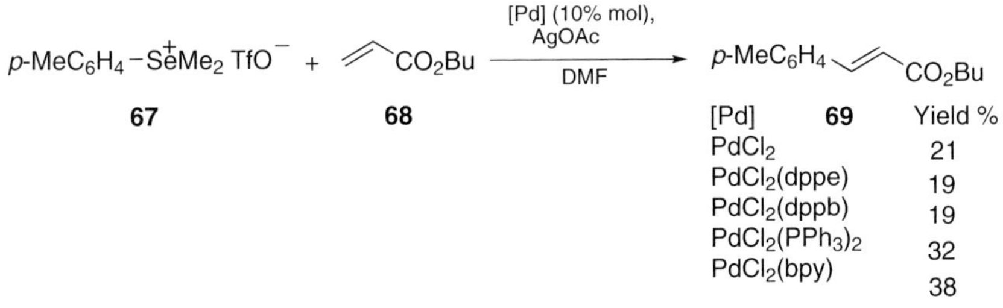

[Pd]	69 Yield %
PdCl$_2$	21
PdCl$_2$(dppe)	19
PdCl$_2$(dppb)	19
PdCl$_2$(PPh$_3$)$_2$	32
PdCl$_2$(bpy)	38

Scheme 5.19

Some selenonium salts **19** are liquids at room temperature and, due to their properties, are selenonium ionic liquids and can be used in reactions as catalysts. Due to their low volatility and ionic structure, this new class of higly polar compounds can be treated as environmentally friendly solvents. A few selenium ionic liquids **72a–d** were prepared by the reaction of diorganyl selenides **70a–d** with alkyl halides **71** in the presence of AgBF$_4$ (Scheme 5.20). All of the synthesized selenonium salts are liquid at room temperature and can be stored for several days at room temperature without decomposition [28].

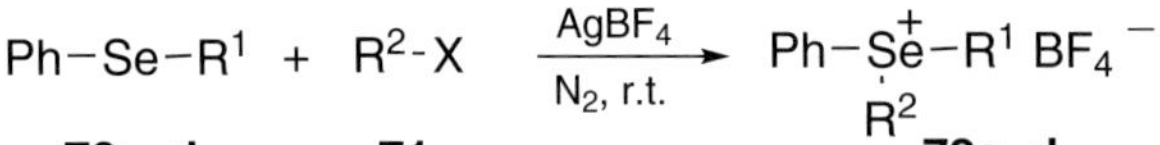

$$\text{Ph-Se-R}^1 + \text{R}^2\text{-X} \xrightarrow[\text{N}_2,\ \text{r.t.}]{\text{AgBF}_4} \text{Ph-}\overset{+}{\underset{\text{R}^2}{\text{Se}}}\text{-R}^1\ \text{BF}_4^-$$

70a–d   **71**   **72a–d**

Salt	R^1	R^2	X	Yields (%)
72a	n-Bu	Et	Br	94
72b	n-Bu	Me	I	97
72c	n-Bu	n-Bu	Br	87
72d	Ph	Et	Br	99

Scheme 5.20

Phenyl butyl ethyl selenonium tetrafluoroborate [pbeSe]BF$_4$ **72a** was used as a catalyst in the preparation of octahydroacridines **75a** and **75b** by the hetero-Diels–Alder reaction involving (R)-citronellal **73** and aniline **74** (Scheme 5.21) [28]. When the reaction of **73** and **74** was carried out at room temperature for 24 h in the absence of supported catalyst, no reaction took place [29].

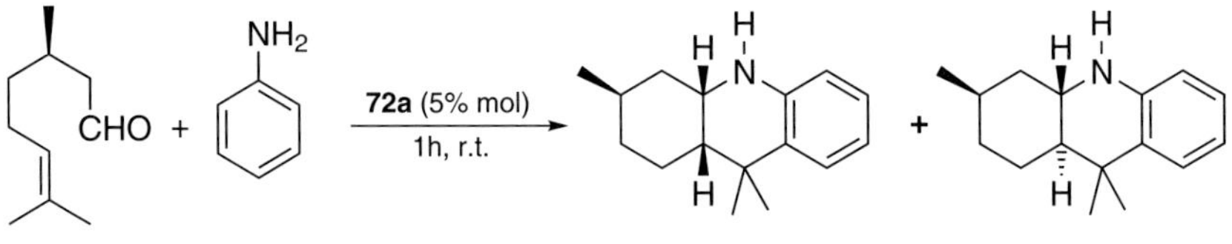

73   **74**   **75a (38%)**   **75b (38%)**

Scheme 5.21

The same salt **72a** was applied for the clean conversion of aromatic, aliphatic, α,β-unsaturated aldehydes and ketones to their corresponding dithioacetals, for example, when benzaldehyde was treated with thiophenol in the presence of 15% of selenonium salt **72a** at room temperature dithiophenylacetal **76** was obtained in 83% yield after 2.5 h [30]. Selenonium salt **72a** was also used as efficient cocatalyst for the Baylis–Hillman reaction of aldehydes and electron-deficient alkenes **77a–d** to give products **78a–d** in good yields. The reaction time was reduced from 24 to 2 h when the selenonium salt **72a** was used (Scheme 5.22) [31].

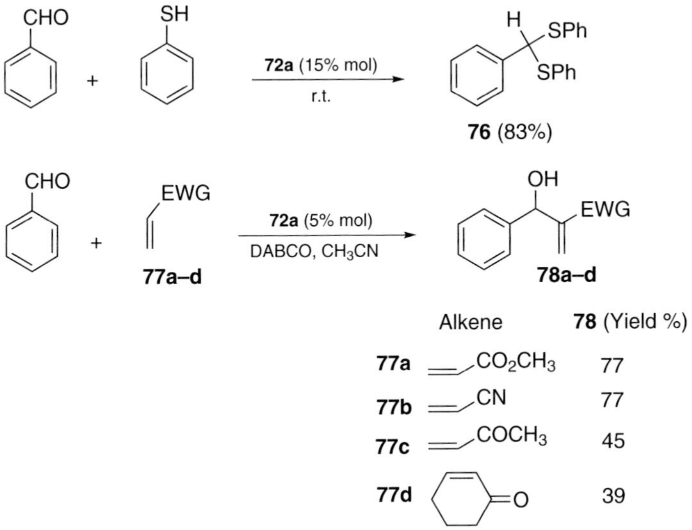
CHO
SH
72a (15% mol)
r.t.
H
SPh
SPh
76 (83%)
CHO
EWG
77a–d
72a (5% mol)
DABCO, CH3CN
OH
EWG
78a–d
Alkene
78 (Yield %)
77a
CO2CH3
77
77b
CN
77
77c
COCH3
45
77d
O
39

Scheme 5.22

Several alkoxyselenonium salts **80a–f** were synthesized via the reaction of alkoxychloroselenuranes **79a–e** with $AgClO_4$ or $AgBF_4$ (Scheme 5.23) [32].

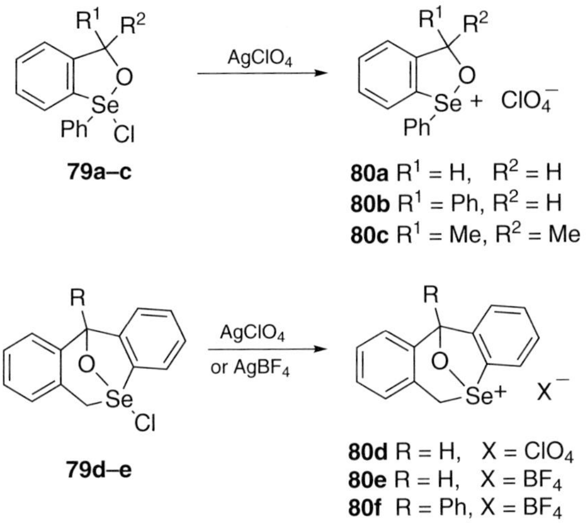
R1
R2
O
Se
Ph
Cl
79a–c
AgClO4
R1
R2
O
Se +
ClO4−
Ph
80a R1 = H, R2 = H
80b R1 = Ph, R2 = H
80c R1 = Me, R2 = Me
R
O
Se
Cl
79d–e
AgClO4
or AgBF4
R
O
Se+
X−
80d R = H, X = ClO4
80e R = H, X = BF4
80f R = Ph, X = BF4

Scheme 5.23

The reactions of selenonium salts **80a–b** and **80d** with organometallic reagents have been studied (Scheme 5.24). Reactions of salts **80a–b** at room temperature gave only benzyl alcohols **81a–b**, and reactions under reflux in THF afforded benzophenone derivatives **82a–b** and benzyl ethers **83a–b** in addition to **81a–b**. The **80d** salt reacted with Grignard reagents at room temprature to give three products **84–86**.

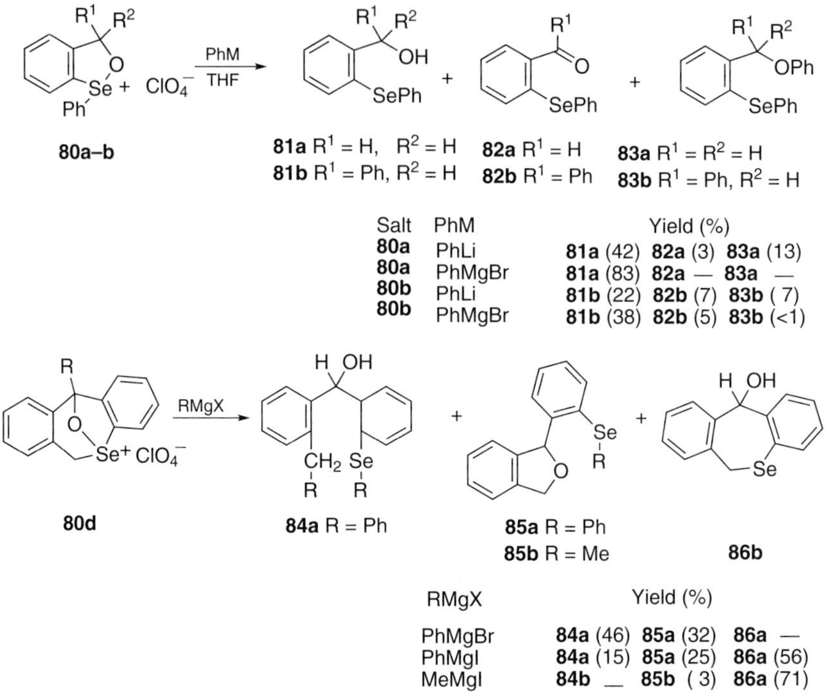

Scheme 5.24

Ylides are well-known synthons in modern organic chemistry, especially due to their use in Wittig-type conversions. In recent years, selenonium ylides **20** have also found some interesting applications in synthetic chemistry. In most cases, they are obtained *in situ* by the reaction of selenonium salts **19** with bases.

Aryl-substituted epoxides **89** have been obtained via the formation of selenonium ylides, by the addition of *t*-BuOK to a mixture of aromatic aldehydes **88** and selenonium salts **87** (Scheme 5.25). Unfortunately, these reactions are limited to nonenolisable carbonyl compounds [33].

Scheme 5.25

$R_2\overset{+}{Se}-CH_2R^1$ X^- + R^2\\$R^3$\\C=O $\xrightarrow{\text{t-BuOK, DMSO}}$ $R^1HC\overset{O}{-\!\!-\!\!-}C(R^2)(R^3)$

87   **88**   **89**

Entry	R	R¹	X	R²	R³	Yield (%)
1	Me	H	BF₄	H	Ph	90
2	Ph	H	BF₄	H	Ph	90
3	Me	H	I	H	p-MeOC₆H₄	75
4	Me	H	I	H	p-NCC₆H₄	54
5	Me	H	I	H	mesityl	79
6	Ph	Me	BF₄	H	Ph	69
7	Ph	H	BF₄	Ph	Ph	90
8	Me	H	I	H	PhCH=CH	71
9	Me	H	I	Ph	PhCH=CH	50

Although the syntheses of optically active selenonium ylides have been well known [34], the first use of chiral selenonium ylides for asymmetric synthesis of epoxides was described only recently. (2R,5R)-2,5-Dimethyltetrahydroselenophene **90**, obtained from (2S,5S)-hexanediol via activation of both hydroxyl groups into mesylates and subsequent double nucleophilic substitution with lithium selenide, proved to be an efficient catalyst (0.2 equivalent) for the benzylidenation of aromatic aldehydes **91**. The enantiomer excess was usually higher than 90%, but equal amounts of *trans*- and *cis*-epoxides **92** were obtained (Scheme 5.26) [35]. When a stoichiometric amount of **90** was used, the *trans*-epoxide was the major isomer. The formation of selenonium salt **93** and selenonium ylide **94** as intermediate was postulated.

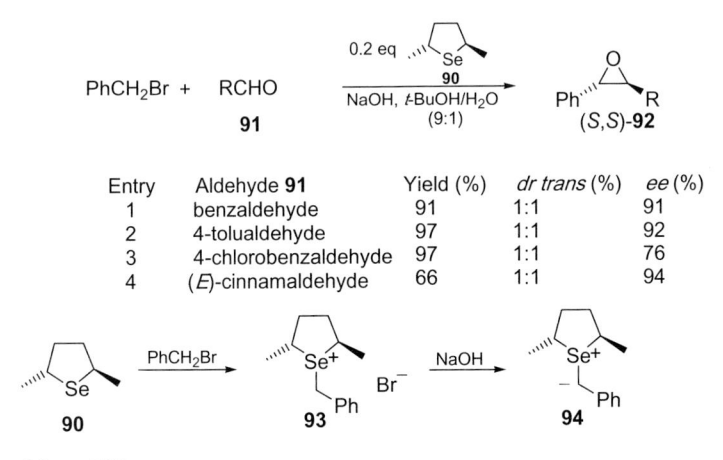

PhCH₂Br + RCHO **91** $\xrightarrow[\text{NaOH, t-BuOH/H}_2\text{O (9:1)}]{\text{0.2 eq } \mathbf{90}}$ Ph$\overset{O}{-\!\!-\!\!-}$R (S,S)-**92**

Entry	Aldehyde **91**	Yield (%)	dr trans (%)	ee (%)
1	benzaldehyde	91	1:1	91
2	4-tolualdehyde	97	1:1	92
3	4-chlorobenzaldehyde	97	1:1	76
4	(E)-cinnamaldehyde	66	1:1	94

90 $\xrightarrow{\text{PhCH}_2\text{Br}}$ **93** (Se⁺–Ph, Br⁻) $\xrightarrow{\text{NaOH}}$ **94** (Se⁺–⁻CHPh)

Scheme 5.26

Recently camphor-derived selenonium ylides were employed to asymmetric synthesis of chiral epoxides in good yields with good stereoselectivities. It was shown that the reactions of the selenonium salts **96** and **98** with aldehydes and t-BuOK gave corresponding chiral *trans*-diaryl epoxides **99** (Scheme 5.27). Selenonium salt **96** was prepared in the reaction of *exo-α*-methylseleno isoborneol **95** with benzyl bromide in 85% yield. The salt **98** could only be formed from *exo-α*-phenylseleno isoborneol **97** with the assistance of silver tetrafluoroborate (82% yield) [36].

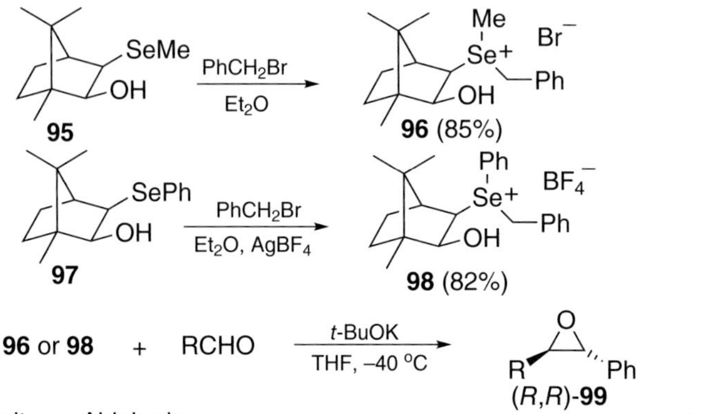

Salt	Aldehyde	Yield (%)	dr trans (%)	ee (%)
96	benzaldehyde	78	86	58
96	4-tolualdehyde	80	85	50
96	4-chlorobenzaldehyde	77	81	55
98	benzaldehyde	85	87	80
98	4-tolualdehyde	86	86	81
98	4-chlorobenzaldehyde	81	91	80

Scheme 5.27

It was found that selenonium ylides can be good precursors for the synthesis of cyclopropane derivatives. Selenonium salt **102**, obtained by the reaction of dimethylselenide **100** with phenacyl bromide **101**, yielded dimethylphenacylselenonium ylide **103** in the reaction with sodium hydroxide. Further reaction of **103** with benzalacetophenone (**104**) gave phenyl-*trans*-dibenzoylcyclopropane **105a**. The "one-pot" reaction of selenonium salt **102** with a substoichiometric amount of sodium hydroxide and **104** gave *cis*-isomer **105b**, in the case of excess sodium hydroxide a mixture of *cis*- and *trans*-isomers **105a–b** were obtained (Scheme 5.28) [37].

An alternative methodology for the synthesis of cyclopropane derivatives is the generation of selenonium ylides by reaction with carbenes. The Cu(acac)$_2$ catalyzed decomposition of α-diazoacetophenones **106** in the presence of

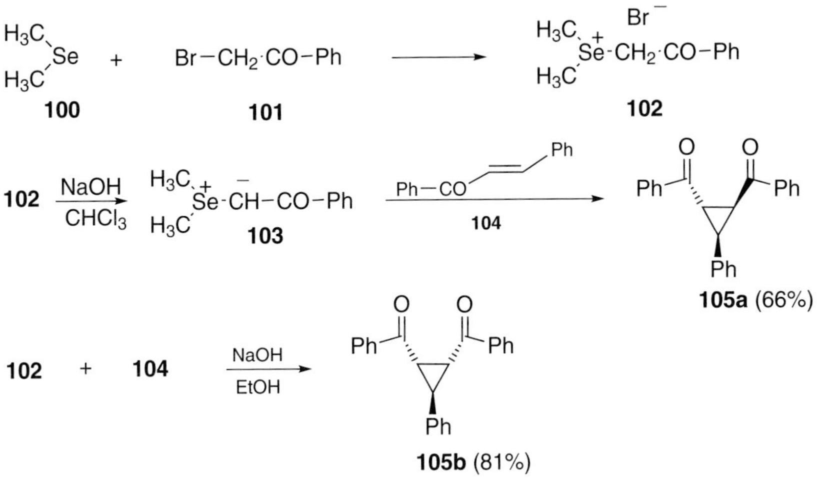

Scheme 5.28

bis(*p*-methoxyphenyl) selenide **107** gave ylide **108**. Further reactions with two equivalents of **106** yielded *cis*-cyclopropane derivatives **109a–h** and selenide **107** (Scheme 5.29) [38]. To confirm the catalytic action of the selenide **107** the reactions were carried out using different amounts of **107**, and even 1/20 amount of **107** to **106** was effective for promotion the reaction.

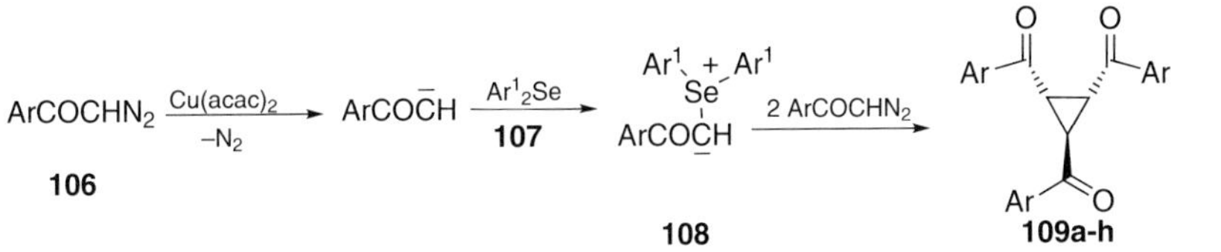

Scheme 5.29

109a Ar = *p*-MeOC$_6$H$_4$ (55%), **109b** Ar = *p*-MeC$_6$H$_4$ (68%)
109c Ar = *m*-MeC$_6$H$_4$ (54%), **109d** Ar = Ph (59%)
109e Ar = *p*-ClC$_6$H$_4$ (54%), **109f** Ar = *m*-ClC$_6$H$_4$ (49%)
109g Ar = *p*-BrC$_6$H$_4$ (48%), **109h** Ar = *p*-NCC$_6$H$_4$ (41%)

Most recently, the first example of the application of chiral selenonium ylides in asymmetric cyclopropanation with good yields, excellent diastereoselectivities, and very high enantioselectivities (up to >99%) was described [39]. Chiral camphor-derived selenonium salts **111a–b** and **113** were prepared via the reaction of selenides **95a–b** and **112** with allyl bromide **110** in very good yields (Scheme 5.30). On

Scheme 5.30

the same way with addition of NaBPh$_4$ selenonium salt **114** was prepared (Scheme 5.31). The reactions of selenonium salts with base, then with various α,β-unsaturated carbonyl compounds gave trisubstituted cyclopropane derivatives, for example, salts **111**, **113**, and **114** reacted with unsaturated esters to afford three stereoisomers of chiral 1,2,3-trisubstututed cyclopropanes **115–117** (Scheme 5.32).

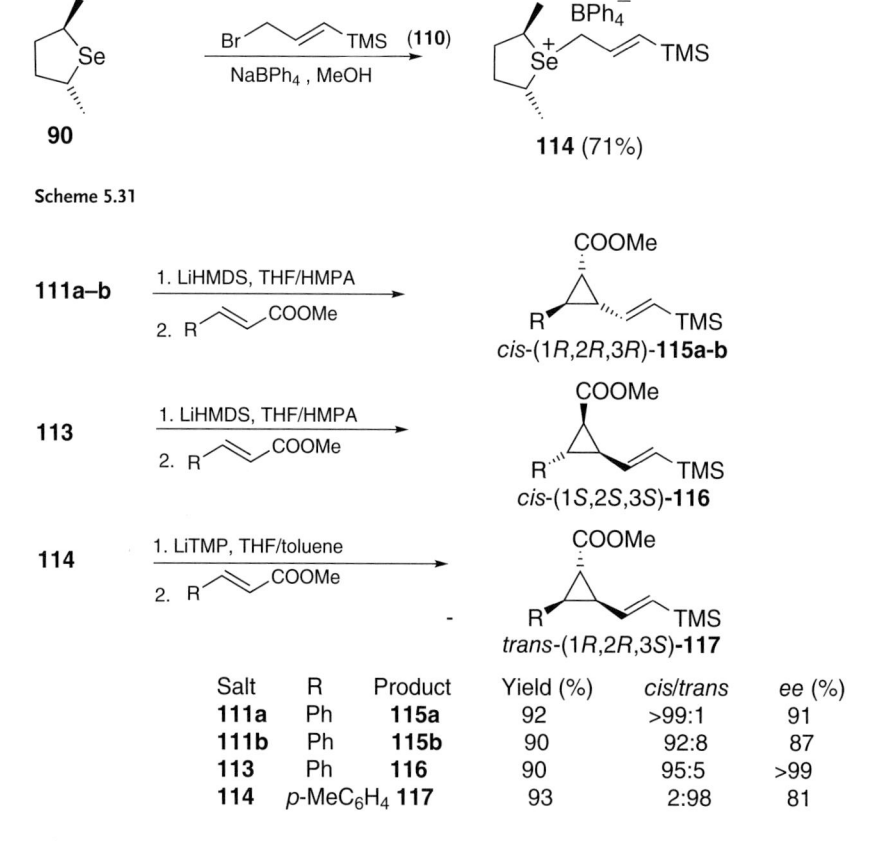

Scheme 5.31

Salt	R	Product	Yield (%)	cis/trans	ee (%)
111a	Ph	115a	92	>99:1	91
111b	Ph	115b	90	92:8	87
113	Ph	116	90	95:5	>99
114	p-MeC$_6$H$_4$	117	93	2:98	81

Scheme 5.32

α,β-Unsaturated ketones, α,β-unsaturated nitriles, and α,β-unsaturated amide with selenonium salt **114** gave the corresponding *trans*-cyclopropanes in good yields, high diastereoselectivities and selectivities. The enantioselectivities of the cyclopropanation with *exo*-salts **111a–b** are opposite to those with *endo*-salt **113**. Therefore, two enantiomers of *cis*-cyclopropane **115** and **116** could be obtained by the choice of *exo*- or *endo*-selenonium salt.

Convenient precursors for the synthesis of epoxide and cyclopropane derivatives can also be vinylselenoniun tetrafluoroborates **119**, obtained via the reaction of vinyl selenides **118** with alkyl iodides and silver tetrafluoroborate. Treatment of selenonium salts **119a–b** with alkoxides gave ylides **120a–b**, and then a reaction with benzaldehyde led to epoxides **121a–b** (Scheme 5.33). The reaction of **119a** with carbanions, for example, derived from dimethyl malonate **122a** or malononitrile **122b** and NaH, yielded the disubstituted cyclopropanes **124a–b** (Scheme 5.34) [40]. The cyclopropane formation goes through the selenonium ylide **123**, and the subsequent intermolecular proton transfer followed by the cyclopane formation with removal of methyl phenyl selenide.

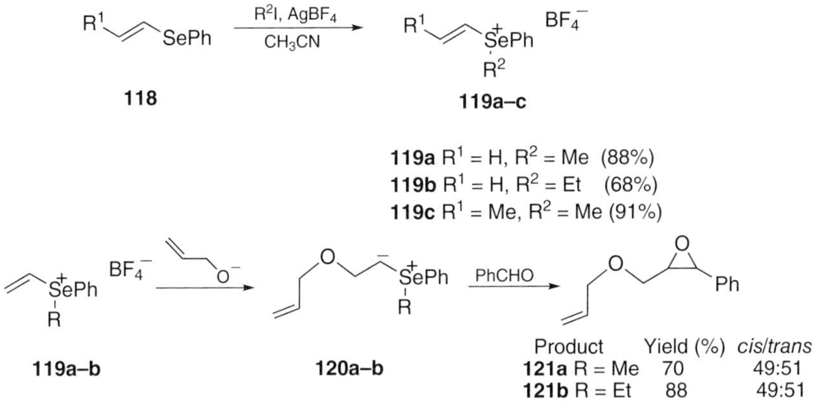

Scheme 5.33

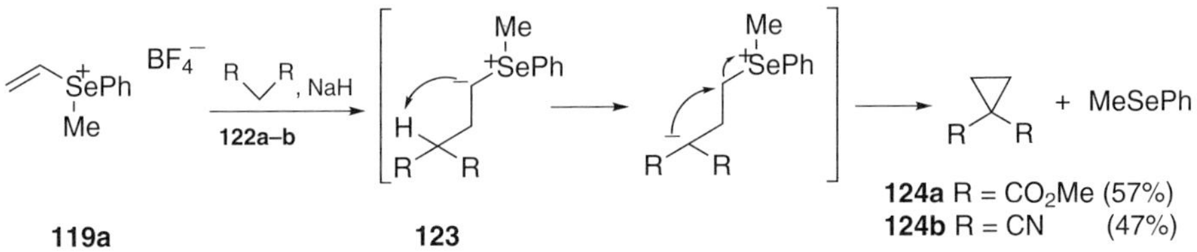

Scheme 5.34

(*E*)-β-Styrylselenonium triflate (**126**) was found to be a good substrate for a novel tandem Michael–Favorski-type reaction [41]. Vinylselenonium salt **126** was prepared via the reaction of trimethylstyrylsilane **125** with diphenylselenoxide in the presence of trifluoromethanesulfonic anhydride [42]. Treatment of active methylene carbanions, prepared by the reaction of NaH and 1,3-dicarbonyl compounds **127**, with **126** gave cyclopropane derivatives **128**. In the first step of a reaction of the selenonium salt **126** with active methylene carbanion the selenonium ylide **129** is formed. The ylide carbanion attacks the more active carbonyl group intermoleculary to give a cyclobutane **129a**, which is followed by the 1,2-migration of the *endo* carbon–carbon bond with the elimination of diphenylselenide to form a cyclopropane derivatives **128** (Scheme 5.35).

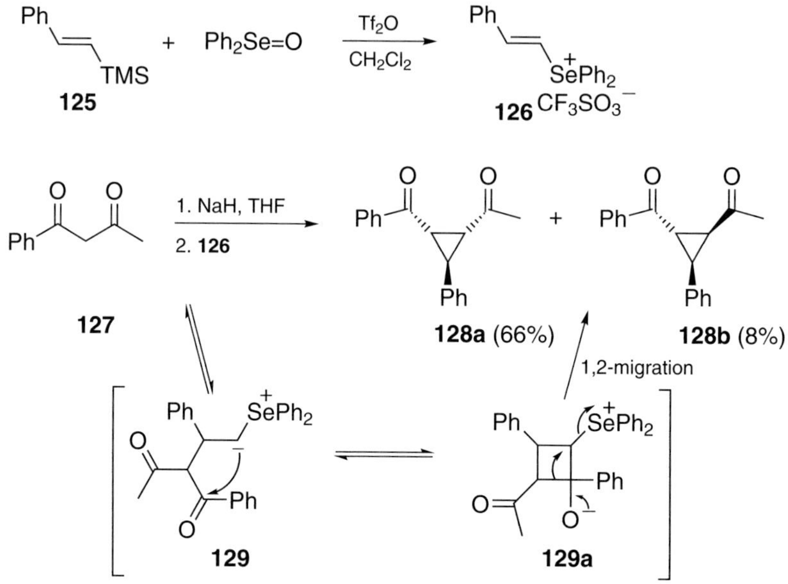

Scheme 5.35

The reaction of (*E*)-vinylselenonium salt **126** with a base such as sodium or potassium hydride gave vinylselenonium ylide **130**, which reacted with aromatic aldehydes **131** to produce the α,β-unsaturated ketones **132a–e** with a retention of configuration (Scheme 5.36) [42].

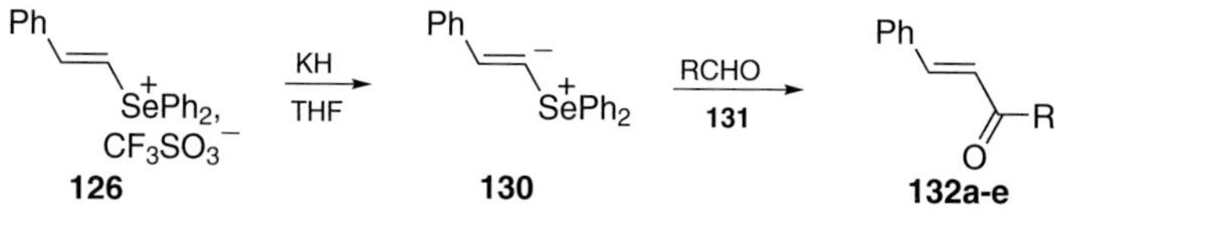

RCHO	Yield (%)
p-O$_2$NC$_6$H$_4$CHO	94
p-ClC$_6$H4CHO	77
p-BrC$_6$H$_4$CHO	76
PhCHO	60
p-MeC$_6$H$_4$CHO	49

Scheme 5.36

Recently, a novel synthesis of ketodiphenylselenonium ylide **133** from alkynylselenonium salt **35** via a Michael-type addition of hydroxide ions was described. Further reaction with aldehydes in the presence of silver triflate and triethylamine gave *trans*-oxiranylketones **134** in good yields (Scheme 5.37). When sodium *p*-toluenesulfonamide was used instead of hydroxide, aziridine **135** derivatives were obtained (Scheme 5.38) [43]. The epoxides **134** obtained from the reactions were pure *trans*-isomers. On the other hand, the compounds **135** were single *cis*-isomers.

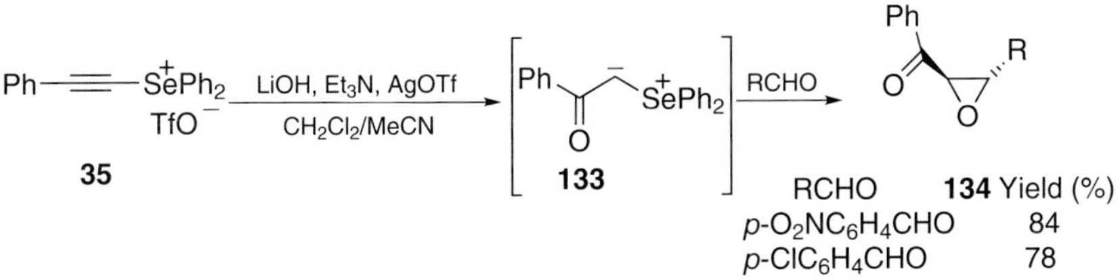

RCHO	**134** Yield (%)
p-O$_2$NC$_6$H$_4$CHO	84
p-ClC$_6$H$_4$CHO	78

Scheme 5.37

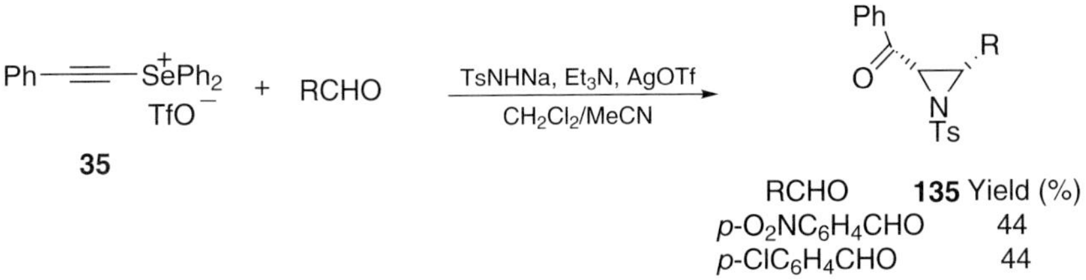

RCHO	**135** Yield (%)
p-O$_2$NC$_6$H$_4$CHO	44
p-ClC$_6$H$_4$CHO	44

Scheme 5.38

It is important to note that except of many applications of selenonium salts, new syntheses of these compounds were also presented [44], and analogues of naturally occurring glycosidase inhibitors were obtained [45].

5.4 Tetravalent, Dicoordinated Derivatives

This class of selenium compounds consists only of two representatives, for example, selenium dioxide **136** and substituted selenodiimides **137**. Selenium dioxide chemistry has been well covered in 2000 by Drabowicz and Mikołajczyk [2f] and some of its reactions have been described by Wirth [2g] in 2007. Therefore, in this chapter, the literature on the chemistry of selenium dioxide since 2000 until 2010 will be covered:

O=Se=O   RN=Se=NR

136   **137**

As for selenodiimides **137**, their chemistry has not been reviewed; therefore, we describe their synthesis and conversions since the beginning.

The most important synthetic application of selenium dioxide is undoubtedly the oxidation of various organic species. Microwave-assisted, oxidative rearrangement of chiral oxazolidines **138** with SeO_2, led to the formation of 1,4-oxazin-2-ones **139**, which was then easily converted to the chiral, fluorinated amino acids, which are highly of interest [46] (Scheme 5.39).

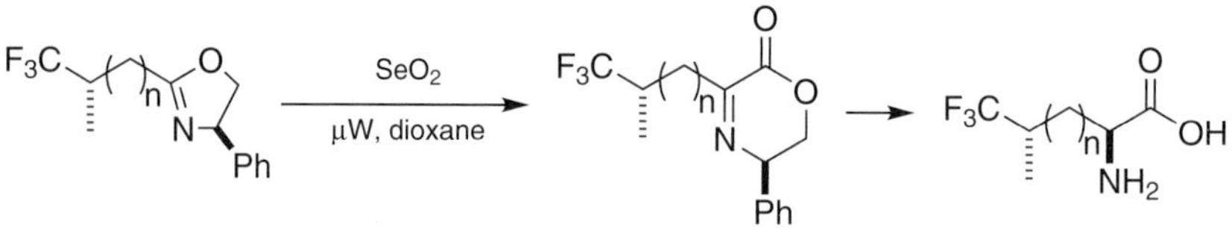

138 n=1,2   **139** n = 0,1

Scheme 5.39

Selenium dioxide was applied to the one-pot synthesis of aryl α-keto esters, where the key step is the oxidation of aryl methyl ketones **140a–f** to aryl α-keto acids **141a–f** [47, 49] (Scheme 5.40).

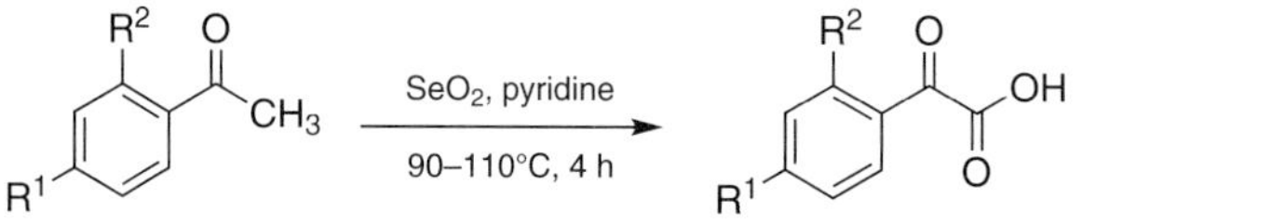

140a–f   **141a–f** (77–99%)

a: R^1=R^2=H; **b**: R^1 = Cl; R^2 = H; **c**: R^1 = NO_2, R^2 = H;
d: R^1 = R^2 = OMe, **e**: R^1 = OPh, R^2 = H;
f: R^1 = OC_6H_4-*p*-OMe, R^2 = H

Scheme 5.40

Similarly, the phenyl methyl ketone **140a** has been converted to phenyl glyoxal monohydrate **142** by the action of selenium dioxide in the dioxane–water mixture [48] (Scheme 5.41).

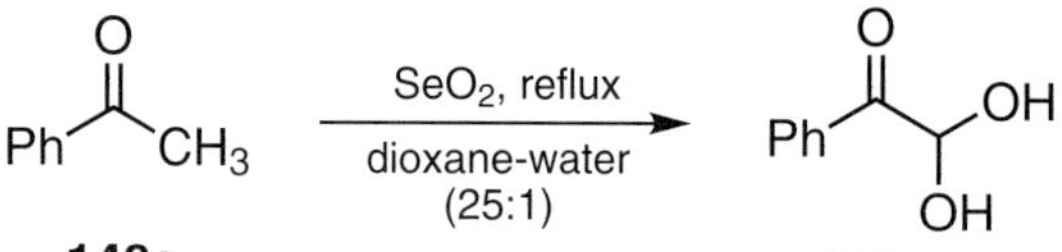

140a   **142**

Scheme 5.41

Selenium dioxide was involved in the methoxyhydroxylation and dihydroxylation of cyclic aryl substituted alkenes [50]. The use of SeO$_2$ together with aqueous hydrogen peroxide in methanol in reaction with 1-methanesulfonyl-4-phenyl-1,2,3,6-tetrahydropyridine (**143**) led to the formation of a mixture of dihydroxy and methoxhydroxy derivatives **144** and **145** in a ratio of 1:3 (Scheme 5.42).

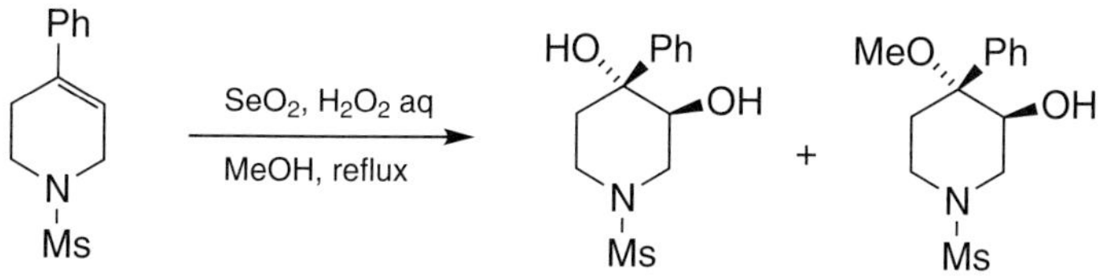

Scheme 5.42

When methanolic solution of hydrogen peroxide was used, the exclusive formation of methoxyhydroxy derivatives was observed, whereas replacing methanol with dioxane allowed us to obtain the dihydroxy compound exclusively [50]. Moreover, the method was extended to other cyclic derivatives, such as pyrane and cyclohexane (Scheme 5.43).

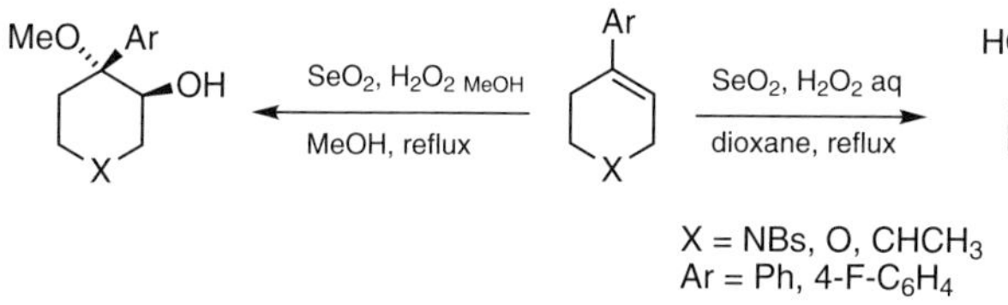

X = NBs, O, CHCH$_3$
Ar = Ph, 4-F-C$_6$H$_4$

Scheme 5.43

Selenium dioxide together with the urea hydrogen peroxide (UHP) was used as the oxidizing agent for the allylic oxidation of two sesquiterpene lactones: dehydrocostus lactone **146** and isoalantolactone **147** [51]. In both cases, products of monohydroxylation **148**, **149**, and **150** have been obtained demonstrating the occurrence of the high selectivity (Scheme 5.44).

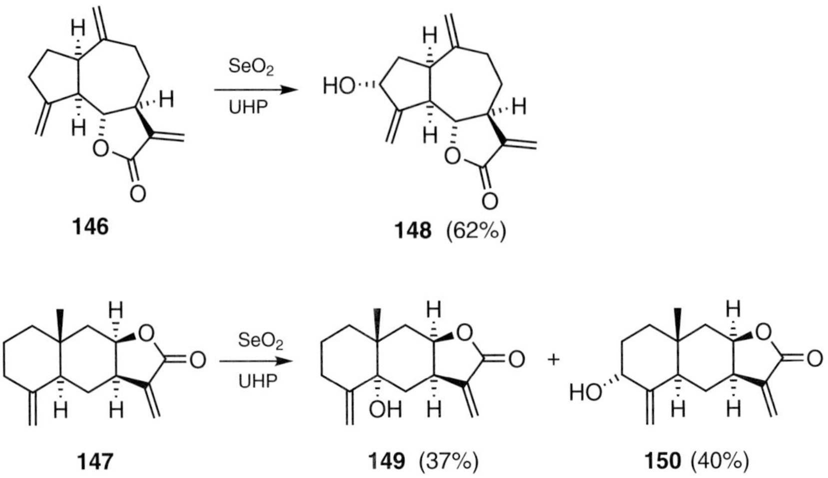

Scheme 5.44

Selenium dioxide was used for the chemoselective oxidation of compounds 151–153 obtained by the 1,4-addition to benzylideneacetophenone, which gave the products of α-oxidation 154, dehydrogenation 155, and cyclization 156a–b [52] (Schemes 5.45 and 5.46).

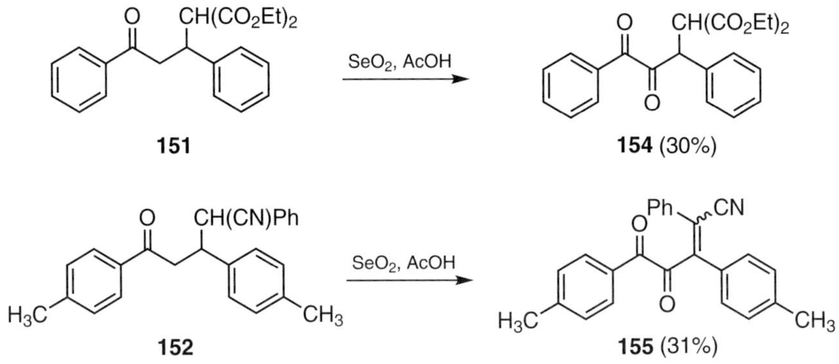

Scheme 5.45

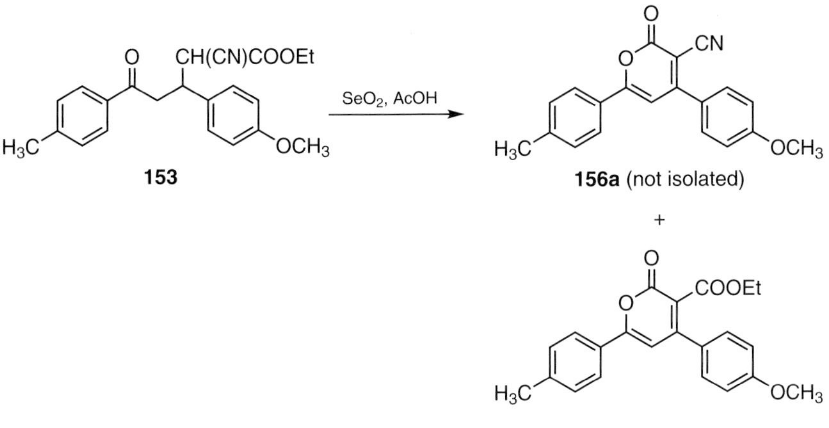

5.4 Tetravalent, Dicoordinated Derivatives | 215

Scheme 5.46

The regioselective oxidation of methyl groups in 2,3,5,6-tetramethylpyrazine **157** with a selenium dioxide–silver nitrate oxidizing system yielded 3,6-dimethylpyrazine-2,5-dicarboxylic acid **158** [53]. The reaction was very efficient, affording the product in 84% yield (Scheme 5.47).

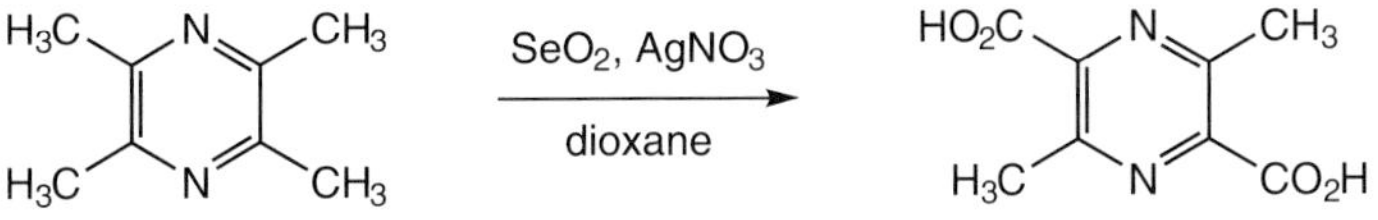

Scheme 5.47

The microwave-assisted oxidation with selenium dioxide in glacial acetic acid of tetralones **159** allowed to afford 1,2-naphthoquinones **160** in reasonable yields [54] (Scheme 5.48). As in many microwave-assisted reactions, the reaction times were very short, oxidations took 1 s with 100% conversion rates.

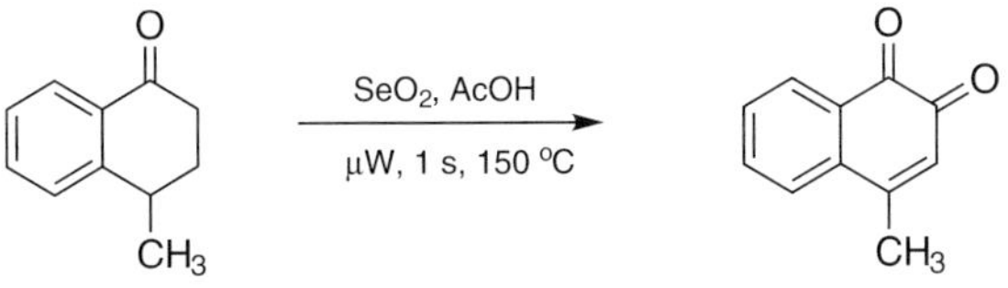

Scheme 5.48

A series of other 2-methylpyridines and 2-methylpyrazines have been converted in good yield by microwave-assisted selenium dioxide oxidation to important heterocyclic aldehydes [55]. In this way 2-pivaloylamino-6-methylpterin **161** gave 2-pivaloylamino-6-formylpterin **162**, and interestingly, 2-methylpyrazine **163** gives 2-pyrazinecarboxylic acid **164** under these conditions (Scheme 5.49).

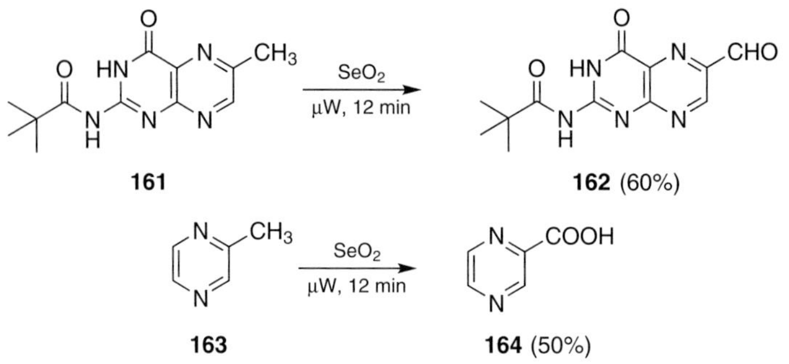

Scheme 5.49

The second important application of selenium dioxide in organic synthesis is the catalysis of various condensations. The condensation between glyoxal monohydrates **142** and β-carbonyl sulfonium salts **165**, [56] catalyzed by selenium dioxide and sodium carbonate, led to the formation of vinyl sulfides **166** (Scheme 5.50).

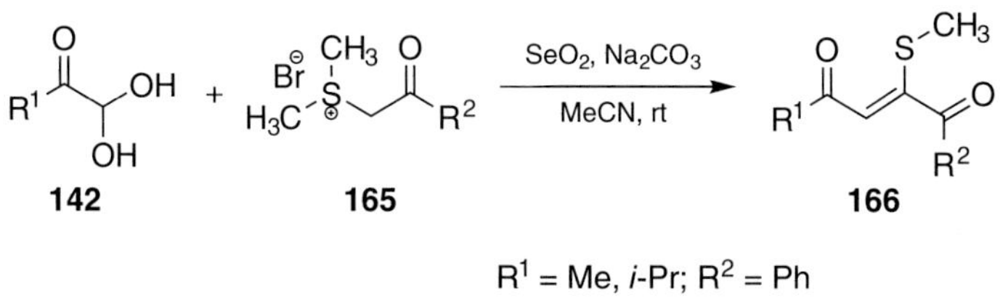

Scheme 5.50

The reaction of ninhydrin **167** with sulfonium salts (**168**, R = Me) catalyzed by selenium oxide and cesium carbonate yielded the mixture of vinyl alcohol **169** and vinyl sulfide **170** in a ratio of 7:3 up to 8:2 [57] (Scheme 5.51). Interestingly, a diizopropyl sulfonium salt (**168**, R = i-Pr) in the similar reaction resulted in the exclusive formation of vinyl alcohol **169** (Scheme 5.51). The reaction of ninhydrin **167** with sulfonium salts (**168**, R = Me), when carried out without the selenium oxide catalyst, led to the occurrence of excess hydroxysulfide **171**, stressing the importance of selenium oxide catalysis [57].

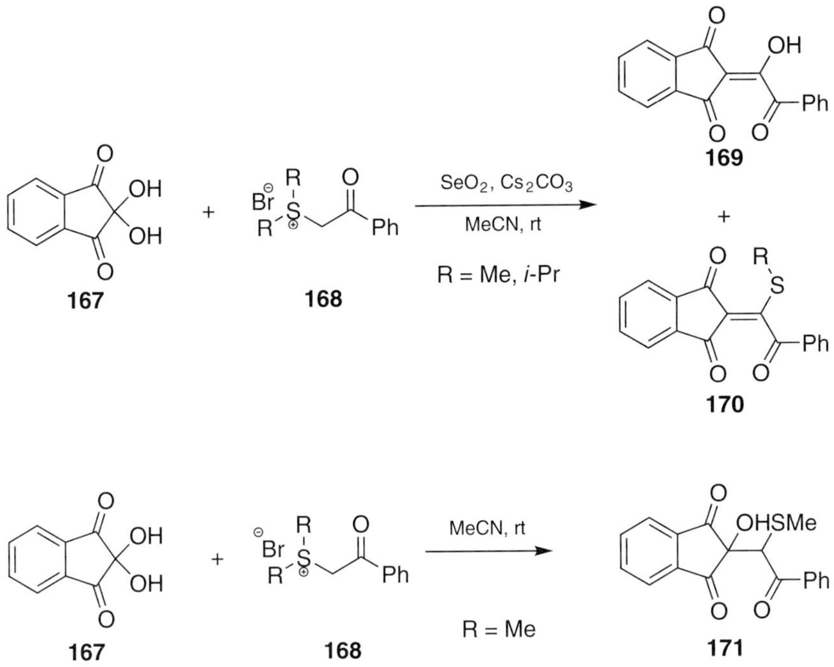

Scheme 5.51

Oxidative coupling of 1-ethoxy-1-oxobenzophosphole-3-one (**172**) catalyzed by selenium dioxide in acetic anhydride led to the formation of the important and valuable phosphoindigo **173** in reasonable yields [58] (Scheme 5.52).

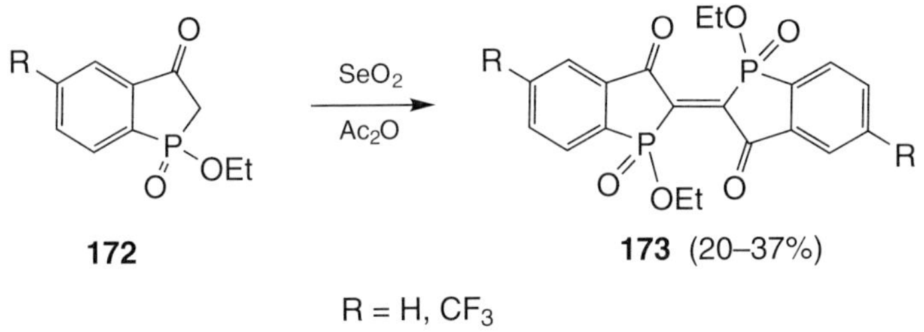

Scheme 5.52

Catalytic value of selenium dioxide has been confirmed also in the series of reactions of 2-methoxy-5-nitropyridine **174** with carbon monoxide and variously substituted aminopyridines **175**, which allowed to synthesize a series of dipyridylurea derivatives **176** [59]. When aminopyrimidines **177** were used, the formation of corresponding urea derivatives **178** was observed (Scheme 5.53). As for the mechanism, it is suggested that soluble SeCO is formed, which reacts with

2-methoxy-5-nitropyridine **174** leading to the production of 5-isocyanato-2-methoxy-pyridine **179**, which easily reacts with the amino component to produce urea derivatives [59] (Scheme 5.53).

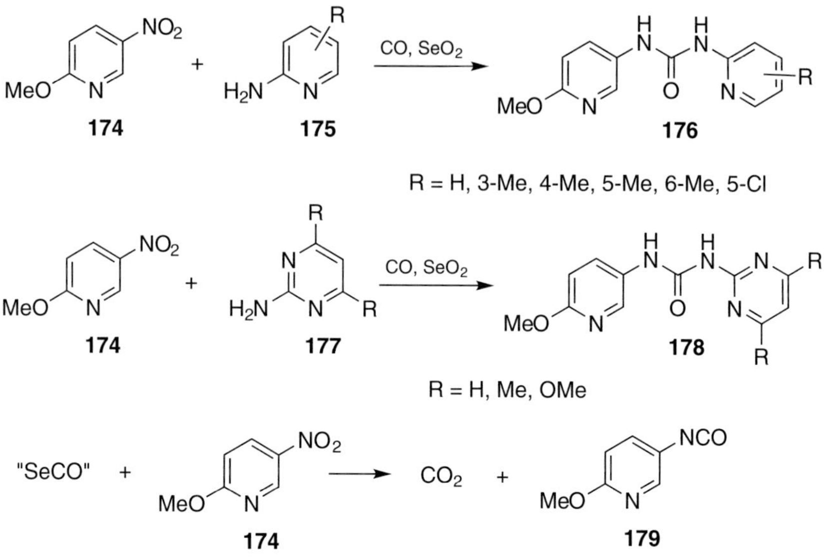

R = H, 3-Me, 4-Me, 5-Me, 6-Me, 5-Cl

R = H, Me, OMe

Scheme 5.53

Selenium dioxide as the oxidation-promoting catalyst was used for the oxidation of methyl *p*-aminobenzoate **180** to *p*-nitrosobenzoate **181** or to azoxybenzene derivatives **182** [60]. When the reaction was carried out in methanol, the exclusive formation of azoxybenzene derivative (**182**) was detected, whereas the use of dichloromethane as a solvent yielded *p*-nitrosobenzoate **181** and 4,4′-dicarbomethoxyazoxybenzene (**182**) in a ratio of 4:1 [60] (Scheme 5.54).

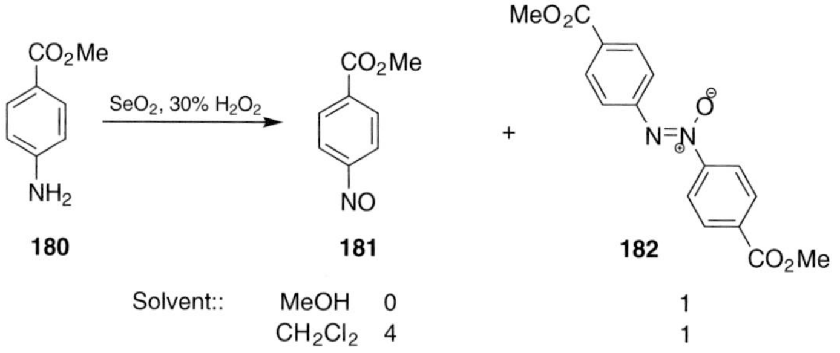

Scheme 5.54

Selenium dioxide also serves in the production of various selenium-bearing heterocyclic compounds. Carbonyl dienes **183** underwent [4+2] cycloaddition with selenium dioxide to give selenophenes **184** in good yields [61, 62] (Scheme 5.55) and addition to symmetrical tetraenes **185a–b** yielded bis-selenophenes **186a–b** (Scheme 5.56).

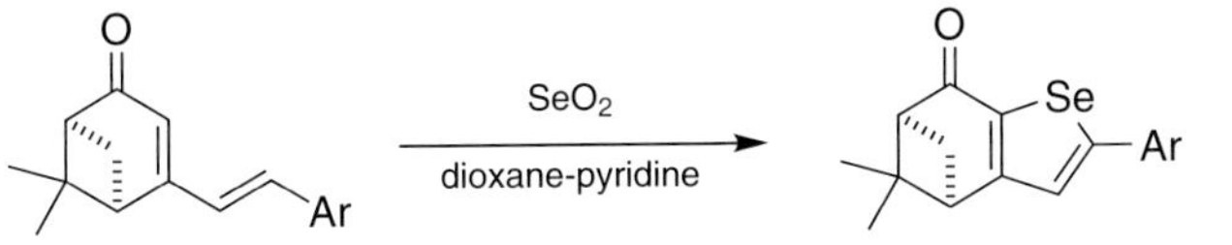

Ar = Ph; 2,6-Cl$_2$C$_6$H$_3$; 2,6-Cl$_2$C$_6$H$_3$; p-ClC$_6$H$_4$; p-MeOC$_6$H$_4$; CH=CH-Ph

Scheme 5.55

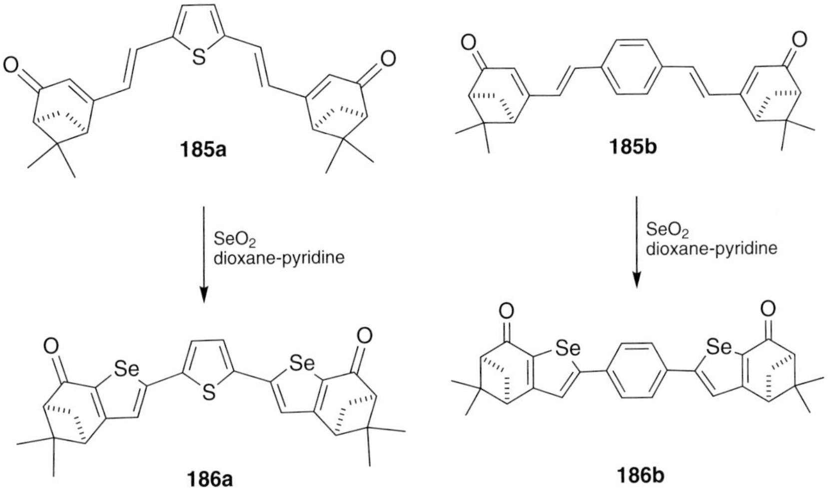

Scheme 5.56

The reaction of selenium dioxide with malononitrile in dimethylsulfoxide or dimethylformamide resulted in the formation of triselenium dicyanide **187**, which after addition of indole yields the corresponding selenocyanate **188** [63]. The reaction with quinoxaline in dioxane gave cyanoquinoxaline **189** [64] (Scheme 5.57).

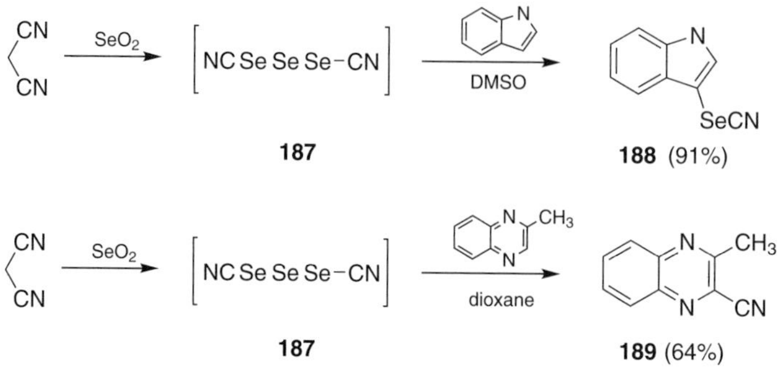

Scheme 5.57

Selenium dioxide reacted with the titanocene compound **190** in a dry benzene–THF mixture to give the organometallic species **191** in 53% yield [65] (Scheme 5.58).

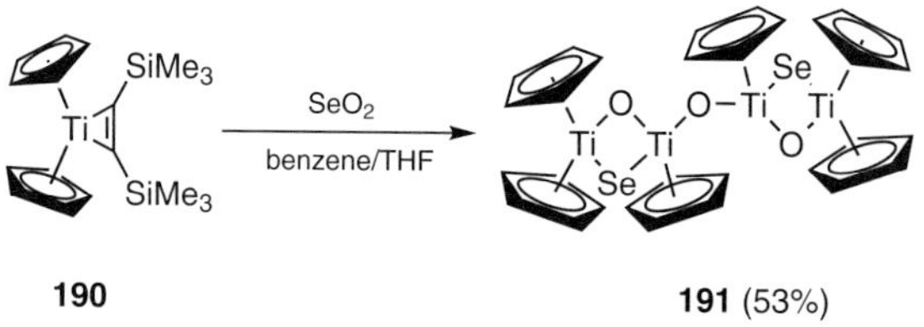

Scheme 5.58

Selenodiimides **137** constitute the second class of tetravalent, dicoordinated selenium compounds. Their chemistry is much less developed then the chemistry of selenium dioxide and its application – definitively more rarely mentioned. To the authors' knowledge, only one paper appeared within the elaborated period [76], but as this class was not yet reviewed, complete coverage of the literature is intended.

A rich series of works presented the ene-reaction or cycloaddition of selenodiimides followed by the [2,3]-sigmatropic rearrangement of the formed product.

The ene-reaction of (+)-2-carene **192** with bis-(p-toluenesulfonyl)selenodiimide **137a** led to the formation of two products **193a** and **193b** of the ene-reaction, which then underwent a [2,3]-sigmatropic rearrangement to give substituted carenes **194a** and **194b**. The simple hydrolysis with aqueous NaOH gave N-carenyl sulfonamides **195c** and **195d** [66] (Scheme 5.59).

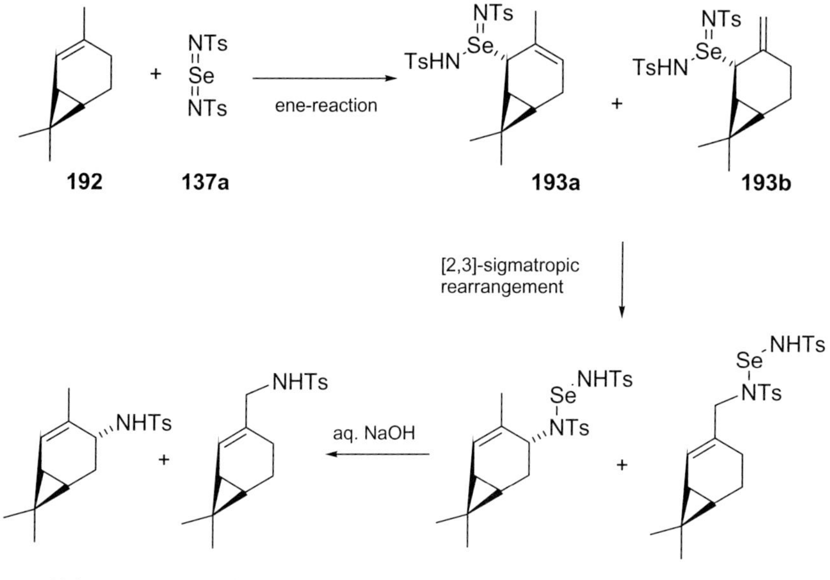

Scheme 5.59

The reaction of bis-(p-toluenesulfonyl)selenodiimide **137a** with (+)-α-phellandrene **195** led to the formation of the product **196** in a [4+2] cycloaddition, which after the reaction with tosyl amide to the organoselenium compound **197** underwent subsequently the spontaneous [2,3]-sigmatropic rearrangement to give substituted phellandrene **198**. Simple hydrolysis with aqueous NaOH gave N,N-(menthene-5,6-diyl)di-p-toluenesulfonamide **199** [67] (Scheme 5.60).

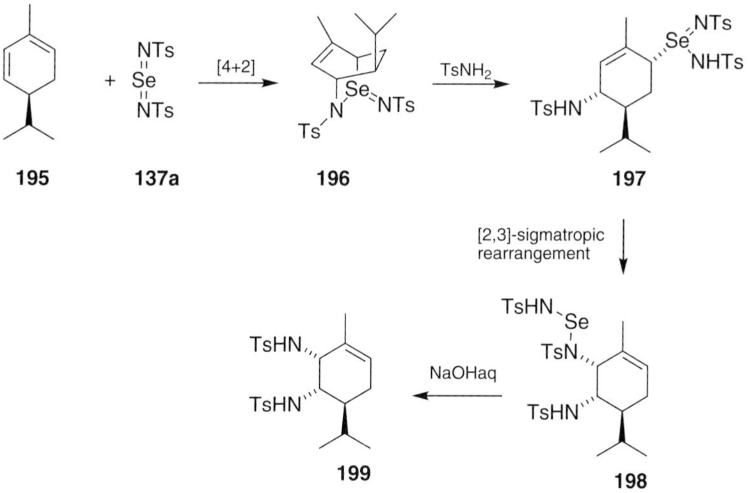

Scheme 5.60

The similar reaction of selenodiimide **137a** was performed with (*S*)-(−)-limonene [68], which led to menthenediyl disulfonamide derivatives.

Variously substituted 1,3-cyclohexadienes reacted with bis-(*p*-toluenesulfonyl) selenodiimide **137a** to give cyclohexenyl disulfonamides **204a–c** [69, 70]. There were cases, where disulfonamide was accompannied by the mono-derivative **205** [69] (Scheme 5.61).

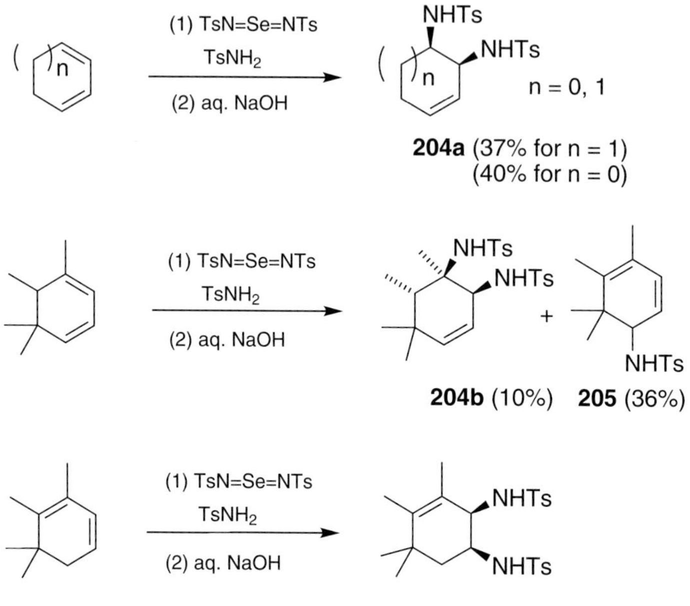

204a (37% for n = 1)
(40% for n = 0)

204b (10%) **205** (36%)

204c (31%)

Scheme 5.61

Similar reactions have been performed for several other cyclic and acyclic conjugated dienes [70] and similar mechanism was suggested for these reactions. Reactions were performed, for example, for 3,4-dimethyl-1,3-butadiene **206** to obtain disulfonamide **207** and for cyclopentadiene to obtain the product (**204a**, *n* = 0) (Scheme 5.62).

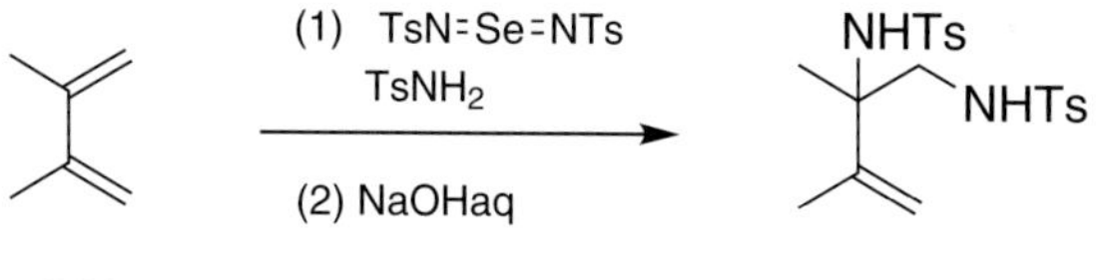

206 **207** (68%)

Scheme 5.62

5.4 Tetravalent, Dicoordinated Derivatives

Sharpless and coworkers [71] studied the reaction of two model selenodiimides **137a** and **137b** with a series of cyclic and acyclic olefins. In this way, a series of N-substituted sulfonamides **208** and **209** or secondary amine **210** were obtained in reasonable yields (Schemes 5.63 and 5.64).

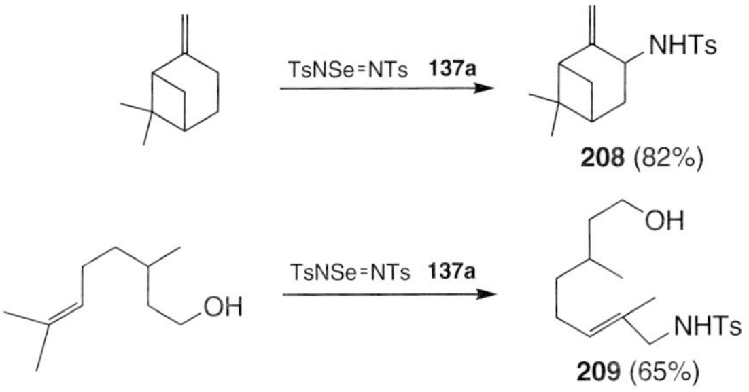

Scheme 5.63

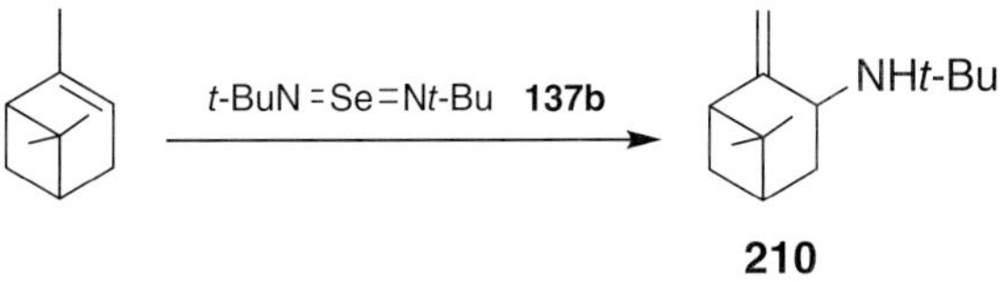

Scheme 5.64

Similar studies to obtain di- and monosulfonamides from conjugated dienes and olefins [72] have been performed, but apart from N-tosyl selenodiimide **137a**, the N-nosyl (p-nitrobenzenesulfonyl) derivative **137c** was used. Products have been synthesized in good yields:

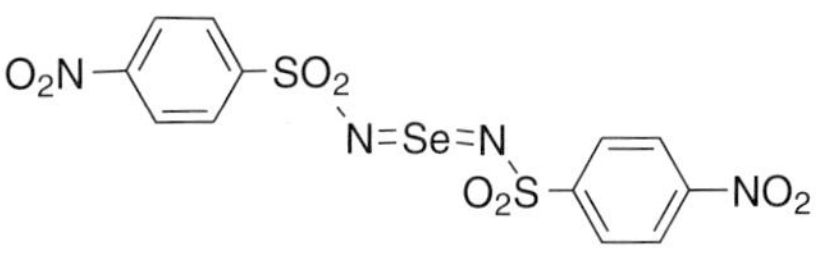

Silyl enol ethers **211** and **212** underwent a similar reaction with bis-(p-toluenesulfonyl)-selenodiimide **137a** to yield the corresponding sulfonamides **213** and **214** [73, 74] (Scheme 5.65).

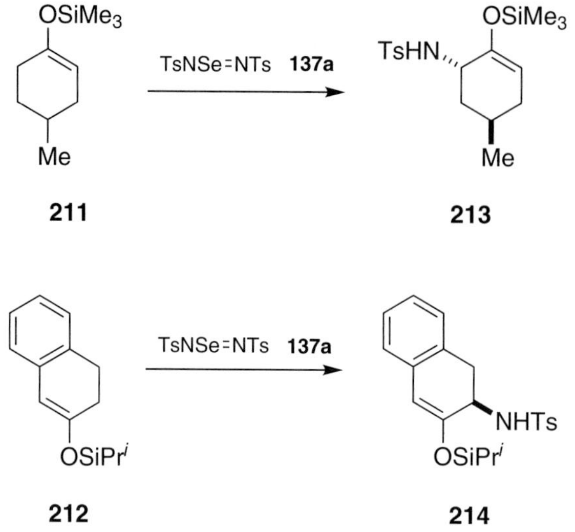

Scheme 5.65

211 → **213** (TsNSe=NTs **137a**)

212 → **214** (TsNSe=NTs **137a**)

The reaction of bis-(N-trimethylsilyl)selenodiimide **137d** reacting with phosphimidoyl-amidine **215** in diethyl ether resulted in the cyclization product **216** [75] (Scheme 5.66).

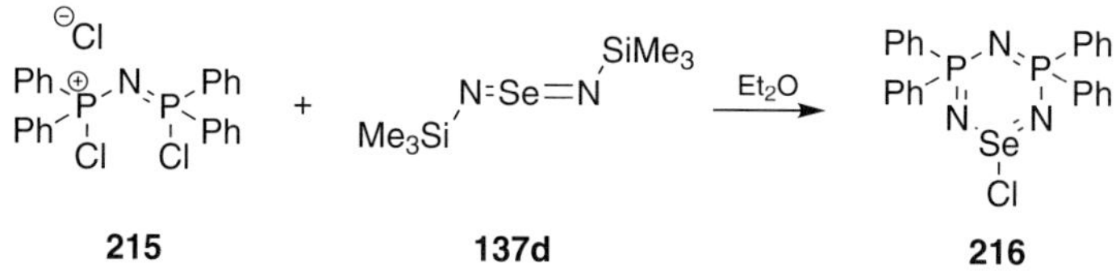

Scheme 5.66

215 + **137d** $\xrightarrow{Et_2O}$ **216**

The synthesis and structural study of two bis-aryl-selenodiimides **137e–f** by the reaction of selenium tetrachloride with lithium arylamides was also performed [76] (Scheme 5.67).

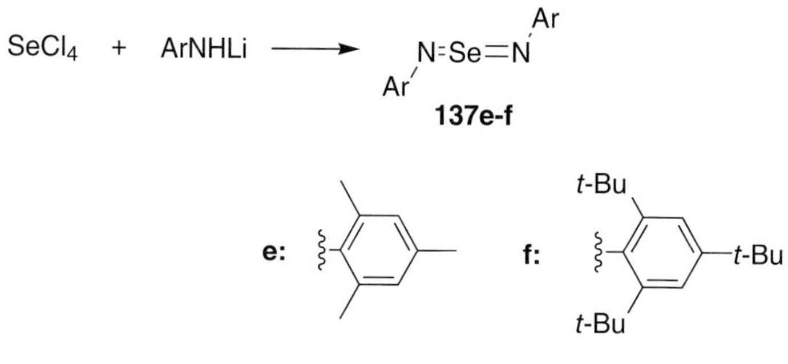

$$SeCl_4 + ArNHLi \longrightarrow \underset{Ar}{\overset{Ar}{N{=}Se{=}N}}$$

137e-f

e: 2,4,6-trimethylphenyl

f: 2,4,6-tri-t-butylphenyl

Scheme 5.67

5.5
Tetravalent, Tricoordinated Derivatives

In this class of compounds, the most important from the synthetic viewpoint are selenoxides **217**, seleninic acids **219**, seleninic acid esters **220**, amides **221**, and seleninic anhydrides **222**. To this group, selenimides **218** should be also included, but their importance is much less than the representatives mentioned above:

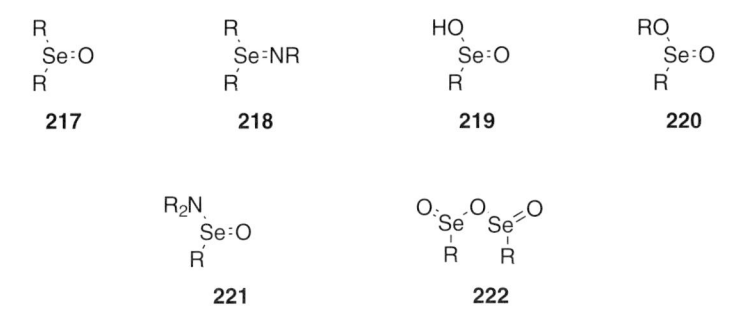

$$\underset{\textbf{217}}{\underset{R'}{\overset{R}{\text{Se}}}\text{:O}} \qquad \underset{\textbf{218}}{\underset{R'}{\overset{R}{\text{Se}}}\text{=NR}} \qquad \underset{\textbf{219}}{\underset{R'}{\overset{HO}{\text{Se:O}}}} \qquad \underset{\textbf{220}}{\underset{R'}{\overset{RO}{\text{Se:O}}}}$$

$$\underset{\textbf{221}}{\underset{R'}{\overset{R_2N}{\text{Se:O}}}} \qquad \underset{\textbf{222}}{\underset{R \quad R}{\overset{O}{\text{Se}}}\overset{O}{\text{Se}}\overset{O}{=}}$$

Among these compounds, selenoxides have found the largest application in chemical synthesis. Therefore, it is no wonder that methods for their preparation are still developed. For example, sodium hypochlorite in DMF was used for the oxidation of dialkyl- or diarylselenides **223a–h** to obtain the corresponding selenoxides **217a–h** in good yields [77a] (Scheme 5.68).

$$\underset{\textbf{223a-h}}{\underset{R'}{\overset{R}{\text{Se}}}} \xrightarrow[\text{DMF, 5 min}]{\text{NaOCl, rt}} \underset{\textbf{217a-h}}{\underset{R'}{\overset{R}{\text{Se:O}}}}$$

	R	yield (%)
a	Ph	77
b	CH_2Ph	69
c	$4\text{-Me-}C_6H_4$	76
d	$4\text{-Cl-}C_6H_4$	84
e	$4\text{-Br-}C_6H_4$	83
f	$4\text{-MeO-}C_6H_4$	80
g	$CH_2(CH_2)_2CH_3$	80
h	$n\text{-}C_{12}H_{25}$	72

Scheme 5.68

Several other oxidizing agents were also used for the oxidation of selenides to selenoxides, amongst them the most important were *m*-chloroperbenzoic acid [77b] or *tert*-butyl hypochlorite [77b].

A very important methodology for synthesis of terminal olefins is proceeding *via* selenoxides. Oxidation of selenide **224** with Dess–Martin periodinane **225** led, after hydrolysis, to the formation of the corresponding selenoxide **226**, which underwent the spontaneous elimination to give an olefin with a terminal double bond **227** in fair yield [77c] (Scheme 5.69).

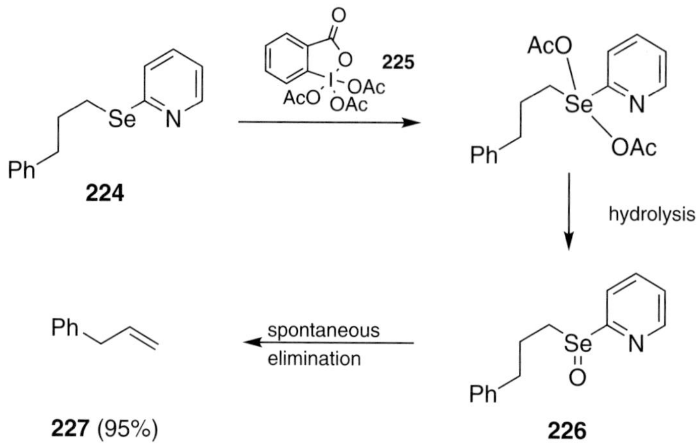

Scheme 5.69

Selenoxides **217** underwent a photochemical degradation on two possible routes [78]. One of them led to the formation of diselenides, alcohols and, in a case of **217** with R = Ph to benzaldehyde. It was suggested that the reaction went through a selenenic acid ester as the rearrangement product **228**. The second possible route allowed us to obtain selenides **223** and the participation of the intermediate **229** or 1,2-diselenetane **230** cannot be excluded [78] (Scheme 5.70).

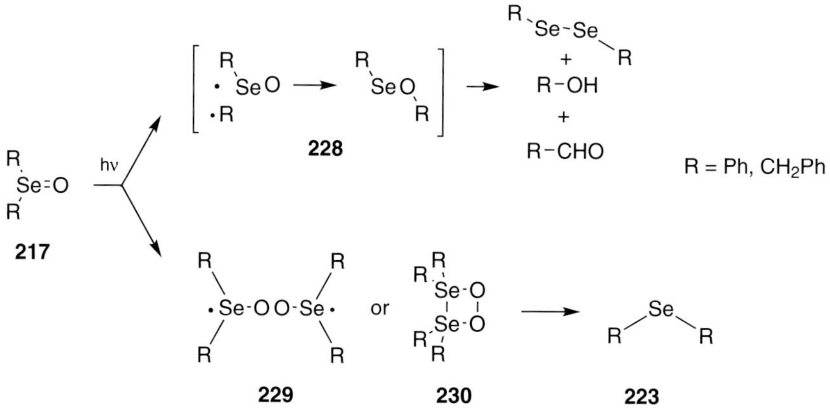

Scheme 5.70

Water-soluble organic selenides, such as 3,3-selenodipropanonic acid (**223**, R = $CH_2CH_2CO_2H$) have been oxidized to the corresponding selenoxides (**217**, R = $CH_2CH_2CO_2H$) with hydrogen peroxide in deuterium oxide [79a]. In the same way, 2-selenanecarboxylic acid **231** was oxidized to 2-selenanecarboxylic acid Se-oxide **232**. These selenoxides, when treated subsequently with dithiothreitol (DTT), underwent reduction to selenides with simultaneous oxidation of DTT to form

3,4-dihydroxy-1,2-dithiane **233**. Similar studies demonstrated that the selenoxides may act as glutathione peroxidase (GPx) antioxidants [79b] (Scheme 5.71).

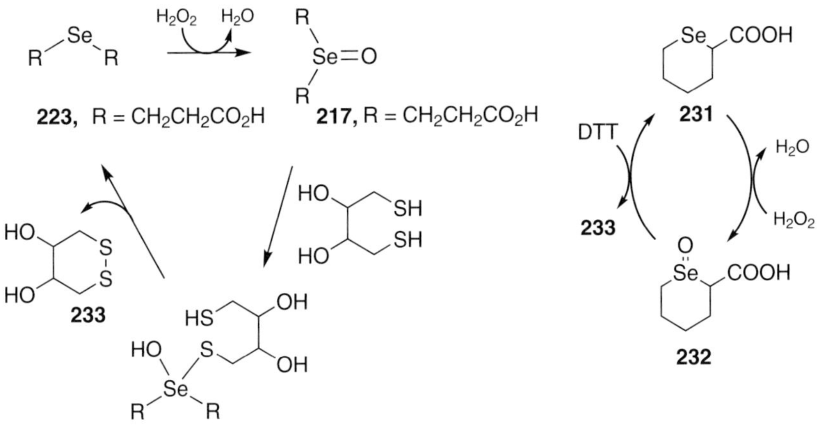

Scheme 5.71

A similar catalytic behavior has been studied in the oxidation of selenides with *tert*-butyl hydroperoxide [80]. The action of *tert*-butyl hydroperoxide on di(3-hydroxypropyl)selenide **234** led to the formation of di(3-hydroxypropyl)selenoxide **235**, which underwent intramolecular dehydration to form spiro-selenurane **236** (Scheme 5.72).

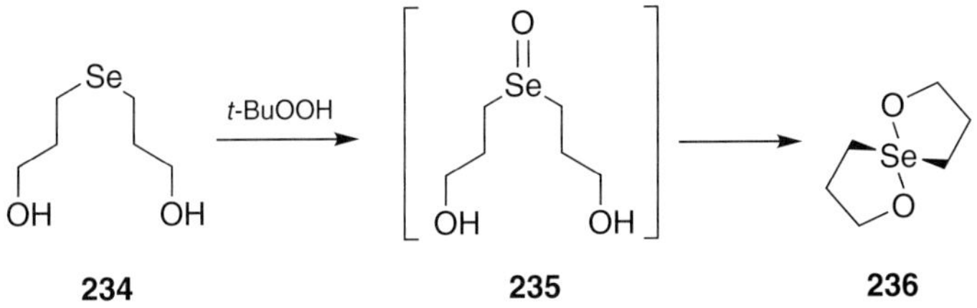

Scheme 5.72

Similarly, selenomethionine **237** was oxidized with hydrogen peroxide to selenomethionine *Se*-oxide **238**, which, upon the reaction with thiol, gave a disulfide with a recovery of selenomethionine **237** [81]. The formation of selenurane **239** was confirmed by the NMR and IR spectroscopy and mass spectrometry (Scheme 5.73).

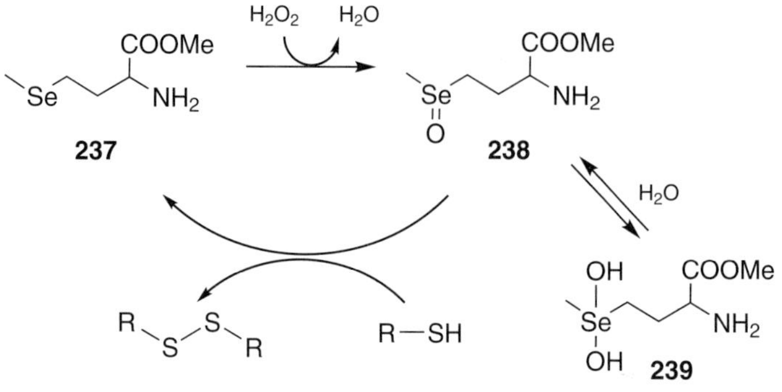

Scheme 5.73

Interestingly, the oxidation of selenomethionine **237** to selenomethionine Se-oxide **238** could also be performed by the action of flavin-containing monooxygenases (FMO) [82].

The action of osmium(VIII) reagent (catalytic amounts of $K_2OsO_2(OH)_4$–$(DHQD)_2PHAL$) on selenides caused a subsequent rearrangement of allyl selenoxides to allyl alcohols [83]. It allowed the transformation of methyl geranyl selenide **240** via selenoxide **241** to linalool **242**. Methyl geranyl selenide **240** and benzyl phenyl selenoxide acted as the two partners where the equilibrium took place and be drawn to completion by the allyl selenoxide-[2,3] sigmatropic rearrangement to produce finally linalool **242** and benzyl phenyl selenide. Methyl citronellyl selenoxide **243** is efficiently transformed to 6,7-dihydroxy citronellyl selenide **244** (Scheme 5.74).

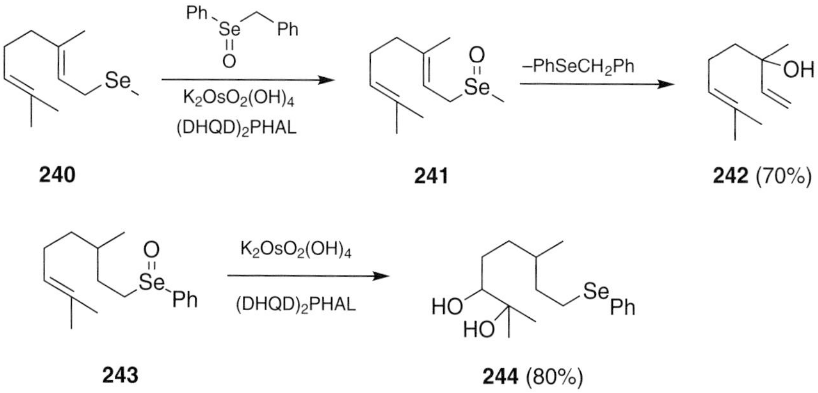

Scheme 5.74

Vinyl benzyl selenides were oxidized with hydrogen peroxide in THF to obtain the corresponding selenoxides in reasonable yields [84]. Oxidation of dimethyl selenide to dimethylselenoxide with peroxynitrite was also studied, and was

5.5 Tetravalent, Tricoordinated Derivatives | 229

demonstrated to be quite efficient [85]. The structure of dimethylselenoxide has been well established [86a] as well as structure and interactions between substituents in 1,8-bis-seleninylnaphthalene **245** [86b]:

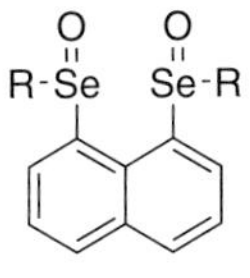

245

R = Me, Et, Ph

Some biochemical studies showed that oxidation of *Se*-methylselenocysteine dipeptide **246** and *Se*-methylselenocysteine gives methaneseleninic acid and dimethyl diselenide [87]. This oxidation occurs *via* selenoxide **247**, which was detected by chromatographic and spectroscopic methods:

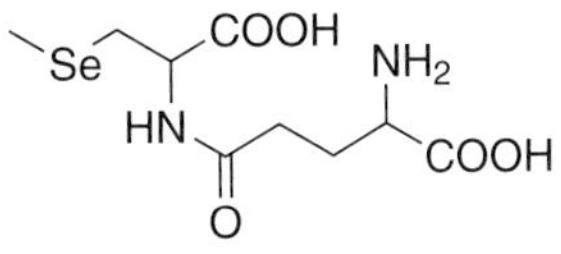

246 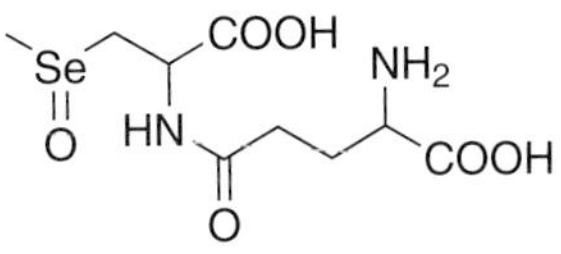   **247**

Selenoxides **217**, deoxygenate by reaction with magnesium in methanol to selenides **223** at room temperature in almost quantitative yields [88]. The most probable mechanism of this reaction is the attack of magnesium on the Se=O oxygen and further liberation of magnesium oxide (Scheme 5.75).

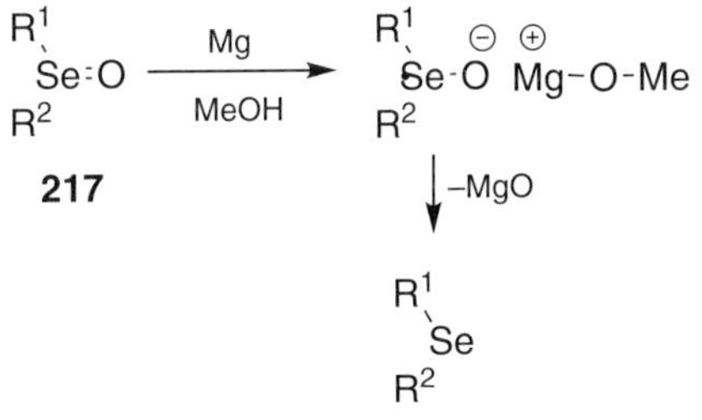

	yield [%]
$R^1 = R^2 = Ph$	80
$R^1 = R^2 = 4\text{-Me-C}_6H_4$	82
$R^1 = R^2 = 4\text{-MeO-C}_6H_4$	82
$R^1 = R^2 = 4\text{-Br-C}_6H_4$	82
$R^1 = R^2 = t\text{-Bu}$	70
$R^1 = Ph$, $R^2 = C_3H_5$	71

Scheme 5.75

Perfluorinated selenides **247** were oxidized with HOF without a solvent to perfluoroalkyl selenoxides **248** in good yields varying from 50 to 90% [89] (Scheme 5.76).

Scheme 5.76

R-Se-R →(HOF) R-Se(=O)-R
 247   248

R =	CF$_3$	91
R =	C$_2$F$_5$	71
R =	n-C$_3$F$_7$	47
R =	n-C$_4$F$_9$	51

yield [%]

Selenoxides appeared also in some biochemical works as their biological significance is rather large. Methylselenocysteine was oxidized to methylselenocysteine selenoxide, which in the presence of cysteine conjugate β-lyase, was converted to unstable methylselenenic acid, which converted into methylselenenylsulfide and finally to methylselenol [90]. Some selenoxides oxidized the remaining thiol groups, which resulted in inhibition of the ryanodine receptor [91]. Some studies on biofortified grains demonstrated that selenomethionine selenoxide 238 is present in this biomaterial [92].

In vivo, selenides are oxidized to selenoxides by FMO and selenoxides regenerate selenides by oxidation of thiols [93]. Phenyl methyl selenoxide (PhSe(O)CH$_3$) and 1-hexynyl methyl selenoxide (C$_4$H$_9$C≡CSe(O)CH$_3$) inhibited δ-aminolevulinic acid dehydratase from liver in rats. This enzyme is inhibited probably by a possible involvement of thiol groups. The inhibitory action of selenoxides was antagonized by dithiotreitol (DTT) [93], which was separately studied in papers [79, 82], which are described above.

Selenimides are not largely represented in the literature; in the studied period, there are notably three methods described for their preparation.

The reaction of aryl benzyl selenides 223 with N-(p-toluenesulfonyl)imino]phenyliodinane 249 in the presence of copper(I) triflate in toluene allowed us to obtain the corresponding N-tosylselenimides 218 in 31–46% yield [94] (Scheme 5.77). Interestingly, the addition of molecular sieves allowed to raise the yield up to 64% and the presence of optically active 4,4-disubstituted bis(oxazoline) 250

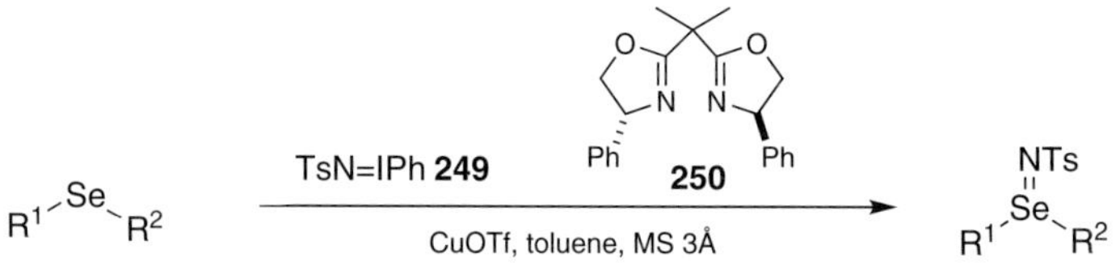

R^1-Se-R^2 →(TsN=IPh 249, 250, CuOTf, toluene, MS 3Å)   R^1-Se(=NTs)-R^2
  223   218

R^1 = Ph; R^2 = CH$_2$Ph
R^1 = 1-naphthyl; R^2 = CH$_2$Ph
R^1 = Ph; R$_2$ = CH$_3$
R^1 = 2,4,6-t-Bu$_3$-C$_6$H$_2$; R^2 = CH$_2$Ph

Scheme 5.77

as a ligand, the imidation turned out to be enantioselective to give optically active *N*-tosylselenimides with enantiomeric excesses of up to 36% [94]. The diastereoselective imidation of diaryl selenides bearing a chiral oxazolinyl moiety with **249** or chloramine–T trihydrate has been successfully carried out to give the corresponding optically active *N*-tosylselenimides **251** in good yields and diastereomeric excess (Scheme 5.78). Noteworthy, chloramine–T trihydrate turned out to be a better reagent for this reaction as yields reached up to 97% and *de* up to 76% [94].

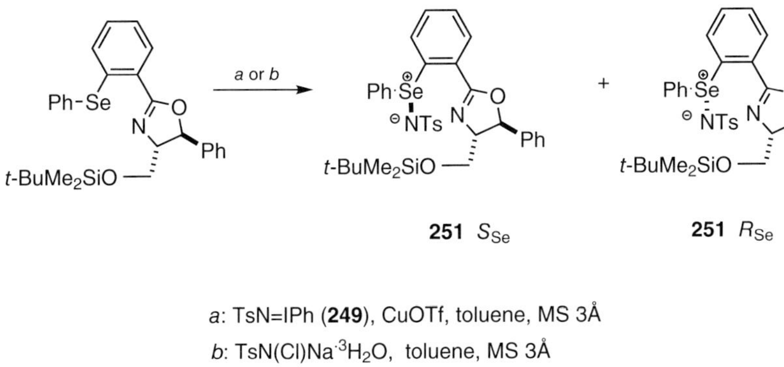
Ph-Se
t-BuMe2SiO
a or b
Ph·Se
NTs
Ph
+
251 SSe
251 RSe
a: TsN=IPh (249), CuOTf, toluene, MS 3Å
b: TsN(Cl)Na·3H2O, toluene, MS 3Å

Scheme 5.78

When this reaction was performed with allyl aryl selenides **252** in the presence of the chiral catalyst (**250**), the reaction occurred in a different way. The initially formed chiral allylic *N*-tosylselenimide **253** underwent a [2,3]-sigmatropic rearrangement to give chiral allylic *N*-tosylamides **254** in moderate to good yields and with up to 30% *ee* [94] (Scheme 5.79). Interestingly, it was shown that the chirality transfer from selenium to carbon atom occurred at the rearrangement step.

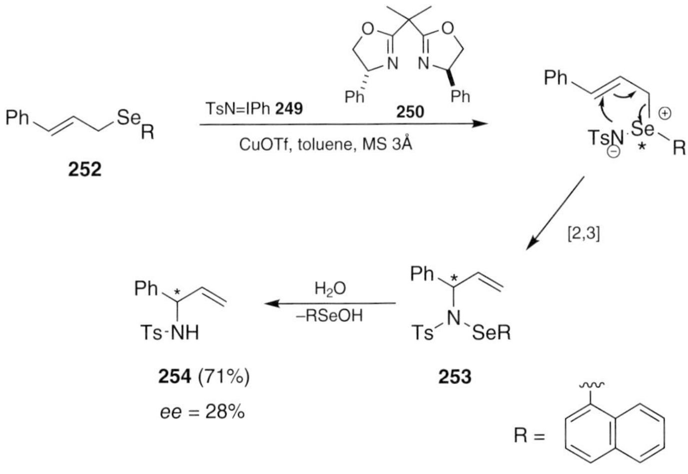
252
TsN=IPh 249
250
CuOTf, toluene, MS 3Å
[2,3]
H2O
–RSeOH
254 (71%)
ee = 28%
253
R =

Scheme 5.79

The reaction of 1,4-phenylene-bis-(N-bromo-phenylsulfinimide) (**255**) with diphenyl selenide led to the bis-N-selenosulfimide, which was subsequently converted to the tetraphenylborate salt **256** [95] (Scheme 5.80).

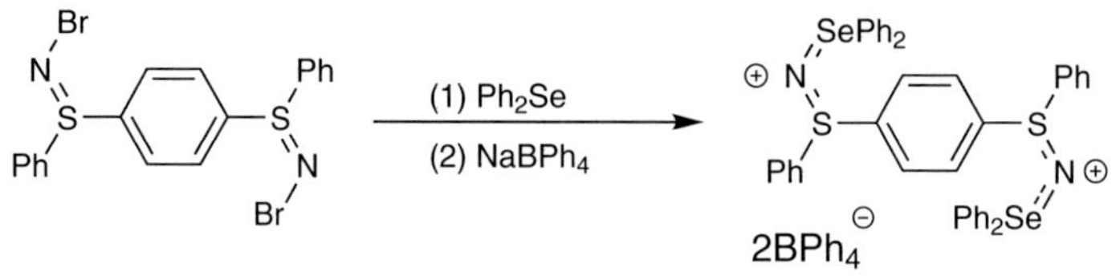

Scheme 5.80

Reaction of o-mesitylsulfonylhydroxylamine **257** with diphenyl selenide yielded the diphenyl selenimidium cation, which after the low-temperature deprotonation with LDA led to diphenyl selenimide **258** [96]. The subsequent bromination with NBS resulted in N-bromodiphenyl selenimide **259**. The final treatment with diphenyl selenide and then with sodium tetraphenylborate generated tetraphenyl diselenimidium tetraphenylborate **260** [96] (Scheme 5.81).

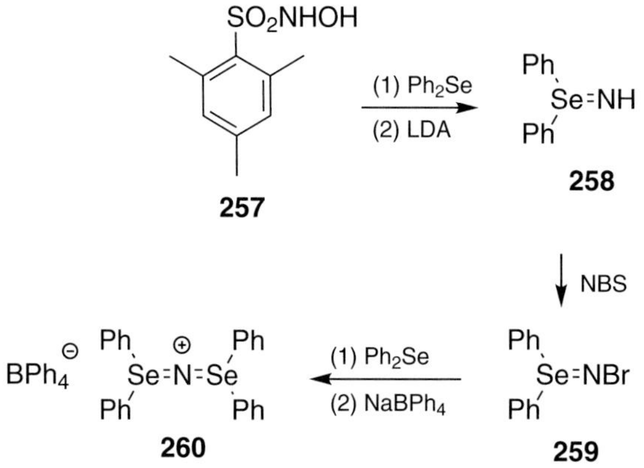

Scheme 5.81

The application of selenininic acid derivatives to various oxidation cycles is well established [2f]. For example, the reaction of alkenes with iodoxybenzene as reoxidant, in the presence of perfluorooctylseleninic acid, which catalyzes the allylic oxidation, gave alkenones in moderate to good yields. After a workup with sodium metabisulfite, the catalyst is recovered in the form of bis(perfluorooctyl) diselenide, which can serve as a convenient catalyst precursor. Using this methodology, cholesteryl benzoate **261** has been oxidized to the corresponding ketone **262** in 92% yield [97] (Scheme 5.82).

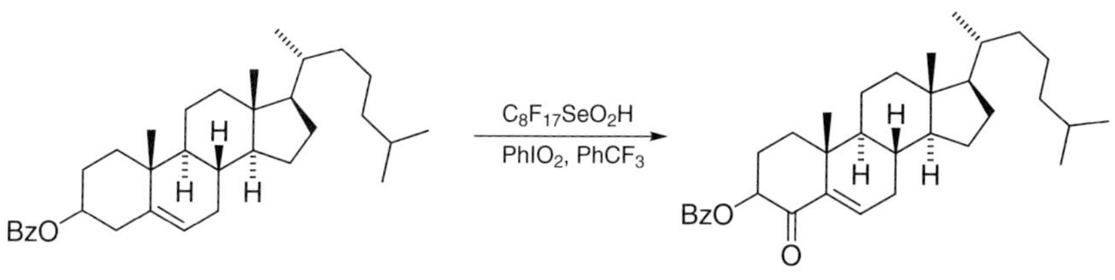

261 **262** (92%)

Scheme 5.82

Allyl hydroxypropyl selenide **263a** and allyl hydroxybutyl selenide **263b** were oxidized with *tert*-butylhydroxyperoxide to form allyl hydroxyalkyl selenoxides [98]. These allyl selenoxides underwent the [2,3]-sigmatropic rearrangement to form allyl selenenates, which was oxidized again to allyl seleninates **264a–b**. Seleninates **264a–b** underwent the immediate intramolecular cyclization to produce 1,2-oxaselenolane *Se*-oxide **265a** and 1,2-oxaselenane *Se*-oxide **265b** [98] (Scheme 5.83).

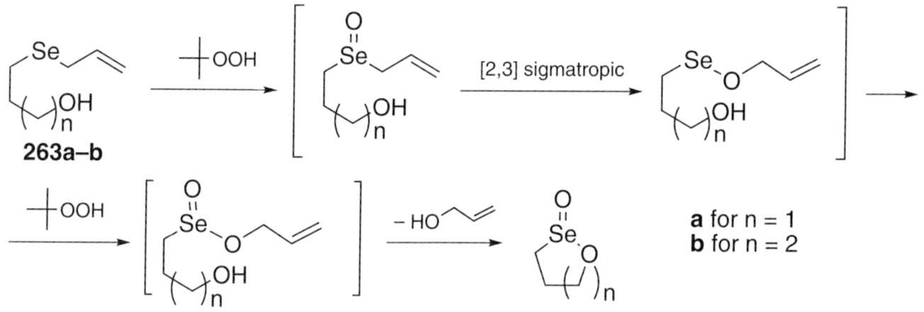

264a–b **265a–b**

Scheme 5.83

Selenyl sulfide **268** was oxidized with *tert*-butylhydroxyperoxide to form seleninyl sulfide **269**, which underwent cyclization to form 1,2-oxaselenolane *Se*-oxide **265a** [98] (Scheme 5.84).

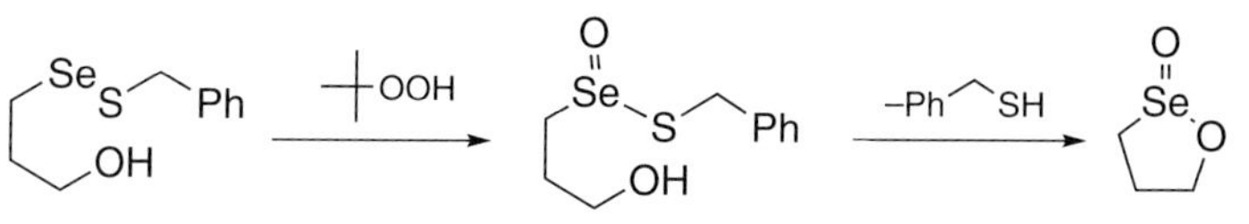

268 **269** **265a**

Scheme 5.84

The oxidation of polyfunctional selenoesters with dimethyldioxirane (DMDO) allowed us to obtain initially seleninic acid and then its selenonic analogue [99]. Thus, the oxidation of Se-glucopyranosyl phenylselenoacetate **270** with DMDO formed 6-(6-deoxy)glucopyranose seleninic acid derivative **271**, which upon further DMDO oxidation, produced the corresponding selenonic acid derivative **272** [99] (Scheme 5.85).

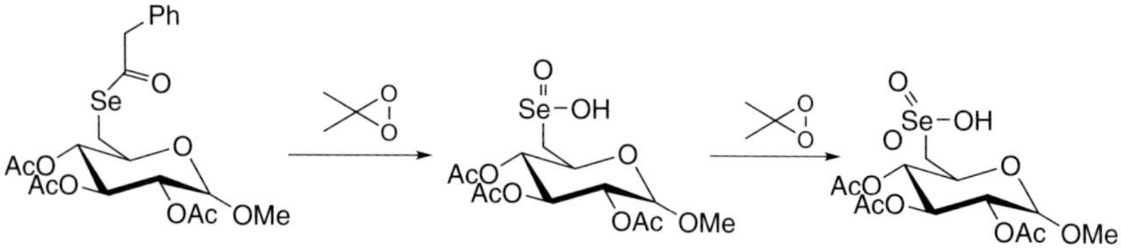

Scheme 5.85

Similarly, Se-[(3,4-di-butyryloxy)butyl]phenylacetate (**273**) was oxidized with DMDO to produce 1-(3,4-di-butyryloxy)butyl seleninic acid (**274**) and finally the corresponding 1-(3,4-di-butyryloxy)butyl selenonic acid (**275**) [99] (Scheme 5.86).

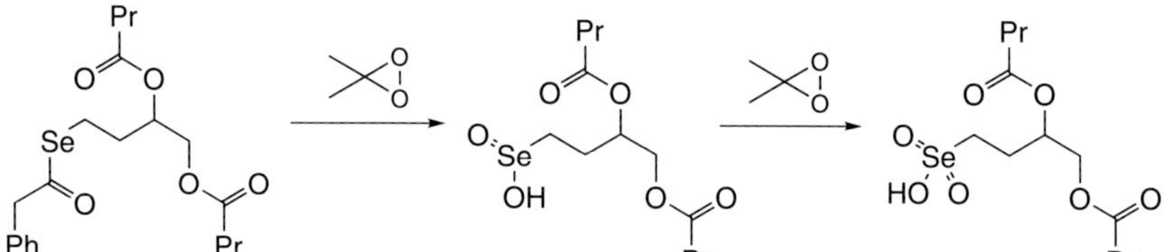

Scheme 5.86

The action of hydrogen peroxide on areneseleninic acids produced arene peroxyseleninic acid, which oxidized bromide to hypobromous acid. This oxidative system was applied to a bromolactonization and to the Baeyer–Villiger oxidation of carbonyl compounds [100] (Scheme 5.87).

Scheme 5.87

The oxidation of perfluorooctaneseleninic acid with a hypervalent iodine oxidant to peroxyseleninic acid allowed to oxidize olefins to α,β-unsaturated ketones [3a, 100] (Scheme 5.88).

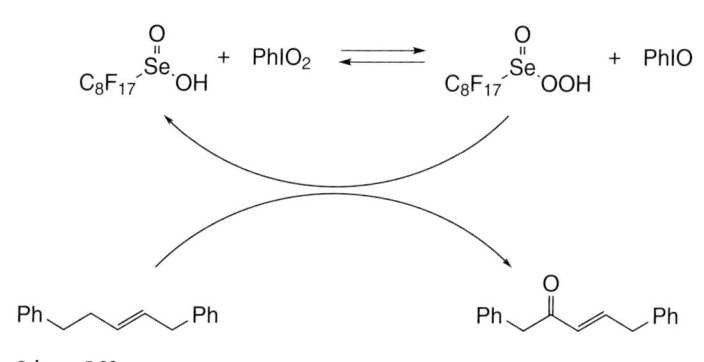

Scheme 5.88

The oxidation of butyl 2,4-di-*tert*-butyl-6-hydroxymethylphenylselenide **276** with hydrogen peroxide allowed us to obtain 3H-benzo[c][1,2]oxaselenole Se-oxide derivative **277** [101]. The cyclic seleninate **277** can be chromatographically resolved and the (R)-enantiomer of benzooxaselenole **277** with 98% ee was isolated [101]. This enantiomer underwent the reaction with methylmagnesium bromide to afford the selenoxide **278** having (S)-configuration and 97% ee [101] (Scheme 5.89).

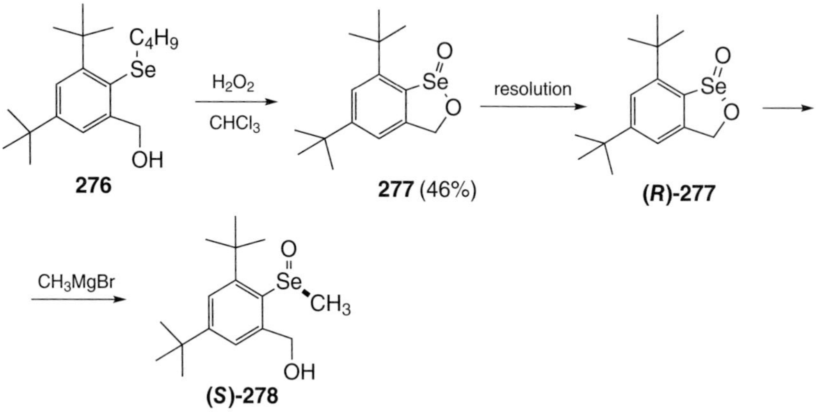

236 | *5 Selenium Compounds with Valency Higher than Two*

Scheme 5.89

Diaryl diselenides **279** have been used for the preparation of a series of racemic areneseleninic acids **280** by ozonization [102] (Scheme 5.90). Racemic mixtures of areneseleninic acids **280** can be resolved on a chiral column using medium-pressure liquid chromatography [102].

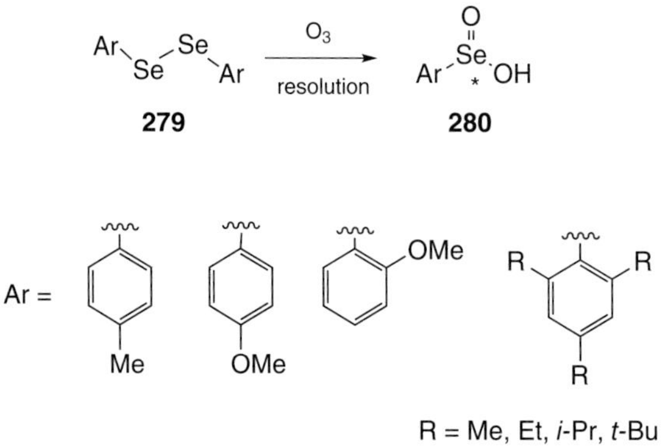

Scheme 5.90

Acid-catalyzed hydrolysis of seleno-seleninate derivatives **281** in a mixture of perchloric acid–dioxane yielded selenenic acids **282** [103]. Alkaline hydrolysis of the seleno-seleninate derivatives **281** gave the corresponding diselenides **283** and seleninic acids **284**. Similar results were obtained when seleno-seleninate derivatives **281** reacted with selenols **285** [103] (Scheme 5.91).

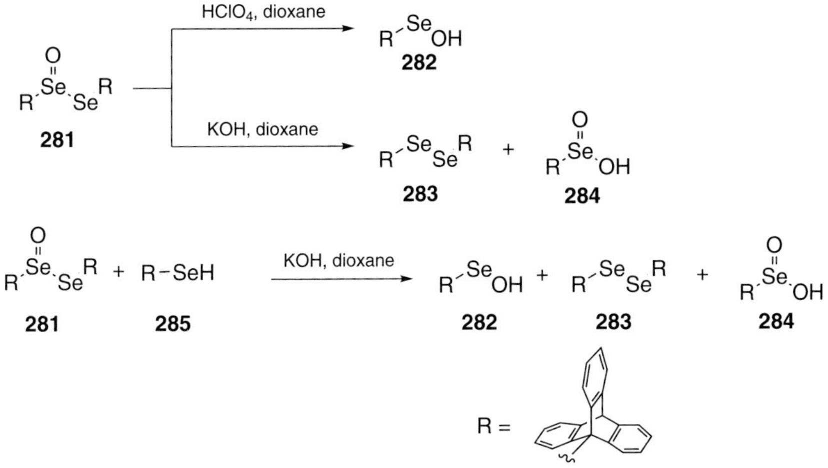

5.5 Tetravalent, Tricoordinated Derivatives

Scheme 5.91

The reaction of N-substituted salicylamide **286** with selenium tetrachloride in THF resulted in 5-(trichloroseleno)salicylamide **287**, which underwent hydrolysis to yield areneseleninic anhydride **288** [104] (Scheme 5.92).

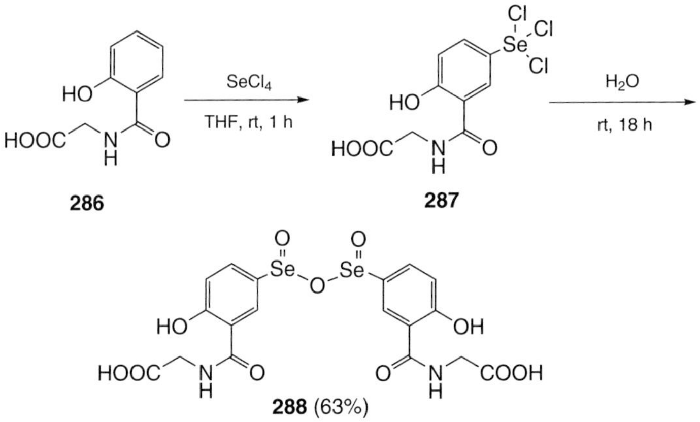

Scheme 5.92

The oxidation of benzyl alcohol to benzaldehyde with benzeneseleninic anhydride **289** was performed and the reaction also resulted in benzeneseleninic acid **219e** and benzeneselenenic acid **290** as side products [105] (Scheme 5.93).

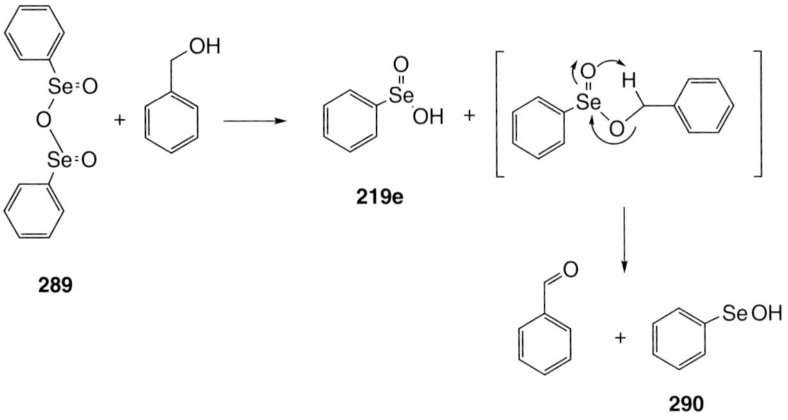

Scheme 5.93

The oxidation of acetaldehyde to glyoxal and glyoxylic acid in the presence of SeO_2 was performed at 90 °C for 12 h, in acetic acid with the observed conversion rate reaching 90% [106]. The yields of this reaction were as follows: glyoxal (33%), glyoxylic acid (13%), and formic and oxalic acids formed in less than 1%. The reaction, when performed under the same experimental conditions, but with propionic acid as solvent demonstrated the same conversion rate and led to the formation of glyoxal (48–50%), glyoxylic acid (10–20%), formic acid (0–3%) and acetic acid (20%) [106] (Scheme 5.94). In all solvents, metallic selenium formed and precipitated.

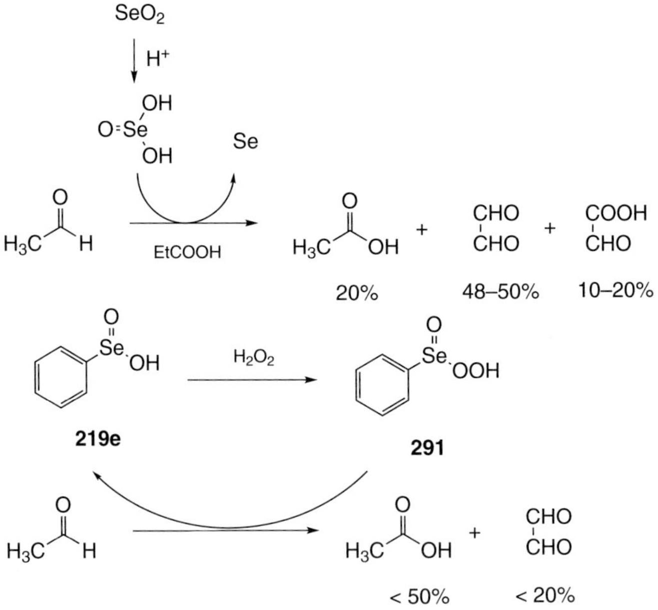

Scheme 5.94

The oxidation of acetaldehyde was also performed with benzeneseleninic acid **219e** or benzeneseleninic peracid **291**, which was generated *in situ* by addition of hydrogen peroxide to **219e** [106]. Both compounds are known to be highly reactive oxidants. In both cases, mostly acetic acid is formed (in up to 50% yield), while less than 20% of glyoxal is obtained (Scheme 5.94).

A series of seleninic acid enantiomers were attempted to be resolved. Each optical isomer of methaneseleninic acid (**219a**) was isolated as chiral crystals by crystallization from the methanol–toluene mixture [107, 108]. The absolute configuration of the enantiomers was determined by x-ray crystallographic analysis, and the relationship between the absolute configuration and the circular dichroism spectra of **219a** was established. The other seleninic acids **219b–e** gave no crystals showing the Cotton effect [107]:

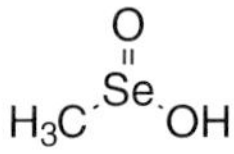

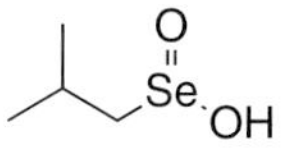

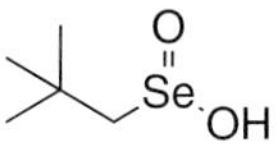

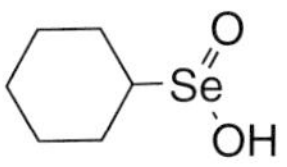

3-Aminopropaneseleninic acid (**292**) and piperidin-4-ylseleninic acid **293** were studied with regard to the agonist potency and efficacy towards heteromeric GABA$_A$ receptors expressed in *Xenopus* oocytes. However, they were not efficient enough, especially the compound **293** [109]. Certain seleninic acid derivatives, such as **294** cause irreversible inhibition of protein tyrosine phosphatases [110].

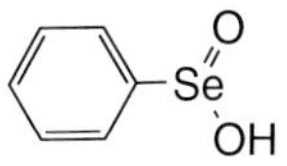

The GPx activities of some seleninic acids were studied with hydrogen peroxide [111]:

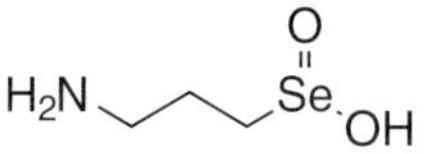

5.6
Pentavalent Derivatives

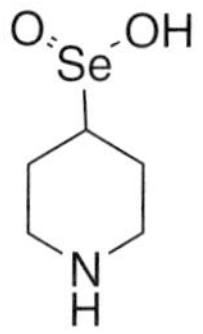

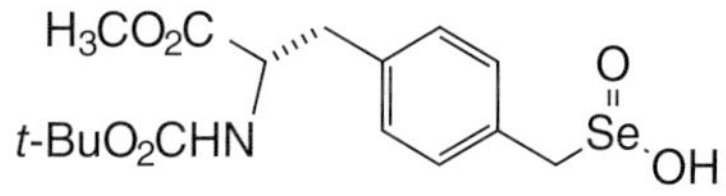

Pentavalent derivatives are divided into two groups. First, the tetracoordinated system, includes selenonium salt oxides and second pentacoordinated systems, where X$_4$SeX$^-$ anions exist. But, it must be admitted that this class of selenium compounds is rather rarely represented in the chemical literature, only Kobayashi et al. [112] reported in late 1980s the synthesis of triaryl selenonium oxide salts (**295**):

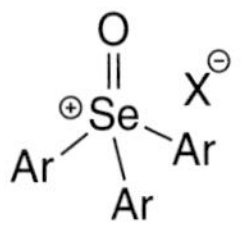

295

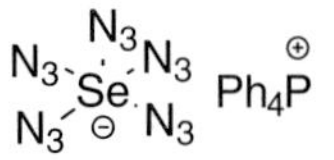

296

As for pentacoordinated selenium anions, Klapoetke et al. [113] described the synthesis of tetraphenylphosphonium pentazaselenite (**296**) by the reaction of tetraphenylphosphonium azide and trimethylsilyl azide with selenium tetrafluoride [113] (Scheme 5.95).

$$(PhP_4)N_3 + 4Me_3SiN_3 + SeF_4 \longrightarrow Se(N_3)_5^{\ominus} Ph_4P^{\oplus} + 4Me_3SiF$$
$$\mathbf{296}$$

Scheme 5.95

5.7
Hexavalent, Tetracoordinated Derivatives

Selenones **297**, and selenonic acids **298** are the best-known hexavalent, tetracoordinated selenonium derivatives. Less common members of this family of compounds such as selenonic esters **299** [114, 115], selenonic amides **300** [115], and selenoximines **301** were also obtained [116]:

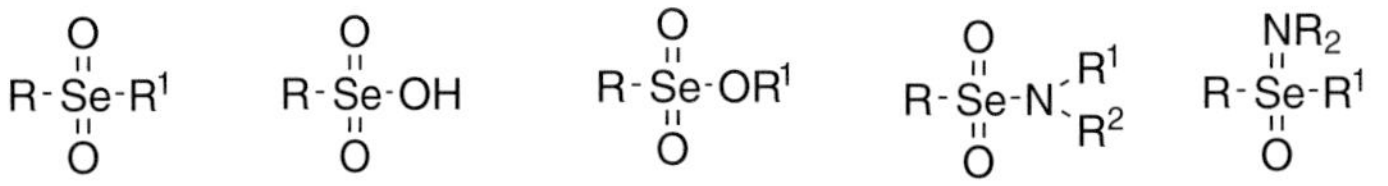

297   298   299   300   301

It is important to note that optically active selenoximines **302a–d** were synthesized and isolated for the first time by optical resolution using chromatography with a chiral column [116c]. Asymmetric selenoximines **302a–d** were obtained in good yields by reaction of the corresponding selenones with N-sulfinyl-p-toluenesulfonamide in the presence of anhydrous magnesium sulfate (Scheme 5.96). No racemization of the optically active selenoximines under neutral, acidic, and basic conditions was observed.

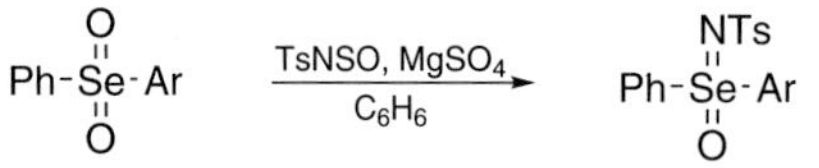

302a Ar = p-MeC$_6$H$_4$ (67%)
302b Ar = p-MeOC$_6$H$_4$ (61%)
302c Ar = p-CF$_3$C$_6$H$_4$ (52%)
302d Ar = 2-Naphthyl (33%)

Scheme 5.96

Till date only selenones **297** have found many interesting synthetic applications and are recognized as excellent leaving groups in intramolecular [117–130] and intermolecular [131–134] nucleophilic substitutions. They can be easily obtained by the oxidation of selenides or selenoxides with standard oxidants, such as hydrogen peroxide, *m*CPBA, or ozone. A new oxidant, aqueous sodium hypochlorite [135], and the photo-oxygenation of selenoxides with addition of dimethylsulfide [136] have also been developed. In the last 10 years, special interest in this group of compounds has been observed.

The oxidation of β-amido selenides **303** with *m*CPBA in THF at −60 °C afforded the corresponding β-amido selenones. *In situ* treatment of the selenones with *t*-BuOK gave *N*-acylaziridines **304** in 75–87% yield (Scheme 5.97) [124]. The formation of selenones was confirmed by ^{77}Se NMR. In the first step of the oxidation of selenides **303** with *m*CPBA the selenoxides were formed. The low temperature of the oxidation reaction was necessary to ensure a sufficiently long-lived selenoxide to enable its further oxidation to the selenone. When the reaction was conducted at higher temperature, the formation of *syn*-elimination products was observed.

303a–d

304a n = 3, (75%)
304b n = 4, (83%)
304c n = 5, (81%)
304d n = 6, (87%)

Scheme 5.97

A similar reaction sequence was described for the synthesis of four-membered heterocyclic compounds. The arenesulfonyl azetidines **307** were obtained from substituted phenylseleno arylsulfonamides **305** via their oxidation with *m*CPBA to selenones **306**, then by the reaction with KOH. Different synthetic methodologies for the synthesis of **305** from α-amino epoxides, γ-lactones, and γ-hydroxyselenides have been described (Scheme 5.98) [125].

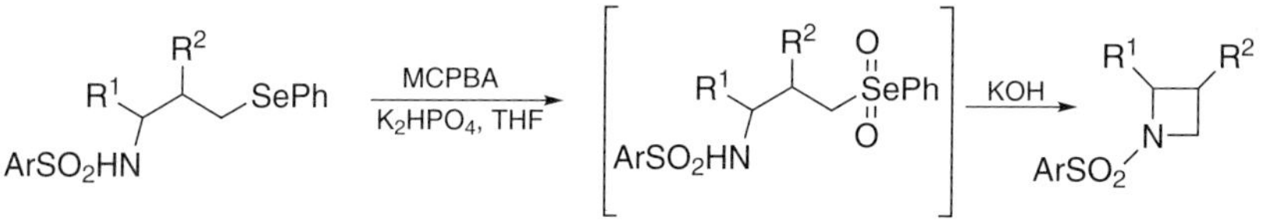

R^1 = H, Me, Et, *n*-Pr, CyCH$_2$, Bn, CH$_2$OH, CO$_2$Me
R^2 = H or THP, Ar = *o*-O$_2$NC$_6$H$_4$, *p*-MeC$_6$H$_4$

Scheme 5.98

Very interesting ring-closure reactions through intermolecular displacement of the phenylselenonyl group by nitrogen nucleophiles leading to various substituted 1,3-oxazolidin-2-ones were presented [126, 127]. For example, the optically active β-phenylseleno alcohol **308** was converted into the corresponding carbamate and oxidized *in situ* with *m*CPBA to generate a selenone group. Intermolecular nucleophilic substitution of the selenone group by the nitrogen atom of the carbamate gave 1,3-oxazolidinone **309** in good yield (Scheme 5.99). A convenient methodology for the synthesis of β-phenylseleno alcohols *via* organocatalytic asymmetric α-selenenylation of aldehydes reduction with NaBH$_4$ was described [128].

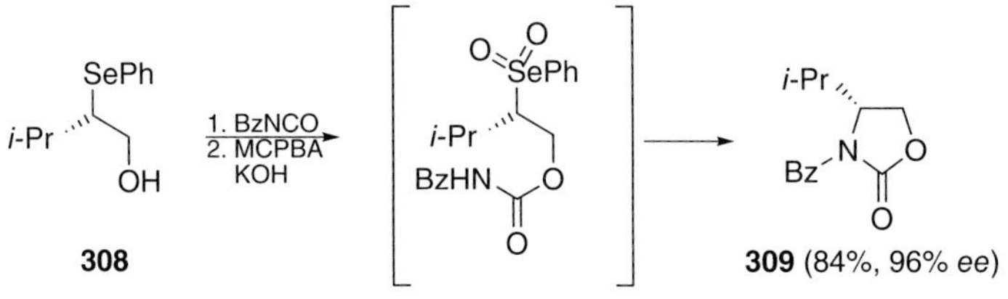

Scheme 5.99

A novel organocatalytic one-pot method for the stereoselective synthesis of highly substituted cyclopropanes **312** was reported. A key step of this procedure is the Michael addition of α-aryl-substituted cyanoacetates **310** to β-substituted vinyl selenones **311** catalyzed by a bifunctional ureidic catalyst **313**, followed de-ethoxydecarbonylation and ring closure reaction (Scheme 5.100) [129]. Highly enantiomerically enriched cyclopropanes using vinyl selenones and chiral nonracemic methylenic compounds such as di-(−)-bornyl malonate or di-(−)-menthyl malonate were also obtained [130].

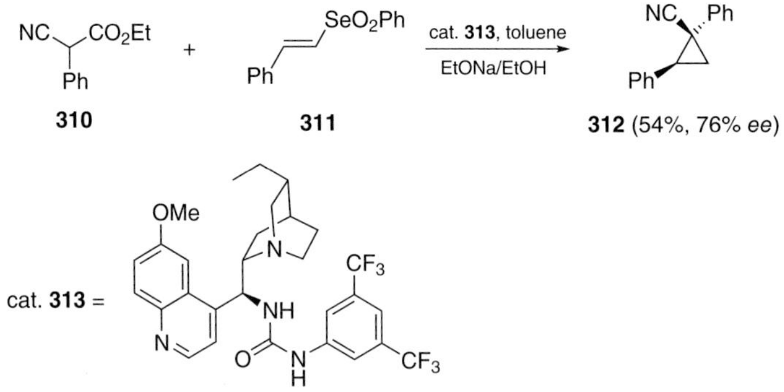

Scheme 5.100

5.7 Hexavalent, Tetracoordinated Derivatives

The intermolecular reaction of selenone **314** with various nucleophiles such as NaN$_3$, KCN, NaSCH$_3$, NaI produced the corresponding azido, cyano, methylthio, and iodo derivatives **315a–d** in good yields (Scheme 5.101) [133]. Using this methodology, other synthetically useful derivatives **316a–d** were also obtained [134]. The compound **316d** was formed by elimination of the intermediate iodide **316c**.

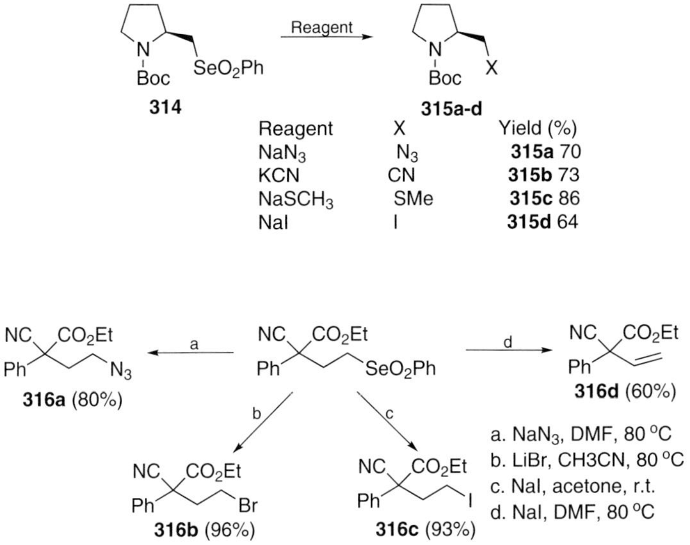

Reagent	X	Yield (%)
NaN$_3$	N$_3$	**315a** 70
KCN	CN	**315b** 73
NaSCH$_3$	SMe	**315c** 86
NaI	I	**315d** 64

a. NaN$_3$, DMF, 80 °C
b. LiBr, CH$_3$CN, 80 °C
c. NaI, acetone, r.t.
d. NaI, DMF, 80 °C

Scheme 5.101

It is important to note, that deoxygenation of a variety of selenones with magnesium in methanol at room temperature in nearly quantitative yields has been reported (Scheme 5.102) [88].

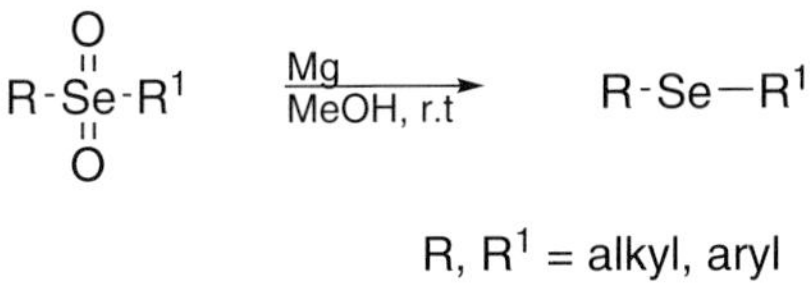

$$R-\overset{\overset{O}{\|}}{\underset{\underset{O}{\|}}{Se}}-R^1 \quad \xrightarrow{\text{Mg}, \text{MeOH, r.t}} \quad R-Se-R^1$$

R, R^1 = alkyl, aryl

Scheme 5.102

Recently, the synthesis of N,N-diethyl-3-(arylselenonyl)-1H-1,2,4-triazole-1-carboxamides **317** and their interesting herbicidal activity was presented [137]:

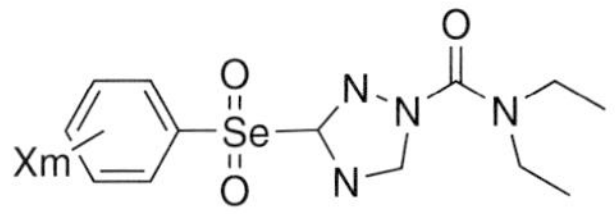

317 Xm = alkyl, chalcogen, alkoxy, phenylate

5.8
Hypervalent Derivatives

To this class, we have included the large range of compounds of various oxidation states, which are characterized by the high coordination number versus valency value. Therefore, we decided to include in one class: tetravalent, tetracoordinated compounds (selenuranes), hexavalent, pentacoordinated systems (selenuranes oxides) and finally, hexavalent, hexacoordinated systems, so-called perselenuranes. Although the word *"hypervalency"* in reference to these compounds is a little exaggeration, the tetravalent selenuranes may be compared to perselenuranes because of their structural and chemical resemblance.

5.8.1
Selenuranes

Selenuranes are tetravalent, tetracoordinated selenium compounds possessing a lone electron pair, which constitutes a phantom ligand. During the past decade, this class of high-coordinated organoselenium compounds was revealed to have the ability to introduce various functional groups into complex chemical systems in a selective manner under mild reaction conditions.

The chemistry of selenuranes has been reviewed rather extensively by Furukawa and Sato [138], who covered the literature until 1999. A little later, Drabowicz [139, 140] has written two reviews updating the topic, and very recently, Allenmark [141] presented a mini-review with discussion on chemistry of chiral selenuranes.

Haloselenuranes are very useful halogenating agents. For example, dichlorodimethyl selane **318** together with triphenylphosphine turned out to be a very powerful chlorinating system [142], which was able to convert alcohols to their chlorides. The chlorination of a chiral alcohol **319** allowed us to obtain a chloro derivative **320** with inversion of configuration (Scheme 5.103).

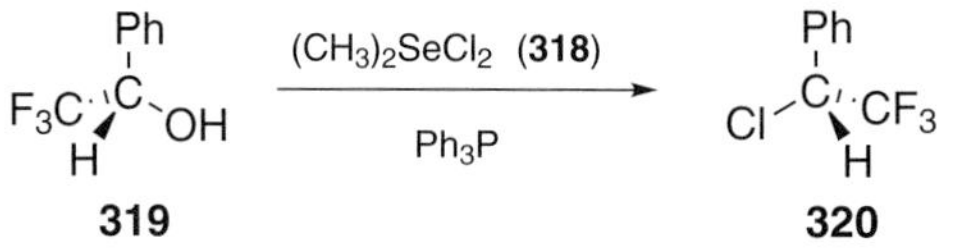

Scheme 5.103

Not only alcohols were efficiently chlorinated [142], but also thiophosphate **321** underwent the reaction with selenurane **318** and triphenylphosphine and was converted into thiophosphorochloridite **322** (Scheme 5.104).

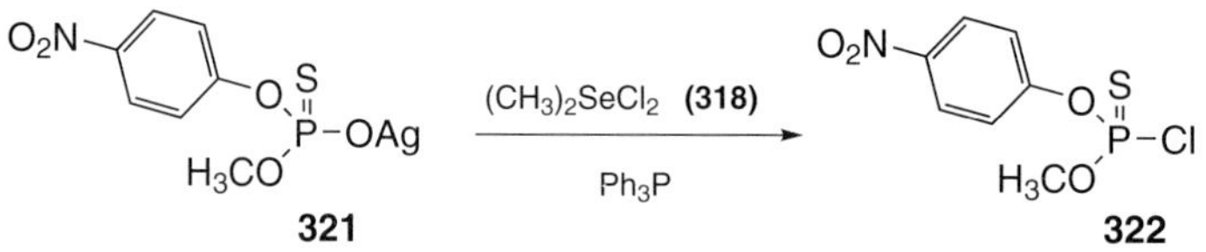

Scheme 5.104

5.8 Hypervalent Derivatives | 245

The optically active cyclic selenurane **323a** stays in equilibrium with an optically active selenoxide form **323b**. When this chiral cyclic selenurane/selenoxide **323** was used together with hydrogen peroxide, this oxidizing agent allowed us to convert enantioselectively sulfides to sulfoxides [142] (Scheme 5.105).

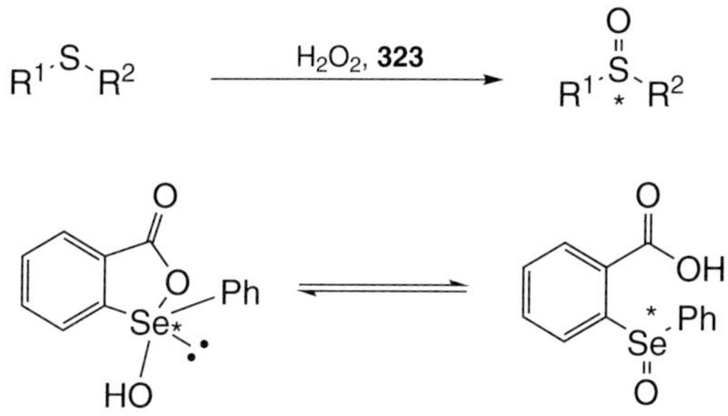

R^1	R^1	yield (%)
p-MeC$_6$H$_5$	Me	75
CH$_2$Ph	Me	69

Scheme 5.105

Selenuranes have been easily transformed to compounds bearing selenium at higher oxidation states. Oxidation of selenurane **324** with m-chloroperbenzoic acid resulted in selenurane oxide **325** [143, 144]. The synthesis of 2,2′-spirobi-3,3-dimethyl-3H-benzo[c] [1, 2] oxaselenole **324** was performed by the reaction of diethyl selenate with organomagnesium compound **326** [145] (Scheme 5.106).

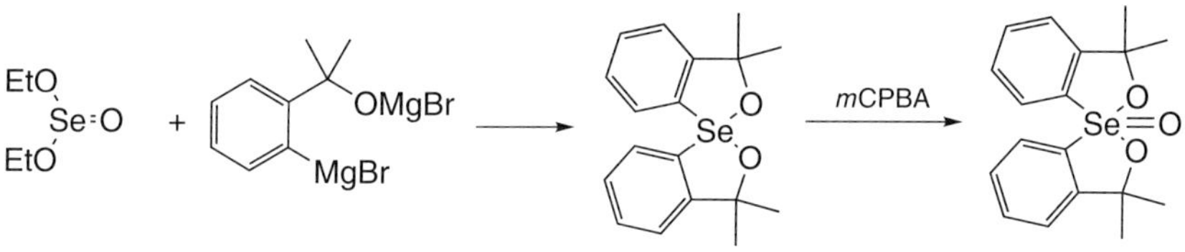

Scheme 5.106

The reaction of selenurane **327** with xenon difluoride allowed for the synthesis of the difluoroperselenurane system **328** [146] (Scheme 5.107).

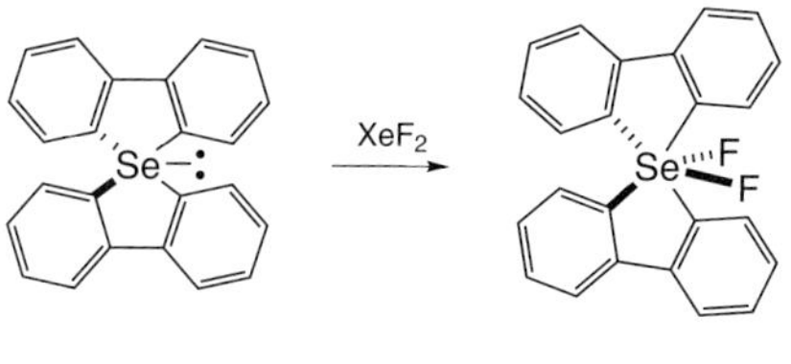

Scheme 5.107

The reaction of selenafluorene Se-oxide 36 with methyllithium and trimethylsilyl triflate at −90 °C led to the unstable Se,Se-dimethyl selenafluorene 329, which decomposed at room temperature to selenafluorene 330 [147] (Scheme 5.108).

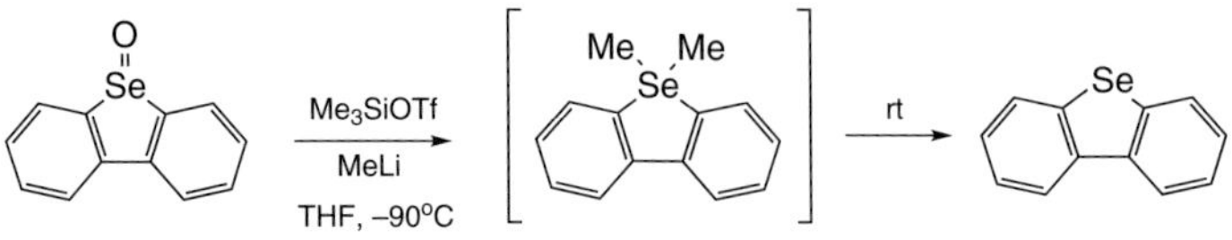

Scheme 5.108

The bipyridyl complex of selenium dichloride (selenurane 331) underwent the reaction with trimethylsilyl triflate to give selenonium salt 332 [148] (Scheme 5.109). A similar complex 333, when reacted with a Grignard reagent, afforded a selane [148] (Scheme 5.109).

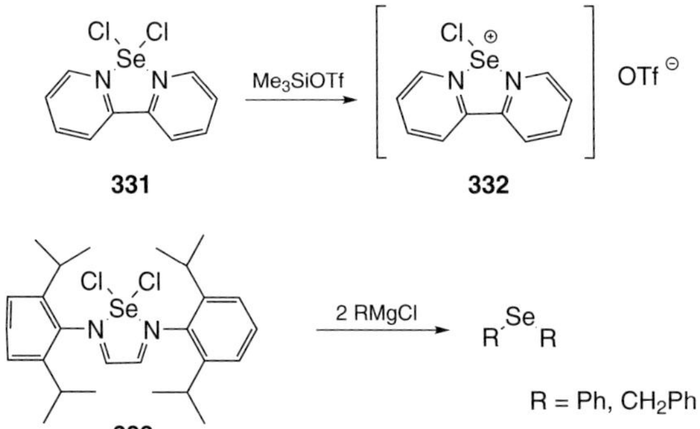

Scheme 5.109

Selenomethionine Se-oxide methyl ester 237 was able to form the amine-based selenurane 1-hydroxy-1-methyl-1λ^4-isoselenazolidine-4-carboxylic acid 334 over a wide pH range, whereas the carboxylic acid-based selenurane 5-amino-2-hydroxy-2-methyl-2λ^4-[1, 2]oxaselenan-6-one 335 was formed at low pH values [81] (Scheme 5.110).

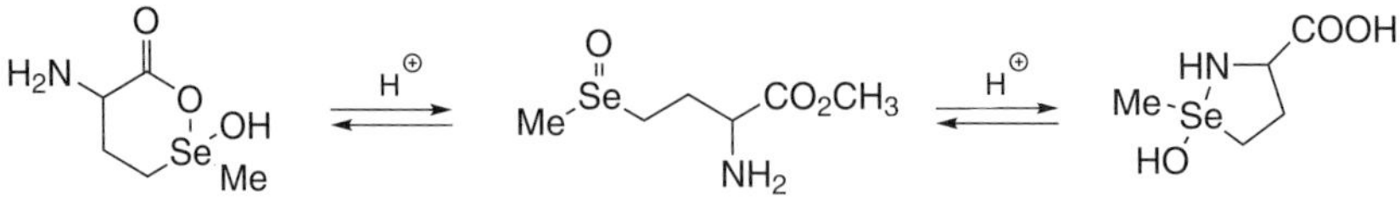

Scheme 5.110

5.8 Hypervalent Derivatives | 247

Selenuranes are very nice objects for structural studies, both theoretically and experimentally. For example, 1,4,6,9-tetraoxa-5λ^4-selena-spiro[4.4]nonane (**336**) was thoroughly investigated by means of B3LYP/6-31G and X-ray studies [149]:

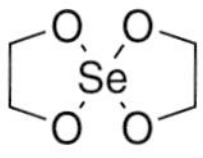

336

The reaction of selenium tetrachloride with N,N′-di-*tert*-butylglyoxaldiimine **337** in THF led to the formation of 2-*tert*-butyl-[1, 2, 5]selenadiazol-2-ium hexachloroselenate **338** [150, 151] (Scheme 5.111).

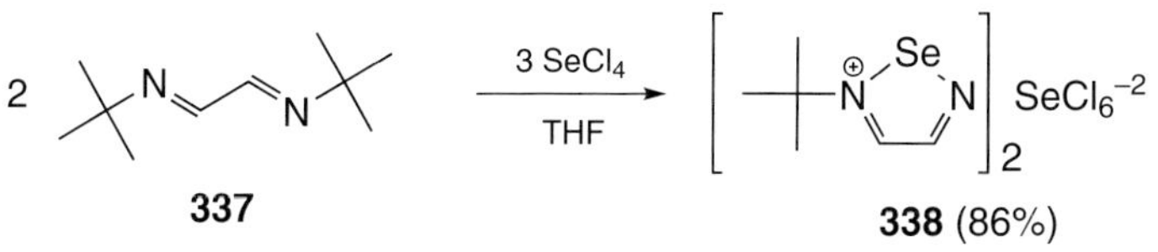

Scheme 5.111

A very similar reaction was performed with N,N′-diarylglyoxaldiimine **339**, where di-N-substituted salt **340** was obtained [152] (Scheme 5.112).

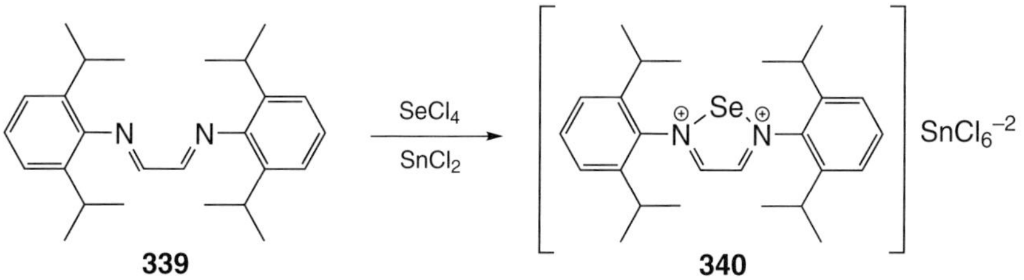

Scheme 5.112

Very interesting conversions of ferrocene-bearing selenuranes **341** and **344** have been reported recently [153]. The reaction of ferrocenyl-*n*-heptyl selenium dibromide **341** with 4-pentenoic acid led to the formation of 5-ferrocenylselanylmethyl-dihydrofuran-2-one **342** and *n*-heptyl bromide **343** [153] (Scheme 5.113).

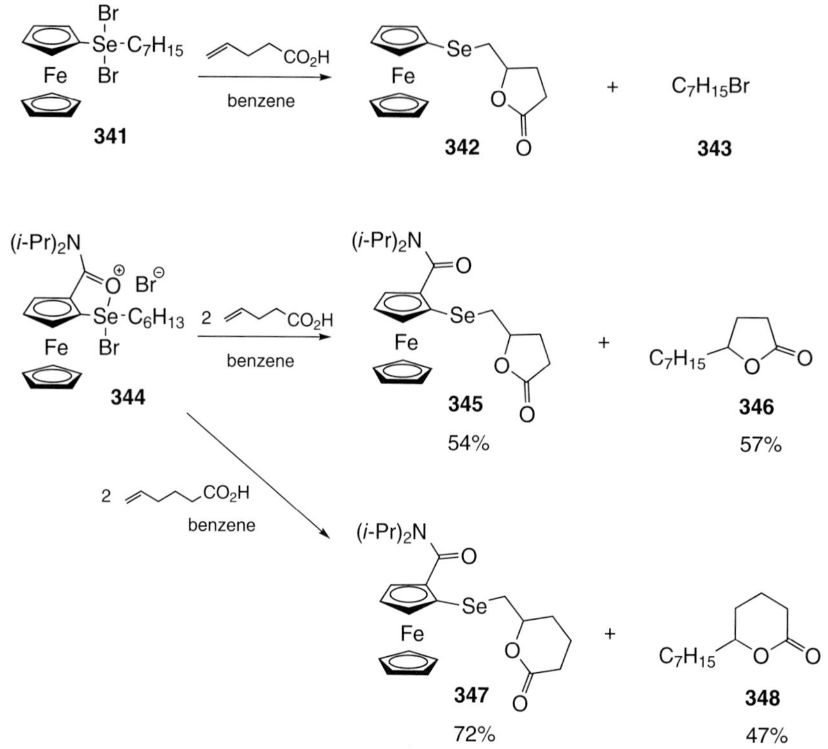

Scheme 5.113

A cyclic selenurane **344** reacting with 4-pentenoic acid gave the substituted 5-ferrocenylselanylmethyl-dihydrofuran-2-one **345** and 5-heptyl-dihydrofuran-2-one **346** in a ratio of 1:1 [153] (Scheme 5.113). The similar reaction of **344** with 5-hexenoic acid led to 6-ferrocenylselanylmethyl-tetrahydro-pyran-2-one **347** and 6-heptyl-tetrahydro-pyran-2-one **348** in a ratio of 3:2 [153] (Scheme 5.113).

Phenylselenium trichloride, in reaction with acetophenone in diethyl ether at room temperature followed by the reduction with sodium disulfite gave 1-phenyl-2-phenylselanyl-ethanone **349** [154] (Scheme 5.114).

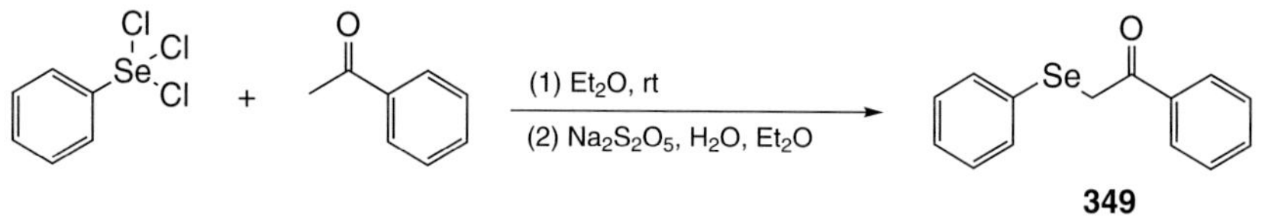

Scheme 5.114

An intriguing conversion, although not from the viewpoint of organic synthesis is the reaction of selenium tetrafluoride with trimethylsilyl azide in sulfur dioxide at −65 °C, which led readily to the formation of selenium tetraazide **350** for the first time [113] (Scheme 5.115).

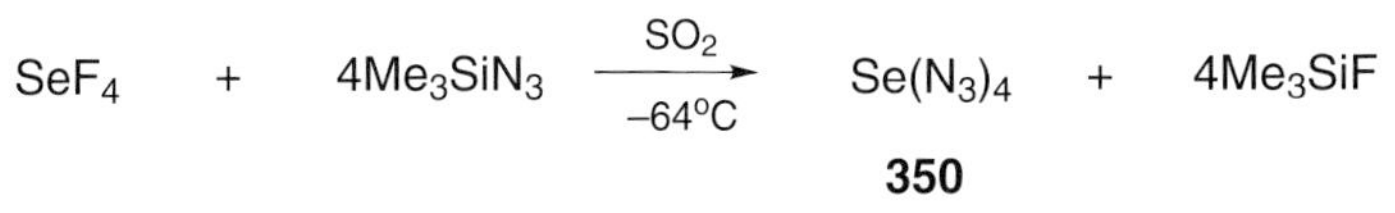

$$SeF_4 \ + \ 4Me_3SiN_3 \ \xrightarrow[-64°C]{SO_2} \ Se(N_3)_4 \ + \ 4Me_3SiF$$
$$\textbf{350}$$

Scheme 5.115

The reactivity of selenium compounds results in their toxicity towards living organisms. Seleninic acids can react with thiols leading to selenurane derivatives **351**, which explained well their toxicity, as they react with cysteine in similar way [155] (Scheme 5.116).

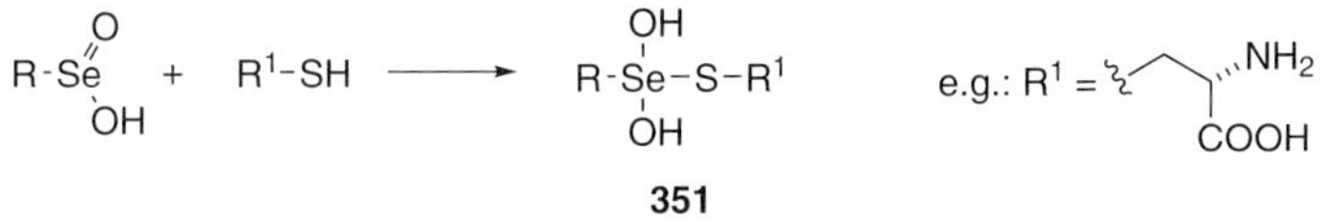

$$R\text{-}\overset{O}{\underset{OH}{Se}} \ + \ R^1\text{-}SH \ \longrightarrow \ R\text{-}\underset{OH}{\overset{OH}{Se}}\text{-}S\text{-}R^1 \quad \text{e.g.: } R^1 = \text{\textasciitilde}\underset{COOH}{\overset{NH_2}{\diagup}}$$
$$\textbf{351}$$

Scheme 5.116

5.8.2
Selenurane Oxides

Selenurane oxides constitute a class of hypervalent selenium compounds that are commonly described as 10-Se-5 species. They have a trigonal bipyramidal geometry and may be chiral if the appropriate arrangement pattern of ligands around the hypervalent selenium atom generate a stereogenic center. However, this class of compounds has not been extensively studied. It is noteworthy that such structures are chiral, due to molecular dissymmetry, even when they are constructed by a pair of the same two-arms ligands, but to date no member of the selenurane oxide family has been prepared in an optically active form.

The first report on the enantiomers of the optically active selenurane oxide, 3,3,3′,3′-tetramethyl-1,1′-spirobi[3H,2,1]-benzoxaselenole oxide **325** [143, 144], described its resolution *via* enantioselective liquid chromatography of the racemate or by spontaneous resolution that occurs during the slow evaporation of an acetonitrile solution or the slow crystallization from the same solvent. This compound was obtained by the procedure discussed in Section 5.8.1.

Another representative of this class has been synthesized in the early 1980's. 2-(Phenylseleno)benzoyl chloride **352** reacted with *tert*-butyl hydroperoxide in the presence of pyridine at −15 °C for 5 days. The mixture of a selenurane **353** and selenurane oxide **354** was purified by chromatography and selenurane oxide **354** was isolated by crystallization from ether as a microcrystalline solid [156] (Scheme 5.117).

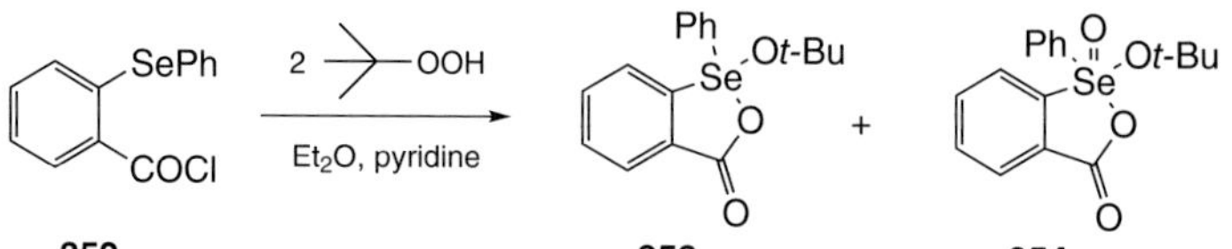

Scheme 5.117

The chemistry of the selenurane oxide, 3,3,3′,3′-tetramethyl-1,1′-spirobi[3H,2,1]-benzoxaselenole oxide (**325**) stayed in a sharp contrast to its sulfur analogue. For example, the selenurane oxide **325** is unexpectedly reduced to the parent selenurane **324** in the presence of HCl [140b] (Scheme 5.118).

Also, the selenurane oxide **325** is converted into the symmetrical hydroxyalkyl selenide **355** by the action of two equivalents of triphenylphosphine in the presence of water, with simultaneous oxidation of triphenylphosphine to the corresponding oxide [140b] (Scheme 5.118).

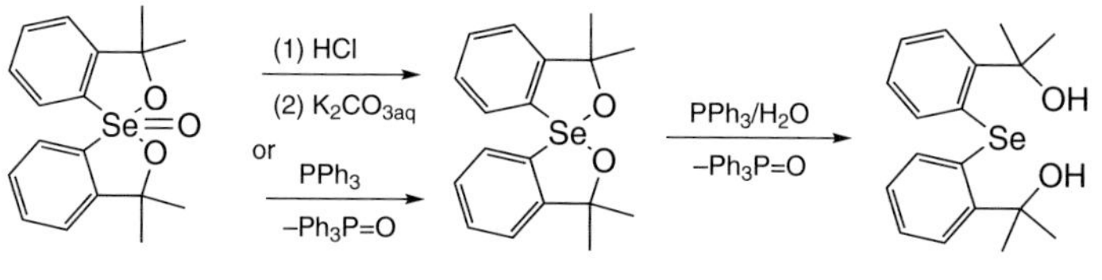

Scheme 5.118

5.8.3
Perselenuranes

Although perselenuranes constitute an interesting class of compounds, their synthetic applications are very scarcely explored. To our knowledge, in the years 2000–2010, no report dealing with this topic has been published. In a previous edition [2f], we did not include a very interesting paper from late eighties [157], where the reactions of two perselenuranes have been reported. Trifluoromethyltetrafluorselenium chloride **356** and trifluoromethylselenium pentafluoride **357** underwent the reaction with bis-trifluoromethyl mercury **358**. The reaction of perselenurane **356** resulted in the formation of chlorotrifluoromethane, trifluoromethaneselenyl chloride **359**, and (trifluoromethyl)selenium trifluoride **360** [157] (Scheme 5.119). In a case of the perselenurane **357** with bis-trifluoromethyl

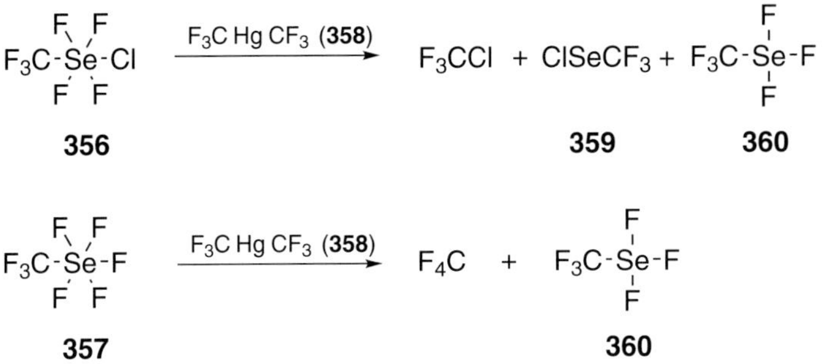

$$F_3C\text{-}\underset{\underset{F}{|}}{\overset{\overset{F}{|}}{Se}}\text{-}Cl \xrightarrow{F_3C\,Hg\,CF_3\ (358)} F_3CCl + ClSeCF_3 + F_3C\text{-}\underset{\underset{F}{|}}{\overset{\overset{F}{|}}{Se}}\text{-}F$$

356   **359**   **360**

$$F_3C\text{-}\underset{\underset{F}{|}}{\overset{\overset{F}{|}}{Se}}\text{-}F \xrightarrow{F_3C\,Hg\,CF_3\ (358)} F_4C + F_3C\text{-}\underset{\underset{F}{|}}{\overset{\overset{F}{|}}{Se}}\text{-}F$$

357   **360**

Scheme 5.119

mercury **358**, their reaction led to carbon tetrafluoride and (trifluoromethyl)selenium trifluoride **360** [157] (Scheme 5.119).

Acknowledgment

The authors wish to dedicate this chapter to Professor Romuald Skowroński from the University of Łódz, Poland.

References

1. (a) Huguet, J.L. (1968) *Adv. Chem. Ser.*, **76**, 345–351; (b) Jones, D.N., Mundy, D., and Whitehouse, R.D. (1970) *J. Chem. Soc., Chem. Commun.*, 86–87; (c) Walter, R. and Roy, J. (1971) *J. Org. Chem.*, **36**, 2561–2563.
2. (a) Paulmier, C. (ed.) (1986) *Selenium Reagents and Intermediates in Organic Synthesis*, Pergamon Press, Oxford; (b) Liotta, D. (ed.) (1987 *Organoselenium Chemistry*, John Wiley & Sons, Inc., New York; (c) Patai, S. and Rappoport, Z. (eds) (1986) *The Chemistry of Organic Selenium and Tellurium Compounds*, vol. 1, John Wiley & Sons, Inc., New York; (d) Patai, S. and Rappoport, Z. (eds) (1987) *The Chemistry of Organic Selenium and Tellurium Compounds*, vol. 2, John Wiley & Sons, Inc., New York; (e) Back, T.G. (ed.) (1999) *Organoselenium Chemistry: A Practical Approach*, Oxford University Press, Oxford; (f) Drabowicz, J. and Mikołajczyk, M. (2000) Selenium at higher oxidation state, in *Organoselenium Chemistry* (ed. T. Wirth), Top. Curr. Chem., **208**, 143–176; (g) Wirth, T. (2007) *Comprehensive Organometallic Chemistry III*, vol. 9 (eds R.H. Crabtree and D.M.P. Mingos), Elsevier, Oxford, pp. 457–500; (h) Wirth, T. (2000) *Angew. Chem. Int. Ed.*, **39**, 3740–3749; (i) Zhu, C. and Huang, Y. (2006) *Curr. Org. Chem.*, **10**, 1905–1920; (j) Braga, A.L., Lüdtke, D.S., and Vargas, F. (2006) *Curr. Org. Chem.*, **10**, 1921–1938; (k) Murai, T. and Kimura, T. (2006) *Curr. Org. Chem.*, **10**, 1963–1973; (l) Coles, M.P. (2006) *Curr. Org. Chem.*, **10**, 1993–2005; (m) Freudendahl, D.M., Shahzad, S.A., and Wirth, T. (2009) *Eur. J. Org. Chem.*, **48**, 1649–1664.
3. (a) Freundendahl, D.M., Santoro, S., Shahzad, S.A., Santi, C., and Wirth, T. (2009) *Angew. Chem. Int. Ed.*, **48**, 8409–8411; (b) Santi, C., Santoro, S., and Battistelli, B. (2010) *Curr. Org. Chem.*, **14**, 2442–2462.
4. Kaboub, L., Fradj, S., and Gouasmia, A. (2006) *Molecules*, **11**, 776–785.

5 Chesney, A., Bryce, M.R., Green, A., Lay, A.K., Yoshida, S., Batsanow, A.S., and Howard, J.A.K. (1998) *Tetrahedron*, **54**, 13257–13266.

6 Binet, L., Fabre, J.M., Montginoul, C., Simonsen, K.B., and Becher, J. (1996) *J. Chem. Soc., Perkin Trans. 1*, 783–788.

7 Binet, L., Fabre, J.M., and Becher, J. (1997) *Synthesis*, 26–28.

8 Bryce, M.R., Chesney, A., Yoshida, S., Moore, A.J., Batsanow, A.S., and Howard, J.A.K. (1997) *J. Mater. Chem.*, **7**, 381–385.

9 Chesney, A., Bryce, M.B., Chalton, M.A., Batsanow, A.S., Howard, J.A.K., Fabre, J.-M., Binet, L., and Chakroune, S. (1996) *J. Org. Chem.*, **61**, 2877–2881.

10 Blum, T., Ermet, J., and Coenen, H.H. (2001) *J. Labelled Cpd. Radiopharm*, **44**, 587–601.

11 Denmark, S.E. and Edwards, M.G. (2006) *J. Org. Chem.*, **71**, 7293–7306.

12 Borodkin, G.L., Chernyak, E.I., Shakirov, M.M., Gatilov, Y.V., Rybalova, T.V., and Shubin, V.G. (1990) *Z. Org. Khim.*, **26**, 1163–1179.

13 (a) Wirth, T. (1999) *Tetrahedron*, **55**, 1–28; (b) Browne, D.M. and Wirth, T. (2006) *Curr. Org. Chem.*, **10**, 1893–1903; (c) Tiecco, M., Testaferri, L., Bagnoli, L., Marini, F., Santi, C., Temperini, A., Scarponi, C., Sternativo, S., Terlizzi, R., and Tomassini, C. (2006) *ARKIVOC*, **vii**, 186–206; (d) Ścianowski, J., Rafiński, Z., and Wojtczak, A. (2006) *Eur. J. Org. Chem.*, 3216–3225; (e) Ścianowski, J., Rafiński, Z., Szuniewicz, A., and Wojtczak, A. (2009) *Tetrahedron*, **65**, 10162–10174; (f) Ścianowski, J., Rafiński, Z., Wojtczak, A., and Burczyński, K. (2009) *Tetrahedron: Asymmetry*, **20**, 2871–2879; (g) Freudendahl, D.M., Iwaoka, M., and Wirth, T. (2010) *Eur. J. Org. Chem.*, 3934–3944.

14 Yamauchi, K., Nakamura, K., and Kinoshita, M. (1979) *Tetrahedron Lett.*, **20**, 1787–1790.

15 Umemoto, T. and Ishihara, S. (1993) *J. Am. Chem. Soc.*, **115**, 2156–2164.

16 Kobayashi, M., Koyabu, K., Shimizu, T., Umemura, K., and Matsuyama, H. (1986) *Chem. Lett.*, 2117–2120.

17 Kataoka, T., Banno, Y., Watanabe, S., Iwamura, T., and Shimizu, H. (1997) *Tetrahedron Lett.*, **38**, 1809–1812.

18 Kataoka, T., Watanabe, S., and Nara, S. (1998) *Phosphorus Sulfur Silicon*, **136**, 497–500.

19 Ochiai, M., Nagaoka, T., Sueda, T., Yan, J., Chen, D.-W., and Miyamoto, K. (2003) *Org. Biomol. Chem.*, **1**, 1517–1521.

20 Kataoka, T., Watanabe, S., Yamamoto, K., Yoshimatsu, M., Tanabe, G., and Muraoka, O. (1998) *J. Org. Chem.*, **63**, 6382–6386.

21 (a) Watanabe, S., Mori, E., Nagai, H., and Kataoka, T. (2000) *Synlett*, (1) 49–52; (b) Watanabe, S., Mori, E., Nagai, H., Iwamura, T., Iwama, T., and Kataoka, T. (2000) *J. Org. Chem.*, **65**, 8893–8898.

22 Kataoka, T., Watanabe, S., and Yamamoto, K. (1999) *Tetrahedron Lett.*, **40**, 931–934.

23 (a) Kataoka, T., Watanabe, S., and Yamamoto, K. (1999) *Tetrahedron Lett.*, **40**, 2153–2156; (b) Watanabe, S., Yamamoto, K., Itagaki, Y., Iwamura, T., Iwama, T., and Kataoka, T. (2000) *Tetrahedron*, **56**, 855–863.

24 (a) Watanabe, S., Yamamota, K., Itagaki, Y., and Kataoka, T. (1999) *J. Chem. Soc., Perkin Trans. 1*, 2053–2055; (b) Watanabe, S., Yamamoto, K., Itagaki, Y., Iwamura, T., Iwama, T., Kataoka, T., Tanabe, G., and Muraoka, O. (2001) *J. Chem. Soc., Perkin Trans. 1*, 239–247.

25 Watanabe, S., Ikeda, T., Kataoka, T., Tanabe, G., and Muraoka, O. (2003) *Org. Lett.*, **5**, 565–567.

26 Watanabe, S., Miura, Y., Iwamura, T., Nagasawa, H., and Katanka, T. (2007) *Tetrahedron Lett.*, **48**, 813–816.

27 Hirabayashi, K., Nara, Y., Yamashita, Y., Kiyota, K., Kamigata, N., and Shimizu, T. (2009) *J. Sulfur. Chem.*, **30**, 346–350.

28 Lenardão, E.J., Mendes, S.R., Ferreira, P.C., Perin, G., Silveira, C.C., and Jacob, R.G. (2006) *Tetrahedron Lett.*, **47**, 7439–7442.

29 Jacob, R.G., Perin, G., Botteselle, G.V., and Lenardão, E.J. (2003) *Tetrahedron Lett.*, **44**, 6809–6812.

30 Lenardão, E.J., Borges, E.L., Mendes, S.R., Perin, G., and Jacob, R.G. (2008) *Tetrahedron Lett.*, **49**, 1919–1921.

31 Lenardão, E.J., Feijo, J.O., Thurow, S., Perin, G., Jacob, R.G., and Silveira, C.C. (2009) *Tetrahedron Lett.*, **50**, 5215–5217.

32 Kataoka, T., Iwamura, T., Tsutsui, H., Kato, Y., Banno, Y., Aoyama, Y., and Shimizu, H. (2001) *Heteroatom Chem.*, **12**, 317–326.

33 (a) Dumont, W., Bayet, P., and Krief, A. (1975) *Angew. Chem. Int. Ed. Eng.*, **13**, 274–275; (b) Takaki, K., Yasumura, M., and Negoro, K. (1981) *Angew. Chem. Int. Ed. Eng.*, **20**, 671–672; (c) Krief, A., Dumont, W., Van Ende, D., Halazy, S., Labar, D., Labourer, J.-L., and Le, T.Q. (1989) *Heterocycles*, **28**, 1203–1228.

34 (a) Sakaki, K. and Oae, S. (1976) *Tetrahedron Lett.*, **41**, 3703–3706; (b) Kamigata, N., Nakamura, Y., Kukuchi, K., Ikemoto, I., Shimizu, T., and Matsuyama, H. (1992) *J. Chem. Soc., Perkin Trans. 1*, 1721–1728.

35 (a) Takada, H., Metzner, P., and Philouze, C. (2001) *Chem. Commun.*, 2350–2351; (b) Briere, J.-F., Takada, H., and Metzner, P. (2005) *Phosphorus Sulfur Silicon*, **180**, 965–968.

36 Li, X.-L., Wang, Y., and Huang, Z.-Z. (2005) *Aust. J. Chem.*, **58**, 749–752.

37 Lotz, W.W. and Gosselck, J. (1973) *Tetrahedron*, **29**, 917–919.

38 Ibata, T. and Kashiuchi, M. (1986) *Bull. Chem. Soc. Jpn.*, **59**, 929–930.

39 Wang, H.-Y., Yang, F., Li, X.-L., Yan, X.-M., and Huang, Z.-Z. (2009) *Chem. Eur. J.*, **15**, 3784–3789.

40 Watanabe, Y., Ueno, Y., and Toru, T. (1993) *Bull. Chem. Soc. Jpn.*, **66**, 2042–2047.

41 Watanabe, S., Nakayama, I., and Kataoka, T. (2005) *Eur. J. Org. Chem.*, 1493–1496.

42 Watanabe, S., Kusumoto, T., Yoshida, C., and Kataoka, T. (2001) *Chem. Commun.*, 839–840.

43 Watanabe, S., Asaka, S., and Kataoka, T. (2004) *Tetrahedron Lett.*, **45**, 7459–7463.

44 (a) Shimizu, T., Urakubo, T., Jin, P., Kondo, M., Kitagawa, S., and Kamigata, N. (1997) *J. Organomet. Chem.*, **539**, 171–175; (b) Miyatake, K., Yamamoto, K., Endo, K., and Tsuchida, E. (1998) *J. Org. Chem.*, **63**, 7522–7524; (c) Wada, M., Nobuki, C., Tenkyuu, Y., Natsume, S., Asahara, M., and Erabi, T. (1999) *J. Organomet. Chem.*, **580**, 282–289; (d) Mullica, D.F., Guziec, F.S., Grant, J.R., Kautz, J.A., and Farmer, J.M. (1999) *J. Mol. Struc.*, **478**, 235–241; (e) Asahara, M., Morikawa, T., Nobuki, S., Erabi, T., and Wada, M. (2001) *J. Chem. Soc., Perkin Trans. 2*, 1899–1903; (f) Sorokin, M.S. and Voronkov, M.G. (2001) *Russ. J. Gen. Chem.*, **71**, 1883–1890; (g) Sorokin, M.S. and Voronkov, M.G. (2006) *Russ. J. Gen. Chem.*, **76**, 461–468; (h) Klapotke, T.M., Krumm, B., and Scherr, M. (2008) *Eur. J. Inorg. Chem.*, 4413–4419.

45 (a) Johnston, B.D., Ghavami, A., Jensen, M.T., Svensson, B., and Pinto, B.M. (2002) *J. Am. Chem. Soc.*, **124**, 8245–8250; (b) Szczepina, M.G., Johnston, B.D., Yuan, Y., Svensson, B., and Pinto, B.M. (2004) *J. Am. Chem. Soc.*, **126**, 12458–12469; (c) Liu, H. and Pinto, B.M. (2005) *J. Org. Chem.*, **70**, 753–755; (d) Liu, H., Nasi, R., Jayakanthan, K., Sim, L., Heipel, H., Rose, D.R., and Pinto, B.M. (2007) *J. Org. Chem.*, **72**, 6562–6572; (e) Nasi, R., Sim, L., Rose, D.R., and Pinto, B.M. (2007) *Carbohydr. Res.*, **342**, 1888–1894.

46 Pigza, J.A., Qauch, T., and Molinski, T.F. (2009) *J. Org. Chem.*, **74**, 5510–5515.

47 Zuang, J., Wang, C., Xie, F., and Zhang, W. (2009) *Tetrahedron*, **65**, 9797–9800.

48 Turbiak, A.J., Kampf, J.W., and Showalter, H.D.H. (2010) *Tetrahedron Lett.*, **51**, 1326–1328.

49 Wadhwa, K., Yang, C., West, P.R., Deming, K.C., Chemburkar, S.R., and Reddy, R.E. (2008) *Synth. Commun.*, **38**, 4434–4444.

50 Chang, M.-Y., Liu, C.-H., and Chen, Y.-L. (2010) *Tetrahedron Lett.*, **51**, 1430–1433.

51 Sabir, H., Muhta, S., and Meenakshi. (2010) *J. Glob. Pharm. Technol.*, **2**, 88–92.

52 Saravanan, S., Purushothaman, S., Amali, I.B., and Muthusubramanian, S. (2009) *Synth. Commun.*, **39**, 2882–2888.

53 Rambaran, V.H., Balof, S., Moody, L.M., VanDerveer, D., and Holder, A.A. (2009) *Cryst. Eng. Commun.*, **11**, 580–582.

54 Gelman, D.M. and Perlmutter, P. (2009) *Tetrahedron Lett.*, **50**, 39–40.

55 Goswami, S., Kumar, A., and Adak, A.K. (2003) *Synthetic Commun.*, **33**, 475–480.

56 Shao, Q. and Li, C. (2008) *Synlett*, 2317–2220.

57 Shao, Q., Shi, W., and Li, C. (2009) *Synlett*, 823–827.

58 Vollbrecht, S., Dobreva, G., Cartis, I., du Mont, W.W., Jeske, J., Ruthe, F., Jones, P.G., Ernst, L., Grahn, W., Papke, U., Marzini, M., and Mayer, H.A. (2008) *Z. Anorg. Allg. Chem.*, **634**, 1321–1325.

59 Chen, J., Ling, G., and Lu, S. (2003) *Tetrahedron*, **59**, 8251–8256.

60 Priewisch, B. and Rueck-Braun, K. (2005) *J. Org. Chem.*, **70**, 2350–2352.

61 Nguyen, T.M. and Lee, D. (2001) *Org. Lett.*, **3**, 3161–3163.

62 Nguyen, T.M., Guzei, I.A., and Lee, D. (2002) *J. Org. Chem.*, **67**, 6553–6556.

63 Kachanov, A.V., Slabko, O.Y., Baranova, O.V., Shilova, E.V., and Kaminskii, V.A. (2004) *Tetrahedron Lett.*, **45**, 4461–4463.

64 Goswami, S., Maity, A.C., Garcia-Grandab, S., and Torre-Fernandez, L. (2007) *Acta Cryst.*, **E63**, o1741–o1742.

65 Theilmann, O., Ruhmann, M., Schulz, A., Villinger, A., and Rosenthal, U. (2010) *Inorg. Chem. Commun.*, **13**, 837–839.

66 Uzarewicz, A. and Ścianowski, J. (1997) *Pol. J. Chem.*, **71**, 48–51.

67 Uzarewicz, A., Wyzlic, I., and Ścianowski, J. (1995) *Pol. J. Chem.*, **69**, 681–684.

68 Uzarewicz, A., Ścianowski, J., and Wyzlic, I. (1995) *Pol. J. Chem.*, **69**, 1153–1157.

69 Uzarewicz, A. and Ścianowski, J. (1998) *Pol. J. Chem.*, **72**, 93–98.

70 Sharpless, K.B. and Singer, S.P. (1976) *J. Org. Chem.*, **41**, 2504–2506.

71 Sharpless, K.B., Hori, T., Truesdale, L.K., and Dietrich, C.O. (1976) *J. Am. Chem. Soc.*, **98**, 269–271.

72 Bruncko, M., Khuong, T.-A.V., and Sharpless, K.B. (1996) *Angew. Chem. Int. Ed. Engl.*, **35**, 454–456.

73 Magnus, P. and Mugrage, B. (1990) *J. Am. Chem. Soc.*, **112**, 462–464.

74 Magnus, P. and Coldham, I. (1991) *J. Am. Chem. Soc.*, **113**, 672–613.

75 Bestari, K., Cordes, A.W., Oakley, R.T., and Young, K.M. (1990) *J. Am. Chem. Soc.*, **112**, 2249–2255.

76 Maaninen, T., Tuononen, H.M., Kosunen, K., Oilunkaniemi, R., Hiitola, J., Lattinen, R., and Chivers, T. (2004) *Z. Anorg. Allg. Chem.*, **630**, 1947–1954.

77 (a) Khurana, J.M., Kandpal, B.M., and Chauhan, Y.K. (2003) *Phosphorus Sulfur Silicon*, **178**, 1369–1375; (b) Sama, T., Shimizu, T., Hirabayashi, K., and Kamigata, N. (2007) *Heteroatom Chem.*, **18**, 301–311; (c) Andreou, T., Bures, J., and Vilarrasa, J. (2010) *Tetrahedron Lett.*, **51**, 1863–1866.

78 Yamazaki, Y., Tsuchiya, T., and Hasegawa, T. (2003) *Bull. Chem. Soc. Jpn.*, **76**, 201–202.

79 (a) Iwaoka, M. and Kumakura, F. (2008) *Phosphorus Sulfur Silicon*, **183**, 1009–1017; (b) Kumakura, F., Mishra, B., Priyadarsini, K.I., and Iwaoka, M. (2010) *Eur. J. Org. Chem.*, 440–445.

80 Back, T.G., Moussa, Z., and Parvez, M. (2004) *Angew. Chem. Int. Ed.*, **43**, 1268–1268.

81 Ritchey, J.A., Davis, B.M., Pleban, P.A., and Bayse, C.A. (2005) *Org. Biomol. Chem.*, **3**, 4335–4342.

82 Krause, R.J., Glocke, S.C., Sicuri, A.R., Ripp, S.L., and Elfarra, A.A. (2006) *Chem. Res. Toxicol.*, **19**, 1643–1649.

83 Krief, A., Destree, A., Durisotti, V., Moreau, N., Smal, C., and Colaux–Castillo, C. (2002) *Chem. Commun.*, 558–559.

84 Block, E., Birringer, M., DeOrazio, R., Fabian, J., Glass, R.S., Guo, C., He, C., Lorance, E., Qian, Q., Schroeder, T.B., Shan, Z., Thiruvazhi, M., Wilson, G.S., and Zhang, X. (2000) *J. Am. Chem. Soc.*, **122**, 5052–5064.

85 Geletii, Y.V., Musaev, D.G., Khavrutskii, L., and Hill, C.L. (2004) *J. Phys. Chem. A*, **108**, 289–294.

86 (a) Filatov, A.S., Block, E., and Petrukhina, M.A. (2005) *Acta Cryst.*, **C61**, o596–o598; (b) Hayashi, S., Nakanishi, W., Furuta, A., Drabowicz, J., Sasamori, T., and Tokitoh, N. (2009) *New. J. Chem.*, **33**, 196–206.

87 Block, E., Birringer, M., Jiang, W., Nakahodo, T., Thompson, H.J., Toscano, P.J., Uzar, H., Zhang, X., and Zhu, Z. (2001) *J. Agric. Food Chem.*, **49**, 458–470.

88 Khurana, J.M., Sharma, V., and Chacko, S.A. (2007) *Tetrahedron*, **63**, 966–969.

89 Gockel, S., Haas, A., Probst, V., Boese, R., and Mueller, I. (2000) *J. Fluor. Chem.*, **102**, 301–311.

90 Sinha, R., Unni, E., Ganther, H.E., and Medina, D. (2001) *Biochem. Pharmacol.*, **61**, 311–317.
91 Xia, R., Ganther, H.E., Egge, A., and Abramson, J.J. (2004) *Biochem. Pharmacol.*, **67**, 2071–2079.
92 Kirby, J.K., Lyons, G.H., and Karkkainen, M.P. (2008) *J. Agric. Food Chem.*, **56**, 1772–1779.
93 Farina, M., Folmer, V., Bolzan, R.C., Andrade, L.H., Zeni, G., Braga, A.L., and Rocha, J.B.T. (2001) *Toxicol. Lett.*, **119**, 27–37.
94 Miyake, Y., Oda, M., Oyamada, A., Takada, H., Ohe, K., and Uemura, S. (2000) *J. Organomet. Chem.*, **611**, 475–487.
95 Aucott, S.M., Bailey, M.R., Elsegood, M.R.J., Gilby, L.M., Holme, K.E., Kelly, P.F., Papageorgiou, M.J., and Pedron–Haba, S. (2004) *New. J. Chem.*, **28**, 959–966.
96 Elsegood, M.R.J., Kelly, P.F., Reid, G., and Staniland, P.M. (2008) *Dalton Trans.*, 3798–3800.
97 Crich, D. and Zou, Y. (2004) *Org. Lett.*, **6**, 775–777.
98 Back, T.G. and Moussa, Z. (2003) *J. Am. Chem. Soc.*, **125**, 13455–13460.
99 Abdo, M. and Knapp, S. (2008) *J. Am. Chem. Soc.*, **130**, 9234–9235.
100 Crich, D. and Zou, Y. (2005) *J. Org. Chem.*, **70**, 3309–3311.
101 Nakashima, Y., Shimizu, T., Hirabayashi, K., Iwasaki, F., Yamasaki, M., and Kamigata, N. (2005) *J. Org. Chem.*, **70**, 5020–5027.
102 Shimizu, T., Nakashima, Y., Watanabe, I., Hirabayashi, K., and Kamigata, N. (2002) *J. Chem. Soc., Perkin Trans. 1*, 2151–2155.
103 Ishii, A., Takahashi, T., and Nakayama, J. (2001) *Heteroatom Chem.*, **12**, 198–203.
104 Yu, S.-C., Borchert, A., Kuhn, H., and Ivanov, I. (2008) *Chem. Eur. J.*, **14**, 7066–7071.
105 Van der Toorn, J.C., Kemperman, G., Sheldon, R.A., and Arends, I.W.C.E. (2009) *J. Org. Chem.*, **74**, 3085–3089.
106 Provendier, H., Santini, C.C., Basset, J.-M., and Carmona, L. (2003) *Eur. J. Inorg. Chem.*, 2139–2144.
107 Nakashima, Y., Shimizu, T., Hirabayashi, K., Yasui, M., Nakazato, M., Iwasaki, F., and Kamigata, N. (2005) *Bull. Chem. Soc. Jpn.*, **78**, 710–714.

108 Nakashima, Y., Shimizu, T., Hirabayashi, K., Kamigata, N., Yasui, M., Nakazato, M., and Iwasaki, F. (2004) *Tetrahedron Lett.*, **45**, 2301–2303.
109 Ebert, B., Mortensen, M., Thompson, S.A., Kehler, J., Wafford, K.A., and Krogsgaard–Larsen, P. (2001) *Bioorg. Med. Chem. Lett.*, **11**, 1573–1577.
110 Abdo, M., Liu, S., Zhou, B., Walls, C.D., Wu, L., Knapp, S., and Zhang, Z.-Y. (2008) *J. Am. Chem. Soc.*, **130**, 13196–13197.
111 (a) Bhabak, K.P. and Mugesh, G. (2008) *Chem. Eur. J.*, **14**, 8640–8651; (b) Mugesh, G., Panda, A., Singh, H.B., Punekar, N.S., and Butcher, R.J. (2001) *J. Am. Chem. Soc.*, **123**, 839–850; (c) Sarma, B.K. and Mugesh, G. (2008) *Chem. Eur. J.*, **14**, 10603–10614.
112 Kobayashi, M. and Mihoya, T. (1986) *Chem. Lett.*, 809–810.
113 Klapoetke, T.M., Krumm, B., Scherr, M., Haiges, R., and Christe, K.O. (2007) *Angew. Chem. Int. Ed.*, **46**, 8686–8690.
114 Boese, R., Haas, A., Herkt, S., and Pryka, M. (1995) *Chem. Ber.*, **128**, 423–428.
115 Haas, A. and Schinkel, K. (1990) *Chem. Ber.*, **123**, 685–689.
116 (a) Derkach, N.Y., Lyapina, T.V., and Parsmurtseva, N.A. (1974) *Zh. Org. Khim.*, **10**, 807–810; (b) Derkach, N.Y., Barashenkov, G.G., and Slyusarenko, E.I. (1982) *Zh. Org. Khim.*, **18**, 70–78; (c) Shimizu, T., Mitsuya, K., Hirabayashi, K., and Kamigata, N. (2004) *Bull. Chem. Soc. Jpn.*, **77**, 375–378
117 Kuwajima, I., Ando, R., and Sugawara, T. (1983) *Tetrahedron Lett.*, **24**, 4429–4432.
118 Ando, R., Sugawara, T., Shimizu, M., and Kuwajima, I. (1984) *Bull. Chem. Soc. Jpn.*, **57**, 2897–2904.
119 Krief, A., Dumont, W., and Denis, J.-N. (1985) *J. Chem. Soc., Chem. Commun.*, 571–572.
120 Krief, A., Dumont, W., and Laboureur, J.L. (1988) *Tetrahedron Lett.*, **29**, 3265–3268.
121 Krief, A., Dumont, W., and De Mahieu, A.F. (1988) *Tetrahedron Lett.*, **29**, 3269–3272.
122 Toshimitsu, A., Hirosawa, C., Tanimoto, S., and Uemura, S. (1992) *Tetrahedron Lett.*, **33**, 4017–4020.

123 Cooper, M.A. and Ward, A.D. (1997) *Aust. J. Chem.*, **50**, 181–188.
124 Ward, V.R., Cooper, M.A., and Ward, A.D. (2001) *J. Chem. Soc., Perkin Trans. 1*, 944–945.
125 Tiecco, M., Testaferri, L., Temperini, A., Terlizzi, R., Bagnoli, L., Marini, F., and Santi, C. (2007) *Org. Biomol. Chem.*, **5**, 3510–3519.
126 Tiecco, M., Testaferri, L., Temperini, A., Bagnoli, L., Marini, F., and Santi, C. (2004) *Chem. Eur. J.*, **10**, 1752–1764.
127 Sheng, S.-R., Luo, H.-R., Huang, Z.-Z., Sun, W.-K., and Liu, X.-L. (2007) *Synth. Commun.*, **37**, 2693–2699.
128 Tiecco, M., Carlone, A., Sternativo, S., Marini, F., Bartoli, G., and Melchiorze, P. (2007) *Angew. Chem. Int. Ed.*, **46**, 6882–6885.
129 Marini, F., Sternativo, S., Del Verme, F., Testaferri, L., and Tiecco, M. (2009) *Adv. Synth. Catal.*, **351**, 1801–1806.
130 Bagnoli, L., Scarponi, C., Testaferri, L., and Tiecco, M. (2009) *Tetrahedron: Asymmetry*, **20**, 1506–1514.
131 Tiecco, M., Chianelli, D., Testaferri, L., Tingoli, M., and Bartoli, D. (1986) *Tetrahedron*, **42**, 4889–4896.
132 Tiecco, M., Chianelli, D., Testaferri, L., Tingoli, M., and Bartoli, D. (1986) *Tetrahedron*, **42**, 4897–4906.
133 Tiecco, M., Testaferri, L., Bagnoli, L., Scarponi, G., Temperini, A., Marini, F., and Santi, C. (2007) *Tetrahedron: Asymmetry*, **18**, 2758–2767.
134 Marini, F., Sternativo, S., Del Verme, F., Testaferri, L., and Tiecco, M. (2009) *Adv. Synth. Catal.*, **351**, 103–106.
135 Khurana, J.M., Kandpal, B.M., and Chauhan, Y. (2003) *Phosphorus Sulfur Silicon*, **178**, 1369–1375.
136 Sofikiti, N. and Stratakis, M. (2003) *ARKIVOC*, **vi**, 30–35.
137 Ma, Y., Liu, R., Gong, X., Li, Z., Huang, Q., Wang, H., and Song, G. (2006) *J. Agric. Food. Chem.*, **54**, 7724–7728.
138 Furukawa, N. and Sato, S. (1999) Structure and reactivity of hypervalent chacogen compounds: selenurane (selane) and tellurane (tellane), in *Chemistry of Hypervalent Compounds* (ed. K. Akiba), Wiley-VCH, New York,.
139 Drabowicz, J. and Halaba, G. (2000) *Rev. Heteroatom Chem.*, **22**, 1–32.
140 Drabowicz, J. (2002) *Heteroatom Chem.*, **13**, 437–442.
141 Allenmark, S. (2008) *Chirality*, **20**, 544–551.
142 Drabowicz, J., Łuczak, J., Łyzwa, P., Kiełbasiński, P., Mikołajczyk, M., Yamamoto, Y., Matsukawa, S., Akiba, K., Wang, F., Polavarapu, P.L., and Wieczorek, M.W. (2005) *Phosphorus Sulfur Silicon*, **180**, 741–753.
143 Drabowicz, J., Łuczak, J., Mikołajczyk, M., Yamamoto, Y., Matsukawa, S., and Akiba, K. (2004) *Chirality*, **16**, 598–601.
144 Petrovic, A.G., Polavarapu, P.L., Drabowicz, J., Zhang, Y., McConnel, O.J., and Duddeck, H. (2005) *Chem. Eur. J.*, **11**, 4257–4262.
145 Drabowicz, J., Łuczak, J., Mikołajczyk, M., Yamamoto, Y., Matsukawa, S., and Akiba, K. (2002) *Tetrahedron Asymmetry*, **13**, 2079–2082.
146 Furukawa, N. and Sato, S. (2002) *Heteroatom Chem.*, **13**, 406–413.
147 Sato, S., Matsuo, M., Nakahodo, T., Furukawa, N., and Nabeshima, T. (2005) *Tetrahedron Lett.*, **46**, 8091–8093.
148 Dutton, J.L., Farrar, G.J., Sgro, M.J., Battista, T.L., and Ragogna, P.J. (2009) *Chem. Eur. J.*, **15**, 10263–10271.
149 Betz, R., Pfister, M., Reichvilser, M.M., and Kluefers, P. (2008) *Z. Anorg. Allg. Chem.*, **634**, 1393–1396.
150 Dutton, J.L., Tindale, J.J., Jennings, M.C., and Ragogna, P.J. (2006) *Chem. Commun.*, 2474–2476.
151 Dutton, J.L., Sutrisno, A., Schurko, R.W., and Ragogna, P.J. (2008) *Dalton Trans.*, 3470–3477.
152 Chivers, T. and Konu, J. (2009) *Angew. Chem. Int. Ed.*, **48**, 3025–3027.
153 Kumar, S., Helt, J.-C.P., Autschbach, J., and Detty, M.R. (2009) *Organometallics*, **28**, 3426–3436.
154 Srivastava, P. and Engman, L. (2010) *Tetrahedron Lett.*, **51**, 1149–1151.
155 Boyse, C.A. (2010) *J. Inorg. Biochem.*, **104**, 1–8.
156 Nakanishi, W., Ikeda, Y., and Iwamura, H. (1982) *J. Org. Chem.*, **47**, 2275–2278.
157 Haas, A. (1986) *J. Fluor. Chem.*, **32**, 415–440.

6
Selenocarbonyls
Toshiaki Murai

6.1
Overview

In principle, oxygen atoms of all types of carbonyl compounds can be replaced with a selenium atom, but the stability and reactivity of the resulting selenocarbonyl compounds are highly dependent on the substitution patterns. The selenocarbonyl compounds are classified into four categories (Figure 6.1) [1]. Those in the category A possess only hydrogen and carbon substituents on the carbon atoms of selenocarbonyl groups. These are called selenoformaldehyde, selenoaldehydes, and selenoketones. These compounds are generally unstable, but their generation and trapping with dienes leading to Diels–Alder adducts were one of the big topics in the last 15 years of the 20th century [2]. In turn, this direction of the research has decreased in this century, and only limited examples are seen. In contrast, several types of molecules belonging to the category A are proposed and have been an object of theoretical studies. The selenocarbonyl compounds belonging to the category B possess one-hetereoatom-containing substituents on the carbon atom of the selenocarbonyl group. Compounds synthesized in this category possess nitrogen, oxygen, chalcogens, and halogens as heteroatoms. The compounds with halogens **VI** [3] must be handled at very low temperatures. In contrast, selenoamides **IV** [4] can be handled at room temperature, in particular, when alkyl substituents are attached to the nitrogen atom. As a result, tremendous studies on syntheses, properties, and reactivity are still growing. Selenoic acid esters **V** are interesting classes of compounds in terms of the stability. Among four types of esters **Va–Vd**, selenothioic acid esters **Vb** have been most widely studied [5]. The stability of the esters **Vb** is also dependent on the substituents, and the introduction of aromatic substituents does not necessarily stabilize the esters. Selenothioacetic acid *S*-alkyl esters **1** are readily prepared as deep purple oils and can be stored in the refrigerator for a long time (Figure 6.2). In contrast, selenothiobenzoic acid *S*-alkyl esters **2** gradually decompose during the purification and cannot be stored for a long time. Selenothiobenzoic acid *S*-aryl esters **3** are the least stable. This trend of stability is not applicable for diselenoic acid esters **Vc**. Aromatic esters **4** are more stable than aliphatic esters **5**. In contrast to these

Organoselenium Chemistry: Synthesis and Reactions, First Edition. Edited by Thomas Wirth.
© 2012 Wiley-VCH Verlag GmbH & Co. KGaA. Published 2012 by Wiley-VCH Verlag GmbH & Co. KGaA.

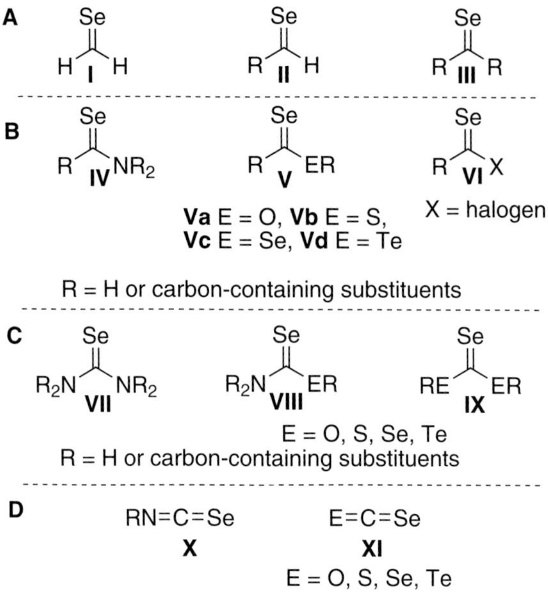

Figure 6.1 Categories of selenocarbonyls.

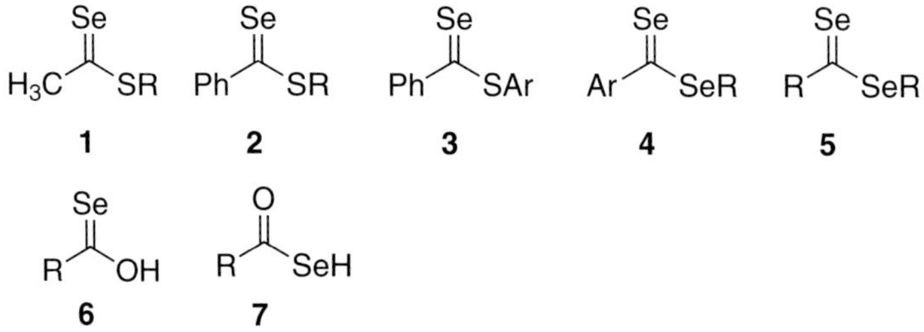

Figure 6.2 Examples of selenoic acid derivatives **V**.

studies, much less attention has been paid to selenoic acid *O*-esters **Va** [6]. No example of selenotelluroic acid *Te*-esters **Vd** has been known. Acids, which can be categorized into **V** in category B, are of interest [7]. In particular, the tautomerization of the acids can lead to two tautomers. One formally involves C=Se, and the other involve Se–H as in **6** and **7**. In an aprotic polar solvent such as THF, C=Se form **6** is observed, but in nonpolar solvents or in the solid state, Se–H form **7** is formed. In this regard, selenoacetic acid **7** (R = CH_3) as a Se–H form is prepared, and the structure and physical properties are disclosed [8].

In the compounds in category C, two-heteroatom-containing substituents are attached to the carbon atom of the selenocarbonyl group [9]. As heteroatoms, nitrogen, oxygen, and chalcogens are attached. The attachment of the heavier elements in groups 13–15 on the periodic table to the selenocarbonyls can be envisioned, but no examples had been reported until very recently. Selenoureas **VII** are well-known compounds and have been utilized for the preparation of selenium-containing heterocycles [10] and metal complexes. Additionally, selenourea **VII** (R = H) is one of the most important selenium sources in the field of inorganic chemistry. It has been treated with cadmium salts to give CdSe for nanoparticles [11], thin films [12], and quantum belts [13]. Cyclic selenodithiocarbonates and their derivatives are used as precursors of tetrathiafluvalene cores [14]. As the compounds in the category D, the carbon atom of the selenocarbonyl group is an sp carbon atom, and two heteroatoms are attached to the carbon atom. Isoselenocyanates **X** [15] are key precursors for selenoureas and selenium-containing heterocycles. Among selenium isologues of carbon dioxide **XI**, carbonyl diselenide XI (E = Se) has been used as starting materials for selenocarbonyls, although highest care is necessary to handle it [16]. The *in situ* generation of carbonyl selenide XI (E = O), which was originally studied by Sonoda and coworkers [17], is used by other researchers to generate H_2Se [18].

As one of characteristic features of selenocarbonyls for the last 10 years, biological activities of many types of selenocarbonyls have been tested [19]. This is because of the high activity of organoselenium compounds as well as because the comparison of biological activities of selenium isologues of biologically active carbonyl compounds are of interest. In the former case, even if some activities are observed, it should be carefully analyzed if the activities are due to the selenocarbonyl compounds used or selenium compounds derived from the decomposition of the original selenocarbonyl compounds. To treat these increasingly expanded fields of selenocarbonyls in this paper, focus has been mainly laid on the results reported after 2005.

6.2
Theoretical Aspects of Selenocarbonyls

Recent progresses on the methods for the theoretical calculations have stimulated many chemists to treat selenocarbonyls as objects of research. One of trends is to depict selenocarbonyls on the basis of the corresponding carbonyl compounds, and structural and electronic structures are compared. In related to the tautomerization between **6** and **7**, molecular orbital calculation of **6** and **7** (R = CH_3) was carried out with dimethyl ether as an aprotic solvent. As a result, the form **6** is stable and the energy barrier between **6** and **7** is lower [20]. The activation energies for the tautomerizm between **6** and **7** without a solvent is also calculated, and the transition structures are optimized [21]. In this paper, the activation energies for a variety of selenocarbonyls **6** (R = H, F, NH_2, OH, CN) are also calculated. Tautomerization of selenium isologues of carbamic acid is also studied [22]. The comparison of the relative energy between C=Se form **8** and SeH form **9** shows that the latter one **9** is more stable, and the activation barrier between **8** and **9** is higher in the case of **8** and **9** with E = O.

Rotational barriers of amides, carbamic acid esters, and urea through carbon–nitrogen single bond are of great interest from the fundamental point of view. Likewise, selenium isologues with C–N single bonds **10** are theoretically studied [23]. The solvent effect for the barriers is elucidated, and the results have suggested that the increased polarity of the solvents enhances the barrier. Rotational barriers of selenoamide **11** and methyleneselenourea **12** are studied in detail combined with NMR experiments [24]. The bond dissociation enthalpies of N–H bond of selenocarbamates **13** [25], bond dissociation energies of C–H bonds in **14** [26], and proton affinities of **14** and related compounds [27] are calculated and compared with those of oxo and thio derivatives.

As other examples of the selenocarbonyls to be theoretically calculated, selenouracils **15** [28], selenothymine **16** [29], selenoketones **17** [30], **18** [31], **19** [32], **20** [33], **21** [33], selenium isologue of quinine [33], selenofluorenone **22** [34], and selenofluroyl fluoride **23** [35] have been reported (Figure 6.3).

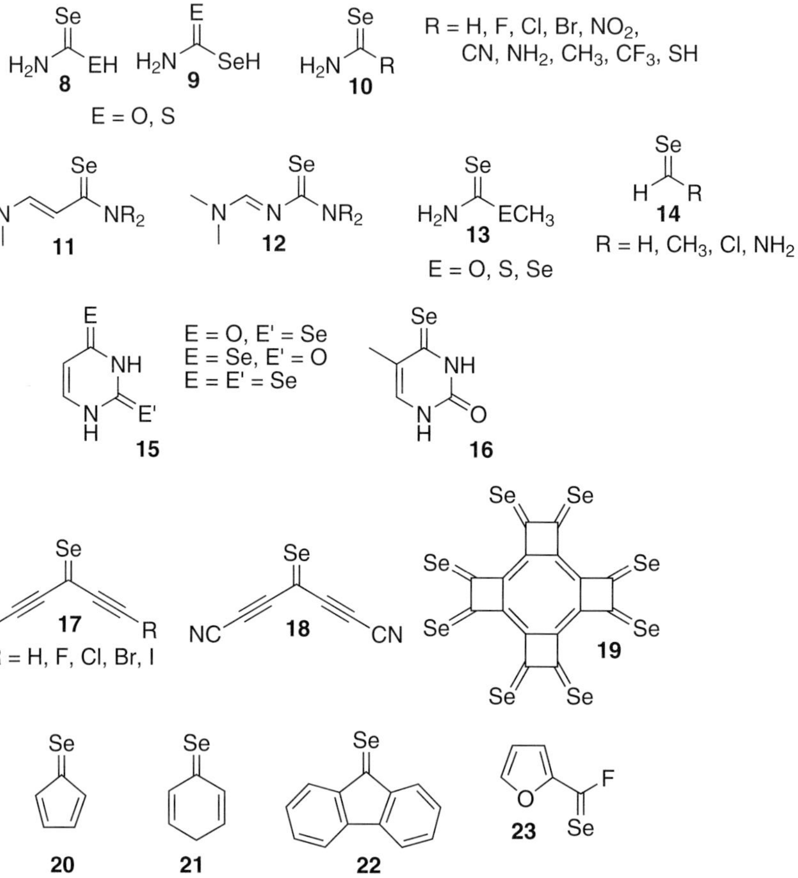

Figure 6.3 Selenocarbonyls of theoretical interest.

6.4 Synthetic Procedures of Selenocarbonyls | 261

The reaction of imidazoline-2-selone **24** with halogens gives charge transfer complexes **25** or T-shaped adducts **26** (Scheme 6.1). The kinetic and thermodynamic aspects of the reaction have been elucidated [36].

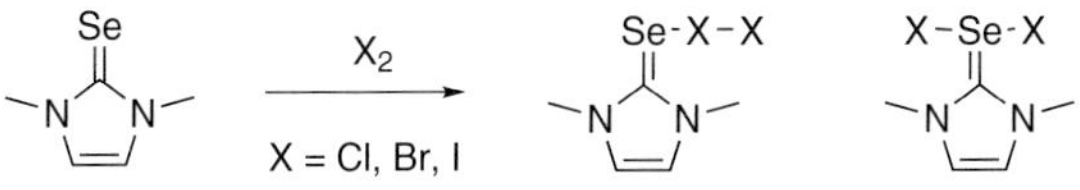

Scheme 6.1

6.3
Molecular Structure of Selenocarbonyls

Molecular structures of a range of selenocarbonyls have mainly been disclosed by the end of last century [1]. Nevertheless, several papers have disclosed new results. First, the structure of *N*-benzyl selenoamide **27** is dissolved, and the length of C=Se bond is reported to be 1.815(5) Å [37]. The X-ray analyses of 12 primary aromatic selenoamides **28** are disclosed, and several types of hydrogen-bonding interactions are discussed [38]. The lengths of C=Se bonds range from 1.822(5) to 1.856(4) Å. The X-ray structure analysis of imidazo-2-selone **29** has shown that **29** is present as a dimeric form through the Se–H hydrogen-bonding interaction, and C=Se bond lengths are around 1.85 Å (Figure 6.4) [39].

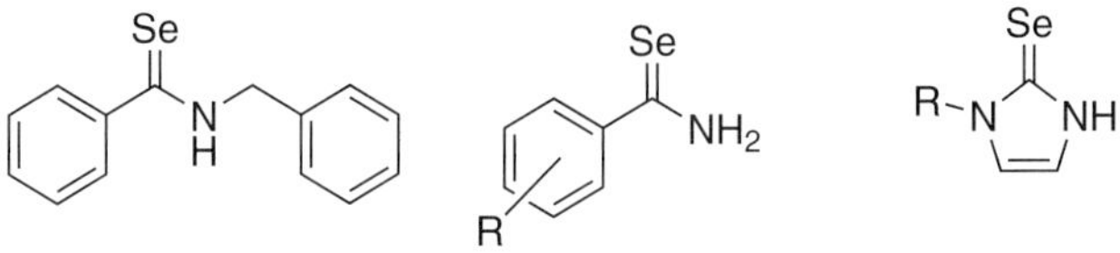

R = 2,4,6-Me$_3$C$_6$H$_2$, Me

Figure 6.4 Some selenocarbonyls for X-ray analyses.

6.4
Synthetic Procedures of Selenocarbonyls

A wide variety of synthetic methods for selenocarbonyls have been established. In all cases, elemental selenium are the original source of C=Se. Insertion of elemental selenium to organometallic reagents followed by the hydrolysis can lead to the desired compounds. As one of other methods, elemental selenium may be reduced to Se^{2-} species, and it is *in situ* handled with carbonyl compounds, haloimines, and nitriles. Selenating agents, which can be handled under air, are developed. Potassium selenocyanate (KSeCN) and selenoisocyanates derived from elemental selenium and icocyanides are precursors of selenocarbonyls. Recent examples of these methods are shown below.

The deprotonation of oxazoline **30** generates lithiated oxazoline **31** (Scheme 6.2) [40]. The THF solution is then transferred to a solution of red selenium activated by sublimation in THF followed by the hydrolysis with citric acid to give cyclic selenocarbamate **32**. Similarly, *N*-mesityl or *N*-methyl imidazole **33** is treated with *n*-BuLi to form lithiated imidazole **34**, and to the reaction mixture is added elemental selenium (Scheme 6.3) [39]. Hydrochloric acid is used for the protonation of the solution to give **29**.

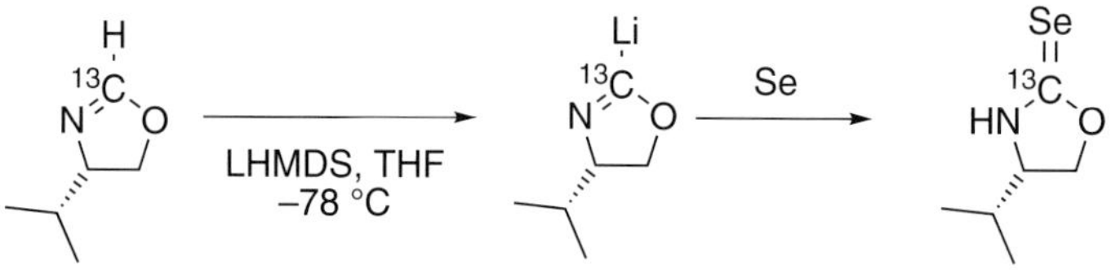

Scheme 6.2

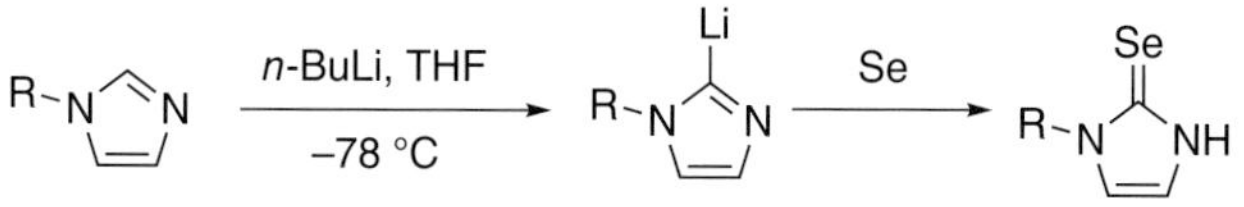

Scheme 6.3

R = 2,4,6-Me$_3$C$_6$H$_2$, Me

Maltol **35** is converted to selenomaltol **36** by reacting with elemental selenium in the presence of red phosphorus under reflux in *m*-xylene, albeit in low yield (Scheme 6.4) [41].

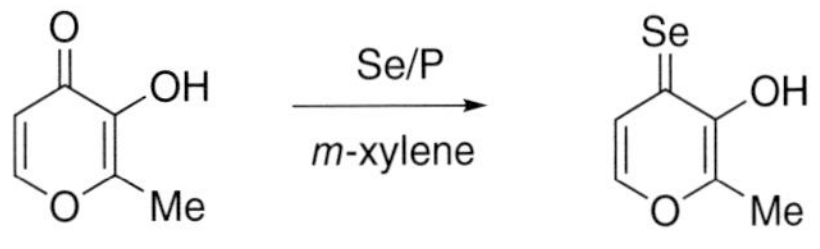

6.4 Synthetic Procedures of Selenocarbonyls

Scheme 6.4

For direct conversion of carbonyls to selenocarbonyls, *in situ* generated or isolated M$_2$Se species is used. The reaction of aldehydes with (Me$_2$Al)$_2$Se generated from (Bu$_3$Sn)$_2$Se and AlMe$_3$ to form selenoaldehydes **37**, which is then trapped with anthracene to give the adduct **38** (Scheme 6.5) [42]. CoCl$_2$·6H$_2$O-mediated reaction of aldehydes with (Me$_3$Si)$_2$Se also generates selenoaldehydes **37** followed by trapping with 2,3-butadiene (Scheme 6.6) [43]. This procedure is applied to acylsilanes, and Diels–Alder adducts **41** are formed (Scheme 6.7). These results have suggested that selenoacylsilanes **40** are generated *in situ* [43]. The combination of elemental selenium, hydrochlorsilanes, and amines can generate a selenating agent, which is reacted with formamides to give selenoformamides **42** (Scheme 6.8) [44]. This system is also applied to sialic acid derivative, and the carbamoyl group is selectively converted to selenocarbamoyl group as in **43** (Scheme 6.9).

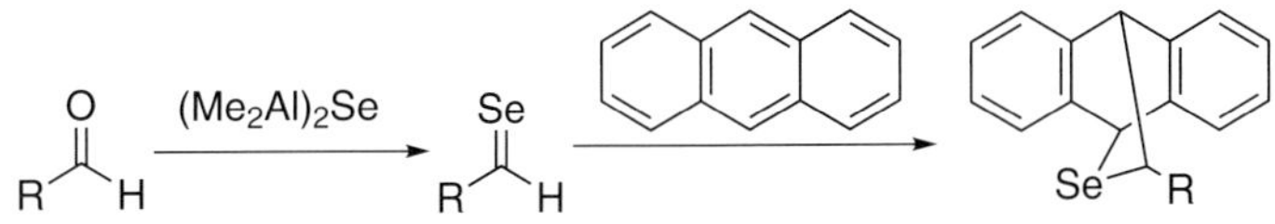

Scheme 6.5

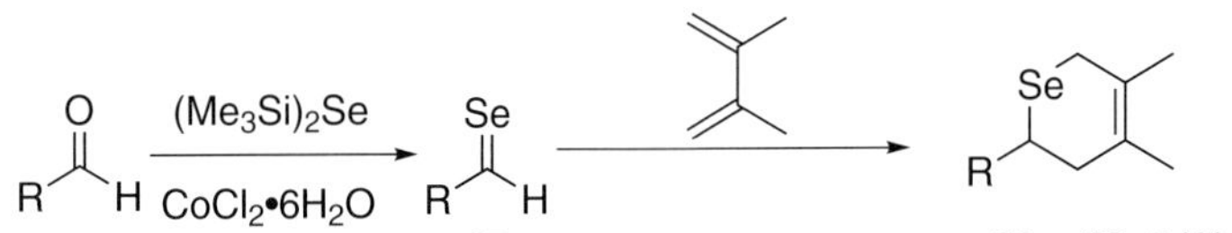

Scheme 6.6

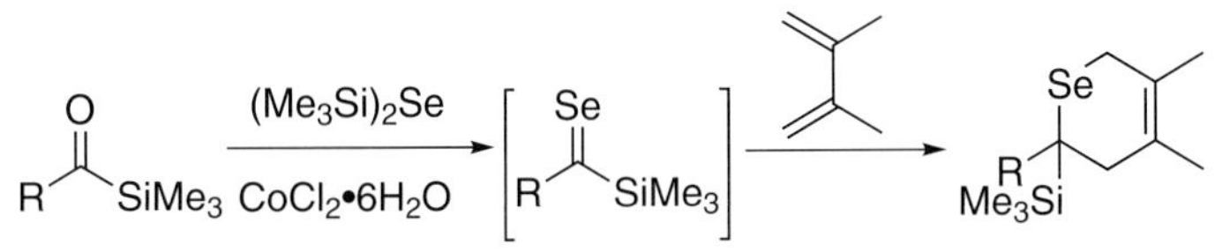

Scheme 6.7

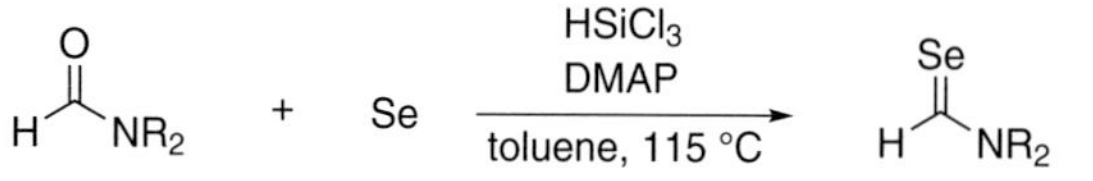

6 Selenocarbonyls

Scheme 6.8

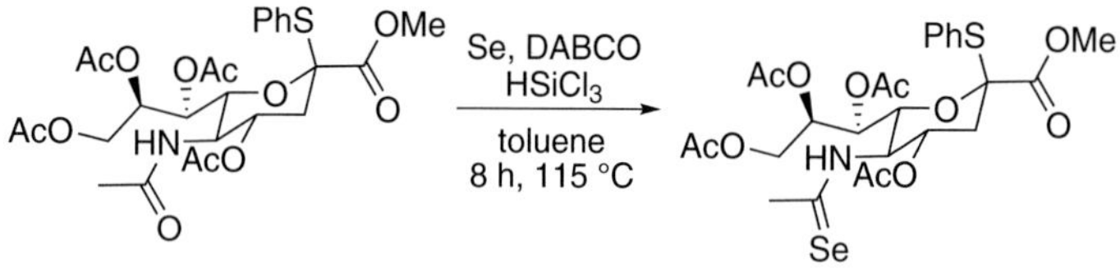

Scheme 6.9

2,4-Bis(phenyl)-1,3-diselenadiphosphetane-2,4-diselenide (**44**) [45], which has been called Woollins' reagent, is used for the selenation of amides [46]. The reagent **44** has been applied to selective formation of **45** from dipeptide (Scheme 6.10) [47]. The addition reaction of H_2Se to nitriles has been known to lead to primary selenoamides **46**. The combination of Woollins' reagent and water is used to the conversion of aromatic nitriles (Scheme 6.11) [48]. The reagent **44** is in equilibrium with diselenaphosphorane **47**. The formation of cyclic compound **48** via [2+2] cyclization of **47** and nitriles has been postulated. Similarly, the use of cyanamides in place of nitriles in the reaction in Scheme 6.11 leads to selenoureas [49]. The addition of H_2Se generated *in situ* from elemental selenium, carbon monoxide, and water to nitriles has been studied [18]. The protocol of this reaction has already been reported in 1985 [50], and the difference is the reaction with or without bases.

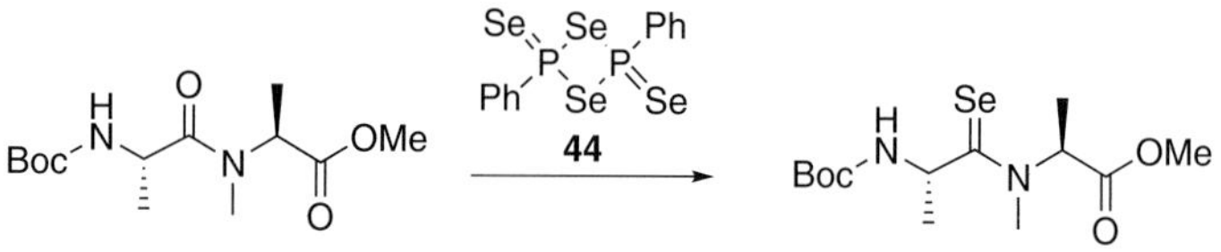

Scheme 6.10

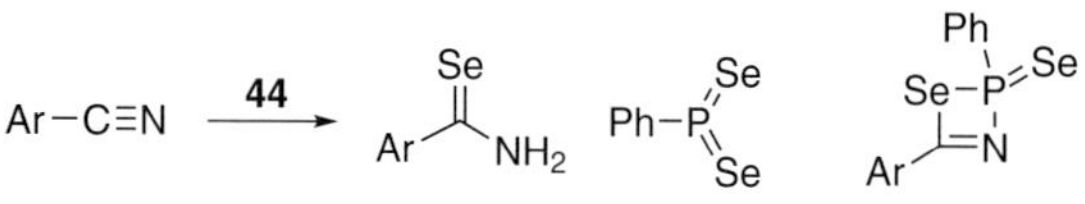

Scheme 6.11

The reduction of elemental selenium with NaBH$_4$ in EtOH at 0 °C for 30 min leads to Se^{2-} species, that is, usually formulated as NaSeH or NaHSe. With this reagent, 2-chloro-3-formylquinoline is converted to 3-selenoquinoline **49** for 1 h under reflux in EtOH (Scheme 6.12) [51]. Similarly, the introduction of selenium atom to 3-chloropyrazine is carried out with NaSeH to give selone **50** in high yield (Scheme 6.13) [52]. Wide applicability of *in situ* generated NaSeH has been proved by applying this to the selenation of iminium salt (Scheme 6.14) [53] and thioimide (Scheme 6.15) [54]. In the former reaction, cyclic diselenocarbonate **51** is obtained as a product, which is then converted to diselenafulvalene **52** by reacting with PPh$_3$. In the latter case, thioimide moiety in **53** is selectively converted to a selenocarbonyl group, and hydroxy and carbonyl groups remains intact under the reaction

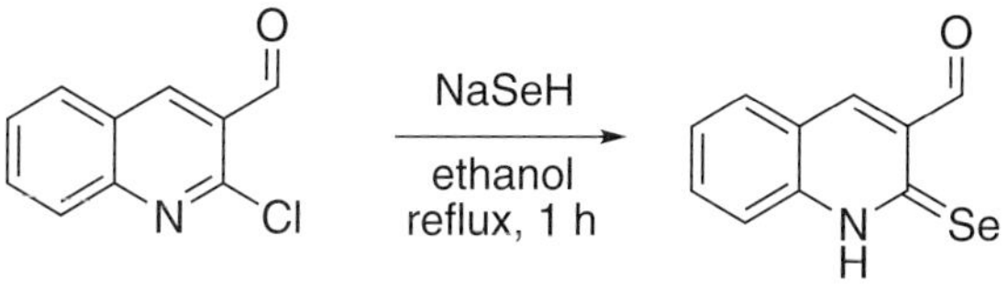

Scheme 6.12

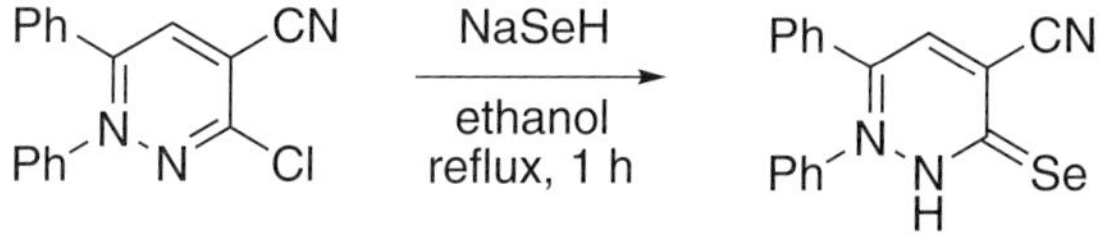

Scheme 6.13

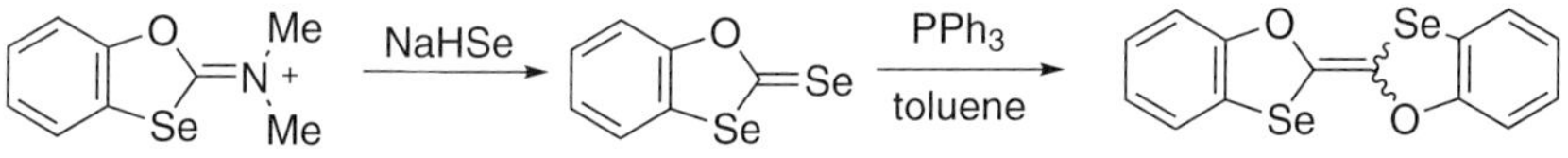

Scheme 6.14

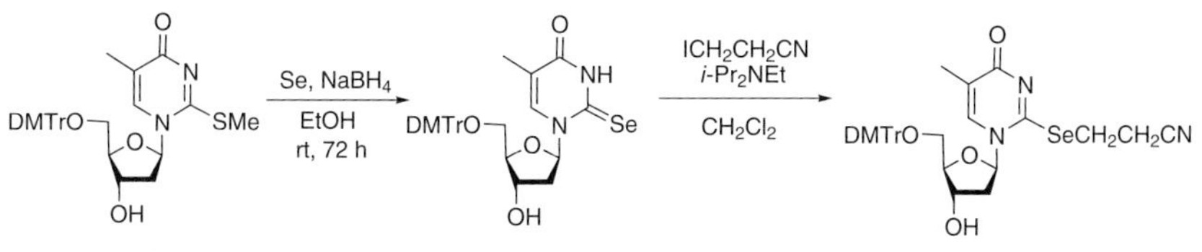

Scheme 6.15

conditions, where **53** was treated with NaSeH at room temperature for 72 h. The product **54** is then cyanoethylated selectively at the selenium atom.

Selenation of Fischer carbene complexes **55** with NaSeH also proceeds smoothly to form pyrimidine selones **56** along with metal complexes **57** (Scheme 6.16) [55]. The formation of free selones **56** is slightly higher yield in the case of chromium complex **55**. X-ray structure analyses of **56** and **57** (M = W) are carried out. The bond length of C=Se in **56** (1.843(6) Å) becomes shorter in tungsten complex **57** (1.835(6) Å). As an alternative method for the selenation of Cr carbene complexes, carbonyl selenide (O=C=Se) can be used [56].

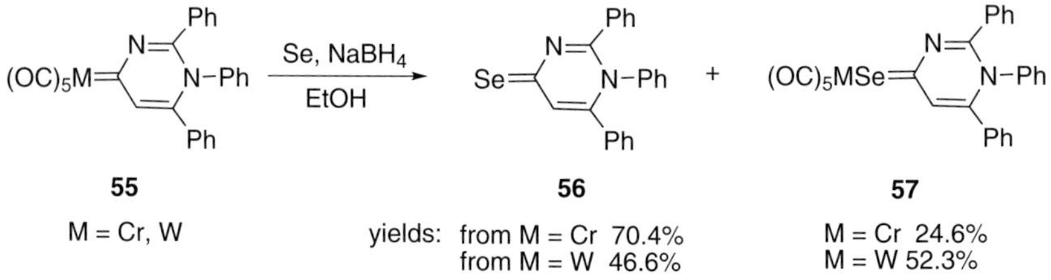

Scheme 6.16

In place of NaBH$_4$, LiAlH$_4$ can be used for the reduction of elemental selenium. In this case, aprotic polar solvent is available, and the species, which may be formulated as LiSeAlH$_2$ rather than as LiAlHSeH proposed by the authors [57], is generated since the generation of gas ascribed to hydrogen has occurred during the reduction. Selenourea **58** is prepared by reacting 1-cyanopiperidine with hydrochloric acid and Se/LiAlH$_4$ (Scheme 6.17). This system enables the conversion of dichloroiminium salt to selenocarbamoyl chloride **59**, which is then reacted with lithium vinylselenolate **60** to form diselenocarbamic acid Se-vinyl ester **61** (Scheme 6.18) [58].

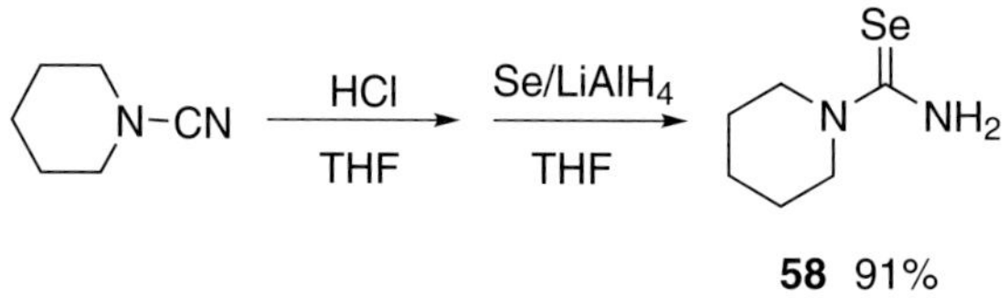

Scheme 6.17

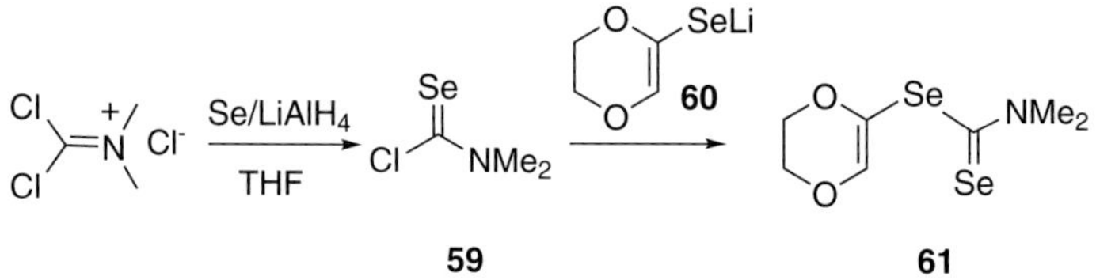

Scheme 6.18

Tetraethyl ammonium tetraselenotungstate **62** has been established as a selenium transfer reagent to iminium salts (Scheme 6.19) [59]. The iminium salts are derived from amides by reacting with oxalyl chloride or phosphorous trichloride and are treated with **62** at −78 °C. The reaction was continued at room temperature for 30 min to give tertiary or secondary selenoamides **63**. Attempts to prepare primary selenoamides and tertiary selenoamides with bulky substituents on the nitrogen atom were not successful. Alkoxylcarbonyl groups in the iminium salts are tolerated by the reaction conditions. The reaction is also applicable to iminium salts derived from lactams. A similar reaction of iminium salt **64** with **62** and secondary amines in the presence of potassium carbonate gives unsymmetrically substituted ureas **65** (Scheme 6.20) [60]. Sulfur isologue of **62** leads to thioureas as isolable compounds. In contrast, some of selenoureas are not isolable, although the general tendency of the stability is not clear.

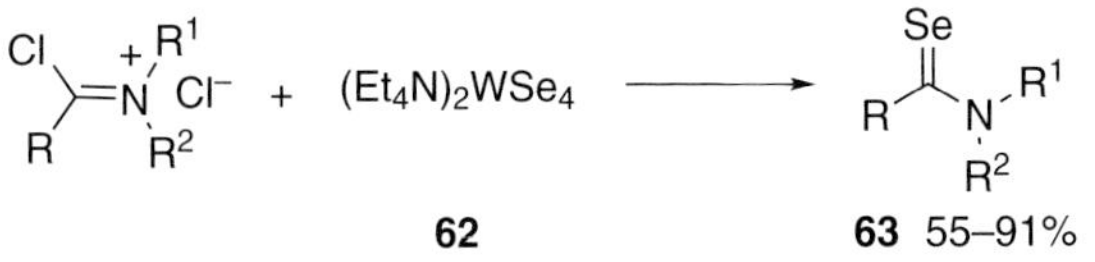

Scheme 6.19

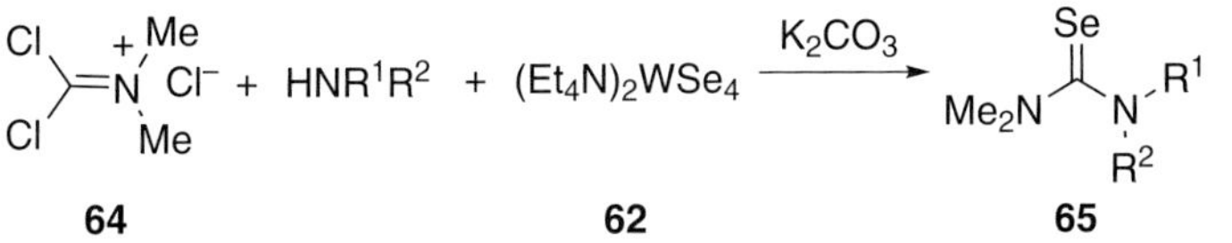

Scheme 6.20

Potassium selenocyanate (KSeCN) has long been known as precursors of selenoisocyanates [16b]. The reaction of KSeCN with aromatic acid chlorides proceeds smoothly, and the resulting N-acyl selenoisocyanates **66** are susceptible to the addition of heteroatom-containing nucleophiles such as amines, alcohols, thiols, and lithium selenolates and tellurolate, which has recently been re-examined, to give N-acyl selenoureas **67** and N-acyl chalcogenolates **68** (Scheme 6.21) [61]. The

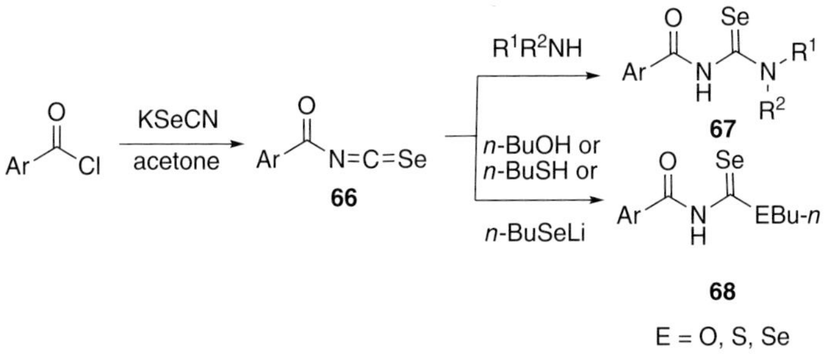

Scheme 6.21

use of aliphatic acid chlorides gives the corresponding products only in low yields. The reaction of **66** with phenylhydrazine [62] and carbodiimide [63] has also been tested to generate selenourea derivatives. Trityl isoselenocyanate has been prepared from KSeCN and trityl chloride [64].

Elemental selenium has been known to readily react with isocyanates to lead to isoselenocyanates [16b]. To the isoselenocyanate **69** are added carbon nucleophiles such as enolate ion **70** and *n*-butyl iodide to form selenoimidates **71** and **72** (Scheme 6.22) [65]. Propargyl alcohols and amines are also used as nucleophiles to selenoisocyanates to give selenium-containing heterocycles. 4-Isocyanopiperidine-1-oxyl **73** is reacted with elemental selenium to form isoselenocyanate **74** followed by the addition of amines to give selenoureas **75** (Scheme 6.23) [66]. A similar reaction of *N*-aryl isoselenocyanates with dimethylamine leading to selenoureas is reported [67]. *N*-protected amino acids are used as a starting material for the synthesis of isocyanates **76**, which are reacted with elemental selenium to give isoselenocyanates **77**, followed by the addition of amino acid *O*-methyl esters to give selenoureido products **78** (Scheme 6.24) [68].

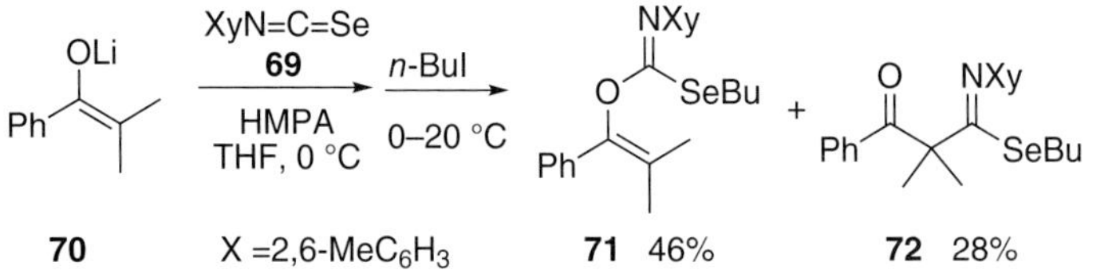

Scheme 6.22

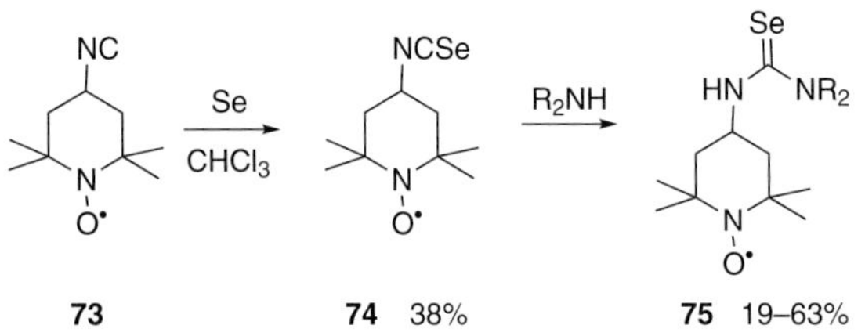

Scheme 6.23

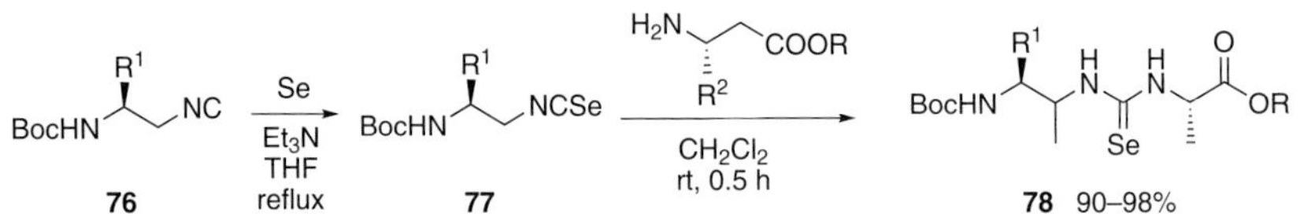

Scheme 6.24

6.4 Synthetic Procedures of Selenocarbonyls

The combination of lithium acetylides, elemental selenium, and CSe$_2$ leads to 1,3-diselenole-2-selones. For example, lithium acetylide **79** is reacted with selenium powder, CSe$_2$, and MeSCN to give the products **80** (Scheme 6.25) [69]. Lithium alkyneselenolate **81** may be initially formed and react with CSe$_2$ to form lithium triselenocarbonate **82** followed by the intramolecular cyclization to generate intermediate **83**.

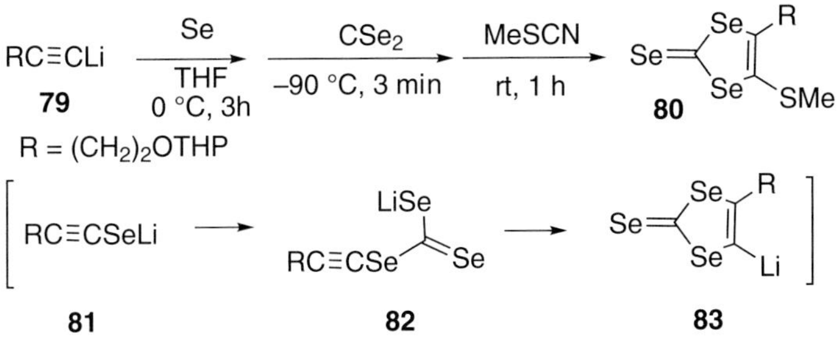

Scheme 6.25

Diselenocarbamoylmethane **85** can be a precursor of selenoaldehydes **37** [70]. Initially, diselenide **84** is reduced with NaH, followed by the reaction with dihaloalkanes to form **85**, which is then treated with SnCl$_4$ in the presence of 2,3-dimethylbutadiene **86** to give Diels–Alder adducts **87** of selenoaldehydes **37** (Scheme 6.26).

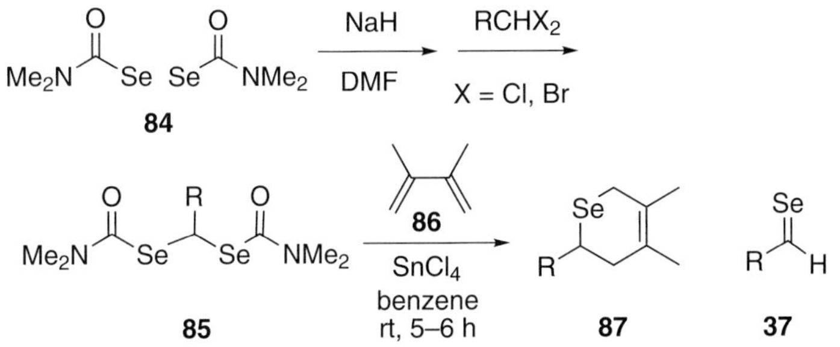

Scheme 6.26

Bis(2-cyanoethylselenide) **88** has been used to introduce the selenocarbonyl group [71]. The reaction of triazolide **89** with sodium 2-cyanoethylselenolate generated from **88** and NaBH$_4$ followed by desilylation forms **90**, which is then treated with several reagents to give selenothymidine derivative **91** [71a] (Scheme 6.27). The conversion of the cyanoethylseleneyl group to the selenocarbonyl group in **91** is achieved by the reaction with K$_2$CO$_3$ in methanol.

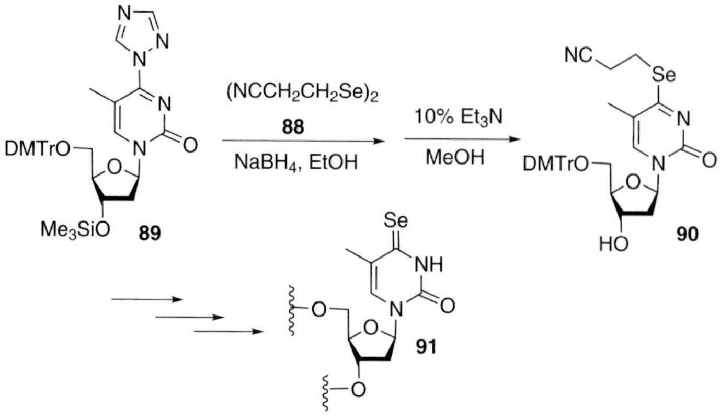

Scheme 6.27

6.5
Manipulation of Selenocarbonyls

In this section, the transformation of selenocarbonyls prepared is described based on the substituents attached to selenocarbonyl.

The aromatic selenoaldehydes **37** generated *in situ* from **38** via a retro Diels–Alder reaction undergo a Diels–Alder reaction with phosphole selenides **92** to give the adducts *endo*-**93** as single stereoisomers [72] (Scheme 6.28). On the basis of the stereochemistry of **93**, selenoaldehydes **37** approach the butadienyl group in **92** from the opposite side of aryl groups on the phosphorus atom. A similar reaction with phosphole sulfides gives four types of the products derived from the disproportionation of sulfur and selenium atoms, although their stereochemistry is identical.

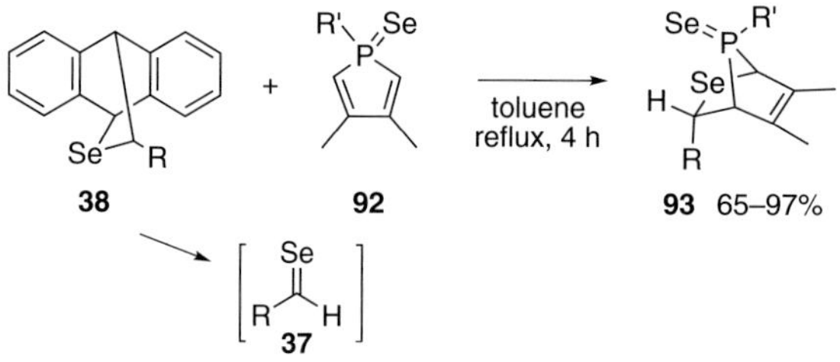

Scheme 6.28

The cycloaddition of pyrrole *N*-oxide **94** with di-*tert*-butyl selenoketone **95** is carried out under microwave irradiation conditions to give selenolactam **96** as a product along with di-*tert*-butyl ketone (Scheme 6.29) [73]. Initially, the attack of the oxygen atom in **94** to the carbon atom of **95** may occur, followed by the transfer of the selenium atom accompanied with the elimination of the ketone.

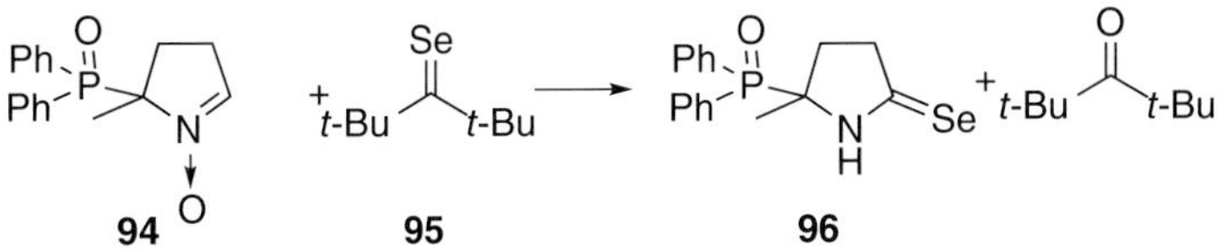

6.5 Manipulation of Selenocarbonyls | 271

Scheme 6.29

Primary selenoamides are one of the most frequently used starting materials leading to heterocycles involving nitrogen and selenium atoms. The reaction of selenoformamide **97** with α-bromoketones in the presence of Amberlite IR-120 or pyridine is carried out to give 1,3-selenazoles **98** (Scheme 6.30) [74]. A similar reaction using α-bromoacetaldehyde proceeds but the corresponding unsubstituted 1,3-selenazole is obtained in only 3% yield. Instead of **97** selenourea has been used in an ionic liquid and water to form 2-amino-1,3-selenazoles in high yields [75]. A combination of primary selenoamides and α-bromo-α,β-unstaturated esters [76a] or acetylene dicarboxylate [76b] has also been investigated.

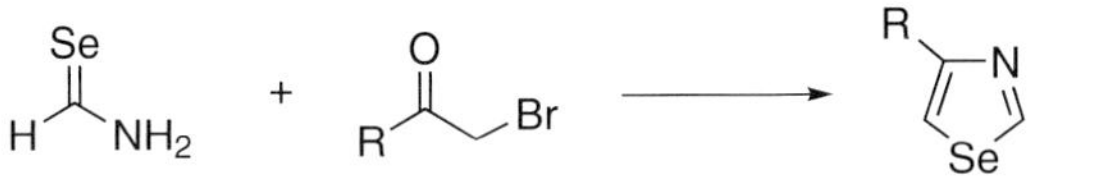

Scheme 6.30

Primary selenoamides have been reacted with propargyl cations generated *in situ* to form selenazoles as a product [77]. For example, the reaction of propargyl alcohol **99** with selenoamide **100** in the presence of a catalytic amount of Sc(OTf)$_3$ gives selenazole **101** (Scheme 6.31). The generation of propargyl cation **102** with the aid of Sc(OTf)$_3$ is postulated, and the selenium atom in **100** may attack the allenyl carbon atom of the resonance form **102**.

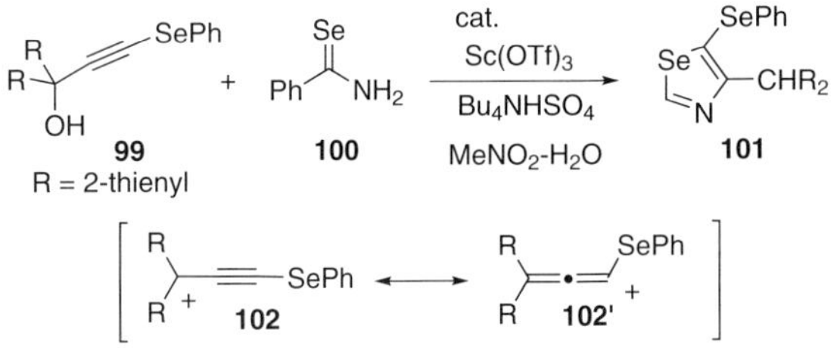

Scheme 6.31

α,β-Unsaturated selenoamides have also been used as precursors of selenium-containing heterocycles [78]. The reaction of selenoamide **103** with primary amines and formaldehyde leads to selenadiadines **104** [78b] (Scheme 6.32). In the product, two molecules of imines derived *in situ* from primary amines and formaldehyde

6 Selenocarbonyls

are incorporated. Alkylation of selenoamides takes place at the selenium atom to form selenoimidates [79].

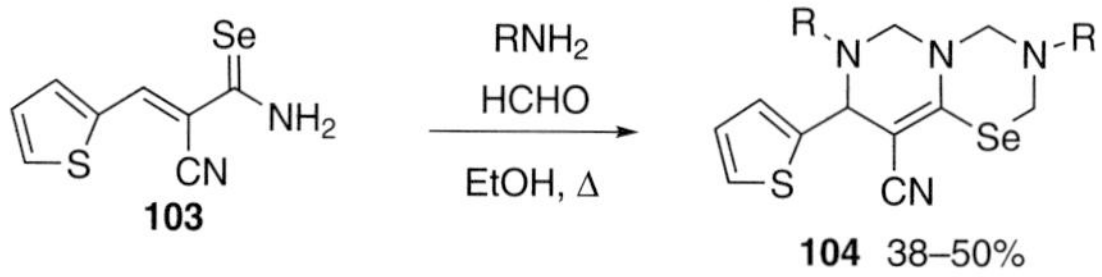

Scheme 6.32

Eschenmoser coupling reaction of thioamides α-haloesters are well known. A similar transformation is carried out with selenolactam **105** [80]. The selenolactam **105** reacts with α-bromoesters **106** under basic conditions to give enaminoesters **107** albeit in low yields (Scheme 6.33).

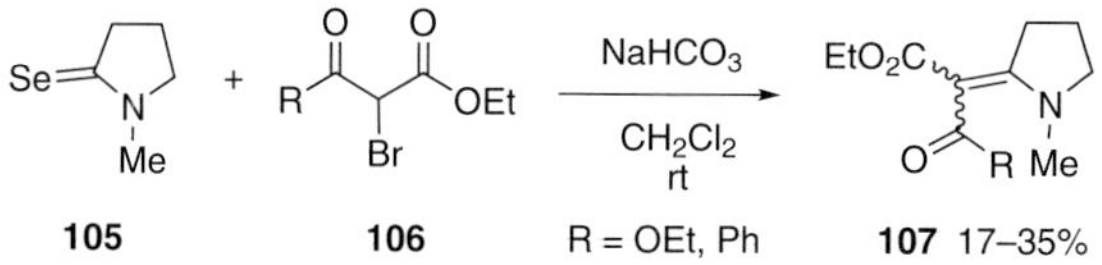

Scheme 6.33

Selenoiminium salts **109** generated from secondary or tertiary selenoamides **108** and MeOTf is subjected to telluration with LiAlH$_4$ and elemental tellurium to give telluroamides **110** (Scheme 6.34) [81a]. The stability of telluroamides is highly dependent on the substituents on the carbon atom α position to the tellurocarbonyl group and the nitrogen atom. The sequential addition reaction of lithium acetylides and Grignard reagents to selenoiminium salts **109** successfully proceeds to give propargylamines **111** having a tetrasubstituted carbon atom next to the nitrogen atom (Scheme 6.35) [81b]. The products derived from the incorporation of two molecules of lithium acetylides or Grignard reagents are not observed. A similar reaction of thioiminium salts is reported, but the reaction of **109** proceeds under milder reaction conditions and with less excess of Grignard reagents.

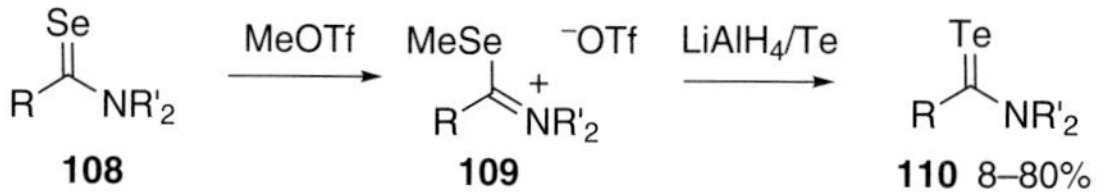

Scheme 6.34

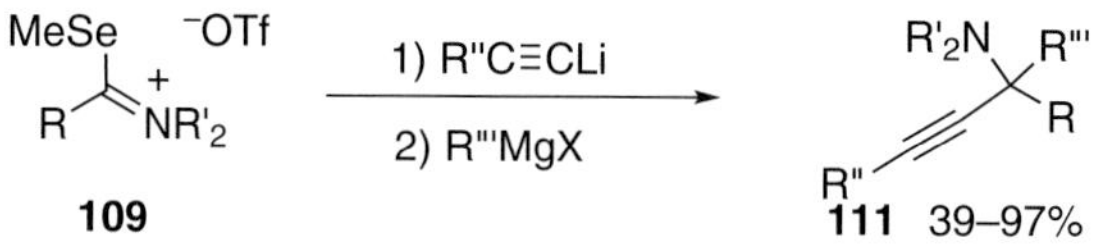

Scheme 6.35

6.5 Manipulation of Selenocarbonyls

The reaction of cyclic selenoamide **112** with α-chloro ketones has been carried out (Scheme 6.36) [82]. Ketoalkylation of **112** initially occurs followed by the intramolecular cyclization to give selenopyrazines **113**. Cyclic selenoamides **114** bearing a formyl group have been reacted with active methylene compounds in the presence of a base under microwave irradiation conditions to give selenolopyranoquinolines **115** as products (Scheme 6.37) [83].

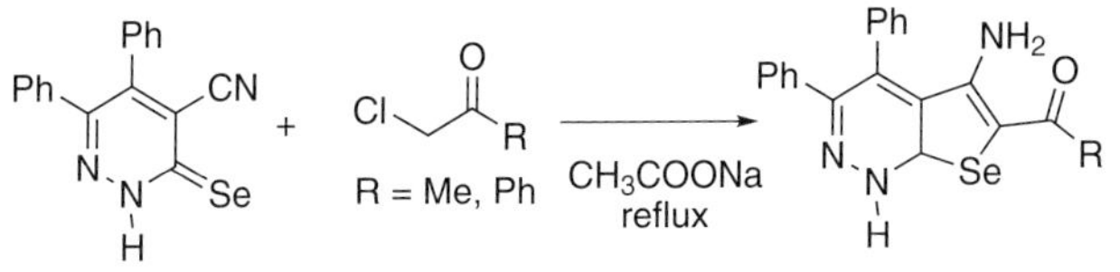

Scheme 6.36

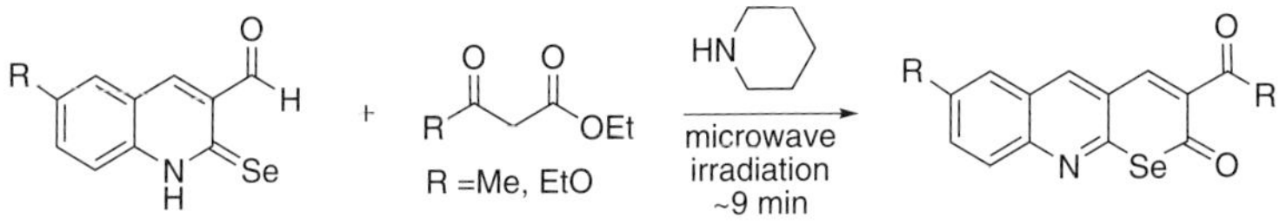

Scheme 6.37

Homocoupling of tertiary selenoamides **108** in the presence of Cu(0) in toluene takes place to nearly quantitatively give enediamines **116** (Scheme 6.38) [84]. The use of terminal alkenes as a solvent produces aminocyclopropanes **117**. α-Aminocarbene species (or α-aminocarbenoids) **118** have been postulated as key intermediates. The Cu(0)-mediated reaction of selenoamides **108** has also been investigated using terminal acetylenes. The products **119**, where α-aminocarbene species are formally inserted into the terminal C–H bond of acetylenes, are obtained [84b].

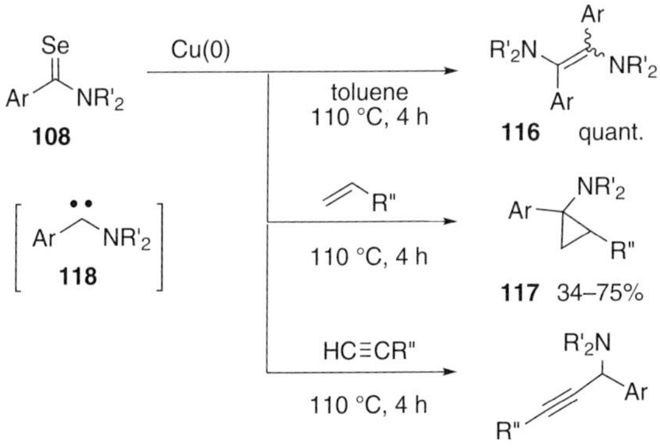

Scheme 6.38

The generation of selenocarbamyllithium **121** has been reported for the first time. The deprotonation from selenoformamide **120** with LDA proceeds in THF at −78 °C for 2 h, which is then trapped with silyl and germyl chlorides to form the first example of selenocarbamoylsilane and –germane **122** as products (Scheme 6.39) [85]. X-ray structures and spectroscopic properties of **122** have been disclosed.

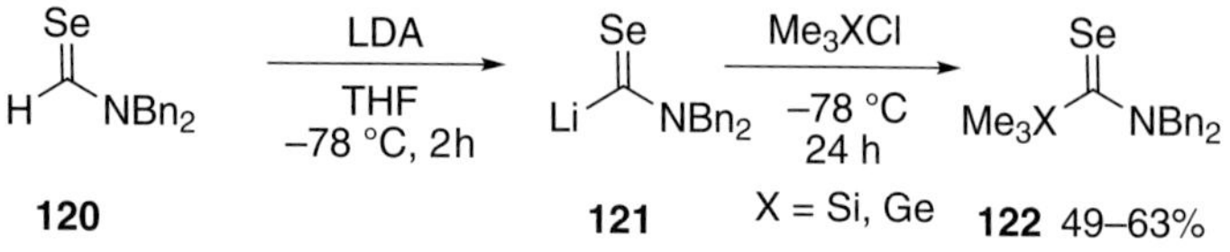

Scheme 6.39

Deselenation of selenoamide **123** with molecular oxygen is achieved in the presence of CuCl (Scheme 6.40) [86].

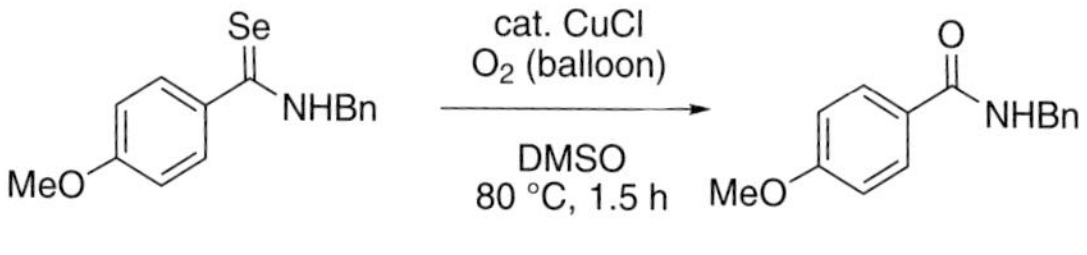

Scheme 6.40

Nucleophilic properties of the selenium atom in selenoureas and selenadienes have also been utilized for the synthesis of 5-membered nitrogen- and selenium-containing heterocycles such as 2-aminoselenazoles [87] and 6-membered heterocycles [88]. The addition reaction of selenourea **124** to divinyl selenide [89a] and benzoylbromoacetylenes has been described [89b]. In the latter case, diselenetane **126** and diselenafulvene **127** are obtained as products in 83% combined yields (Scheme 6.41). In the reaction, benzoyl selenoketene **125** may initially be formed

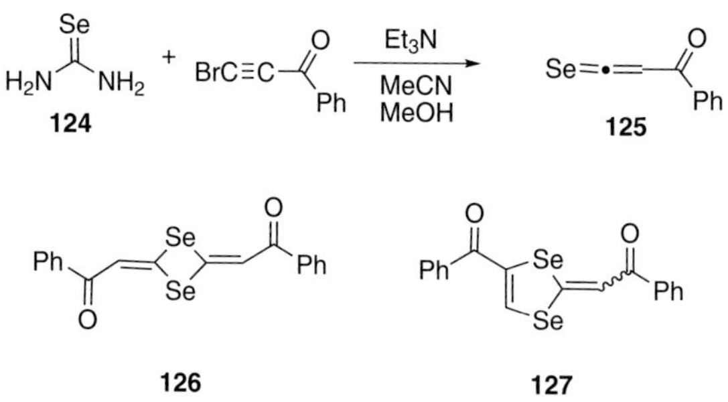

Scheme 6.41

6.5 Manipulation of Selenocarbonyls

and undergo dimerization. Selenourea **128** has been successfully used as a reagent to remove a chloroacetyl group [57]. The glucoside **129** is treated with **128** to give glucoside **130** in high yield along with selenazolone **131** (Scheme 6.42). Initially, the selenium atom in **128** nucleophilically may attack the carbon atom bearing a chloride atom to generate selenoimidate **132**, which may then undergo intramolecular cyclization accompanied with the elimination of the alcoholate to lead to **130** and **131**. N-allyl selenoureas **133** is subjected to acid-mediated cyclization and iodocyclization to give selenazines **134** and selenazolidines **135** (Scheme 6.43) [90]. N-selenocarbamoyl β-lactams bearing propargyl **136** and allenyl groups **137** have also been used as starting materials for iodocyclizations (Scheme 6.44) [91]. The compounds **136** selectively undergo 6-*exo* cyclization to give 3-selena-1-dethiacephem **138**. In contrast, iodocyclization of **137** proceeds both in 6-*exo* and 7-*endo* fashions depending on the substituents on the allenyl group. The use of **137** (R″ = H) selectively gives **139**, whereas the reaction of **137** (R″ = *n*-C$_5$H$_{11}$ and 1-naphthyl) leads to **140** with high selectivity.

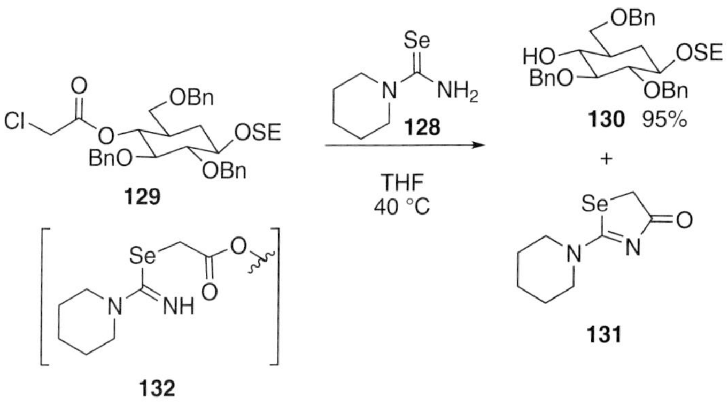

Scheme 6.42

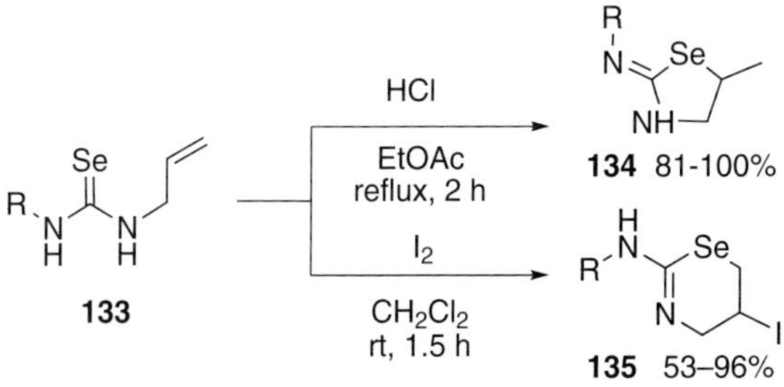

Scheme 6.43

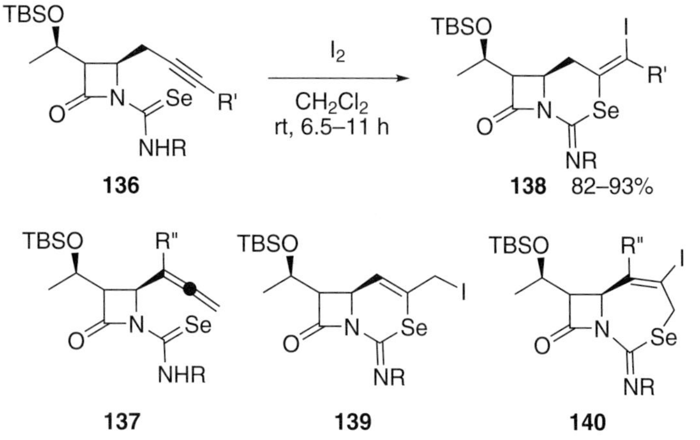

276 | *6 Selenocarbonyls*

Scheme 6.44

O-Allyl selenocarbamates **141** are also subjected to iodocyclization (Scheme 6.45) [92]. The reaction proceeds in a 5-*exo* fashion to give 1,3-oxaselenolanes. The use of NIS instead of I_2 generally gives the products **142** in better yields. The Ti-mediated aldol reaction of *N*-acetyl selenocarbonates **143** with aldehydes has been carried out (Scheme 6.46) [93]. The yields of the aldol adducts **144** are generally high, whereas the diastereoselectivity depends on the aldehydes used. The reaction of aromatic and α,β-unsaturated aldehydes shows higher selectivity when compared with those of aliphatic aldehydes.

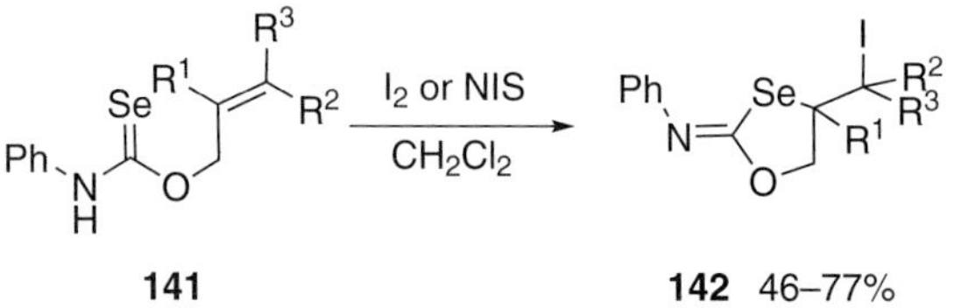

Scheme 6.45

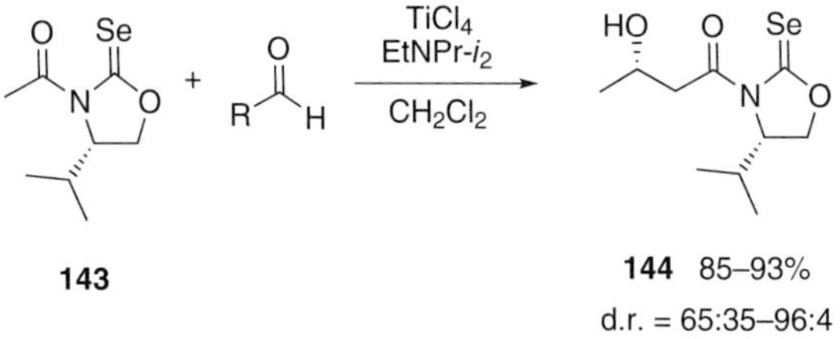

Scheme 6.46

N-Alkyl, N-aryl [94], and N-acryloyl isoselenocyanates [95] have been continuously used to provide nitrogen and selenium-containing heterocycles. As a reaction partner, the compounds bearing electrophilic and nulceophilic centers in the molecules such as β-haloethylamines, α,β-unsaturated carbonyl compounds, and α-aminoacid esters are used. For example, o-ethynylanilines **145** have been reacted with isoselenocyanates in xylene or under solvent-free conditions to give 4-methylene-3-selenaquinolines **146** (Scheme 6.47) [94d].

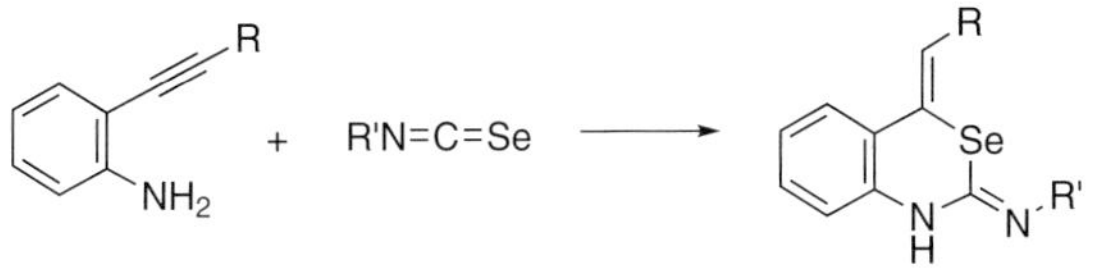

Scheme 6.47

The addition of amines to isoselenocyanates has been known to lead to selenoureas. From the biological point of view, primary amines with carbohydrate skeletons are used to obtain sugar-derived selenoureas [96]. The use of carbon nucleophiles in the reaction of selenoisocyanates can be a synthetic way to selenoamides, but only limited examples are reported. The 2-cyanomethylbenzimidazole **147** has been reacted with isoselenocyanate to form α,β-unsaturated selenoamide **148** (Scheme 6.48) [97]. Thiazole carbenes **150** generated from thiazolium salts **149** with a base is added to selenoisocyanates to form inner salts **151**. The salts **151** are further reacted with dimethyl acetylenedicarboxylate (DMAD) to form 1:2 adducts **152** of **151** and DMAD (Scheme 6.49) [98]. Secondary diselenocarbamate **153** is generated *in situ* by the addition of NaSeH to isoselenocyanates (Scheme 6.50) [99]. The acryloylation of **153** leads to 1,3-selenazine-4-ones **154** as a product. The use of N-(O-methylphenyl)selenoisocyanate and cinnamoyl chlorides gives the corresponding products only in low yields.

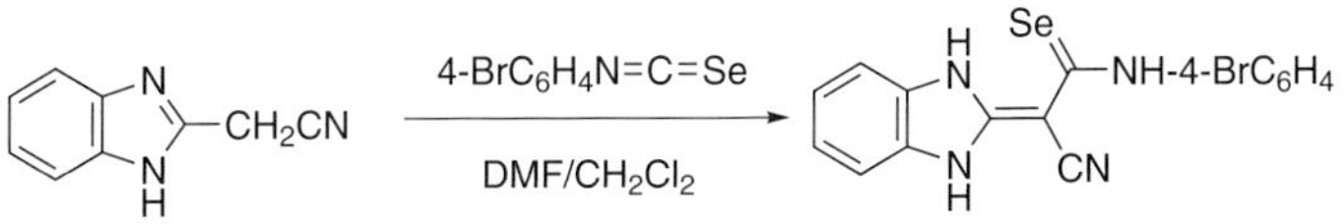

Scheme 6.48

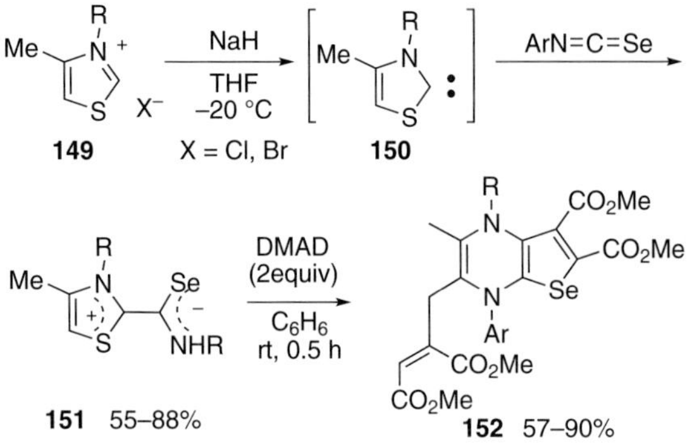

278 | 6 Selenocarbonyls

Scheme 6.49

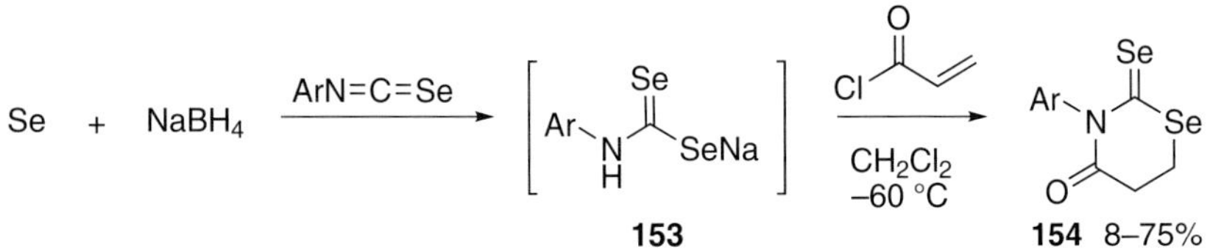

Scheme 6.50

6.6
Metal Complexes of Selenocarbonyls

The complexation of several selenocarbonyls toward transition metals has been studied. The reaction of selenoketones **95**, **155**, and **156** with [Pt(η-C_2H_4)(PPh$_3$)$_2$] takes place at room temperature, and the reaction is complete within 10 min to give Pt complexes **157–159** in high yields (Figure 6.5) [100]. X-ray structure analysis of **158** has clearly shown that the selenocarbonyl group in **158** coordinates to Pt in a η^2 mode. The structures of **157** and **159** are determined by comparing their spectroscopic data to those of **158**.

The reaction of selenoureas with Me$_3$PAuCl [101a] and HgCl$_2$ [101b] is carried out to form metal complexes of selenoureas. For gold complexes, the coordination of the selenium atom to gold is estimated by the NMR spectra. X-ray structure analysis of a mercury complex has indicated the formation of 1:2 complex **160** of HgCl$_2$ and selenourea, and two selenium atoms coordinate to mercury (Figure 6.6).

New types of selenosemicarbazones **161** are synthesized and their coordination ability toward transition metals has been tested (Figure 6.7). Bis(selenosemicarbazone) **162** is reacted with Ni, Zn, and Cd acetates to form the corresponding metal complexes [102a]. In the reaction with Zn and Cd acetates, the formation of 1:1 complexes of **162** and these metals has been confirmed by elemental analyses and NMR spectra. In contrast, the formation of Ni complex

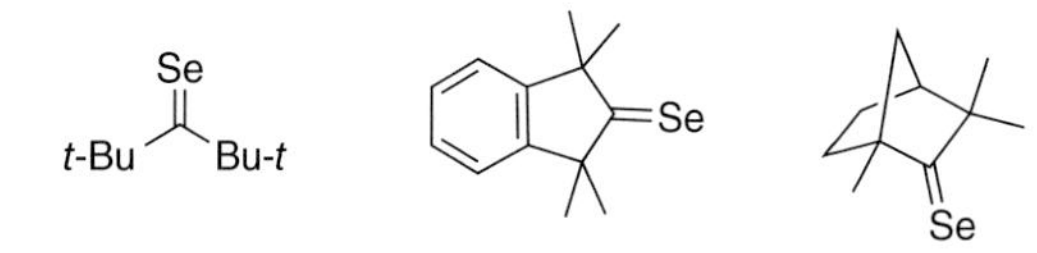

6.6 Metal Complexes of Selenocarbonyls

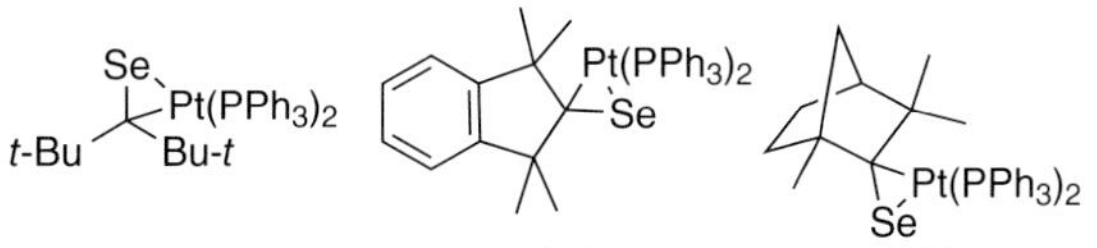

Figure 6.5 Selenocarbonyls and their Pt complexes.

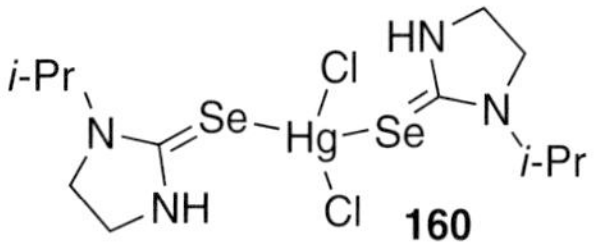

Figure 6.6 Hg complex of selenourea.

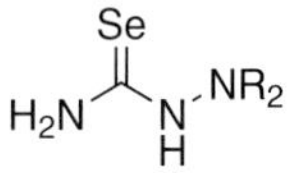

Figure 6.7 Selenosemicarbazones.

163 is accompanied with the elimination of H_2Se (Scheme 6.51). The structure of **163** was established by X-ray analysis. Selenosemicarbazone with a 2-quinolyl group **164** is also used as a ligand to Ni, Zn, and Cd [102b]. Unlike the complexation of **162**, the skeleton of **164** does not collapse to form 1:1 complexes. Pharmacological properties of Cd, Zn, Ni, Pt, and Pd complexes with **162** and **164** have been disclosed [102c, d]. Selenosemicarbazone bearing a pyridyl group **165** is used for the complexation of Gd to elucidate the cytotoxicity (Figure 6.8) [103].

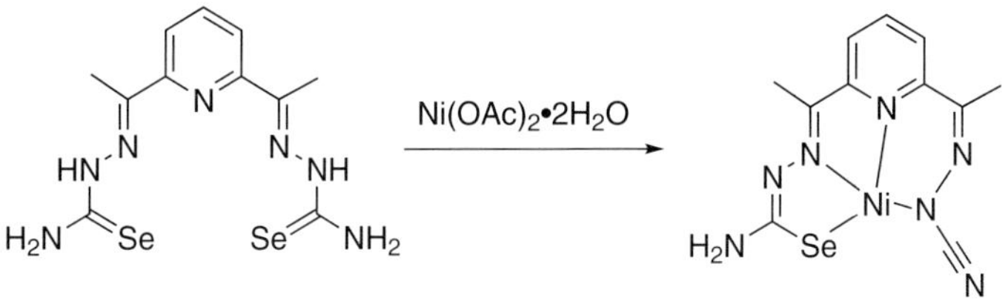

Scheme 6.51

Imidazoline-2-selones **166** have been synthesized and used for the complexation with Ni(II), Co(II), Rh(III), and Ir(III) (Figure 6.9) [104]. The structures of the

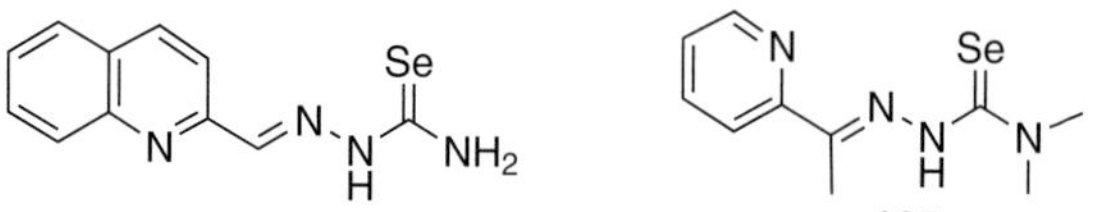

Figure 6.8 Selenosemicarbazones.

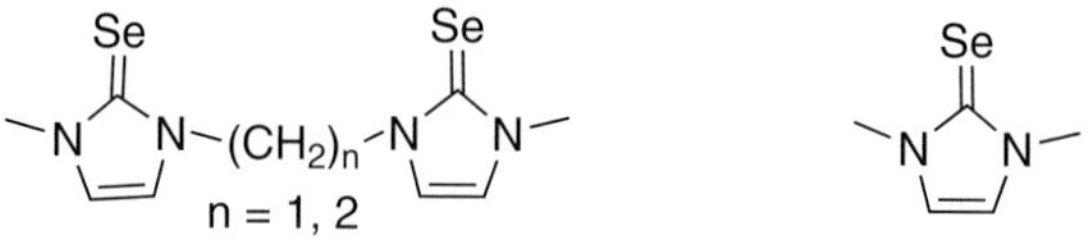

Figure 6.9 Imidazoline-2-selones.

resulting complexes are determined by X-ray analyses. For example, in the Ni complex, Ni adopts a tetrahedral structure, and two selenium atoms and two bromine atoms derived from $NiBr_2$ are coordinated to Ni. Similarly, the formation of 1:1 complexes of Rh(III) and Ir(III) and selones **166** is confirmed. In contrast, the reaction of **166** with $CoCl_2$ gives coordination polymers. N,N-Dimethyl imidazoline-2-selone **24** is treated with HCl, and the resulting compound is used for the selective extraction of Pt ions on the silica surface [105].

N-Phenyl selenocarbamate **167** has been used as a ligand to Au(I) [106]. Gold chlorides bearing several mono- and bidentate phosphine ligands are used as a precursor. The reaction of **167** with gold chlorides and monodentate phosphines gives 1:1 complexes **168**, whereas the use of those with bidentate ligands gives 2:2 complexes **169** (Figure 6.10).

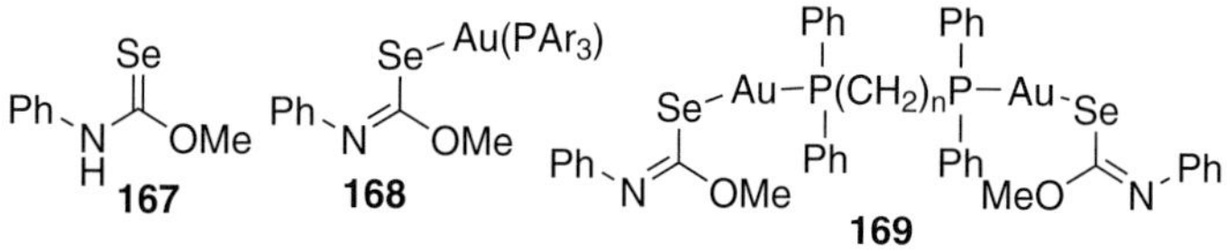

Figure 6.10 Selenocarbamate and Au complexes.

6.7
Future Aspects

Selenocarbonyls have been compounds attracting fundamental organic chemists. This has dramatically changed in the last 10 years. The progresses on computational methods have enabled not only theoretical chemists but also experimental chemists to handle selenocarbonyl in the computers. More precise static and dynamic behaviors of known selenocarbonyls will be investigated. Imaginary designed selenocarbonyls can be research topics, and their properties, possibility of their actual syntheses, and their stability will be predicted with great accuracy. Plenty of selenium isologues have appeared to be introduced, but still limited amounts of derivatives are actually prepared. Intense colors of selenoaldehydes, ketones, and esters can be sensors to selectively detect heavy metals and toxic compounds, but feasible molecules have yet to be introduced. The use of the sequence of the conversion between carbonyl to selenocarbonyl and vice versa allows for the labeling of carbonyl groups in biologically important molecules. Tremendous amounts of polymers having amide moieties have been known. In contrast, these selenium isologues have not been the topics of the research yet. High reactivity of selenocarbonyls can provide new synthetic transformations, and allow the synthesis of new heterocycles as some of them are shown in this review. Studies on metal complexes of selenocarbonyls will increase in relation to inorganic and material chemistry, and biological interest. On the basis of these aspects, fruitful chemistry on selenocarbonyls will be extended in future.

References

1 Murai, T. (2000) *Topics in Current Chemistry*, vol. 208 (ed. T. Wirth), Springer, Berlin, p. 177.
2 Guziec, L.J. and Guziec, F.S., Jr. (2005) *Comprehensive Organic Functional Group Transformations II*, vol. 3 (eds A.R. Katritzky, R.J.K. Taylor, and K. Jones),Elsevier Pergamon, Amsterdam, p. 397.
3 Wirth, T. (2005) *Sci. Synth.*, **22**, 181.
4 (a) Murai, T. (2005) *Topics in Current Chemistry*, vol. 251 (ed. S. Kato), Springer, Berlin, p. 247; (b) Moore, A.J. (2005) *Comprehensive Organic Functional Group Transformations II*, vol. 5 (eds A.R. Katritzky, R.J.K. Taylor, and R.C.F. Jones), Elsevier Pergamon, Amsterdam, p. 519; (c) Flynn, C. and Haughton, L. (2005) *Comprehensive Organic Functional Group Transformations II*, vol. 5 (eds A.R. Katritzky, R.J.K. Taylor, and R.C.F. Jones), Elsevier Pergamon, Amsterdam, p. 571.
5 (a) Murai, T. (2005) *Synlett*, 1509; (b) Ishii, A. and Nakayama, J. (2005) *Topics in Current Chemistry*, vol. 251 (ed. S. Kato), Springer, Berlin, p. 227; (c) Murai, T. (2005) *Comprehensive Organic Functional Group Transformations II*, vol. 5 (eds A.R. Katritzky, R.J.K. Taylor, and R.C.F. Jones), Elsevier Pergamon, Amsterdam, p. 493.
6 Ishii, A., and Nakayama, J. (2005) *Comprehensive Organic Functional Group Transformations II*, vol. 5 (eds A.R. Katritzky, R.J.K. Taylor, and R.C.F. Jones), Elsevier Pergamon, Amsterdam, p. 459.
7 Kato, S. (2005) *Topics in Current Chemistry*, vol. 251 (ed. S. Kato), Springer, Berlin, p. 1.
8 Castano, J.A.G., Romano, R.M., Beckers, H., Willner, H., Boese, R., and Védova, C.O.D. (2008) *Angew. Chem. Int. Ed.*, **47**, 10114.
9 Schmidt, J. and Silks, L.A. (2005) *Sci. Synth.*, **18**, 969.
10 (a) Koketsu, M. and Ishihara, H. (2006) *Current Org. Synth.*, **3**, 439; (b) Koketsu, M. and Ishihara, H. (2007) *Handbook of Chalcogen Chemistry*, (ed. F.A. Devillanova), RSC Publishing, Cambridge, p. 145.
11 Williams, J.V., Kotov, N.A., and Savage, P.E. (2009) *Ind. Eng. Chem. Res.*, **48**, 4316.
12 Henriquez, R., Grez, P., Munoz, E., Gómez, H., Badán, J.A., Marotti, R.E., and Dalchiele, E.A. (2010) *Thin Solid Films*, **518**, 1774.

13 Liu, Y.-H., Wayman, V.L., Gibbons, P.C., Loomis, R.A., and Buhro, W.E. (2010) *Nano Lett.*, **10**, 352.
14 Abashev, G.G., Shklyaeva, E.V., Syutkin, R.V., Lebedev, K.Y., Osorgina, I.V., Romanova, V.A., and Bushueva, A.Y. (2008) *Solid State Sci.*, **10**, 1710.
15 Aitken, R.A. (2005) *Comprehensive Organic Functional Group Transformations II*, vol. 5 (eds A.R. Katritzky, R.J.K. Taylor, and R.C.F. Jones), Elsevier Pergamon, Amsterdam, p. 975.
16 (a) Ogura, F. and Takimiya, K. (1999) *Organoselenium Chemistry A Practical Approach*, (ed. T.G. Back), Oxford University Press, Oxford, p. 257; (b) Braverman, S., Cherkinsky, M., and Birsa, M.L. (2005) *Sci. Synth.*, **18**, 65.
17 Fujiwara, S., Kambe, N., and Sonoda, N. (1999) *Organoselenium Chemistry A Practical Approach*, (ed. Back, T.G.), Oxford University Press, Oxford, p. 223.
18 Chen, Y., Tian, F., Song, M., and Lu, S. (2008) *Heteroatom Chem.*, **19**, 211.
19 (a) Roy, G. and Mugesh, G. (2005) *J. Am. Chem. Soc.*, **127**, 15207; (b) Roy, G. and Mugesh, G. (2008) *Phosphorus Sulfur Silicon*, **183**, 908; (c) Sharma, A.K., Sharma, A., Desai, D., Madhunapantula, S.V., Huh, S.J., Robertson, G.P., and Amin, S. (2008) *J. Med. Chem.*, **51**, 7820; (d) Bhabak, K.P. and Mugesh, G. (2010) *Chem. Eur. J.*, **16**, 1175.
20 Li, Q.-G., Liang, G.-M., Wang, X., Chu, S.-Y., and Ren, Y. (2007) *Int. J. Quantum Chem.*, **107**, 921.
21 Huang, G., Xia, Y., and Li, Y. (2009) *Theochem.*, **896**, 80.
22 Kaur, D., Kaur, R.P., and Kaur, P. (2006) *Bull. Chem. Soc. Jpn.*, **79**, 1869.
23 Kaur, D., Sharma, P., Bharatam, P.V., and Dogra, N. (2006) *J. Mol. Chem.*, **759**, 41.
24 Kleinpeter, E., Schulenburg, A., Zug, I., and Hartmann, H. (2005) *J. Org. Chem.*, **70**, 6592.
25 Kaur, D. and Kaur, R.P. (2008) *Theochem.*, **858**, 94.
26 Kaur, D., Kaur, R.P., and Kaur, R. (2007) *Theochem.*, **803**, 95.
27 Kaur, D., Sharma, P., Kaur, R.P., Kaur, M., and Bharatam, P.V. (2007) *Theochem.*, **805**, 119.
28 (a) Lamsabhi, A.M., Mó, O., Yanez, M., and Boyd, R.J. (2008) *J. Chem. Theory Comput.*, **4**, 1002; (b) Trujillo, C., Mó, O., and Yanez, M. (2008) *Chem. Phys. Chem.*, **9**, 1715.
29 Vázquez-Mayagoitia, A., Huertas, O., Brancolini, G., Migliore, A., Sumpter, B.G., Orozco, M., Luque, F.J., Felice, R.D., and Fuentes-Cabrera, M. (2009) *J. Phys. Chem. B*, **113**, 14465.
30 (a) Ramasami, P. (2006) *Theochem.*, **767**, 19; (b) Ramasami, P. (2006) *Theochem.*, **775**, 87; (c) Ramasami, P. (2008) *J. Sulfur Chem.*, **29**, 475.
31 Ramasami, P. (2009) *Heteroatom Chem.*, **20**, 208.
32 Zbou, X., Hovat, D.A., and Borden, W.T. (2010) *J. Phys. Chem. A*, **114**, 3683.
33 Zborowski, K.K. and Proniewicz, L.M. (2009) *Polish J. Chem.*, **83**, 477.
34 Song, P. and Ma, F. (2010) *J. Phys. Chem. A*, **114**, 2230.
35 Abdallah, H.H. and Ramasami, P. (2009) *Theochem.*, **913**, 157.
36 Aragoni, M.C., Arca, M., Devillanova, F.A., Grimaldi, P., Isaia, F., Lelj, F., and Lippolis, V. (2006) *Eur. J. Inorg. Chem.*, 2166.
37 Mutoh, Y. and Murai, T. (2004) *Organometallics*, **23**, 3907.
38 Li, Y., Hua, G.-X., Slawin, A.M.Z., and Woollins, J.D. (2009) *Molecules*, **14**, 884.
39 Landry, V.K., Minoura, M., Pang, K., Buccella, D., Kelly, B.V., and Parkin, G. (2006) *J. Am. Chem. Soc.*, **128**, 12490.
40 Olivault-Shiflett, M., Kimball, D.B., and Silks, L.A. (2004) *J. Org. Chem.*, **69**, 5150.
41 Tejchman, W., Zborowski, K., Lasocha, W., and Proniewicz, L.M. (2008) *Heterocycles*, **75**, 1931.
42 Segi, M., Zhou, A., and Honda, M. (2005) *Phosphorus Sulfur Silicon*, **180**, 1045.
43 Degl'Innocenti, A., Capperucci, A., Accia, M., and Tiberi, C. (2009) *Phosphorus Sulfur Silicon*, **184**, 1621.
44 Shibahara, F., Sugiura, R., and Murai, T. (2009) *Org. Lett.*, **11**, 3064.

45 Hua, G. and Woollins, J.D. (2009) *Angew. Chem. Int. Ed.*, **48**, 1368.
46 Bethke, J., Karaghiosoff, K., and Wessjohann, L.A. (2003) *Tetrahedron Lett.*, **44**, 6911.
47 Huang, Y., Jahreis, G., Lücke, C., Wildemann, D., and Fischer, G. (2010) *J. Am. Chem. Soc.*, **132**, 7578.
48 Hua, G., Li, Y., Slawin, A.M.Z., and Woolins, J.D. (2006) *Org. Lett.*, **8**, 5251.
49 Hua, G., Zhang, Q., Li, Y., Slawin, A.M.Z., and Woolins, J.D. (2009) *Tetrahedron*, **65**, 6074.
50 Ogawa, A., Miyake, J., Karasaki, Y., Murai, S., and Sonoda, N. (1985) *J. Org. Chem.*, **50**, 384.
51 (a) Raghavendra, M., Naik, H.S.B., and Shergara, B.S. (2008) *Phosphorus Sulfur Silicon*, **183**, 1501; (b) Naik, T.R.R., Naik, H.S.B., Naik, H.R.P., Raghavendra, M., and Ramesha, S.S. (2008) *Phosphorus Sulfur Silicon*, **183**, 1968.
52 Abdel-Hafez, S.H. (2008) *Eur. J. Med. Chem.*, **43**, 1971.
53 Kojima, T., Tanaka, K., Ishida, T., and Nogami, T. (2004) *J. Org. Chem.*, **69**, 9319.
54 Hassan, A.E.A., Sheng, J., Zhang, W., and Huang, Z. (2010) *J. Am. Chem. Soc.*, **132**, 2120.
55 Zheng, Z., Chen, J., Yu, Z., and Han, X. (2006) *J. Organomet. Chem.*, **691**, 3679.
56 Zheng, Z., Chen, J., Luo, N., Yu, Z., and Han, X. (2006) *Organometallics*, **25**, 5301.
57 Sogabe, S., Ando, H., Koketsu, M., and Ishihara, H. (2006) *Tetrahedron Lett.*, **47**, 6603.
58 Imakubo, T., Shirahara, T., and Kibune, M. (2004) *Chem. Commun.*, 1590.
59 Saravanan, V., Mukherjee, C., Das, S., and Chandrasekaran, S. (2004) *Tetrahedron Lett.*, **45**, 681.
60 Sivapriya, K., Suguna, P., Banerjee, A., Saravanan, V., Rao, D.N., and Chandrasekaran, S. (2007) *Bioorg. Med. Chem. Lett.*, **17**, 6387.
61 Koketsu, M., Yamamura, Y., Aoki, H., and Ishihara, H. (2006) *Phosphorus Sulfur Silicon*, **181**, 2699.
62 Koketsu, M., Yamamura, Y., and Ishihara, H. (2006) *Heterocycles*, **68**, 1191.
63 Koketsu, M., Yamamura, Y., Ando, H., and Ishihara, H. (2006) *Heterocycles*, **68**, 1267.
64 Andaloussi, M.B.D. and Mohr, F. (2010) *J. Organomet. Chem.*, **695**, 1276.
65 Asanuma, Y., Fujiwara, S., Shin-ike, T., and Kambe, N. (2004) *J. Org. Chem.*, **69**, 4845.
66 Zakrzewski, J. and Krawczyk, M. (2008) *Heteroatom Chem.*, **19**, 549.
67 Zakrezewski, J. and Krawczyk, M. (2009) *Phosphorus Sulfur Silicon*, **184**, 1880.
68 Chennakrishnareddy, G., Nagendra, G., Hemantha, H.P., Das, U., Row, T.N.G., and Sureshbabu, V.V. (2010) *Tetrahedron*, **66**, 6718.
69 Takimiya, K., Kodani, M., Murakami, S., Otsubo, T., and Aso, Y. (2006) *Heterocycles*, **67**, 655.
70 (a) Shimada, K., Gong, Y., Nakamura, H., Matsumoto, R., Aoyagi, S., and Takikawa, Y. (2005) *Tetrahedron Lett.*, **46**, 3775; (b) Gong, Y., Shimada, K., Nakamura, H., Fujiyama, M., Kodama, A., Otsuki, M., Matsumoto, R., Aoyagi, S., and Takigawa, Y. (2006) *Heteroatom Chem.*, **17**, 125.
71 (a) Salon, J., Sheng, J., Chen, J.J.G., Caton-Williams, J., and Huang, Z. (2007) *J. Am. Chem. Soc.*, **129**, 4862; (b) Salon, J., Jiang, J., Sheng, J., Gerlits, O.O., and Huang, Z. (2008) *Nuleic Acid Res.*, **36**, 7009.
72 Segi, M., Kawaai, K., Honda, M., and Fujinami, S. (2007) *Tetrahedron Lett.*, **48**, 3349.
73 (a) Shioji, K., Matsumoto, A., Takao, M., Kurauchi, Y., Shigetomi, T., Yokomori, Y., and Okuma, K. (2007) *Bull. Chem. Soc. Jpn.*, **80**, 743; (b) Okuma, K., Mori, Y., Shigetomi, T., Tabuchi, M., Shioji, K., and Yokomori, Y. (2007) *Tetrahedron Lett.*, **48**, 8311.
74 Below, H., Pfeiffer, W.-D., Geisler, K., Lalk, M., and Langer, P. (2005) *Eur. J. Org. Chem.*, 3637.
75 Potewar, T.M., Ingale, S.A., and Srinivasan, K.V. (2008) *Arkivoc*, 117.
76 (a) Huang, X., Yu, L., and Chen, Z.-H. (2005) *Synth. Commun.*, **35**, 1253; (b) Koketsu, M., Sasaki, T., Ando, H., and Ishihara, H. (2007) *J. Heterocyclic Chem.*, **44**, 231.

77 (a) Yoshimatsu, M., Yamamoto, T., Sawa, A., Kato, T., Tanabe, G., and Muraoka, O. (2009) *Org. Lett.*, **11**, 2952; (b) Yoshimatsu, M., Matsui, M., Yamamoto, T., and Sawa, A. (2010) *Tetrahedron*, **66**, 7975.

78 (a) Dyachenko, V.D. (2005) *Russ. J. Gen. Chem.*, **75**, 447; (b) Frolov, K.A., Krivokolysko, S.G., Dotsenko, V.V., and Litvinov, V.P. (2009) *Chem. Heterocylic Compd.*, **45**, 255.

79 (a) Takido, T., Toriyama, M., Yamashita, K., Suwa, T., and Seno, M. (2003) *Phosphorus Sulfur Silicon*, **178**, 319; (b) Nakajima, N. and Ubukata, M. (2005) *Sci. Synth.*, **22**, 367.

80 Hussaini, S.H. and Hammond, G.B. (2008) *Arkivoc*, 129.

81 (a) Mutoh, Y., Murai, T., and Yamago, S. (2007) *J. Organomet. Chem.*, **692**, 129; (b) Murai, T., Nogawa, S., and Mutoh, Y. (2007) *Bull. Chem. Soc. Jpn.*, **80**, 2220.

82 Abdel-Hafez, S.H.H. (2008) *Eur. J. Med. Chem.*, **43**, 1971.

83 Raghavendra, M., Naik, H.S.B., and Sherigara, B.S. (2008) *Phosphorus Sulfur Silicon*, **183**, 2086.

84 (a) Mitamura, T., Nomoto, A., Sonoda, M., and Ogawa, A. (2008) *Tetrahedron*, **64**, 9983; (b) Mitamura, T. and Ogawa, A. (2009) *Org. Lett.*, **11**, 2045.

85 Murai, T., Hori, R., Maruyama, T., and Shibahara, F. (2010) *Organometallics*, **29**, 2400.

86 Shibahara, F., Suenami, A., Yoshida, A., and Murai, T. (2007) *Chem. Commun.*, 2354.

87 (a) Ramazani, A., Morsali, A., Ganjeie, B., Kazemizadeh, A.R., and Ahmadi, E. (2005) *Phosphorus Sulfur Silicon*, **180**, 2439; (b) Koketsu, M., Kogami, M., Ando, H., and Ishihara, H. (2006) *Synthesis*, 31; (c) Koketsu, M., Kanoh, K., Ando, H., and Ishihara, H. (2006) *Heteroatom Chem.*, **17**, 88; (d) Koketsu, M., Kanoh, K., and Ishihara, H. (2006) *Heterocycles*, **68**, 2647; (e) Koketsu, M., Kanoh, K., and Ishihara, H. (2006) *Heterocycles*, **68**, 2145; (f) Koketsu, M., Kanoh, K., and Ishihara, H. (2006) *Heterocycles*, **68**, 2627; (g) Koketsu, M., Kanoh, K. and Ishihara, H. (2006) *Heterocycles*, **68**, 2645; (h) Narender, M., Reddy, M.S., Kumar, V.P., Srinivas, B., Sridhar, R., Venkata, Y., Nageswar, D., and Rao, K.R. (2007) *Synthesis*, 3469; (i) Narender, M., Reddy, M.S., Kumar, V.P., Reddy, V.P., Nageswar, Y.V.D., and Rao, K.R. (2007) *J. Org. Chem.*, **72**, 1849; (j) Kano, K., Ishihara, H., and Koketsu, M. (2007) *Heterocycles*, **74**, 1009; (k) Fleischhauer, J., Beckert, R., Günther, W., Kluge, S., Zahn, S., Weston, J., Berg, D., and Görls, H. (2007) *Synthesis*, 2839; (l) Madhav, J.V., Kuarm, B.S., and Rajtha, B. (2008) *Synth. Commun.*, **38**, 3514; (m) Salvador, M.A., Reis, L.V., Almeida, P., and Santos, P.F. (2008) *Arkivoc*, 90; (n) Atanassov, P.K., Linden, A., and Heimgartner, H. (2010) *Helv. Chim. Acta*, **93**, 395.

88 (a) Amosova, S.V., Volkova, K.A., Penzik, M.V., Nakhmanovich, A.S., Albanov, A.I., and Potapov, V.A. (2007) *Russ. J. Org. Chem.*, **43**, 1087; (b) Kanoh, K., Ishihara, H., and Koketsu, M. (2007) *Synthesis*, 2617.

89 (a) Amosova, S.V., Volkova, K.A., Penzik, M.V., Albanov, A.I., and Potapov, V.A. (2008) *Russ. J. Gen. Chem.*, **78**, 503; (b) Amosova, S.V., Elokhina, V.N., Nakhmanovich, A.S., Larina, L.I., Martynov, A.V., Steele, B.R., and Potatov, V.A. (2008) *Tetrahedron Lett.*, **49**, 974.

90 Koketsu, M., Kiyokuni, T., Sakai, T., Ando, H., and Ishihara, H. (2006) *Chem. Lett.*, **35**, 626.

91 (a) Garud, D.R. and Koketsu, M.M. (2008) *Org. Lett.*, **10**, 3319; (b) Garud, D.R., Ninomiya, M., and Koketsu, M. (2010) *Heterocycles*, **81**, 2439.

92 Garud, D.R., Makimura, M., Ando, H., Ishihara, H., and Koketsu, M. (2007) *Tetrahedron Lett.*, **48**, 7764.

93 Silks, L.A. III., Kimball, D.B., Hatch, D., Ollivault-Shiflett, M., Michalczyk, R., and Moody, E. (2009) *Synth. Commun.*, **39**, 641.

94 (a) Favero, F., Sommen, G.L., Linden, A., and Heimgartner, H. (2006) *Heterocycles*, **67**, 749; (b) Sommen, G.L., Linden, A., and Heimgartner, H. (2007) *Helv. Chim. Acta*, **90**, 641; (c) Koketsu, M., Takahashi, A., and Ishihara, H. (2007) *J. Heterocyclic Chem.*, **44**, 79; (d) Sashida, H., Pan, C., Kaname, M., and Minoura, M. (2010) *Synthesis*, 3091;

(e) Kaname, M., Minoura, M., and Sashida, H. (2011) *Tetrahedron Lett.*, **52**, 505; (f) Maeda, H., Takashima, M., Sakata, K., Watanabe, T., Honda, M., and Segi, M. (2011) *Tetrahedron Lett.*, **52**, 415.

95 Koketsu, M., Yamamura, Y., and Ishihara, H. (2006) *Synthesis*, 2738.

96 (a) Somsák, L., Felföldi, N., Kónya, B., Hüse, C., Telepó, K., Bokor, E., and Czifrák, K. (2008) *Carbohydrate Res.*, **343**, 2083; (b) López, O., Maza, S., Ulgar, V., Maya, I., and Fernández-Bolanos, J.G. (2009) *Tetrahedron*, **65**, 2556.

97 Atanassov, P.K., Linden, A., and Heimgartner, H. (2006) *J. Sulfur Chem.*, **27**, 181.

98 Zhang, J.-H. and Cheng, Y. (2009) *Org. Biomol. Chem.*, **7**, 3264.

99 Garud, D.R., Tanahashi, N., Ninomiya, M., and Koketsu, M. (2009) *Tetrahedron*, **65**, 4775.

100 Shigetomi, T., Shioji, K., Okuma, K., and Yokomori, Y. (2007) *Bull Chem. Soc. Jpn.*, **80**, 395.

101 (a) Isab, A.A. and Ahmad, S. (2006) *Transition Metal Chem.*, **31**, 500; (b) Isab, A.A., Wazeer, M.I.M., Fettouhi, M., Ahmad, S., and Ashraf, W. (2006) *Polyhedron*, **25**, 2629.

102 (a) Todorovic, T.R., Bacchi, A., Pelizzi, G., Juranic, N.O., Sladic, D.M., Brceski, I.D., and Andelkovic, K.K. (2006) *Inorg. Chem. Commun.*, **9**, 862; (b) Todorovic, T.R., Bacchi, A., Juranic, N.O., Sladic, D.M., Pelizzi, G., Bozic, T.T., Filipovic, N.R., and Andelkovic, K.K. (2007) *Polyhedron*, **26**, 3428; (c) Gligorijevic, N., Todorovic, T., Radulovic, S., Sladic, D., Sladic, D., Filipovic, N., Godevac, D., Jeremic, D., and Andelkovic, K. (2009) *Eur. J. Med. Chem.*, **44**, 1623; (d) Todorovic, T.R., Bacchi, A., Sladic, D.M., Todorovic, N.M., Bozic, T.T., Radanovic, D.D., Filipovic, N.R., Pelizzi, G., and Andelkovic, K.K. (2009) *Inorg. Chim. Acta*, **362**, 3813.

103 Kowol, C.R., Eichinger, R., Jakupec, M.A., Galanski, M., Arion, V.B., and Keppler, B.K. (2007) *J. Inorg. Bio.*, **101**, 1946.

104 (a) Jia, W.-G., Huang, Y.-B., Lin, Y.-J., Wang, G.-L., and Jin, G.-X. (2008) *Eur. J. Inorg. Chem.*, 4063; (b) Jia, W.-G., Huang, Y.-B., Lin, Y.-J., and Jin, G.-X. (2008) *Dalton Trans.*, 5612.

105 Choi, J., Ko, J.H., Jung, I.G., Yang, H.Y., Ko, K.C., Lee, J.Y., Lee, S.M., Kim, H.J., Nam, J.H., Ahn, J.R., and Son, S.U. (2009) *Chem. Mater.*, **21**, 2571.

106 (a) Gallenkamp, D., Tiekink, E.R.T., and Mohr, F. (2008) *Phosphorus Sulfur Silicon*, **183**, 1050; (b) Gallenkamp, D., Porsch, T., Molter, A., Tiekink, E.R.T., and Mohr, F. (2009) *J. Organomet. Chem.*, **694**, 2380.

7
Selenoxide Elimination and [2,3]-Sigmatropic Rearrangement

Yoshiaki Nishibayashi and Sakae Uemura

7.1
Introduction

Organoselenium compounds are recognized as useful intermediates to introduce novel functional groups into organic compounds under mild conditions. For example, alkyl selenides are easily oxidized by a variety of oxidants to the corresponding alkyl selenoxides, where, if a β-hydrogen atom is present, *syn*-elimination occurs easily even at room temperature to give the corresponding alkenes together with selenenic acids (Scheme 7.1: selenoxide elimination). In the case of allylic selenides, on the other hand, [2,3]-sigmatropic rearrangement from the produced allylic selenoxides precedes selenoxide elimination to afford the corresponding allylic alcohols instead of conjugated dienes via selenenic esters (selenenates) (Scheme 7.2: [2,3]-sigmatropic rearrangement). The [2,3]-sigmatropic rearrangement proceeds quite smoothly even at low temperatures such as −78 °C.

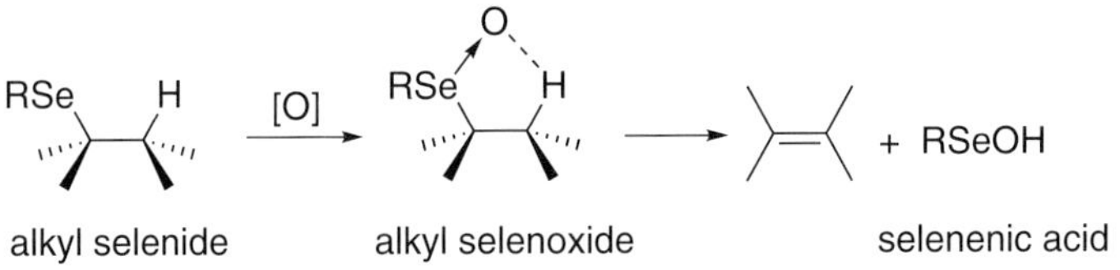

Scheme 7.1 Selenoxide elimination leading to alkenes.

The selenoxide elimination and the [2,3]-sigmatropic rearrangement are very useful synthetic methods for organic chemistry and are now often used for preparation of many synthetically useful compounds including natural products. Several review articles on these topics from a synthetic viewpoint have appeared so far [1]. Compared to the corresponding reactions of organosulfur compounds, the advantages of organoselenium compounds are that the activation energy for the above

Organoselenium Chemistry: Synthesis and Reactions, First Edition. Edited by Thomas Wirth.
© 2012 Wiley-VCH Verlag GmbH & Co. KGaA. Published 2012 by Wiley-VCH Verlag GmbH & Co. KGaA.

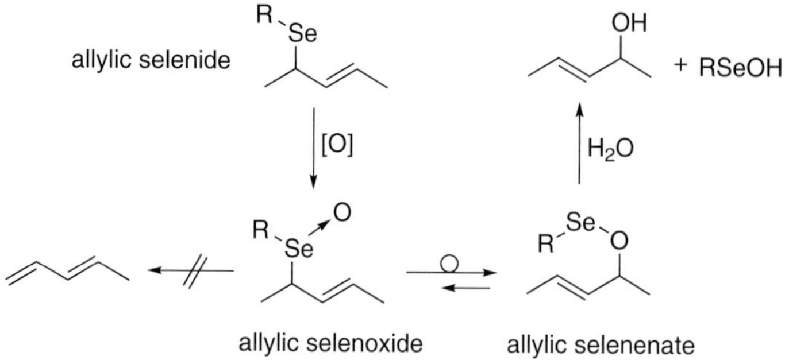

Scheme 7.2 [2,3]-Sigmatropic rearrangement leading to allylic alcohols.

two reactions to proceed is much lower because of a weaker bond energy of the C–Se bond than the C–S bond as well as the favorable conformation of the transition state. This leads to the high potential of organoselenium compounds for performing stereoselective reactions. Although the usefulness of optically active organosulfur compounds has long been known for asymmetric synthesis, the corresponding selenium chemistry was undeveloped until recently in spite of its high potential. In this chapter, we focus on the recent advances of asymmetric version of these two reactions, selenoxide elimination and [2,3]-sigmatropic rearrangement, where optically active organoselenium compounds work as key intermediates. The results so far disclosed indicate that the following features are essential for achieving a high asymmetric induction in the above both reactions: (i) a highly stereoselective formation of organoselenium compounds bearing a chiral center at the selenium atom; (ii) no or slow racemization of the resulting chiral organoselenium compounds; (iii) an efficient chirality transfer from the chiral selenium atom center to the products. Typically, asymmetric selenoxide elimination of the chiral vinyl selenoxides affords the corresponding allenes and cyclohexylidenes with high enantioselectivity. Asymmetric [2,3]-sigmatropic rearrangement of chiral allylic selenoxides, selenimides, and selenium ylides leads to the corresponding allylic alcohols, allylic amines, and homoallylic selenides, respectively, with good to high enantioselectivities.

7.2
Preparation and Properties of Chiral Selenoxides

The preparation of the highly optically active selenoxides should be essential to achieve a high asymmetric induction in both selenoxide elimination and [2,3]-sigmatropic rearrangement. At first, the preparation and the properties of chiral selenoxides which work as key intermediates in both reactions are briefly surveyed. The chemistry of chiral selenoxides has been little studied until recently and the first example of the preparation of chiral selenoxides was reported by Davis and coworkers in 1983 [2]. This is in sharp contrast to the chiral sulfoxides which have

7.2 Preparation and Properties of Chiral Selenoxides

been known since the mid-1920s and have played important roles in the study of the origin of molecular recognition and asymmetric organic synthesis using them as chiral auxiliaries [3]. The principal difficulty in studying and preparing chiral selenoxides with a high enantiomeric purity has been considered to be due to their configurational lability. In earlier studies [4], it was shown that a chiral alkyl aryl selenoxide quite easily racemizes in the presence of moisture via the formation of an achiral hydrate, the racemization being much accelerated by the addition of acid, but not in basic media (Scheme 7.3). Bulky ortho-substituents on an aryl group were shown to slow down the rate of racemization by inhibiting sterically the formation of the achiral hydrate. Subsequent detailed studies by Kamigata and coworkers revealed that the rate-determining step is the protonation on the selenoxide oxygen [5]. They also succeeded in preparing the air stable and enantiomerically pure (−)-4-(methoxycarbonyl)phenyl 2,4,6-triisopropylphenyl selenoxide (1) by fractional crystallization of the corresponding diastereomeric menthyl ester (2) followed by removal of the chiral auxiliary [6] (Scheme 7.4). The absolute configuration of the selenoxide 1 was confirmed by X-ray crystallographic analysis of its diastereomer [6]. These results indicate that the isolation of the chiral selenoxides is possible in the absence of water and acid even though the activation barrier for the racemization of the chiral selenoxides is considerably lower than that of

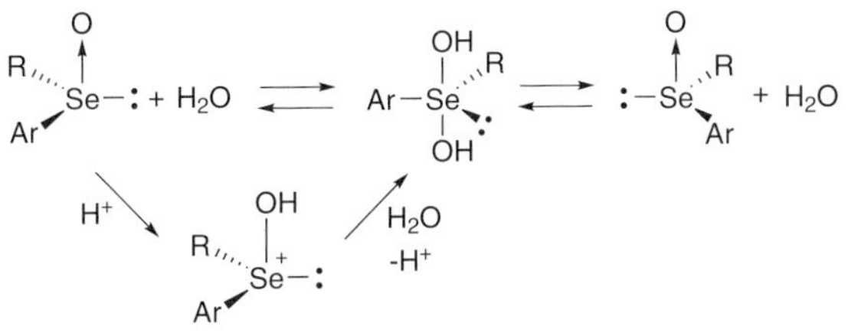

Scheme 7.3 Pathway for the racemization of chiral selenoxides.

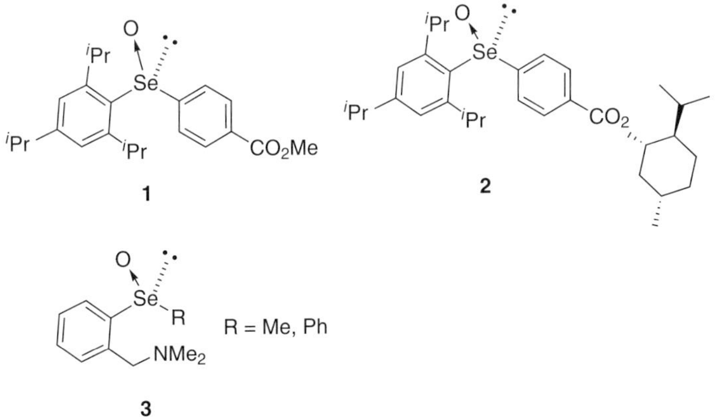

R = Me, Ph

Scheme 7.4 Isolation of enantio- and diastereomerically pure selenoxides.

the chiral sulfoxides. Kamigata and coworkers further succeeded in isolation of the optically pure selenoxides stabilized by intramolecular coordination with an amino group (**3**) by using chromatographic separation of a racemic mixture of selenoxides with optically active columns [7].

Another interesting example is the optical resolution of the chiral selenoxides by complexation of racemic selenoxides with an optically active binaphthol [8]. The resolution proceeded very efficiently due to the facile racemization of the selenoxides.

There are two methods for obtaining chiral selenoxides by direct oxidation of the corresponding selenides. One is the enantioselective oxidation of prochiral selenides, and the other is the diastereoselective oxidation of selenides bearing a chiral moiety (Scheme 7.5). In the former case, almost complete stereoselective oxidation to the chiral selenoxides was accomplished by Davis and coworkers in 1992 [9]. The Davis oxidant (**4**) was found to be the most efficient reagent for the enantioselective oxidation of a variety of prochiral selenides (Scheme 7.6). In fact, asymmetric oxidation was accomplished by treatment of the selenides with 1 molar equivalent of the Davis oxidant **4** at 0 °C affording the corresponding chiral selenoxides in quantitative yields with high enantioselectivities (up to 95% *ee*). The oxidation of methyl phenyl selenide was complete within 1 min, whereas that of alkyl phenyl selenide bearing bulkier alkyl substituents on a phenyl moiety required a few hours. On the other hand, the Sharpless oxidant, which is well known as a reagent for the highly enantioselective oxidation of prochiral sulfides to the corresponding chiral sulfoxides, worked less effectively than the Davis oxidant for the enantioselective oxidation of selenides [10]. In fact, oxidation of achiral selenides using *tert*-butylhydroperoxide (TBHP) in the presence of a tartrate ester and a titanium(IV) alkoxide gave the corresponding chiral selenoxides with moderate enantioselectivities (up to 40% *ee*) (Scheme 7.7). As to the

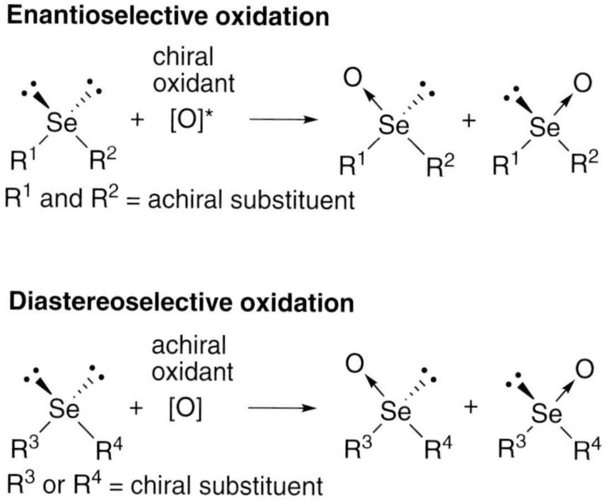

Enantioselective oxidation

R^1–Se–R^2 + [O]* (chiral oxidant) ⟶ O=Se(R^1)(R^2) + O=Se(R^1)(R^2)

R^1 and R^2 = achiral substituent

Diastereoselective oxidation

R^3–Se–R^4 + [O] (achiral oxidant) ⟶ O=Se(R^3)(R^4) + O=Se(R^3)(R^4)

R^3 or R^4 = chiral substituent

Scheme 7.5 Enantio- and diastereo selective oxidation of selenides.

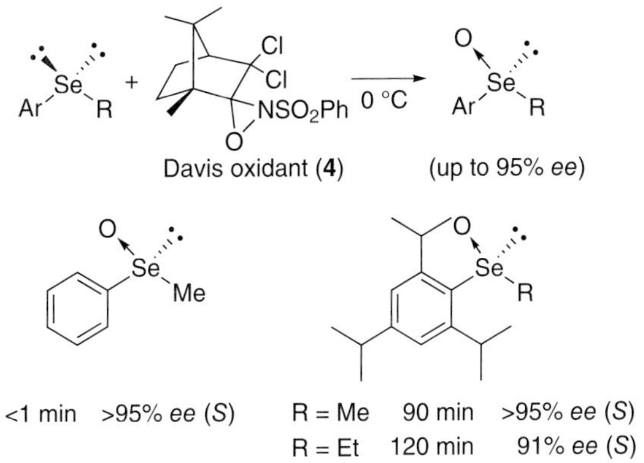

Scheme 7.6 Enantioselective oxidation of prochiral selenides with the Davis oxidant.

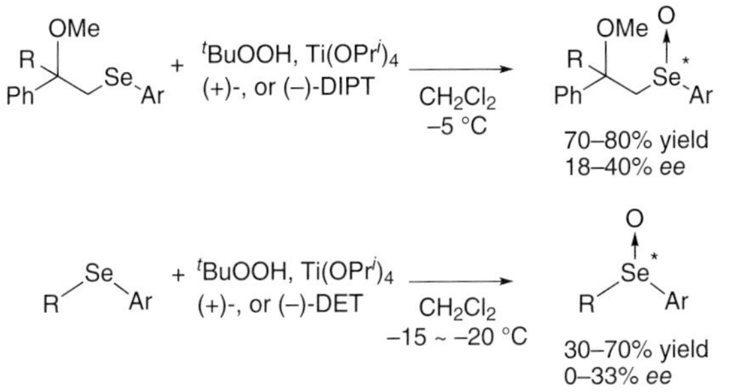

Scheme 7.7 Enantioselective oxidation of selenides with the Sharpless oxidant.

diastereoselective oxidation, Reich and Yelm first reported the preparation of the chiral selenoxides with a high diastereoselectivity by oxidation of the nonracemic [2.2]paracyclophane-substituted selenides as shown in Scheme 7.8 [11]. In fact, oxidation of methyl [2.2]paracyclophanyl selenide with *m*-chloroperbenzoic acid (*m*CPBA) at −60 °C for 3 min gave the corresponding chiral selenoxide, which had a new chiral center at the selenium atom, with 64% *de*.

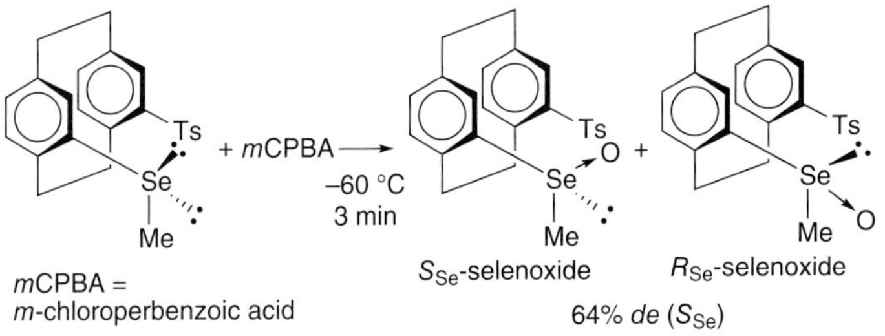

Scheme 7.8 Diastereoselective oxidation of methyl [2.2]paracyclophanyl selenide.

7.3 Selenoxide Elimination

As synthetic methods for introducing an unsaturated carbon–carbon bond into organic compounds, selenoxide elimination reaction has been shown to be quite useful because of its simple procedure and characteristic regioselectivity. Jones and coworkers, who discovered the first example of selenoxide elimination, proposed an intramolecular mechanism via a five-membered ring intermediate to explain its *syn*-elimination [12]. This mechanism was supported by Sharpless and coworkers who applied the method utilized by Cram to determine the stereochemistry of elimination in amine oxides [13]. Thus, oxidation of *erythro*-selenide afforded only *Z*-alkene, while that of *threo*-selenide gave only *E*-alkene (Scheme 7.9).

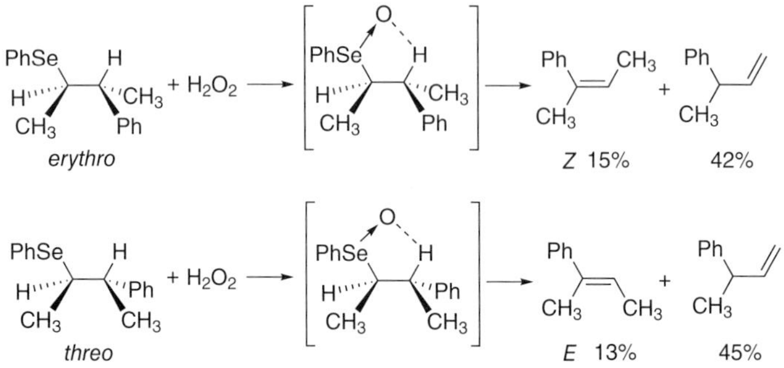
PhSe
H
H
CH3
CH3
Ph
+ H2O2
erythro
O
PhSe
H
Z 15%
42%
threo
E 13%
45%

Scheme 7.9 Reaction pathway for selenoxide *syn*-elimination.

The regioselectivity in selenoxide elimination affording either allylic or vinylic products depends much on the nature of the β-substituent Y in the selenides as shown in Scheme 7.10. Uemura and coworkers carried out theoretical studies using *ab initio* MO calculations [14]. The regioselectivity in the elimination was investigated by taking a simplified β-substituted selenoxide system, $CH_2(Y)CH(Se(O)CH_3)CH_3$ (Y = OCH_3, OH, CN, and Cl), as model compounds. The elimination occurs *via* a five-membered ring intermediate and the breaking of the

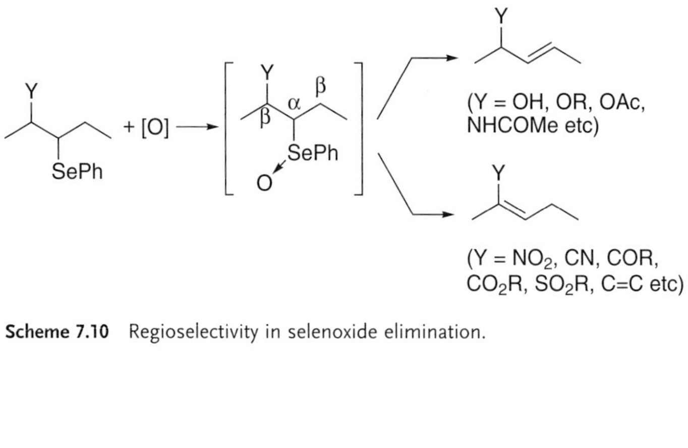
Y
SePh
+ [O]
α
β
O
(Y = OH, OR, OAc,
NHCOMe etc)
(Y = NO2, CN, COR,
CO2R, SO2R, C=C etc)

Scheme 7.10 Regioselectivity in selenoxide elimination.

C_β–H bond takes place earlier than that of the Se–C_α bond. The theoretical results were supported by the experimental results obtained for the elimination of selenoxides. In addition to other theoretical studies on the selenoxide elimination [15], Kingsbury reported quite recently more detailed theoretical study on the transition state of this reaction by using DFT calculations [16].

In 1991, Sonoda and coworkers found the first example of a selenoxide *anti*-elimination on oxidation of (Z)-1,2-bis(phenylseleno)-1-alkenes [17]. Thus, oxidation of (Z)-1,2-bis(phenylseleno)-1-octene **5** with *m*CPBA and the subsequent work-up afforded the corresponding acetylenic phenyl selenoxide **6** in 67% yield (Scheme 7.11). This result indicates that the *anti*-elimination took place from (Z)-1,2-bis(phenylseleno)-1-alkenes. Although further investigations are required for understanding the precise mechanism of this *anti*-elimination, the participation by the neighboring seleninyl group may be conceivable for this novel selenoxide *anti*-elimination.

Scheme 7.11 *anti*-Selenoxide elimination.

Huang and coworkers have developed a novel synthetic approach to the highly selective formation of alkenes bearing functional groups via selenoxide elimination by using polymer-supported selenides [18]. After the selenoxide elimination, the corresponding polymer-supported selenenic acids were recovered and reused as substrates after the conversion into original selenides. This method has significant advantages such as easy separation and good purity of products, and freedom from odors.

If selenoxide elimination from the optically active selenoxides could give the optically active alkenes, the reaction may provide a new methodology for asymmetric elimination to form a carbon–carbon double bond under mild reaction conditions. In this section, typical examples of enantio- and diastereoselective selenoxide eliminations are described.

7.3.1
Enantioselective Selenoxide Elimination Producing Chiral Allenes and α,β-Unsaturated Ketones

In 1992, Uemura and coworkers reported the first successful example of asymmetric selenoxide elimination [19]. When vinyl selenides prepared by the method originally reported by Back and coworkers [20] were oxidized with either the Davis (**4, 7, 8**), the Sharpless (**9, 10**), or the modified Sharpless (**11**) oxidants at 0–30 °C for 10–19 days, the corresponding chiral allenic sulfones were obtained in high chemical yields with a moderate enantioselectivity [19] (Scheme 7.12). The nature of an aryl group of the selenides has a remarkable effect on the enantioselectivity of the products. The o-nitrophenyl group was most effective for this asymmetric induction, the effectiveness being followed by the o,p-dinitrophenyl and o-methoxyphenyl groups, while the phenyl and 2'-pyridyl groups were not effective at all. This fact indicates that an *ortho*-substituent may stabilize the chiral selenoxide intermediate sterically and the electron-withdrawing substituent may accelerate the elimination step preventing the racemization of the produced chiral selenoxide intermediate [21]. As the oxidant, the Sharpless reagents (**9, 10**) were better than the Davis (**4, 7, 8**) and the modified Sharpless (**11**) reagents in enantiomeric excess of the chiral allenes.

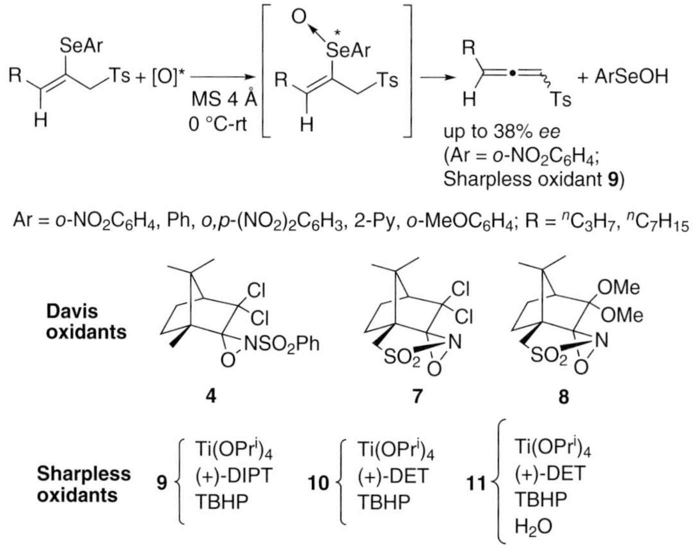

Scheme 7.12 Enantioselective selenoxide elimination for chiral allenic sulfones.

As an extensive work of enantioselective selenoxide elimination, Uemura and coworkers applied this reaction to the preparation of axially chiral alkyl and aryl cyclohexylidenemethyl ketones starting from substituted cyclohexyl methyl and phenyl selenides [22]. Chiral cyclohexylidene compounds have attractive chiroptical properties, and their highly stereoselective preparation has been reported. For example, a variety of benzylidenecyclohexane or alkylidenecyclohexane derivatives were prepared highly stereoselectively only using chiral Wittig reagents [23]. Furthermore, for the stereoselective preparation of chiral cyclohexylidenemethyl carbonyl compounds, the carboxylic acid derivatives were prepared either by optical resolution or by asymmetric dehydrohalogenation with the chiral lithium amides [24], while the ketone, ester, and aldehyde derivatives were mainly prepared from the corresponding chiral carboxylic acids [25]. The stereoselective synthesis of the chiral ketones from cyclohexylideneacetic acids, however, was difficult because of a facile isomerization to the thermodynamically more stable *endo* alkenes. Therefore, the novel approach using enantioselective selenoxide elimination may provide a new and useful method for the synthesis of the chiral cyclohexylidenemethyl ketones (Scheme 7.13).

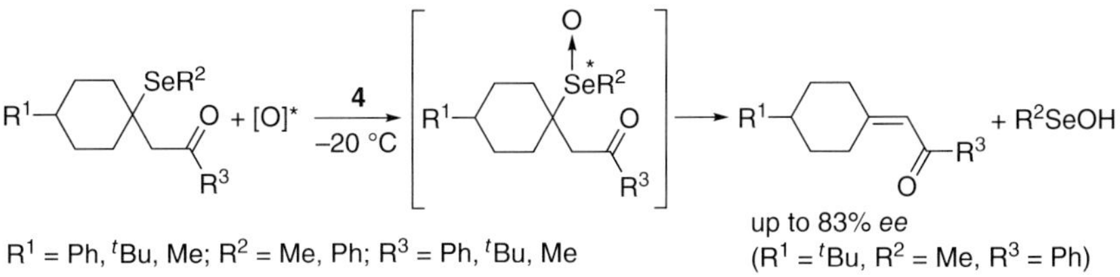

R^1 = Ph, ^{t}Bu, Me; R^2 = Me, Ph; R^3 = Ph, ^{t}Bu, Me

up to 83% *ee*
(R^1 = ^{t}Bu, R^2 = Me, R^3 = Ph)

Scheme 7.13 Enantioselective selenoxide elimination for chiral cyclohexylidenemethyl ketones.

When cyclohexyl selenides, prepared from the corresponding 4-substituted cyclohexanone *via* selenoketal, were oxidized with a variety of Davis and Sharpless oxidants, the chiral 4-substituted cyclohexylidenemethyl ketones were obtained with high enantioselectivities. In this enantioselective induction, R^2 of the substrate selenide and the oxidant employed were revealed to show a remarkable effect upon the enantioselectivity of the product. Thus, the Davis oxidant **4** was revealed to be an oxidant of choice and the use of a methyl moiety as R^2 instead of a phenyl moiety gave a high enantioselectivity (up to 83% *ee*), probably due to a large steric difference between the two groups bonded to the selenium atom of the substrate.

7.3.2
Diastereoselective Selenoxide Elimination Producing Chiral Allenecarboxylic Esters

Asymmetric selenoxide elimination *via* a diastereomeric chiral selenoxide as a key intermediate was also achieved by Uemura and coworkers in 1994 [26]. Novel optically active diferrocenyl diselenides were prepared from the corresponding chiral ferrocenes and the optically active ferrocenyl group was used as a chiral aryl moiety since an arylselenium moiety can be easily introduced into organic compounds. Oxidation of the chiral ferrocenyl vinyl selenides, prepared from the above diferrocenyl diselenides and ethyl propiolate derivatives, with 1 molar equivalent of *m*CPBA under various conditions afforded the corresponding chiral selenoxides. The chiral selenoxides suffered *in situ* selenoxide elimination to afford the axially chiral allenecarboxylic esters with high enantioselectivities (Scheme 7.14) [26b]. The reaction temperature had a remarkable effect on the enantioselectivity and lower temperatures gave better results. The addition of molecular sieves 4 Å

(MS 4Å) to the reaction system improved the enantioselectivity. Asymmetric selenoxide elimination provides a new method for the preparation of chiral allene-carboxylic esters which have so far been prepared only by optical resolution of the corresponding racemic acids.

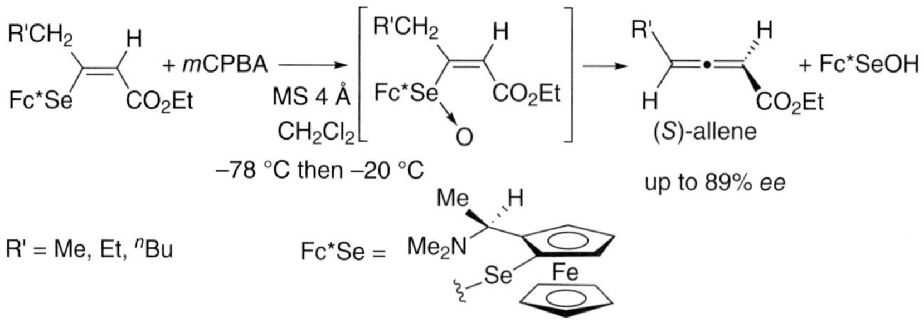

Scheme 7.14 Diastereoselective selenoxide elimination for the synthesis of chiral allenecarboxylic esters.

The highly diastereoselective oxidation of the chiral ferrocenyl vinyl selenides was supported by the oxidation of the corresponding chiral p-tolyl ferrocenyl sulfide **12** with mCPBA at low temperature. An almost completely diastereoselective oxidation occurred at −78 °C to give the corresponding chiral sulfoxide **13** (Scheme 7.15). Oxidation of ethyl 1-(N,N-dimethylamino)ethylphenyl sulfide **14** with mCPBA was carried out separately, but in this case the corresponding sulfoxide **15** showed only a low diastereoselectivity. This result indicates that the high diastereoselectivity observed here is not due to the chirality of the substituent on the ferrocene ring, but rather to the planar chirality of the ferrocene.

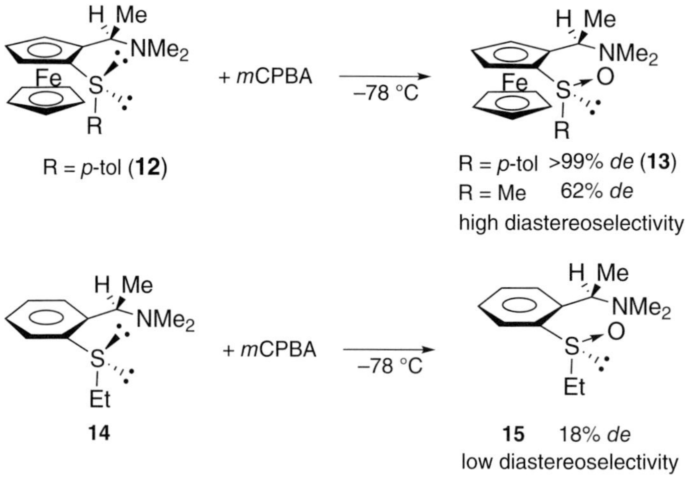

Scheme 7.15 Diastereoselective oxidation of chiral sulfides as model reactions.

7.4
[2,3]-Sigmatropic Rearrangement via Allylic Selenoxides

A typical [2,3]-sigmatropic rearrangement involves a transfer of oxygen from the selenium atom to the carbon atom of allylic selenoxide to produce the corresponding allylic alcohol after hydrolysis of allylic selenenate. As described in Section 7.1 of this chapter, the [2,3]-sigmatropic rearrangement of the allylic selenoxides proceeds much faster than that of the allylic sulfoxides due to the lower activation energy of the former process. Moreover, the equilibrium favors the selenenate in the Se (selenium) series, but the sulfoxide in the S (sulfur) series (Scheme 7.16) [27]. Detailed kinetic and thermodynamic studies of [2,3]-sigmatropic rearrangements of allylic selenoxides have been reported by Reich and coworkers [27].

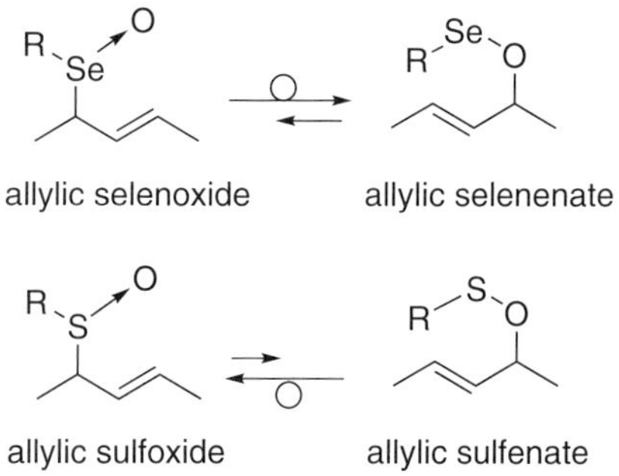

Scheme 7.16 [2,3]-Sigmatropic rearrangement of allylic selenoxides and sulfoxides.

When chiral allylic selenoxides were obtained by enantioselective or diastereoselective oxidation of allylic selenides, the formation of the corresponding chiral allylic alcohols is expected after the hydrolysis of the intermediate chiral allylic selenenates obtained by chirality transfer in the rearrangement step. In fact, several asymmetric reactions *via* this process were reported for the preparation of the chiral allylic alcohols. In this section, typical examples of enantio- and diastereoselective [2,3]-sigmatropic rearrangements *via* the chiral allylic selenoxides affording the corresponding chiral allylic alcohols are described.

7.4.1
Enantioselective [2,3]-Sigmatropic Rearrangement Producing Chiral Allylic Alcohols

The first successful example of an enantioselective [2,3]-sigmatropic rearrangement *via* allylic selenoxides was reported by Davis and coworkers in 1985 [4]. They generated *in situ* a chiral cinnamyl phenyl selenoxide by oxidation of the corresponding prochiral selenide with the Davis oxidant **16** to give

7 Selenoxide Elimination and [2,3]-Sigmatropic Rearrangement

1-phenyl-2-propen-1-ol after hydrolysis with only a low enantioselectivity (less than 10% *ee*). However, the enantiomeric excess of the allylic alcohol product increased to 60% *ee* using the more effective Davis oxidant **4** (Scheme 7.17) [9]. Oxidation of (*E*)- and (*Z*)-cinnamyl phenyl selenides **17** was complete within 5 min at −60 °C whereas the more hindered selenides required higher reaction temperature (0 °C) for several hours. As the rearrangement of allylic selenoxides to the allylic alcohols is extremely fast, it is reasonable to assume that the racemization of the chiral allylic selenoxides did not occur. In fact, it was also clarified that the presence of water, which accelerates the racemization of chiral selenoxides *via* hydration, had no influence on the chirality transfer of the chiral allylic selenoxides to the corresponding selenenates [9]. The selectivity of the produced allylic alcohols may be dependent on the free energy difference of the two possible transition states (*exo* and *endo*), from which the allylic alcohols with opposite configuration are obtained as shown in Scheme 7.18. Here, rational modification of the aryl part of the allylic selenides is required for a significant improvement in the enantioselectivity of the produced allylic alcohols.

17a
Davis oxidant **16**: 10% *ee* (Ar = Ph)
Davis oxidant **4**: 35% *ee* (Ar = Ph)
Davis oxidant **4**: 40% *ee* (Ar = 2,4,6-triisopropylphenyl)

17b
Davis oxidant **4**: 40% *ee* (Ar = Ph)
Davis oxidant **4**: 60% *ee* (Ar = 2,4,6-triisopropylphenyl)

Ar' = 2-chloro-5-nitrophenyl
Davis oxidant **16** (1985)

Davis oxidant **4**

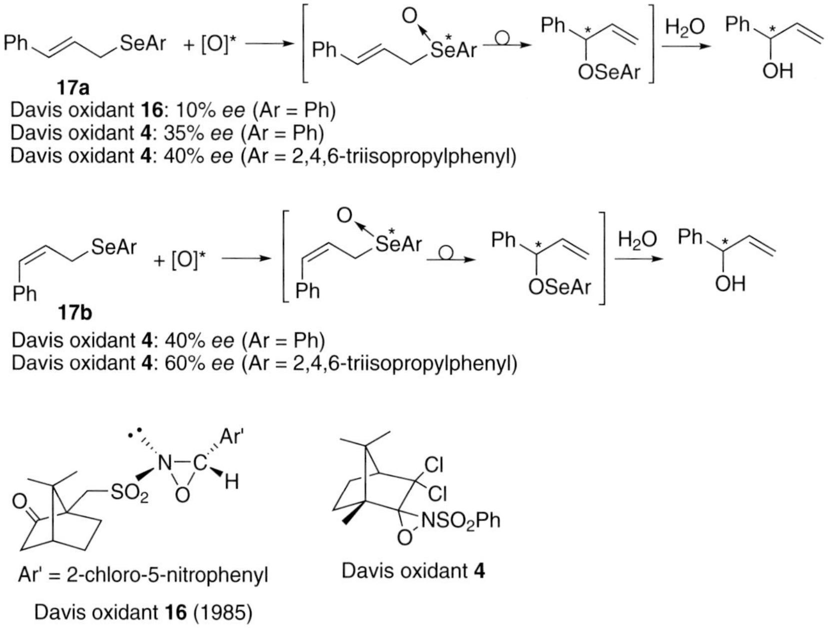

Scheme 7.17 Enantioselective [2,3]-sigmatropic rearrangement of allylic selenoxides using Davis oxidants.

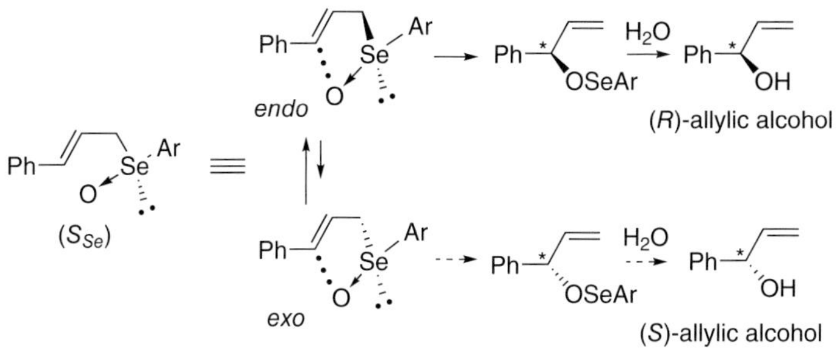

Scheme 7.18 Two possible transition states in [2,3]-sigmatropic rearrangements.

The enantioselective oxidation of several aryl cinnamyl selenides has also been investigated by Uemura's group using the Sharpless oxidant **9** instead of the Davis oxidant [28] (Scheme 7.19). A strong correlation between the aryl moiety and the enantioselectivity of the produced allylic alcohol was shown. The introduction of an *o*-nitro group to an arylselenium moiety remarkably enhanced the enantioselectivity, while the use of either phenyl, 2′-pyridyl, or ferrocenylselenium moiety resulted in only moderate to low enantioselectivities. The large steric and/or electronic effect of the *o*-nitro group to stabilize the chiral selenoxide intermediate may be important to achieve the high enantioselectivity. Namely, a large energy difference between the two possible transition states (*exo* and *endo*) should be formed by the introduction of an *o*-nitro group into the phenyl ring of the selenide.

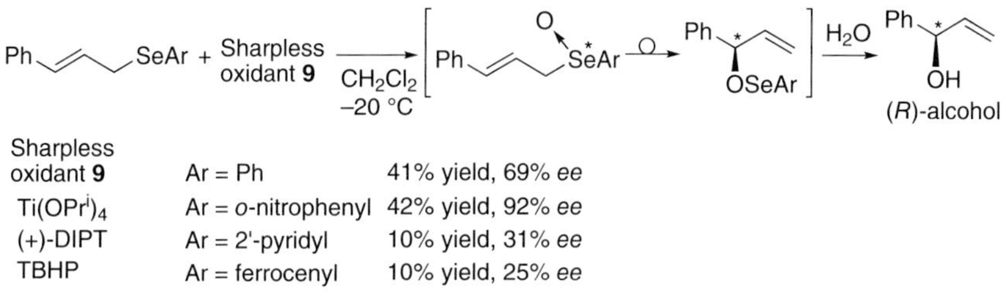

Sharpless oxidant **9**
Ti(OPr^i)$_4$
(+)-DIPT
TBHP

Ar = Ph   41% yield, 69% ee
Ar = *o*-nitrophenyl   42% yield, 92% ee
Ar = 2′-pyridyl   10% yield, 31% ee
Ar = ferrocenyl   10% yield, 25% ee

Scheme 7.19 Enantioselective [2,3]-sigmatropic rearrangement of allylic selenoxides using Sharpless oxidant.

7.4.2
Diastereoselective [2,3]-Sigmatropic Rearrangement Producing Chiral Allylic Alcohols

In 1991, Reich and Yelm reported the diastereoselective oxidation of the optically active geranyl [2.2]paracyclophanyl selenide with *m*CPBA at −60 °C to give (*S*)-linalool with 66% *ee* via the [2,3]-sigmatropic rearrangement of the corresponding selenoxide [11] (Scheme 7.20). The result indicates that the [2,3]-sigmatropic rearrangement of the allylic selenoxide proceeds through an *endo* transition state. Since an almost similar selectivity (64% *de*) was observed by oxidation of methyl [2.2]paracyclophanyl selenide as shown previously (Section 7.2), a chirality transfer should occur at the rearrangement step of the chiral allylic selenoxide.

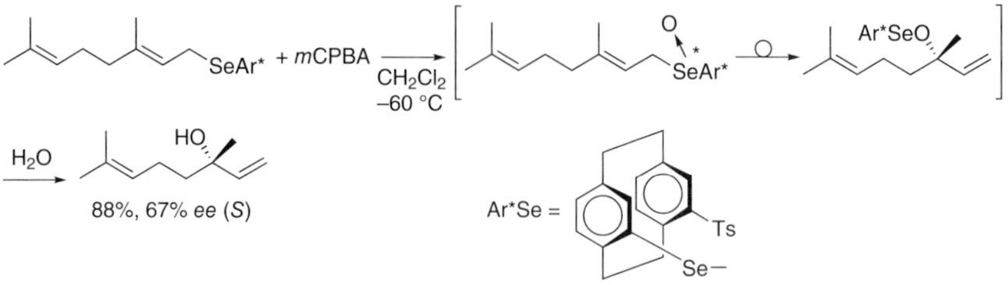

Scheme 7.20 Diastereoselective [2,3]-sigmatropic rearrangement of optically active allylic selenoxides ([2.2]paracyclophanyl moiety as a chiral auxiliary).

The optically active diferrocenyl diselenide has also been applied to asymmetric [2,3]-sigmatropic rearrangements to produce the chiral allylic alcohols [26]. Thus, the oxidation of cinnamyl selenide bearing a chiral ferrocenyl group with *m*CPBA afforded the optically active 1-phenyl-2-propen-1-ol in high enantioselectivity (Scheme 7.21). The chiral linalool was also obtained with a higher enantioselectivity than that obtained by Reich's system using [2.2]paracyclophanyl selenide. The fact that methanol could be used for obtaining the high enantioselectivity indicates the occurrence of a fast [2,3]-sigmatropic rearrangement of the intermediate chiral ferrocenyl selenoxides, since it is known that the racemization of the optically active selenoxides occurs readily in methanol. In this case, the *endo*-transition state leading to (*R*)-allylic alcohol should be more stable than the *exo*-transition state because the steric repulsion between the styryl and chiral ferrocenyl moieties is larger in the *exo*-transition state.

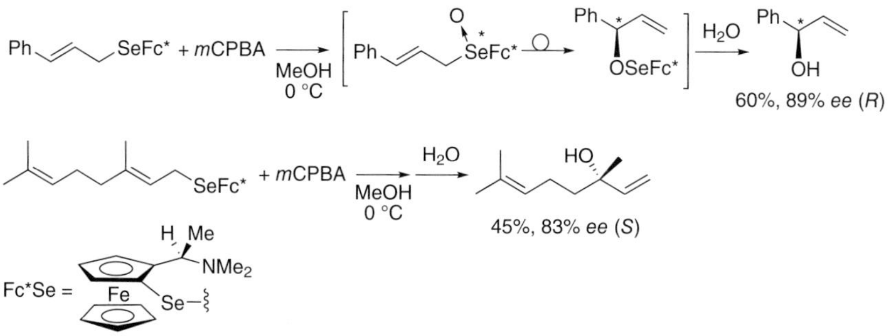

Scheme 7.21 Diastereoselective [2,3]-sigmatropic rearrangement of optically active allylic selenoxides (ferrocenyl moiety as a chiral auxiliary).

Similarly, the diastereoselective [2,3]-sigmatropic rearrangement of several other allylic aryl selenides was reported by Fujita and coworkers [29]. However, unfortunately only a moderate enantioselectivity was observed [29] (Scheme 7.22). Oxidation of aryl cinnamyl and geranyl selenides bearing a chirality derived from L-prolinol at an aryl moiety with mCPBA afforded the chiral 1-phenyl-2-propen-1-ol and linalool, respectively, with moderate to good enantioselectivities.

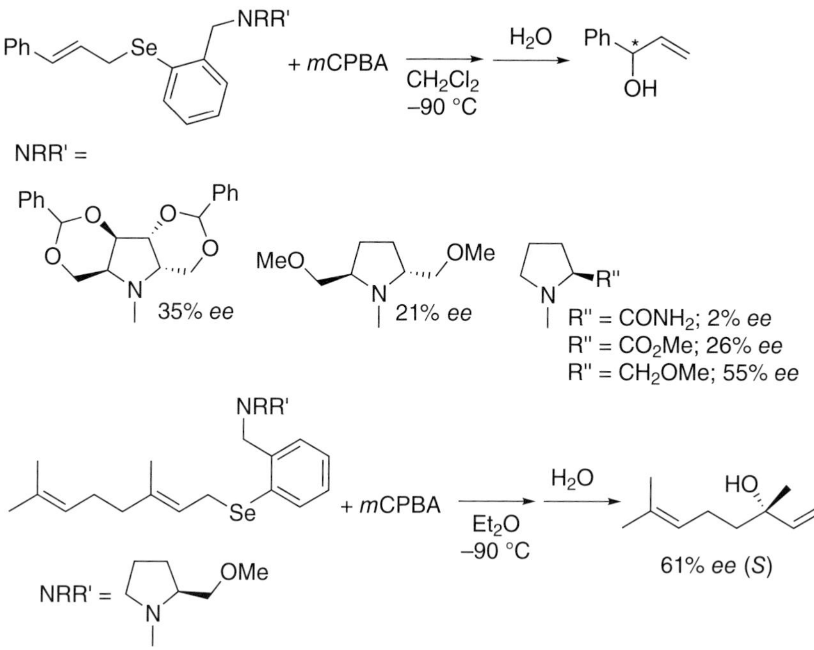

Scheme 7.22 Diastereoselective [2,3]-sigmatropic rearrangement of optically active allylic selenoxides (pyrrolidines as chiral auxiliaries).

The most important step of the [2,3]-sigmatropic rearrangement is the enantio- or diastereoselective oxidation of the allylic selenides and the transfer of the chirality of the selenium atom to C-3 of the resulting allylic alcohols. In other approach to obtain the chiral selenoxides, Koizumi and coworkers used selenuranes (tetrasubstituted selenium[IV]) bearing a 2-*exo*-hydroxyl-10-bornyl group as a chiral auxiliary [30]. An optically pure selenoxide was prepared by the following two methods using 2-*exo*-hydroxyl-10-bornyl phenyl selenide 18 (Scheme 7.23). One is the direct diastereoselective oxidation of the selenide with mCPBA. The other is the nucleophilic substitution of the chloroselenurane, which is readily obtained by the diastereoselective oxidation of the selenide with *tert*-BuOCl. The alkaline hydrolysis of the (R_{Se})-chloroselenurane proceeded with retention of configuration to give the (R_{Se})-selenoxide as the sole product. Oxidation of the selenides with *tert*-BuOCl was complete within 10 min at 0 °C to give the chloroselenurane as a single diastereomer in a high yield. The addition of aqueous NaHCO$_3$ to produce the chloroselenurane at 0 °C instantaneously resulted in a complete hydrolysis, leading to the selenoxide as a single diastereomer in a high yield. This selenoxide is stable at room temperature due to the bulkiness of the bornyl group as well as an intramolecular hydrogen bond between the seleninyl oxygen and the secondary hydroxyl group.

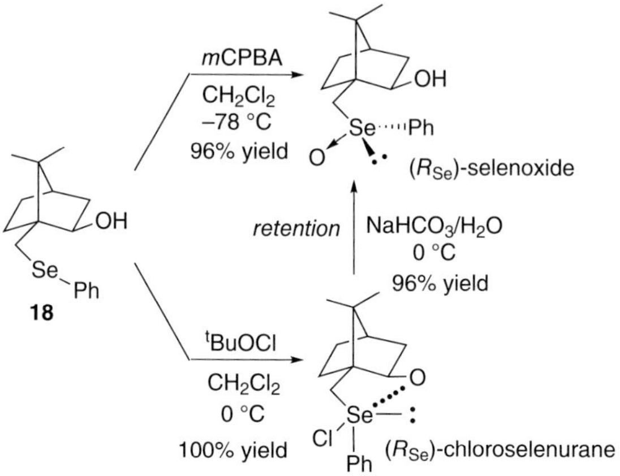

Scheme 7.23 Diastereoselective oxidation of chiral bornyl phenyl selenide.

The stereoselective alkaline hydrolysis of the selenurane to afford a single diastereomer of the selenoxide was applied to the diastereoseletive [2,3]-sigmatropic rearrangement of an allylic selenoxide [31] (Scheme 7.24). Treatment of the allylic chloroselenuranes with aqueous NaHCO$_3$ gave the corresponding allylic alcohols with a moderate to high enantioselectivity (up to 88% ee). The absolute configuration of the resulting allylic alcohols suggests that the [2,3]-sigmatropic rearrangement of allylic selenoxides proceeds predominantly *via* the *endo*-transition state.

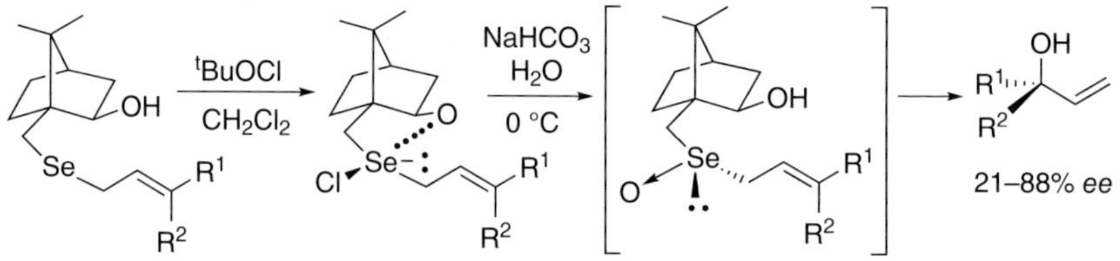

R^1 = Ph, R^2 = H; R^1 = cyclohexyl, R^2 = H; R^1 = ^{n}hexyl, R^2 = H;
R^1 = H, R^2 = Ph; R^1 = H, R^2 = cyclohexyl; R^1 = H, R^2 = ^{n}hexyl

Scheme 7.24 Diastereoselective [2,3]-sigmatropic rearrangement of optically active allylic selenuranes (bornyl moiety as a chiral auxiliary).

In 2004, Carter and coworkers reported the vanadium-catalyzed diastereoselective oxidation of allylic selenides bearing an optically active oxazoline moiety to give the corresponding allylic alcohols with moderate enantioselectivities (Scheme 7.25) [32]. In this reaction system, the vanadium complex is first coordinated to the oxazoline moiety and then the diastereoselective oxidation occurs to give the corresponding allylic selenoxides with moderate diastereoselectivities.

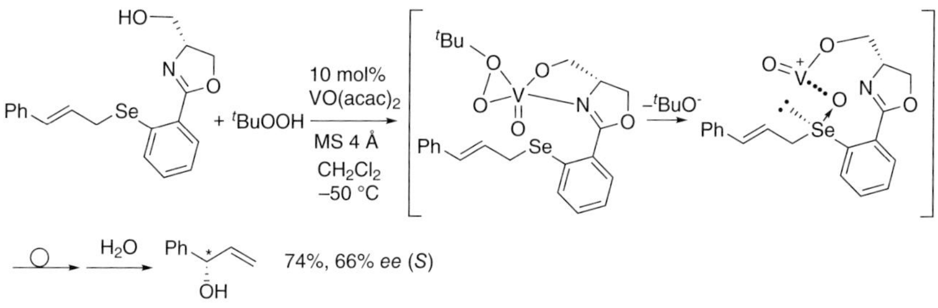

Scheme 7.25 Vanadium-catalyzed diastereoselective [2,3]-sigmatropic rearrangement of optically active allylic selenoxides.

The chirality transfer of selenides bearing a chiral center at the allylic position via [2,3]-sigmatropic rearrangement of allylic selenoxides was achieved by several research groups. It is well known that the introduction of an arylseleno group into the allylic position of alcohols proceeds to give the corresponding allylic selenides with inversion of configuration at the allylic position. This method realizes the chirality transfer of the allylic position in the optically active allylic alcohols. In fact, Portoghese and coworkers applied this system to prepare pseudocodeine, where the diastereoselective oxidation of an optically active allylic selenide afforded the corresponding chiral allylic alcohol after hydrolysis (Scheme 7.26) [33]. In 2007, Martin and coworkers reported the preparation of solandelactone E by using the chirality transfer of an optically active allylic selenide via [2,3]-sigmatropic rearrangement to give the corresponding chiral allylic alcohol after hydrolysis without loss of optical purity (Scheme 7.27) [34]. The selective substitution at the 12-position occurred because of the steric hindrance. In 2010, Posner and Hess reported the efficient three-step preparation of α-hydroxy-(E)-β,γ-unsaturated esters with high enantioselectivities [35] (Scheme 7.28). Enantioselective α-selenenylation of aldehydes in the presence of a catalytic amount of optically active proline derivatives as organocatalysts is the key step for the preparation of optically active allylic selenides [36]. The utility of this method for the total synthesis of natural products such as (+)-symbioramide has been shown [36]. In these cases, the absolute configuration at the selenium atom of the produced allylic selenoxides was not determined, but the highly diastereoselective oxidation of the optically active allylic selenides bearing a chiral center at the allylic position may occur to form the corresponding allylic selenoxides as reactive key intermediates. Thus, the chirality transfer at the selenium atom in the allylic selenoxides is considered to take place

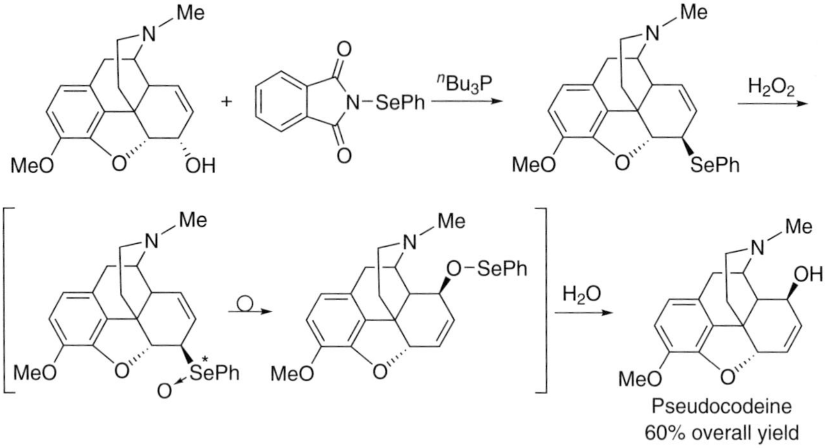

Scheme 7.26 Chirality transfer via [2,3]-sigmatropic rearrangement by Portoghese.

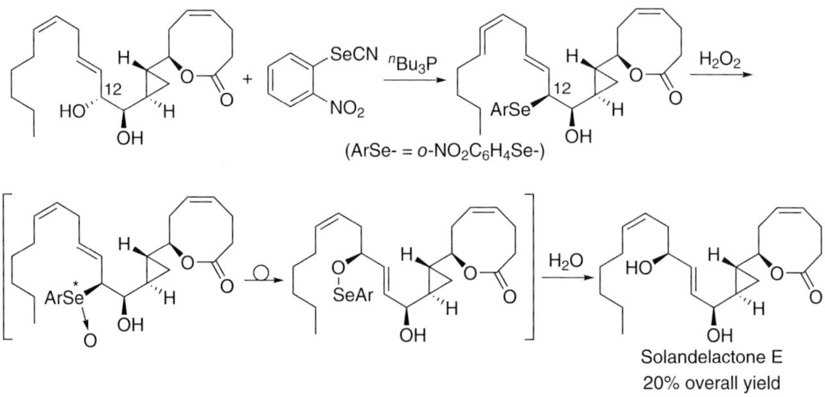

Scheme 7.27 Chirality transfer via [2,3]-sigmatropic rearrangement by Martin.

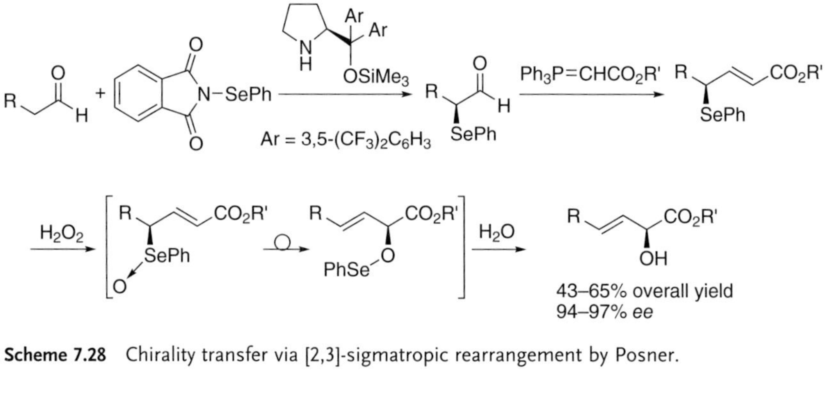

Scheme 7.28 Chirality transfer via [2,3]-sigmatropic rearrangement by Posner.

via [2,3]-sigmatropic rearrangement to give the corresponding chiral allylic alcohols after hydrolysis. In fact, some other groups also reported the preparation of optically active products by using a similar method [37, 38].

7.5
[2,3]-Sigmatropic Rearrangement via Allylic Selenimides

The nitrogen analogues of the selenoxides, the selenimides, should in principle be capable of similar [2,3]-sigmatropic transformations as the selenoxides (Scheme 7.29). Three decades ago, Sharpless and Hori reported the reaction of 10-(phenylseleno)-β-pinene with chloramine T (TsNClNa) to afford the corresponding allylic amine [39] (Scheme 7.30). Later, anhydrous chloramine T in methanol was revealed to be a highly effective reagent for the conversion of allylic selenides to the corresponding rearranged N-allylic-p-toluenesulfonamides (N-tosyl allylic amines) [40]. This reaction presumably proceeds through the

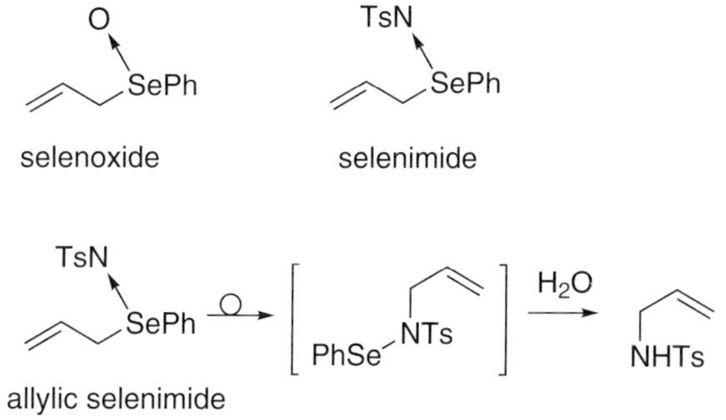

Scheme 7.29 [2,3]-Sigmatropic rearrangement via allylic selenimides.

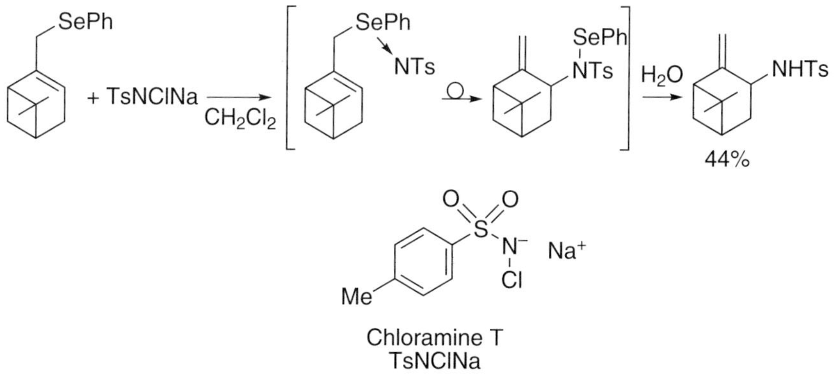

Scheme 7.30 The first example of a [2,3]-sigmatropic rearrangement via an allylic selenimide.

corresponding allylic selenimide intermediate, which undergoes [2,3]-sigmatropic rearrangement to the selenamide, followed by methanolysis, to afford the corresponding allylic amine. In place of hazardous anhydrous chloramine T, which may decompose violently if heated above 130 °C, a mixture of N-chlorosuccinimide (NCS) and an amine can also be used for a similar purpose [41]. A one-step procedure for the preparation of *tert*-butoxycarbonyl (Boc)- or carbobenzyloxy (Cbz)-protected primary allylic amines using NCS is shown in Scheme 7.31, where the reaction starting from the optically active allylic selenides produced the chiral allylic amines with high enantioselectivities [42]. In 2008, Tunge and Waetzig reported a similar transformation of an optically active selenide (96% *ee*) via [2,3]-sigmatropic rearrangement into the corresponding allylic amine (92% *ee*) without loss of optical purity of the starting selenide [43] (Scheme 7.32). Interestingly, in the absence of the additional amine, the corresponding allylic chloride was obtained with 84% *ee* as a mixture of two diastereoisomers (16:1) [43].

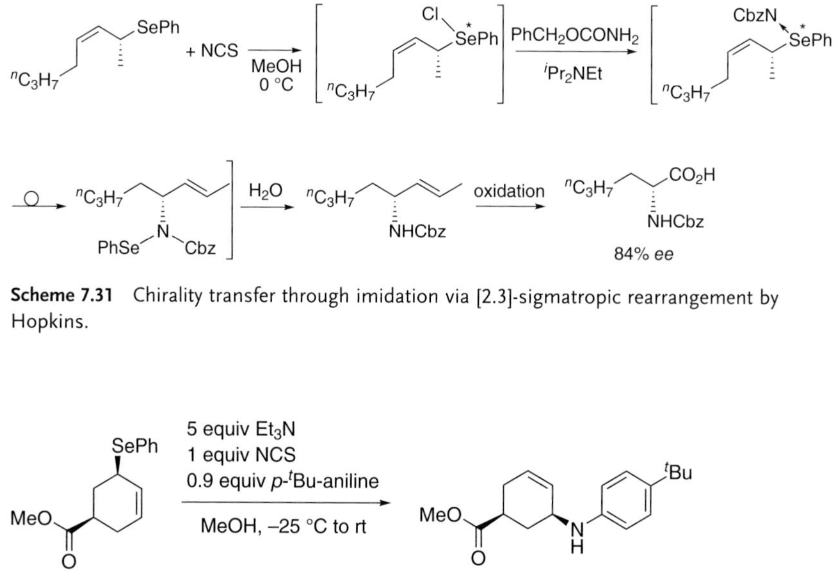

Scheme 7.31 Chirality transfer through imidation via [2.3]-sigmatropic rearrangement by Hopkins.

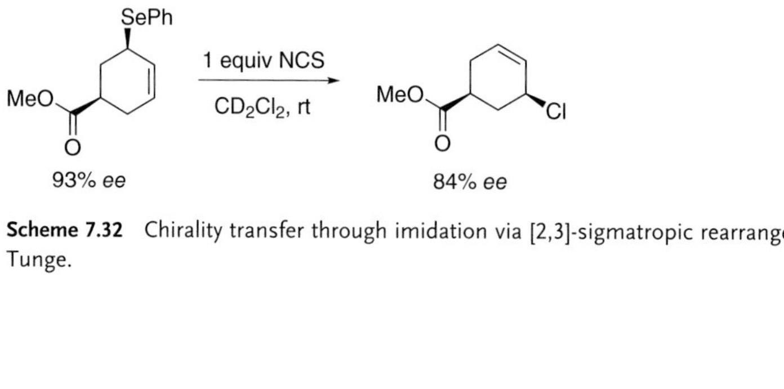

Scheme 7.32 Chirality transfer through imidation via [2,3]-sigmatropic rearrangement by Tunge.

As an imidation reagent, N-(p-tolylsulfonyl)imino(phenyl)iodinane (PhI = NTs) was also found to be convenient for conversion of the sulfides [44] and the selenides [45] to the corresponding sulfimides and selenimides, respectively.

As described previously (Sections 7.4.1 and 7.4.2), the asymmetric [2,3]-sigmatropic rearrangement occurs via chiral selenoxides, prepared by diastereoselective oxidation or enantioselective oxidation of the prochiral selenides. If the chiral selenimides were obtained as the intermediates by imidation of the selenides enantioselectively or diastereoselectively, the chirality transfer should in principle be expected to occur in the intermediates. Compared with the well-established preparative methods for the synthesis of optically active allylic alcohols, access to optically active allylic amines, which are very important compounds in organic synthesis, is still quite limited [46]. In this section, some examples of [2,3]-sigmatropic rearrangements via the chiral allylic selenimides to produce the chiral allylic amines as well as the preparation of chiral selenimides are described.

7.5.1
Preparation and Properties of Chiral Selenimides

Selenimides are the tricoordinate tetravalent compounds and expected to be isolable in the optically active form since the optically active sulfonium imides are known. In 1981, Krasnov and coworkers reported the first synthesis of the optically active selenimides by starting from dialkyl- and diarylselenium dichlorides, but the scope of this reaction was not fully developed probably because of low yields of the products as well as their quite low optical purity [47]. In 1994, Kamigata and coworkers isolated an optically pure selenimide by the optical resolution of a diastereomeric selenimide, and studied its stereochemistry and the kinetics of epimerization by pyramidal inversion [48]. Detailed studies showed that the selenimide is more stable than the corresponding sulfimide bearing the same substituents toward epimerization by pyramidal inversion. Accordingly, a higher reaction temperature or a higher activation energy is required for the pyramidal inversion of the selenimide than that of the sulfimide. Furthermore, they reported an example of the conversion of a chiral selenoxide, obtained by the optical resolution of a diastereomeric mixture, into the corresponding enantiomerically pure selenimide, ascertaining the detailed stereochemistry of this compound [49] (Scheme 7.33). The transformation of the selenoxide into the selenimide

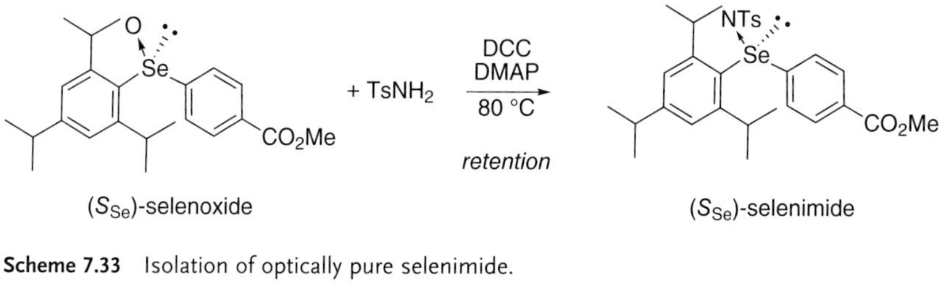

Scheme 7.33 Isolation of optically pure selenimide.

proceeded with an overall retention of configuration at the selenium center in the presence of dicyclohexylcarbodiimide (DCC). More recently, they reported another successful example of the isolation of optically pure selenimides by using chromatographic separation of a racemic mixture of selenimides with optically active columns [50].

In 1998, Uemura and coworkers reported the direct enantioselective imidation of prochiral selenides with PhI = NTs in the presence of a catalytic amount of CuOTf and a chiral bis(oxazoline) to produce the corresponding chiral selenimides with low to moderate enantioselectivities [45] (Scheme 7.34). In the absence of molecular sieves, no enantioselectivity was observed at all. This fact may suggest the interference of the rapid selenimide–selenoxide equilibrium by removal of water present in the reaction mixture. The enantioselectivity of produced chiral selenimides is not high (up to 36% ee), but this direct catalytic imidation of prochiral selenides may be a potentially useful method for the preparation of optically active chiral selenimides.

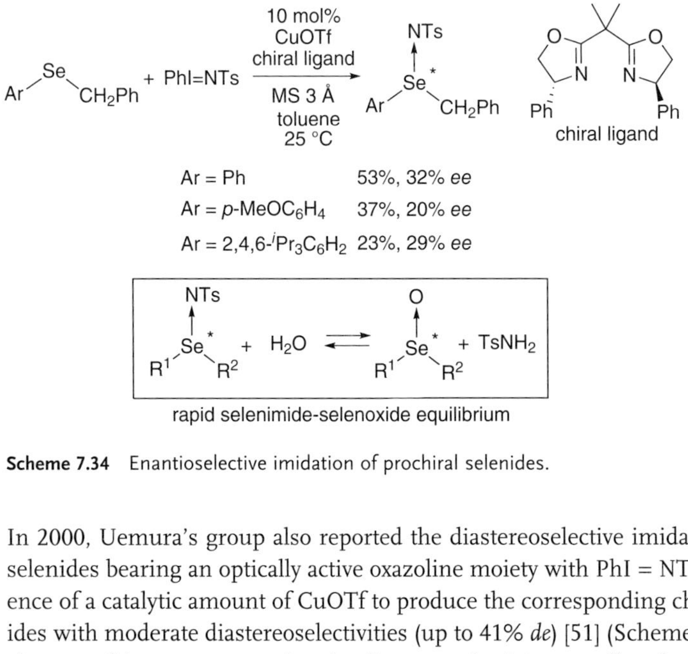

Scheme 7.34 Enantioselective imidation of prochiral selenides.

In 2000, Uemura's group also reported the diastereoselective imidation of chiral selenides bearing an optically active oxazoline moiety with PhI = NTs in the presence of a catalytic amount of CuOTf to produce the corresponding chiral selenimides with moderate diastereoselectivities (up to 41% de) [51] (Scheme 7.35). In the absence of the copper complex, the diastereoselectivity as well as the product yield decreased. When chloramine T trihydrate was used instead of PhI = NTs, the diastereoselective imidation of optically active selenides proceeded smoothly even in the absence of copper to give the corresponding chiral selenimides with a higher diastereoselectivity (up to 76% de). The absolute configuration at the selenium atom was confirmed by X-ray analysis. An ionic reaction pathway involving a chloroselenonium ion as a key intermediate was proposed for this diastereoselective imidation as shown in Scheme 7.36.

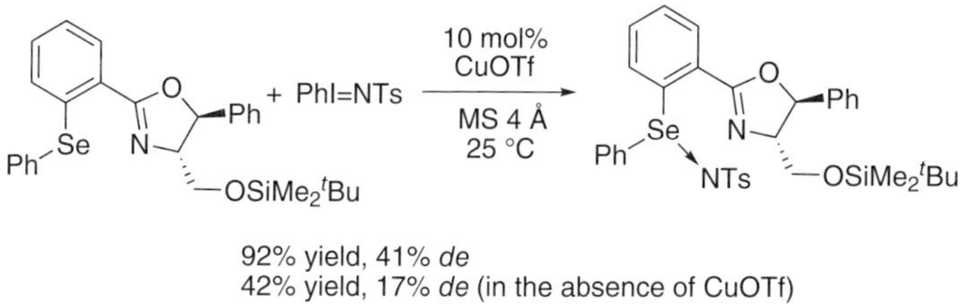

Scheme 7.35 Diastereoselective imidation of chiral selenides using PhI = NTs.

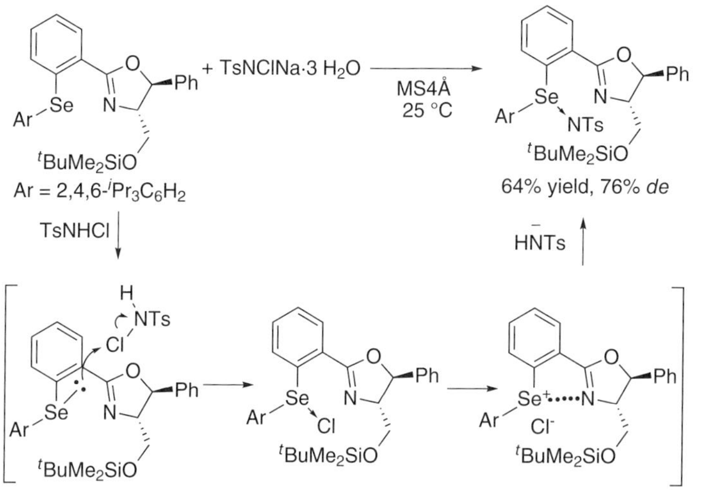

Scheme 7.36 Diastereoselective imidation of chiral selenides using chloramine T.

7.5.2
Enantioselective [2,3]-Sigmatropic Rearrangement Producing Chiral Allylic Amines

The enantioselective [2,3]-sigmatropic rearrangement of allylic selenimides was achieved by Uemura and coworkers as an application of the enantioselective synthesis of selenimides [45]. When the reaction was applied to a variety of aryl cinnamyl selenides, the expected optically active allylic amines were obtained with low to moderate enantioselectivities *via* the [2,3]-sigmatropic rearrangement of the intermediate chiral allylic selenimides (Scheme 7.37). Almost the same selectivity as that of the selenimides obtained by direct imidation of prochiral selenides as shown previously (7.5.1) may indicate that the chirality transfer is not lost in the [2,3]-sigmatropic rearrangement step.

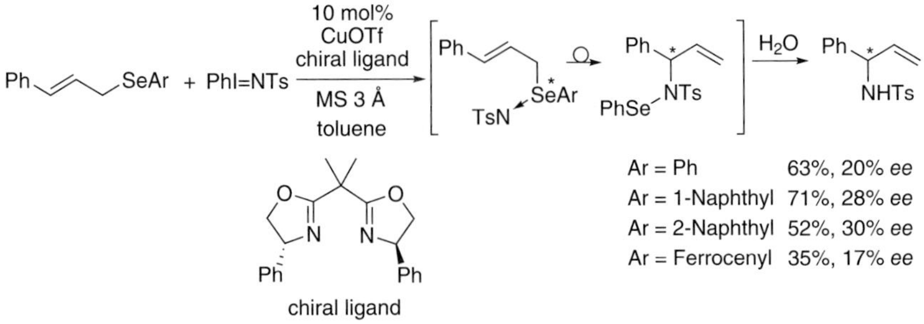

7 Selenoxide Elimination and [2,3]-Sigmatropic Rearrangement

Scheme 7.37 Enantioselective [2,3]-sigmatropic rearrangement of allylic selenimides.

Results for Scheme 7.37:
- Ar = Ph   63%, 20% ee
- Ar = 1-Naphthyl 71%, 28% ee
- Ar = 2-Naphthyl 52%, 30% ee
- Ar = Ferrocenyl 35%, 17% ee

7.5.3
Diastereoselective [2,3]-Sigmatropic Rearrangements Producing Chiral Allylic Amines

In 1995, Uemura and coworkers succeeded in the diastereoselective imidation of optically active allylic ferrocenyl selenides to give the corresponding allylic amines with high enantioselectivities [52]. Thus, treatment of optically active cinnamyl ferrocenyl selenides with PhI = NTs or TsNClNa at 0 °C afforded the allylic amines with high enantioselectivities (up to 87% ee) (Scheme 7.38). PhI = NTs is a more effective reagent than TsNClNa in both product selectivity and enantioselectivity. The result of the high ee value of the produced allylic amines provides the basis for the following speculation concerning this asymmetric reaction: (i) the initial imidation step proceeds with high diastereoselectivity; (ii) the chirality transfer via [2,3]-sigmatropic rearrangement occurs almost without loss of optical purity. Compared with the case of the chiral selenoxides, the epimerization of chiral selenimides is known to be quite slow [48].

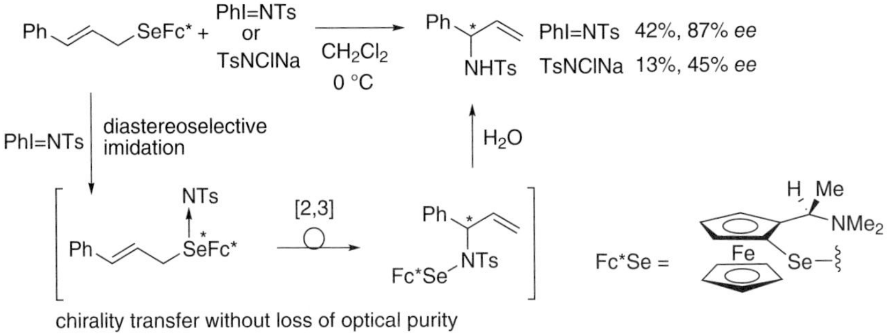

Scheme 7.38 Diastereoselective [2,3]-sigmatropic rearrangement of optically active allylic selenimides by Uemura.

Results for Scheme 7.38:
- PhI=NTs   42%, 87% ee
- TsNClNa   13%, 45% ee

Koizumi and coworkers used optically active allylic chloroselenuranes bearing a 2-*exo*-hydroxyl-10-bornyl moiety to produce optically active allylic amines [53]. Treatment of the allylic selenides with *tert*-BuOCl gave the corresponding allylic chloroselenuranes as exclusive products. Benzylcarbamate (CbzNH$_2$), *tert*-butyl-carbamate (BocNH$_2$), *p*-tosylamide, and diphenylphosphinamide were selected as the N-protected amines for selenimide formation. The nucleophilic reaction of allylic chloroselenuranes with N-protected lithium amides afforded optically active allylic selenimides *in situ* with retention of configuration. The [2,3]-sigmatropic rearrangement of allylic selenimides gave the N-protected allylic amines with low to high enantioselectivities (up to 93% *ee*) (Scheme 7.39). From the absolute configuration of the resulting N-protected allylic amines as well as from their high enantiomeric excess, it was concluded that the [2,3]-sigmatropic rearrangement of allylic selenimides proceeded predominantly *via* the *endo*-transition state.

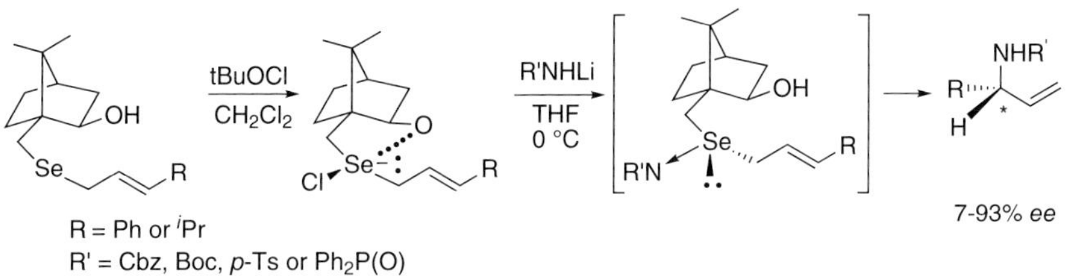

Scheme 7.39 Diastereoselective [2,3]-sigmatropic rearrangement of optically active allylic selenimides by Koizumi.

7.6
[2,3]-Sigmatropic Rearrangement via Allylic Selenium Ylides

The [2,3]-sigmatropic rearrangement of allylic chalcogen ylides provides a useful method for carbon–carbon bond formation (Scheme 7.40). The first successful asymmetric [2,3]-sigmatropic rearrangement involving sulfur ylides appeared in 1973 [54]. Treatment of the resolved and enantiomerically pure sulfonium salt with *tert*-butoxide afforded the corresponding homoallylic sulfide with 94% *ee*, where the chiral information has been transferred from the sulfur atom to the carbon atom with an excellent control (Scheme 7.41). Another approach to generate the chiral sulfur ylide was the enantioselective deprotonation of the diallylic sulfonium salt with an optically active base in an optically active solvent [55]. The reaction produced the corresponding sulfide *via* the chiral allylic sulfur ylide, but with only low enantioselectivities (up to 12% *ee*) (Scheme 7.42). In sharp contrast to the well-studied [2,3]-sigmatropic rearrangement of the allylic sulfur ylides [56], the study of the allylic selenium ylides has been quite limited [57]. Typical results of the preparation of optically active selenium ylides as well as their application to [2,3]-sigmatropic rearrangement are described in this section.

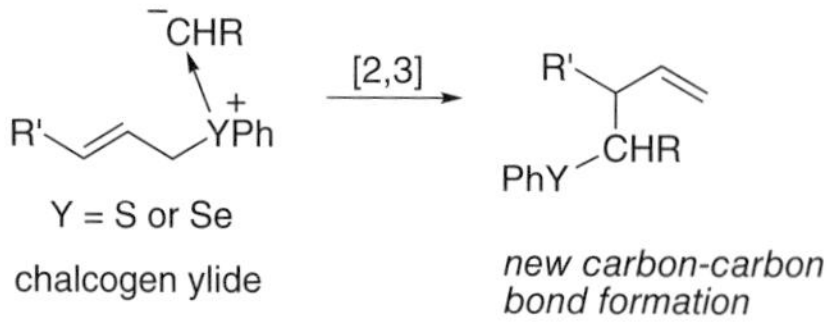

7 Selenoxide Elimination and [2,3]-Sigmatropic Rearrangement

Scheme 7.40 [2,3]-Sigmatropic rearrangement *via* allylic chalcogen ylides.

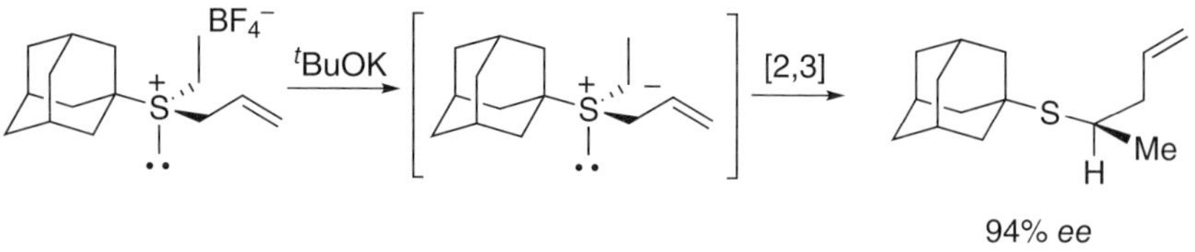

Scheme 7.41 Chirality transfer from sulfur to carbon via [2,3]-sigmatropic rearrangement.

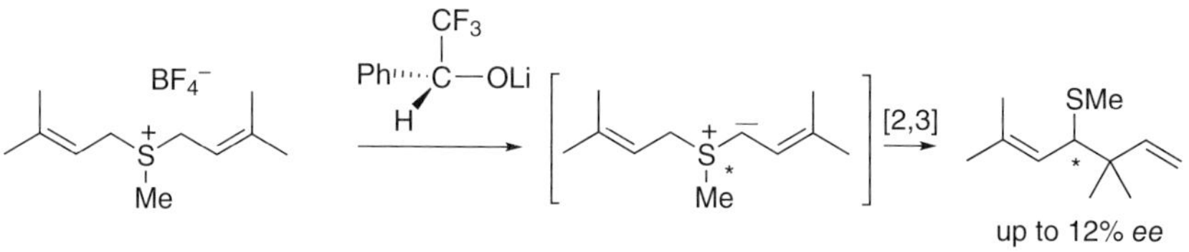

Scheme 7.42 Enantioselective [2,3]-sigmatropic rearrangement via allylic sulfur ylide.

7.6.1
Preparation and Properties of Optically Active Selenium Ylides

After the first preparation of a selenium ylide by Hughes and a coworker in 1935 [58], only a few reports on the preparation of the optically active selenium ylides have appeared until now (Scheme 7.43) [59]. The preparation of an optically pure selenium ylide by fractional recrystallization of single diastereomeric selenium ylides was first reported by Kamigata and coworkers in 1992 [60]. They designed a selenium ylide bearing an electron-withdrawing group on the selenium atom to stabilize the chirality of the selenium center thermodynamically. The fractional recrystallization of the diastereomeric selenium ylides possessing a menthyl group as a chiral source gave the optically pure selenium ylide as stable crystals (Scheme 7.44). The absolute configuration around the selenium atom was confirmed by X-ray crystallographic analysis. The rate of racemization of the optically active selenium ylide was studied by heating of a solution of the optically active selenium ylide. The result indicated that the selenium ylide is not simply racemized thermally but also it decomposes at temperatures higher than 100 °C. The conclusion is that the selenium ylide is extremely stable toward thermal inversion (racemization by pyramidal inversion) in comparison with similar sulfur ylides.

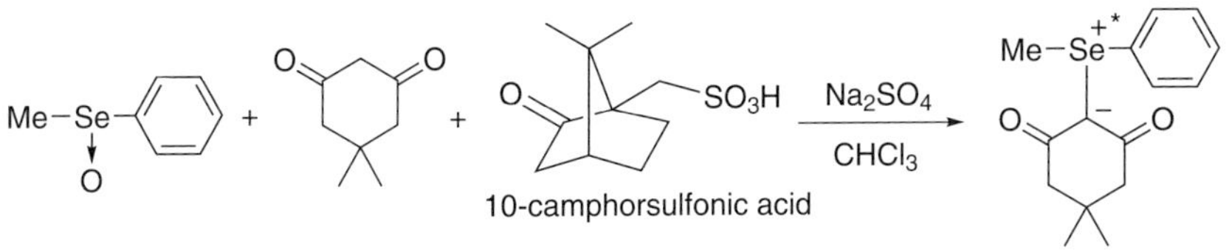

Scheme 7.43 Asymmetric synthesis of optically active selenium ylide.

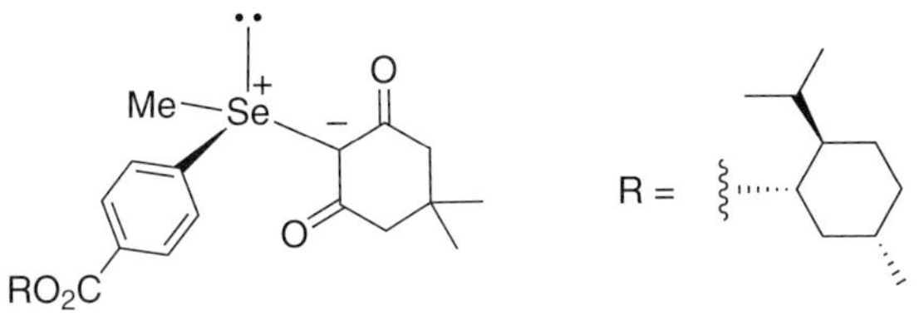

Scheme 7.44 Isolation of optically pure selenium ylide.

7.6.2
Enantioselective [2,3]-Sigmatropic Rearrangements via Allylic Selenium Ylides

The first example of an enantioselective [2,3]-sigmatropic rearrangement *via* allylic selenium ylides appeared in 1995 [61], where the enantioselective catalytic carbenoid addition to chalcogen atoms such as selenium and sulfur was used as the first step. The reaction of cinnamyl phenyl selenide or sulfide with ethyl diazoacetate in the presence of a catalytic amount of Cu(I)-bis(oxazoline) or $Rh_2(5S\text{-MEPY})_4$ afforded a diastereomeric mixture (ca. 3:2 ~ 2:1) of ethyl 2-phenylchalcogeno-3-phenylpent-4-enoates with moderate enantioselectivities (up to 41% *ee*) (Scheme 7.45). The enantiomeric excess in the reaction of the

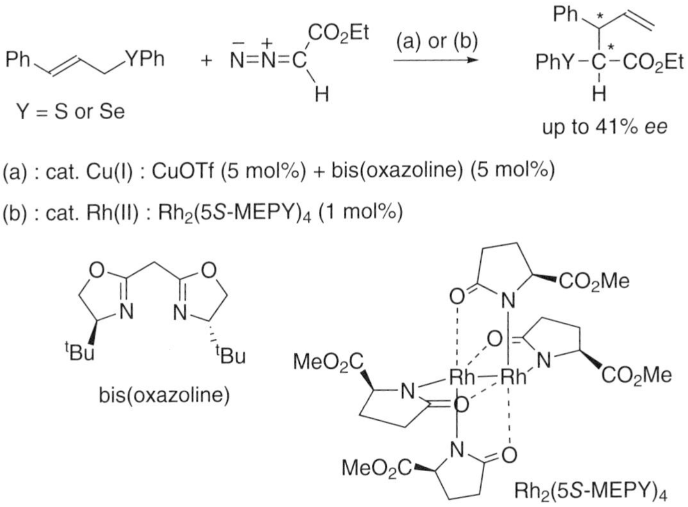

(a) : cat. Cu(I) : CuOTf (5 mol%) + bis(oxazoline) (5 mol%)
(b) : cat. Rh(II) : $Rh_2(5S\text{-MEPY})_4$ (1 mol%)

Scheme 7.45 Enantioselective [2,3]-sigmatropic rearrangement of allylic selenium ylides.

cinnamyl selenide was slightly higher than that of the cinnamyl sulfide under the same reaction conditions. The plausible reaction pathway of this catalytic reaction is shown in Scheme 7.46. Since it is known that the chirality transfer from the sulfur to the carbon in a sulfur ylide via [2,3]-sigmatropic rearrangement proceeds without loss of optical purity and also the racemization does not occur at a chiral selenium ylide, the observed enantioselectivity of the products might depend on the initial step of carbenoid addition to the chalcogen atom. This reaction may provide a new methodology for the carbon–carbon bond formation with asymmetric induction at both carbon centers.

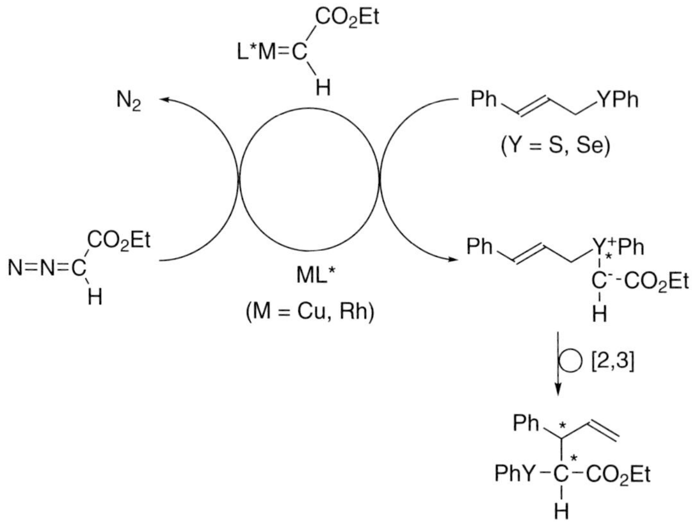

Scheme 7.46 Proposed reaction pathway for enantioselective [2,3]-sigmatropic rearrangement of allylic selenium ylides.

After the first successful example of an asymmetric [2,3]-sigmatropic rearrangement reported by Uemura and coworkers [61], several research groups investigated enantioselective [2,3]-sigmatropic rearrangements of allylic and propargylic sulfur ylides by using a variety of transition metal complexes and optically active ligands as catalysts [62–64]. Typical examples are shown in Schemes 7.47 [63a] and 7.48 [63b]. The development of a highly enantioselective [2,3]-sigmatropic rearrangement of allylic and propargylic selenium ylides is still ongoing because a successful example has not been reported until now.

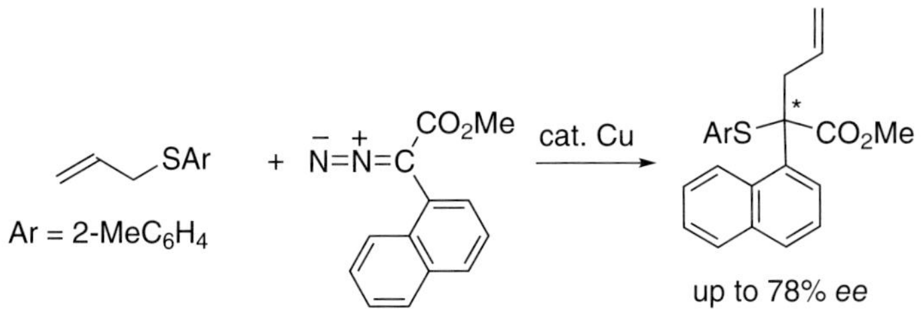

cat. Cu: $[Cu(MeCN)_4]PF_6$ (10 mol%) + bis(oxazoline) (11 mol%)

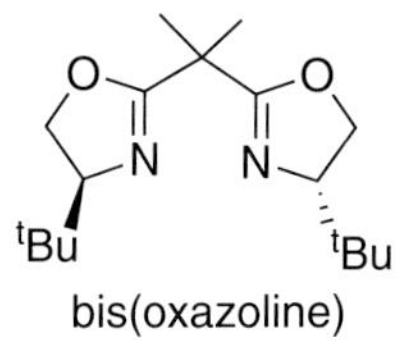

Scheme 7.47 Enantioselective [2,3]-sigmatropic rearrangement of allylic sulfur ylides by Wang [63a].

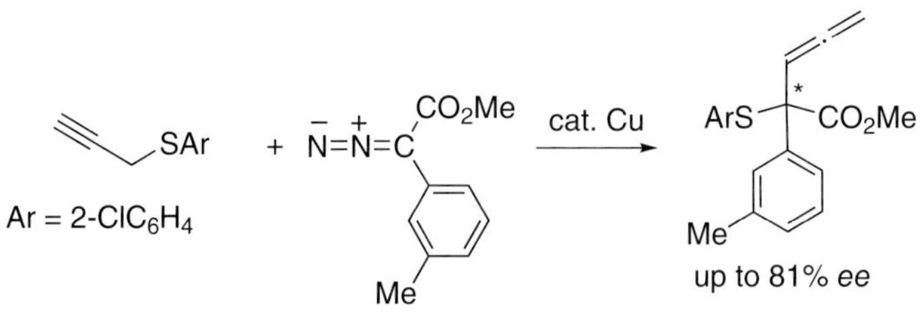

cat. Cu: $[Cu(MeCN)_4]PF_6$ (10 mol%) + bis(oxazoline) (11 mol%)

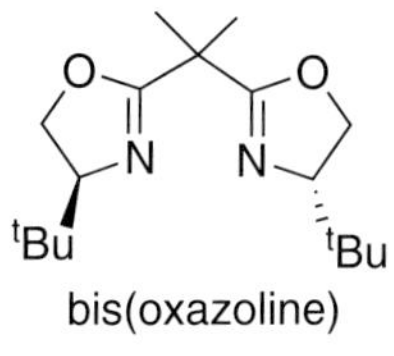

Scheme 7.48 Enantioselective [2,3]-sigmatropic rearrangement of propargylic sulfur ylides by Wang [63b].

7.6.3
Diastereoselective [2,3]-Sigmatropic Rearrangement via Allylic Selenium Ylides

Koizumi and coworkers used optically active allylic chloroselenuranes bearing a 2-*exo*-hydroxyl-10-bornyl moiety to produce the chiral allylic selenium ylides *in situ* [65]. The nucleophilic reaction of the corresponding chiral chloroselenuranes and selenoxides with an active methylene compound occurs in a highly stereoselective manner to give the corresponding chiral selenium ylides with retention of configuration [66] (Scheme 7.49). Treatment of the allylic chloroselenuranes with (phenylsulfonyl)acetonitrile and triethylamine at −20 °C for 0.5–1 h afforded the corresponding homoallylic selenides as a mixture of two diastereoisomers [65] (Scheme 7.50). The result shows that the [2,3]-sigmatropic rearrangement of the resulting allylic selenium ylides proceeded in a highly stereoselective manner to give the corresponding homoallylic selenides. From the absolute configuration of the resulting homoallylic selenides as well as their high diastereomeric excess, the [2,3]-sigmatropic rearrangement of the intermediate allylic selenium ylides was revealed to proceed predominantly *via* the *endo*-transition state.

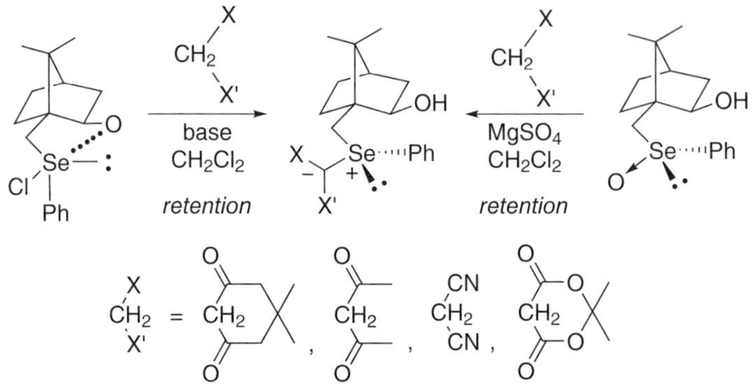

Scheme 7.49 Diastereoselective formation of allylic selenium ylides.

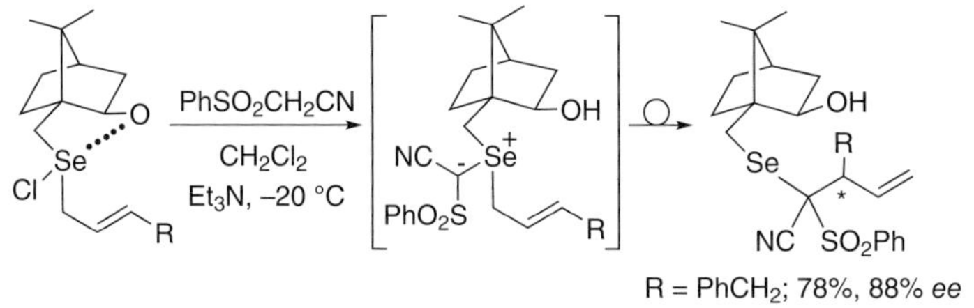

Scheme 7.50 Diastereoselective [2,3]-sigmatropic rearrangement of allylic selenium ylides.

Another example (our unpublished results) of the diastereoselective [2,3]-sigmatropic rearrangement *via* a chiral selenium ylide is the reaction between an optically active allylic ferrocenyl selenide with ethyl diazoacetate in the presence of a catalytic amount of $Rh_2(OAc)_4$. Unfortunately, the diastereoselectivity of the ethyl 2-ferrocenylseleno-3-phenylpent-4-enoates was not high (up to 25% *de*) (Scheme 7.51).

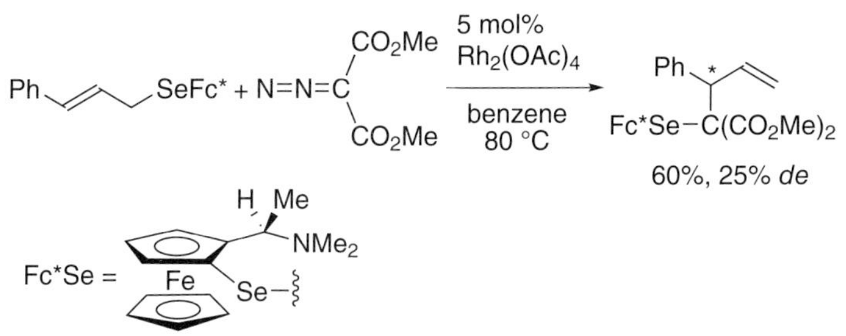

Scheme 7.51 Diastereoselective [2,3]-sigmatropic rearrangement of allylic selenium ylides.

7.7
Summary

This chapter reviews recent advances of the following four types of asymmetric reactions using organoselenium compounds: selenoxide elimination, [2,3]-sigmatropic rearrangement via selenoxides, [2,3]-sigmatropic rearrangement via selenimides, [2,3]-sigmatropic rearrangement via selenium ylides. In all the reactions, the preparation or the intervention of optically active organoselenium compounds bearing a chiral center on the selenium atom such as selenoxides, selenimides, and selenium ylides is considered to be one of the most important steps to obtain the optically active organic compounds with high stereoselectivities.

Although a significant development in asymmetric reactions using organoselenium compounds via selenoxide elimination and [2,3]-sigmatropic rearrangement as key steps has been achieved in the last two decades, the authors believe and hope that new methodology as well as synthetic applications in this field will appear in the near future.

References

1 For selected reviews, see (a) Paulmier, C. (1986) *Selenium Reagents and Intermediates in Organic Synthesis*, Pergamon Press, Oxford; (b) Patai, S. (ed.) (1987) *The Chemistry of Organic Selenium and Tellurium Compounds*, John Wiley & Sons, Inc., New York; (c) Back, T.G. (ed.) (1999) *Organoselenium Chemistry, A Practical Approach*, Oxford University Press, Oxford; (d) Nishibayashi, Y. and Uemura, S. (1996) *Rev. Heteroatom Chem.*, **14**, 83; (e) Uemura, S. (1998) *Phosphorus Sulphur Silicon*, **136–138**, 219; (f) Wirth, T. (1999) *Tetrahedron*, **55**, 1; (g) Wirth, T. (2000) *Angew. Chem. Int. Ed.*, **39**, 3740; (h) Wirth, T. (ed.) (2000) *Organoselenium Chemistry, Modern Developments in Organic Synthesis*, Springer, Berlin, Heidelberg; (i) Uemura, S. (2001) *Phosphorus Sulphur Silicon*, **171**, 13; (j) Uemura, S. (2005) *Phosphorus Sulfur Silicon*, **180**, 721; (k) Freudendahl, D.M., Shahzad, S.A. and Wirth, T. (2009) *Eur. J. Org. Chem.*, 1649; (l) Mukherjee, A.J., Zade, S.S., Singh, H.B., and Sunoj, R.B. (2010) *Chem. Rev.*, **110**, 4357.

2 Davis, F.A., Billmers, J.M., and Stringer, O.D. (1983) *Tetrahedron Lett.*, **24**, 3191.

3 Harrison, P.W.B., Kenyon, J., and Phillips, H. (1926) *J. Chem. Soc.*, 2079.
4 Davis, F.A., Stringer, O.D., and McCauley, J.P. (1985) *Tetrahedron*, **41**, 4747.
5 Shimizu, T., Yoshida, M., and Kamigata, N. (1988) *Bull. Chem. Soc. Jpn.*, **61**, 3761.
6 Shimizu, T., Kikuchi, K., Ishikawa, Y., Ikemoto, I., Kobayashi, M., and Kamigata, N. (1989) *J. Chem. Soc., Perkin Trans. 1*, 597.
7 (a) Shimizu, T., Enomoto, M., Taka, H., and Kamigata, N. (1999) *J. Org. Chem.*, **64**, 8242; (b) Taka, H., Matsumoto, A., Shimizu, T., and Kamigata, N. (2000) *Chem. Lett.*, 726; (c) Taka, H., Matsumoto, A., Shimizu, T., and Kamigata, N. (2001) *Heteroatom. Chem.*, **12**, 227; (d) Kamigata, N. (2001) *Phosphorus Sulfur Silicon*, **171**, 207; (e) Soma, T., Shimizu, T., Hirabayashi, K., and Kamigata, N. (2007) *Heteroatom Chem.*, **18**, 301.
8 Toda, F. and Mori, K. (1986) *J. Chem. Soc., Chem. Commun.*, 1357.
9 Davis, F.A. and Reddy, R.T. (1992) *J. Org. Chem.*, **57**, 2599.
10 (a) Tiecco, M., Tingoli, M., Testaferri, L., and Bartoli, D. (1987) *Tetrahedron Lett.*, **28**, 3849; (b) Shimizu, T., Kobayashi, M., and Kamigata, N. (1989) *Bull. Chem. Soc. Jpn.*, **62**, 2099.
11 Reich, H.J. and Yelm, K.E. (1991) *J. Org. Chem.*, **56**, 5672.
12 Jones, D.N., Mundy, D., and Whitehouse, R.D. (1970) *J. Chem. Soc., Chem. Commun.*, 86.
13 Sharpless, K.B., Young, M.W., and Lauer, R.F. (1973) *Tetrahedron Lett.*, **22**, 1979.
14 Kondo, N., Fueno, H., Fujimoto, H., Makino, M., Nakaoka, H., Aoki, I., and Uemura, S. (1994) *J. Org. Chem.*, **59**, 5254.
15 (a) Bayse, C.A. and Allison, B.D. (2007) *J. Mol. Model.*, **13**, 47; (b) Macdougall, P.E., Smith, N.A., and Schiesser, C.H. (2008) *Tetrahedron*, **64**, 2824; (c) Bayse, C.A. and Antony, S. (2009) *Molecules*, **14**, 3229.
16 Kingsbury, C.A. (2010) *J. Phys. Org. Chem.*, **23**, 513.
17 Ogawa, A., Sekiguchi, M., Shibuya, H., Kuniyasu, H., Takami, N., Ryu, I., and Sonoda, N. (1991) *Chem. Lett.*, 1805.
18 (a) Huang, X. and Xu, W. (2002) *Tetrahedron Lett.*, **43**, 5495; (b) Huang, X. and Xu, W.M. (2003) *Org. Lett.*, **5**, 4649; (c) Xu, W.M. and Huang, X. (2004) *Synthesis*, 2094; (d) Tang, E., Huang, X., and Xu, W.M. (2004) *Tetrahedron*, **60**, 9963; (e) Gian, H. and Huang, X. (2006) *Synlett*, 1547.
19 (a) Komatsu, N., Nishibayashi, Y., Sugita, T., and Uemura, S. (1992) *J. Chem. Soc., Chem. Commun.*, 46; (b) Komatsu, N., Murakami, T., Nishibayashi, Y., Sugita, T., and Uemura, S. (1993) *J. Org. Chem.*, **58**, 3697.
20 Back, T.G. and Krishna, M.V. (1987) *J. Org. Chem.*, **52**, 4265; (b) Back, T.G., Krishan, M.V., and Muralidharan, K.R. (1989) *J. Org. Chem.*, **54**, 4146.
21 Sayama, S. and Onami, T. (2000) *Tetrahedron Lett.*, **41**, 5557.
22 Komatsu, N., Matsunaga, S., Sugita, T., and Uemura, S. (1993) *J. Am. Chem. Soc.*, **115**, 5847.
23 (a) Hanessian, S., Delorme, D., Beaudoin, S., and Leblanc, Y. (1984) *J. Am. Chem. Soc.*, **106**, 5754; (b) Hanessian, S. and Beaudoin, S. (1992) *Tetrahedron Lett.*, **33**, 7655; (c) Denmark, S.E. and Chen, C.T. (1992) *J. Am. Chem. Soc.*, **114**, 10674.
24 (a) Duhamel, L., Ravard, A., and Plaquevent, J.C. (1990) *Tetrahedron Asymmetry*, **1**, 347; (b) Vardecard, J., Plaquevent, J.C., Duhamel, L., and Duhamel, P. (1993) *J. Chem. Soc., Chem. Commun.*, 116; (c) Vadecard, J., Plaquevent, J.C., Duhamel, L., Duhamel, P., and Toupet, L. (1994) *J. Org. Chem.*, **59**, 2285; (d) Amadji, M., Vadecard, J., Cahard, D., Duhamel, L., Duhamel, P., and Plaquevent, J.C. (1998) *J. Org. Chem.*, **63**, 5541.
25 Solladie, G., Zimmermann, R., Bartsch, R., and Warborsky, H.M. (1985) *Synthesis*, 662.
26 (a) Nishibayashi, Y., Singh, J.D., Fukuzawa, S., and Uemura, S. (1994) *Tetrahedron Lett.*, **35**, 3115; (b) Nishibayashi, Y., Singh, J.D., Fukuzawa, S., and Uemura, S. (1995) *J. Org. Chem.*, **60**, 4114.
27 (a) Reich, H.J., Reich, I.L., and Wollowitz, S. (1978) *J. Am. Chem. Soc.*, **100**, 5981; (b)

Reich, H.J., Shah, S.K., Gold, P.M., and Olson, R.E. (1981) *J. Am. Chem. Soc.*, **103**, 3112; (c) Reich, I.L., and Reich, H.J. (1981) *J. Org. Chem.*, **46**, 3721; (d) Reich, H.J. and Wollowitz, S. (1982) *J. Am. Chem. Soc.*, **104**, 7051; (e) Reich, H.J., Yelm, K.E., and Wollowitz, S. (1983) *J. Am. Chem. Soc.*, **105**, 2503.

28 Komatsu, N., Nishibayashi, Y., and Uemura, S. (1993) *Tetrahedron Lett.*, **34**, 2339.

29 Fujita, K., Kanakubo, M., Ushijima, H., Oishi, A., Ikeda, Y., and Taguchi, Y. (1998) *Synlett*, 987.

30 Takahashi, T., Kurose, N., Kawanami, S., Arai, Y., Koizumi, T., and Shiro, M. (1994) *J. Org. Chem.*, **59**, 3262.

31 (a) Kurose, N., Takahashi, T., and Koizumi, T. (1997) *Tetrahedron*, **53**, 12115; (b) Zhang, J., and Koizumi, T. (2000) *Phosphorus Sulfur Silicon*, **157**, 225.

32 Bourland, T.C., Carter, R.G., and Yokochi, A.F.T. (2004) *Org. Biomol. Chem.*, **2**, 1315.

33 Kshirsagar, T.A., Moe, S.T., and Portoghese, P.S. (1998) *J. Org. Chem.*, **63**, 1704.

34 (a) Davoren, J.E. and Martin, S.F. (2007) *J. Am. Chem. Soc.*, **129**, 510; (b) Davoren, J.E., Harcken, C., and Martin, S.F. (2008) *J. Org. Chem.*, **73**, 391.

35 Hess, L.C. and Posner, G.H. (2010) *Org. Lett.*, **12**, 2120.

36 (a) Tiecco, M., Carlone, A., Sternativo, S., Marini, F., Bartoli, G., and Melchiorre, P. (2007) *Angew. Chem. Int. Ed.*, **46**, 6882; (b) Sundén, H., Rios, R., and Córdova, A. (2007) *Tetrahedron Lett.*, **48**, 7865.

37 Zanoni, G., Castronovo, F., Perani, E., and Vidari, G. (2003) *J. Org. Chem.*, **68**, 6803.

38 Bradshaw, B., Etxebarria-Jardí, G., and Bonjoch, J. (2010) *J. Am. Chem. Soc.*, **132**, 5966.

39 Hori, T. and Sharpless, K.B. (1979) *J. Org. Chem.*, **44**, 4208.

40 Fankhauser, J.E., Peevey, R.M., and Hopkins, P.B. (1984) *Tetrahedron Lett.*, **25**, 15.

41 Shea, R.G., Fitzner, J.N., Fankhauser, J.E., and Hopkins, P.B. (1984) *J. Org. Chem.*, **49**, 3647.

42 Shea, R.G., Fitzner, J.N., Fankhauser, J.E., Spaltenstein, A., Carpino, P.A., Peevey, R.M., Pratt, D.V., Tenge, B.J., and Hopkins, P.B. (1986) *J. Org. Chem.*, **51**, 5243.

43 Waetzig, S.R. and Tunge, J.A. (2008) *Chem. Commun.*, 3311.

44 Takada, H., Nishibayashi, Y., Ohe, K., Uemura, S., Baird, C.P., Sparey, T.J., and Taylor, P.C. (1997) *J. Org. Chem.*, **62**, 6512.

45 Takada, H., Oda, M., Miyake, Y., Ohe, K., and Uemura, S. (1998) *Chem. Commun.*, 1557.

46 Johannsen, M. and Jørgensen, K.A. (1998) *Chem. Rev.*, **98**, 1689.

47 Krasnov, V.P., Naddaka, V.I., and Minkin, V.I. (1981) *Zh. Org. Khim.*, **17**, 445.

48 Kamigata, N., Taka, H., Matsuhisa, A., Matsuyama, H., and Shimizu, T. (1994) *J. Chem. Soc., Perkin Trans. 1*, 2257.

49 (a) Shimizu, T., Seki, N., Taka, H., and Kamigata, N. (1996) *J. Org. Chem.*, **61**, 6013; (b) Taka, H., Shimizu, T., Iwasaki, F., Yasui, M., and Kamigata, N. (1999) *J. Org. Chem.*, **64**, 7433; (c) Shimizu, T., Kamigata, N., and Ukuta, S. (1999) *J. Chem. Soc., Perkin Trans. 2*, 1469.

50 Soma, T., Kamigata, N., Hirabayashi, K., and Shimizu, T. (2007) *Bull. Chem. Soc. Jpn.*, **80**, 2389.

51 Miyake, Y., Oda, M., Oyamada, A., Takada, H., Ohe, K., and Uemura, S. (2000) *J. Organomet. Chem.*, **611**, 475.

52 Nishibayashi, Y., Chiba, T., Ohe, K., and Uemura, S. (1995) *J. Chem. Soc., Chem. Commun.*, 1243.

53 Kurose, N., Takahashi, T., and Koizumi, T. (1996) *J. Org. Chem.*, **61**, 2932.

54 Trost, B.M. and Hammen, R.F. (1973) *J. Am. Chem. Soc.*, **95**, 962.

55 Trost, B.M. and Biddlecom, W.G. (1973) *J. Org. Chem.*, **38**, 3438.

56 For reviews, see (a) Padwa, A., and Hornbuckle, S.F. (1991) *Chem. Rev.*, **91**, 263; (b) Li, A.H., Dai, L.X., and Aggarwal, V.K. (1997) *Chem. Rev.*, **97**, 2341; (c) Aggarwal, V.K. (1998) *Synlett*, 329.

57 (a) Reich, H.J. and Cohen, M.L. (1979) *J. Am. Chem. Soc.*, **101**, 1307; (b) Giddings, P.J., John, D.I., and Thomas, E.J. (1980) *Tetrahedron Lett.*, **21**, 395; (c) Kosarych, Z. and Cohen, T. (1982) *Tetrahedron Lett.*, **23**, 3019.

58 Hughes, E.D. and Kuriyan, K.I. (1935) *J. Chem. Soc.*, 1609.
59 Sakaki, K. and Oae, S. (1976) *Tetrahedron Lett.*, 3703.
60 Kamigata, N., Nakamura, Y., Kikuchi, K., Ikemoto, I., Shimizu, T., and Matsuyama, H. (1992) *J. Chem. Soc., Perkin Trans. 1*, 1721.
61 Nishibayashi, Y., Ohe, K., and Uemura, S. (1995) *J. Chem. Soc., Chem. Commun.*, 1245.
62 (a) Fukuda, T. and Katsuki, T. (1997) *Tetrahedron Lett.*, **38**, 3435; (b) Fukuda, T., Irie, R., and Kastuki, T. (1999) *Tetrahedron*, **55**, 649; (c) Murakami, M., Uchida, T., Saito, B., and Kastuki, T. (2003) *Chirality*, **15**, 116.
63 (a) Zhang, X., Qu, Z., Ma, Z., Shi, W., Jin, X., and Wang, J. (2002) *J. Org. Chem.*, **67**, 5621; (b) Zhang, X., Ma, M., and Wang, J. (2003) *Tetrahedron Asymmetry*, **14**, 891; (c) Ma, M., Peng, L., Li, C., Zhang, X., and Wang, J. (2005) *J. Am. Chem. Soc.*, **127**, 15016; (d) Zhang, Y. and Wang, J. (2010) *Coord. Chem. Rev.*, **254**, 941.

64 (a) Carter, D.S. and Vranken, D.L.V. (1999) *Tetrahedron Lett.*, **40**, 1617; (b) Aggarwal, V.K., Ferrara, M., Hainz, R., and Spey, S.E. (1999) *Tetrahedron Lett.*, **40**, 8923; (c) Kitagaki, S., Yanamoto, Y., Okubo, H., Nakajima, M., and Hashimoto, S. (2000) *Heterocycles*, **54**, 623; (d) McMillen, D.W., Varga, N., Reed, B.A., and King, C. (2000) *J. Org. Chem.*, **65**, 2532; (e) Wee, A.G., Shi, Q., Wang, Z., and Hatton, K. (2003) *Tetrahedron Asymmetry*, **14**, 897; (f) Bacci, J.P., Greenman, K.L., and Vranken, D.L.V. (2003) *J. Org. Chem.*, **68**, 4955; (g) Armstrong, A. and Emmerson, D.P.G. (2009) *Org. Lett.*, **11**, 1547.
65 Kurose, N., Takahashi, T., and Koizumi, T. (1997) *J. Org. Chem.*, **62**, 4562.
66 Takahashi, T., Kurose, N., Kawanami, S., Nojiri, A., Arai, Y., Koizumi, T., and Shiro, M. (1995) *Chem. Lett.*, 379.

8
Selenium Compounds as Ligands and Catalysts
Fateh V. Singh and Thomas Wirth

8.1
Introduction

After the discovery of the selenoxide elimination in the early 1970s [1], organoselenium chemistry has been developed as an important tool in synthetic organic chemistry. Many different organoselenium reagents have been introduced since that time and chemo-, regio-, stereoselective, as well as stereospecific reactions in various synthetic transformations such as selenenylations, selenocyclizations, selenoxide eliminations, and 2,3-sigmatropic rearrangements have been described [2]. In the past years, organoselenium reagents have received particular attention due to their use in catalysis [3]. There are several books [4], book chapters [5], and review articles [6] which have appeared and describe various aspects of organoselenium chemistry in the recent past. This chapter highlights the developments of organoselenium reagents as ligands and catalysts in various organic transformations.

8.2
Selenium-Catalyzed Reactions

Organoselenium reagents have been used as ligands and catalysts in various organic transformations. Although the selenium-catalyzed reaction was initially reported in 1975 by Sonoda [7], Uemura and coworkers more widely investigated the utility of chiral selenium-based ligands in asymmetric catalysis [8]. After these initial reports, selenium-based ligands and catalysts have been used with great success in a wide range of catalytic approaches.

8.2.1
Stereoselective Addition of Diorganozinc Reagents to Aldehydes

8.2.1.1 Diethylzinc Addition

The stereoselective addition of diethylzinc to aldehydes in the presence of chiral catalysts is a widely used reaction in asymmetric catalysis [9]. It has become an important tool for the synthesis of various chiral secondary alcohols and a number of catalysts have been employed successfully in this reaction [10]. Wirth has reported the first selenium-catalyzed stereoselective addition reaction of diethylzinc to aldehydes in 1995 [11a]. They have achieved stereoselective addition of diethylzinc to aldehydes using chiral diselenides of type **1** as catalyst (1 mol%). The chiral secondary alcohols **2** were obtained in very high yields in up to 98% enantiomeric excess (Scheme 8.1) [11]. Both aromatic and aliphatic aldehydes can be employed in this reaction but aromatic aldehydes showed higher selectivity than the aliphatic ones.

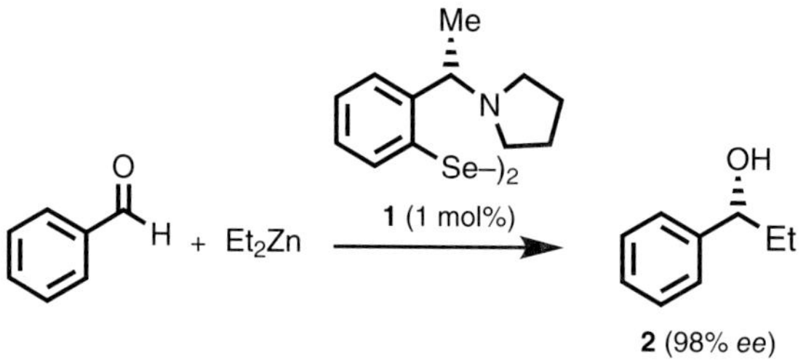

Scheme 8.1

Some other diselenides such as **3a–3d**, based on a similar framework with different substituents, have been synthesized by the same group and were evaluated in the stereoselective addition of diethylzinc addition to benzaldehyde (Figure 8.1). The presence of an electron-withdrawing moiety in diselenide **3a** decreases both yield and selectivity. Catalyst **3b** containing an additional stereogenic center in the five-membered ring is found to be a more efficient catalyst and the product was obtained with 97% enantiomeric excess [11].

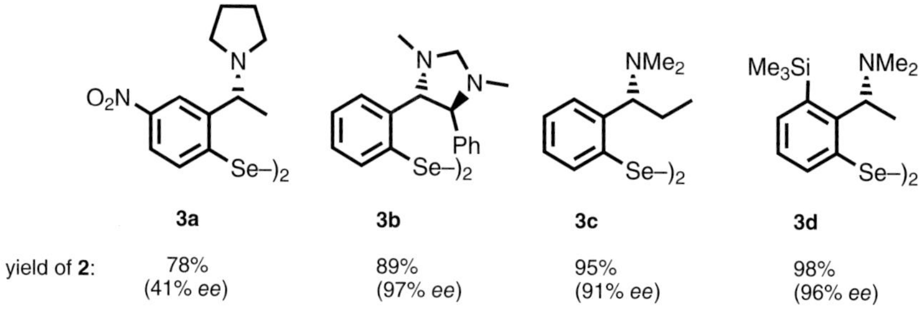

	3a	3b	3c	3d
yield of **2**:	78% (41% ee)	89% (97% ee)	95% (91% ee)	98% (96% ee)

Figure 8.1 Diselenides used as catalysts for diethylzinc addition.

8.2 Selenium-Catalyzed Reactions

Braga and coworkers reported the synthesis of aliphatic diselenides such as **4** [12] and selenides with oxazolidine moieties such as **5** [13], which were employed as chiral catalysts (0.5 mol%) in the stereoselective addition of diethylzinc to benzaldehyde. Chiral diselenide **4** was found to be a more efficient catalyst than the selenide **5**. The chiral alcohol **2** was obtained with up to 99% enentiomeric excess; however, the selenide containing oxazolidine derivative **5** showed a maximum of 95% enantiomeric excess (Scheme 8.2) [12c].

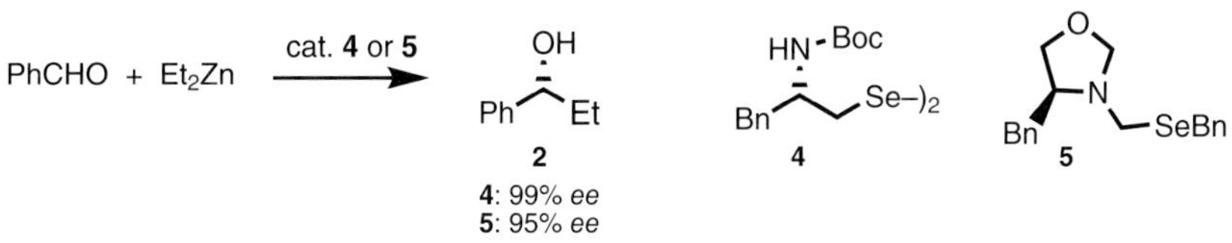

Scheme 8.2

8.2.1.2 Diphenylzinc Addition

The enantioselective arylation of aldehydes has been achieved by the addition of diarylzinc to aldehydes using selenium-based chiral catalysts. Bolm and coworkers synthesized chiral oxazolinyl-substituted ferrocenyl diselenide **6**, which was used as a catalyst (2.5 mol%) in the asymmetric arylation of aldehydes **7** using a zinc reagent prepared by the mixture of diphenylzinc and diethylzinc in a 1 : 2 ratio. The chiral secondary alcohols **8** were obtained up to 84% enantiomeric excess (Scheme 8.3) [14].

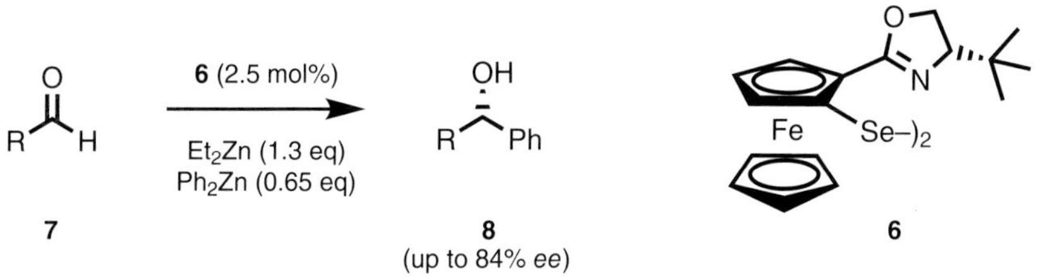

Scheme 8.3

Recently, a new class of selenium-based reagents has been developed and chiral catalysts of type **9** have been used in the enantioselective arylation of aromatic aldehydes using arylboronic acids with up to 91% enantiomeric excess in the products (Scheme 8.4) [15]. Various aromatic aldehydes bearing different functionalities in the aromatic ring can be used in this reaction.

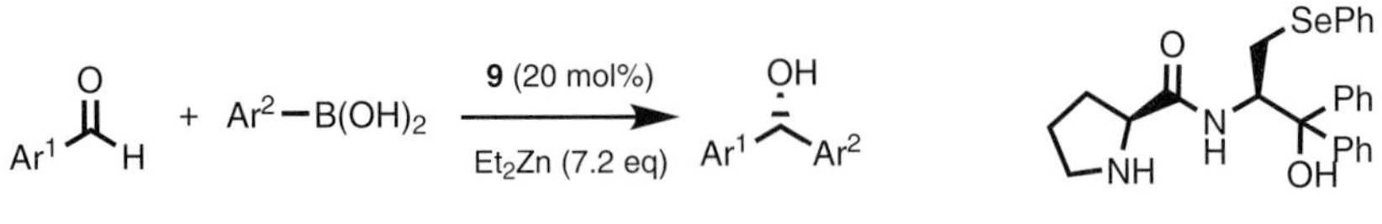

Scheme 8.4

Ar^1-CHO (**7**) + $Ar^2-B(OH)_2$ → [**9** (20 mol%), Et_2Zn (7.2 eq)] → $Ar^1-CH(OH)-Ar^2$ (**8**, up to 91% ee)

9: pyrrolidine-based ligand with SePh, Ph, Ph, OH groups

8.2.2
Selenium-Ligated Transition Metal-Catalyzed Reactions

Various selenium-containing ligands can form complexes with different transition metals and these selenium-ligated metal complexes have been used to catalyze various synthetic transformations.

8.2.2.1 Selenium-Ligated Stereoselective Hydrosilylation of Ketones

Some diselenides have been developed as promising ligands in transition metal-catalyzed asymmetric hydrosilylation or transfer hydrogenations of ketones. The rhodium(I)-catalyzed stereoselective hydrosilylation of alkyl aryl ketones **10** was successfully achieved using diferrocenyl diselenide **11** as chiral ligand. The chiral alcohols **8** were obtained in up to 88% enantiomeric excess (Scheme 8.5) [16]. Ruthenium(II)-catalyzed stereoselective transfer-hydrogenation of ketones have been also developed by using the same diselenide **11** as chiral ligand. However, the stereoselectivities obtained are lower in most of the reported examples (up to 48% ee) [17].

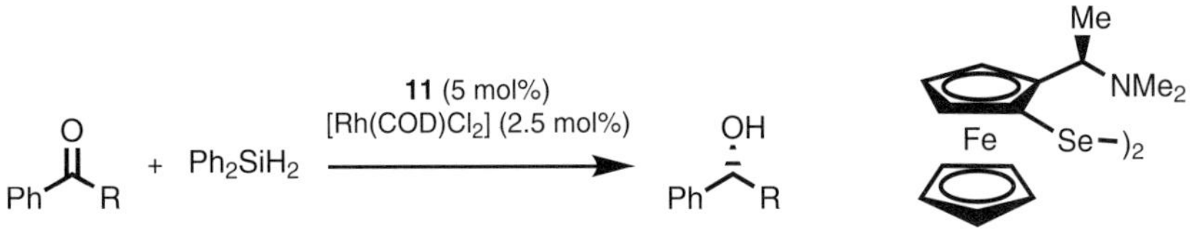

Scheme 8.5

Ph-CO-R (**10**) + Ph_2SiH_2 → [**11** (5 mol%), $[Rh(COD)Cl_2]$ (2.5 mol%)] → Ph-CH(OH)-R (**8**, up to 88% ee)

R = Me, Et, t-Bu

11: ferrocenyl-CH(Me)(NMe$_2$)-Se-)$_2$

The proposed catalytic cycle includes the formation of a tetracoordinated rhodium(I)-diselenide complex **A**, oxidative addition of the silane to Rh(I) to give complex **B**, and the coordination of a carbonyl group to Rh(III). The subsequent reductive elimination from Rh leads to the hydrosilylated compound (Scheme 8.6).

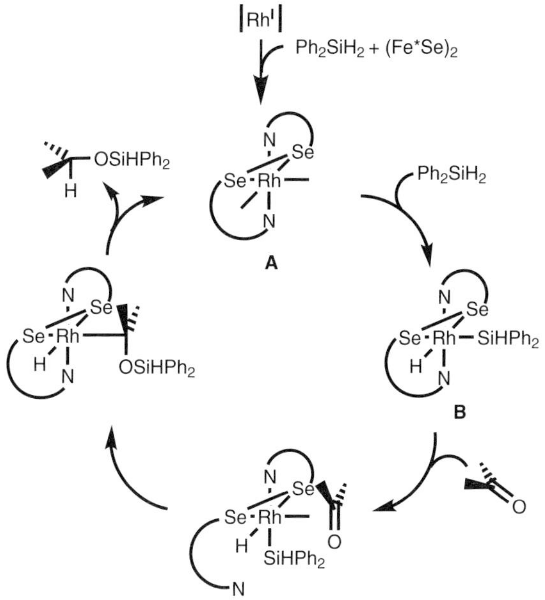

Scheme 8.6

Rhodium(I)-catalyzed stereoselective hydrogenation of enamides **12** also proceed smoothly using the same chiral ligand **11** to give the corresponding amides **13** with up to 69% enantiomeric excess (Scheme 8.7) [9].

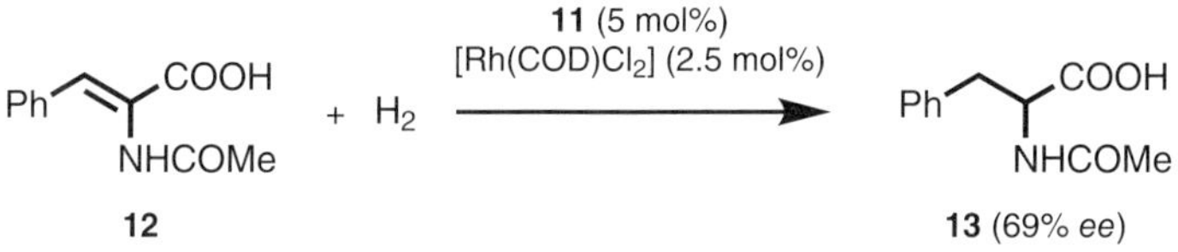

Scheme 8.7

8.2.2.2 Selenium-Ligated Copper-Catalyzed Addition of Organometallic Reagents to Enones

The conjugate addition of carbon nucleophiles to α,β-unsaturated carbonyl compounds is an important synthetic approach for the construction of carbon–carbon bonds. In 2001, Woodward and coworkers reported the synthesis of axially chiral 1,1′-bi-2-naphthol (BINOL)-cored methyl selenide **15** used as ligand in copper-catalyzed asymmetric 1,4-addition of organozinc compounds to enones **14**. The addition products **16** were obtained in good yields with up to 79% enantiomeric excess (Scheme 8.8) [18].

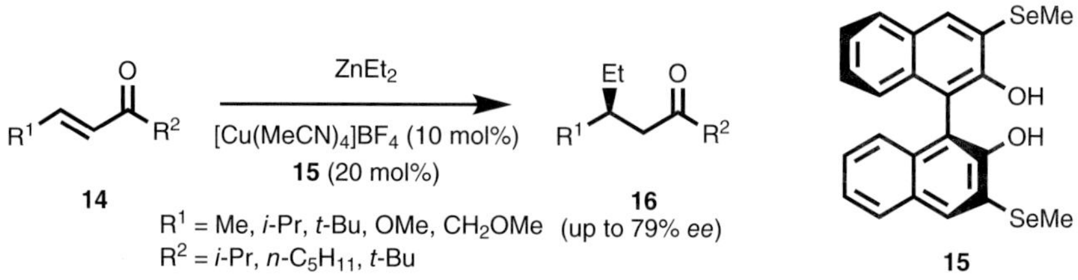

Scheme 8.8

Braga and coworkers developed the stereoselective conjugate addition of Grignard reagents **19** to enones **17** using a combination of an oxazolidine ligand **18** with copper(I) as catalyst. The addition products **20** were obtained in good yields with up to 85% enantiomeric excess as shown in Scheme 8.9 [19]. Recently, Zhang and coworkers investigated a new copper(I)-based catalyst with an axially chiral binaphthyl selenophosphoramide ligand **21** and employed this catalyst in the enantioselective conjugate addition of diethylzinc to enones **17**. The Michael products of type **22** were obtained in high yields with excellent enantiomeric excess for both cyclic and acyclic enones (Scheme 8.9) [20].

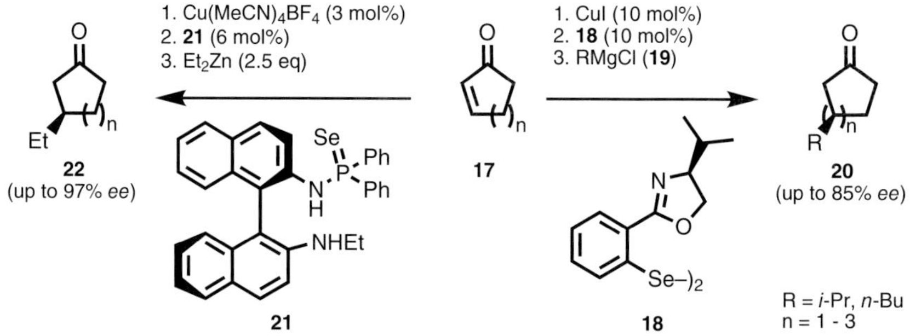

Scheme 8.9

8.2.2.3 Selenium-Ligated Palladium-Catalyzed Asymmetric Allylic Alkylation

Enantioselective palladium-catalyzed allylic alkylation is one of the well-investigated tools for stereoselective carbon–carbon bond-forming reactions [21]. Structurally diverse chiral selenium reagents have been investigated as promising ligands in palladium-catalyzed asymmetric allylic substitution reactions [22].

Helmchen and coworkers reported the first palladium-catalyzed asymmetric allylic substitution of (*E*)-1,3-diphenylallyl acetate **23** with dimethyl malonate **24** using the chiral oxazolidine containing selenide **25** (4 mol%) as chiral ligand. The chiral allylic alkylatated products **26** were obtained in good to excellent yields with up to 98% enantiomeric excess (Scheme 8.10) [22].

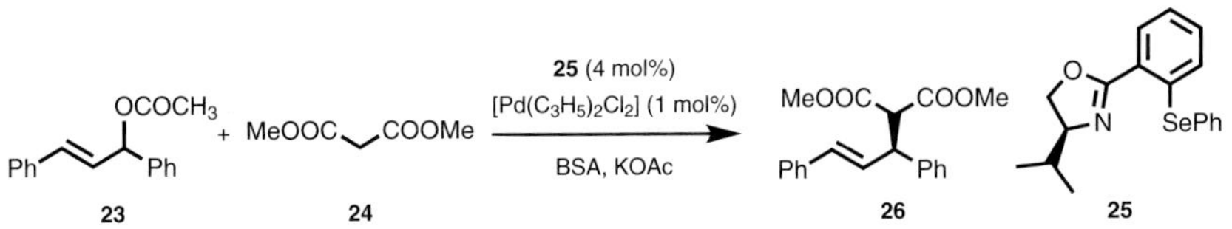

Scheme 8.10

After initial investigations of selenium-ligands in palladium-catalyzed allylic substitution reactions, different classes of chiral selenium ligands have been synthesized and used. Hiroi and coworkers synthesized (*S*)-proline-derived enantiopure selenium-based chiral ligands **27–30** and used them in the palladium-catalyzed asymmetric allylic alkylation shown in Scheme 8.10 leading to the product **26** with 36–86% enantiomeric excess; the values are shown in Figure 8.2 after the compound numbers [23].

The chiral ferrocenyl-oxazoline containing selenides **31a–31d** were synthesized as novel chiral ligands and successfully employed in palladium-catalyzed asymmetric allylic substitution reactions. Although the product **26** was obtained in poor to moderate yields, the enantioselectivities (up to 99% *ee*) were excellent as shown in Figure 8.3 [24].

The same research group developed the synthesis of chiral *N,Se*-ligands **32a–32c** based on [2.2]paracyclophane and have applied these ligands in palladium-catalyzed asymmetric allylic substitutions (Scheme 8.10). Product **26** was obtained in excellent yields with a maximum enantiomeric excess of 93% using **32c** as chiral ligand (Figure 8.4) [25]. Recently, some new chiral organoselenides, containing cinchona alkaloid moieties [26a] and *N*-trifluoracyl β-chalcogeno amides [26b], have been

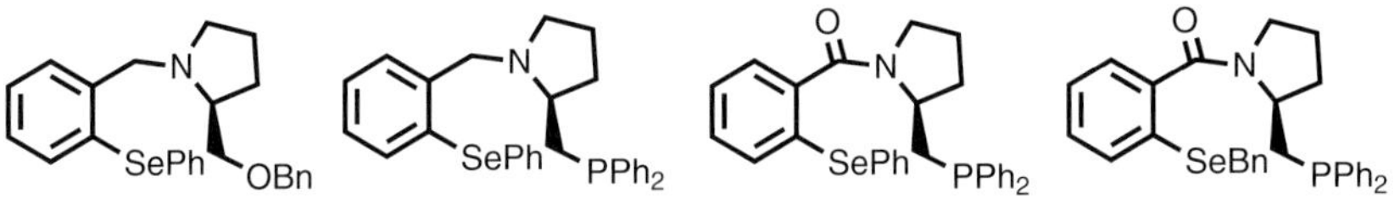

	27	28	29	30
26:	(36% *ee*)	(85% *ee*)	(74% *ee*)	(86% *ee*)

Figure 8.2 Selenides **27–30** employed as chiral ligands.

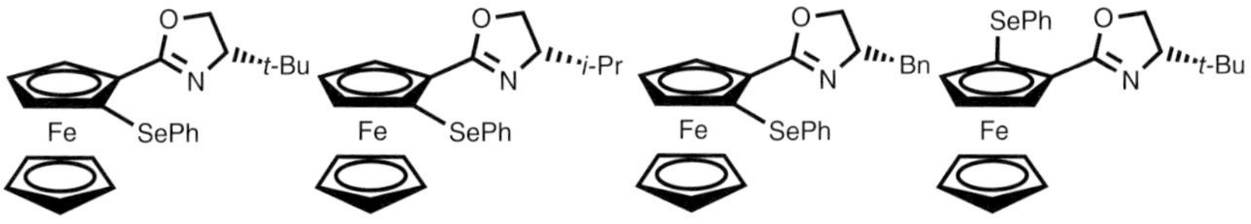

	31a	31b	31c	31d
26:	(99% *ee*)	(90% *ee*)	(92% *ee*)	(92% *ee*)

Figure 8.3 Chiral ferrocenyl-oxazoline containing selenides **31a–31d** for the asymmetric palladium-catalyzed allylic substitution.

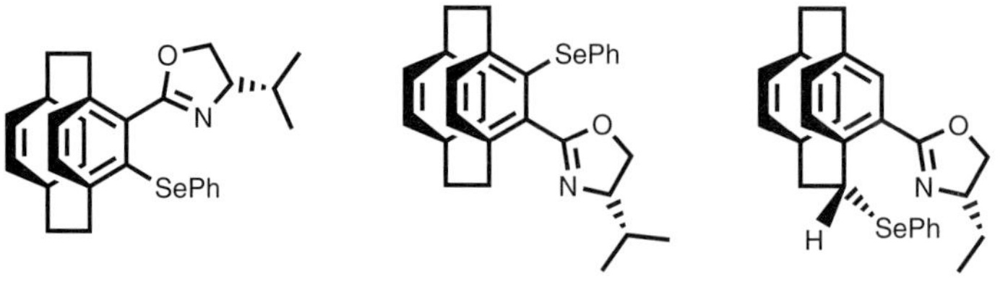

32a
26: (57% ee)

32b
(73% ee)

32c
(93% ee)

Figure 8.4 Chiral N,Se-ligands **32a–32c** based on [2.2] paracyclophane.

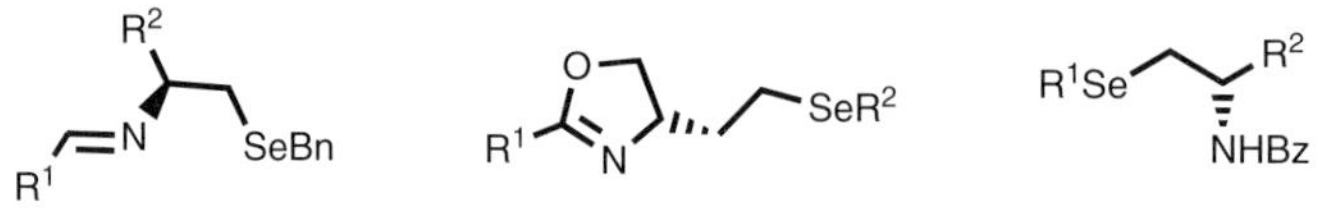

33
26: (up to 97% ee)

34
(up to 91% ee)

35
(up to 98% ee)

R^1 = aryl
R^2 = alkyl

R^1 = aryl, Me, t-Bu
R^2 = aryl, t-Bu

R^1 = alkyl
R^2 = aryl

Figure 8.5 Seleno imines **33**, oxazoline-based selenides **34** and β-seleno amides **35** as chiral ligands.

successfully employed in enantioselective carbon–carbon bond-forming alkylations [26].

Braga and coworkers have developed various classes of selenium-based chiral ligands such as seleno imine derivatives **33** [27], oxazoline-based selenides **34** [15], and β-seleno amides **35** (Figure 8.5) [14]. All chiral selenium ligands have been used in stereoselective palladium-catalyzed asymmetric allylations. The chiral β-seleno amides **35** were found to be the most efficient ligands in term of yields and selectivity.

Various chiral β-selenium-, sulfur-, and tellurium amides have been synthesized by ring-opening reactions of 2-oxazolines. Furthermore, all these compounds were employed as chiral ligands for the palladium-catalyzed asymmetric allylic alkylation of racemic 1,3-diphenyl-2-propenyl acetate with malonates. The alkylated product **26** was obtained with up to 98% ee using chiral β-selenium amides as ligand. It has also been demonstrated that selenium has higher ability to complex with palladium compared to other chalcogens and selenium analogues have shown slightly better performance than sulfur and tellurium analogues [28].

8.2.2.4 Selenium-Ligands in Palladium-Catalyzed Mizoroki–Heck Reactions

Transition metal-catalyzed C–C bond-forming reactions are important transformations in synthetic organic chemistry. Recently, some organometallic systems have been framed amending the stability and efficiency of the palladium-based catalysts. In this respect, diverse palladacycles and a series of electron-rich and

8.2 Selenium-Catalyzed Reactions

sterically hindered phosphane palladium complexes have been synthesized and applied as catalysts in C–C bond-forming reactions. In the Mizoroki–Heck reaction, weak σ-donating phosphane ligands are privileged.

Yao and his colleagues synthesized selenium-ligated Pd(II) complexes **36–38**, which were used as catalyst in Mizoroki–Heck reactions of various aryl bromides **39** including activated, deactivated, and heterocyclic ones with terminal alkenes **40**. The coupling products **41** were obtained in good yields (up to 97%) as shown in Scheme 8.11 [29]. The catalytic activity of the selenide-based Pd(II) complexes not only rivals, but vastly outperforms that of the corresponding phosphorus and sulfur analogues. Practical advantages of the selenium-based catalysts include their straightforward synthesis and high activity in the absence of any additives as well as the enhanced stability of the selenide ligands toward air oxidation.

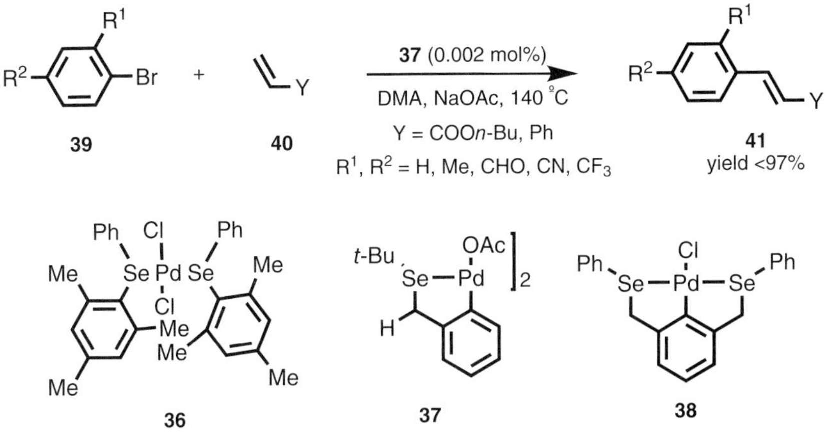

Scheme 8.11

Mingzhong and coworkers reported the synthesis of new silica-supported poly-γ-methylselenopropylsiloxane palladium complexes from γ-chloropropyltriethoxysilane via immobilization on silica, followed by reacting with sodium methylselenolate and then the reaction with palladium chloride. Furthermore, it was used as catalyst in the amidation and butoxycarbonylation of aryl halides under atmospheric pressure of carbon monoxide [30]. The same research group developed another kind of silica-supported poly-ω-(methylseleno)undecylsiloxane palladium(0) complex and employed as catalyst in similar Mizoroki–Heck reactions. The complexes were not only showing the high activity and stereoselectivity for arylation of conjugate alkenes **40** with aryl halides **39** (Scheme 8.12), but offer practical advantages such as ease of handling, separation from the reaction mixture, and reuse. The arylation of butyl acrylate and acrylamide with aryl iodides or bromides catalyzed by "Si-Se-Pd(0)"-complexes provides a better and practical procedure for the synthesis of *trans*-butyl cinnamates and *trans*-cinnamamides [31].

Scheme 8.12

Recently, palladium(II) complexes containing 4- or 6-[(phenylseleno)methyl]pyridine [32] and 2,6-bis[(phenylseleno)methyl]pyridine [33] selenium ligands have been synthesized and successfully used as catalysts in Mizoroki–Heck reactions of aryl halides with styrene or n-butyl acrylate. The coupling products were obtained in high yields.

8.2.2.5 Selenium-Ligands in Palladium-Catalyzed Phenylselenenylation of Organohalides

In recent years, some of the selenium-ligated palladium complexes have been developed as pincer-type complexes and received considerable attention for their use in catalysis. Szabó and Wallner explored the utility of pincer-type complex **38** as a catalyst in the substitution of a phenylselenyl group with various organohalides **43** such as propargyl-, allyl-, benzyl-, and benzoyl halides (Scheme 8.13) [34]. These reactions occur under mild reaction conditions using only 2 mol% of the catalyst and trimethylstannyl phenylselenide **42** as selenenylating agent. The selenenylated products **44** were obtained in good to excellent yields and reaction showed functional group tolerance with both electron-withdrawing and electron-donating functionalities (Scheme 8.13).

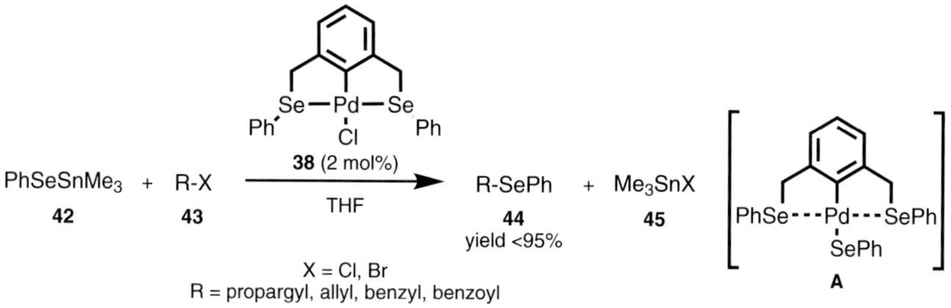

Scheme 8.13

The mechanistic studies indicate that the catalytic cycle is initiated by the formation of intermediate **A** through transmetallation of **42** with **38**, in which the phenylselenyl group is transferred to palladium, followed by an S_N2-type of displacement of the halide-leaving group in **43** by the lone-pair electrons of selenium, regenerating catalyst **38**.

8.2.2.6 Selenium-Ligands in Palladium-Catalyzed Substitution Reactions

In 2005, Szabo and coworkers investigated the palladium pincer-complex **38** as catalyst in substitution reactions of vinyl cyclopropanes **46**, vinyl aziridines **47**, and allyl acetates **48** with tetrahydroxydiboron **49** to produce allylboronic acids, which subsequently react with aqueous KHF_2 to achieve the corresponding potassium trifluoro(allyl)boronate derivates **50**. The potassium trifluoro(allyl)boronate derivatives **50** were obtained in good to excellent yields (up to 98%) (Scheme 8.14) [35]. The reaction is regio- and stereospecific, proceeds under mild reaction conditions, and tolerates various functionalities such as Br, COOEt, $ArSO_2(NH)$, OAc, and $SiRMe_2$.

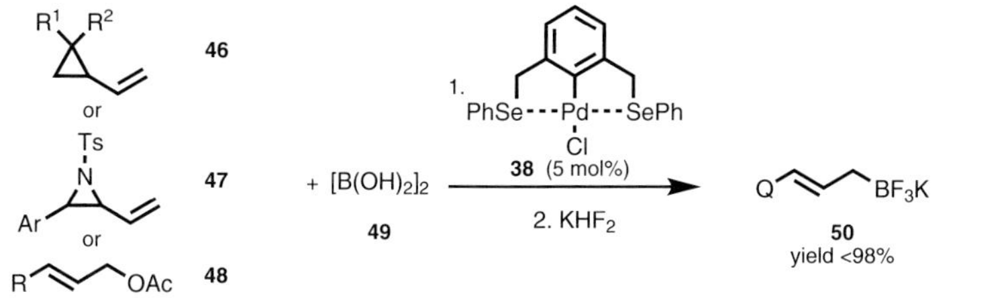

R^1 = COOEt, COOMe, SO_2Ph; R^2 = COOEt, COOMe, SO_2Ph
Ar = Ph, p-BrC_6H_5; R = Ar, $AcOCH_2$, H

Q = $CH_2CHR^1R^2$, CH(NHTs)Ar, R

Scheme 8.14

The same research group synthesized various methoxy-substituted pincer complexes in order to study the substituent effects on the catalytic activity in palladium-catalyzed opening of vinyl epoxides and boronation of cinnamyl alcohol. The results clearly indicate that a methoxy substitution at the *para*-position of the pincer complex leads to up to fourfold acceleration in the catalytic reactions, while substitution of the side-arms does not change the activity of the complex or leads to a slight deceleration of the catalytic processes [36].

8.2.2.7 Selenium-Ligands in the Palladium-Catalyzed Allylation of Aldehydes

Recently, similar pincer complexes have been proven to be important catalysts for the nucleophilic allylation of aldehydes. In 2006, Yao and coworkers achieved the catalyzed allylation of various aromatic and aliphatic aldehydes **51** with allyltributyltin **52** using pincer-complex **37**. The allylated products **53** were isolated in good to excellent yields (Scheme 8.15) [37].

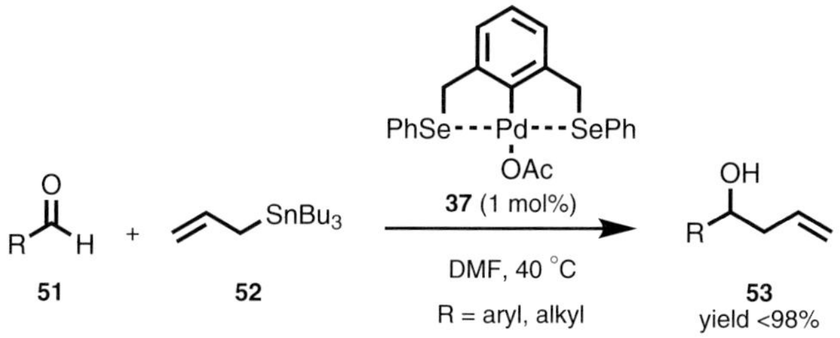

Scheme 8.15

The possible catalytic cycle is depicted in Scheme 8.16; the catalytic cycle is initiated by the reaction of **37** with **52** to generate the allyl-Pd(II) intermediate **A**. This allyl-Pd species then undergoes nucleophilic addition with aldehyde **51**, probably via a cyclic transition state with the metal concurrently activating the carbonyl group. The resulting Pd(II)-alkoxide intermediate **B** undergoes an exchange reaction with tributylstannyl acetate to liberate the homoallyl alcohol product in the form of stannyl ether **C** and to regenerate catalyst **37**. Additionally or alternatively, transmetallation may occur at the stage of intermediate **B** with **52**. Stannyl ether **C** is hydrolyzed to **53** during workup.

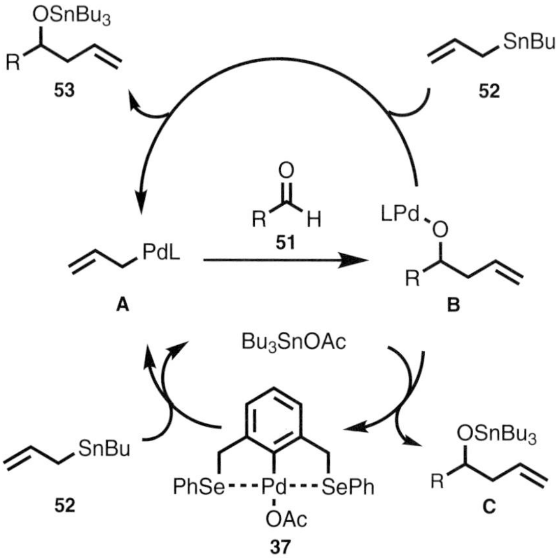

Scheme 8.16

8.2.2.8 Selenium-Ligands in Palladium-Catalyzed Condensation Reactions

Recently, the pincer complexes have been used to catalyze condensation reactions of sulfonimines with isocyanoacetates under mild reaction conditions. In 2007, Szabó and his coworkers have developed palladium-catalyzed condensation of sulfonimines **54** with isocyanoacetates **55** using various palladium pincer complexes, containing phosphorous, sulfur, selenium, and nitrogen atoms in the ligand system [38]. All the reactions proceed quickly at room temperature using only 1 mol% of pincer-complex **56** affording 2-imidazoline scaffolds **57** in excellent yields (Scheme 8.17). The stereoselectivity of the process is strongly dependent on the catalyst. Using electron-deficient and relatively bulky palladium pincer complexes such as **56a–c**, the major product is the *syn*-isomer **57**. However, the diastereoselectivity is reversed by employing the pincer complexes containing a selenium atom in the ligand system such as **56d** and **56e**.

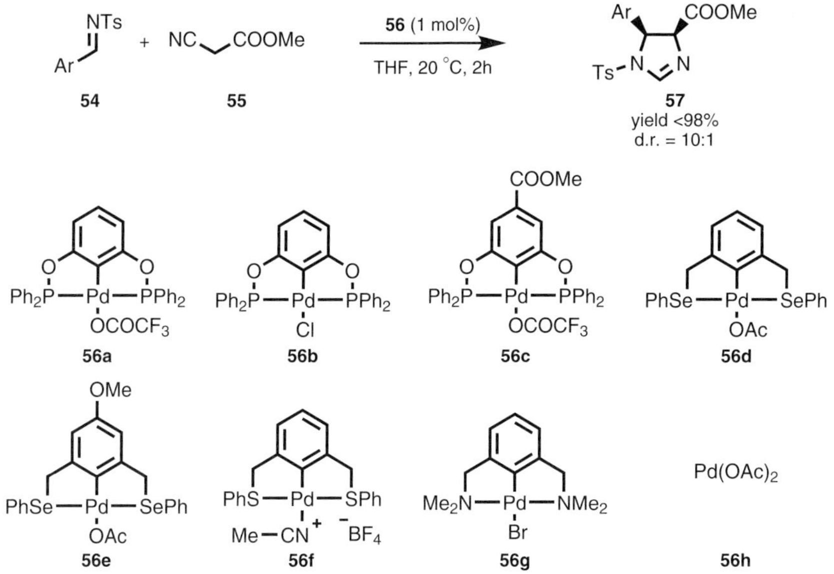

Scheme 8.17

8.2.2.9 Ruthenium-Catalyzed Substitution Reactions

In past decade, Nishibayashi, Uemura, and coworkers reported the synthesis of various chalcogenolate (S,Se,Te)-bridged diruthenium complexes and explored their catalytic activity in the propargylic substitution reactions of propargylic alcohol with carbon, nitrogen, and oxygen nucleophiles. The nucleophilic substitution of propargylic alcohol **58** was achieved with various nucleophiles **59** in high yields using 5 mol% of catalyst **60** and 10 mol% of NH_4BF_4 (Scheme 8.18) [39]. The results clearly indicate that only sulfur- and selenium-bridged complexes are effective catalysts while tellurium-bridged complexes show extremely poor catalytic behavior toward propargylic substitution reactions. The same research group also developed the catalytic propargylation of acetone by using 5 mol% of same selenium-ligated diruthenium complexes **60**. The substitution products were obtained good to excellent yields [40].

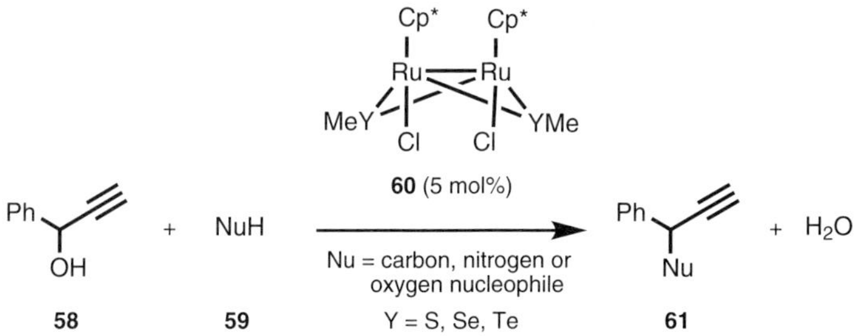

Scheme 8.18

8.2.2.10 Selenium-Ligands in Zinc-Catalyzed Intramolecular Hydroaminations

Recently, Roesky, Blechert, and coworkers have reported the synthesis of some selenenylated aminotroponiminate zinc complexes **63** along with the other functionalities (S and Te) and investigated as catalysts in the intramolecular hydroamination of nonactivated alkenes **62**. The cyclized products **64** were obtained up to 97% yield by using selenenylated aminotroponiminate zinc complex **63** as shown in Scheme 8.19 [41].

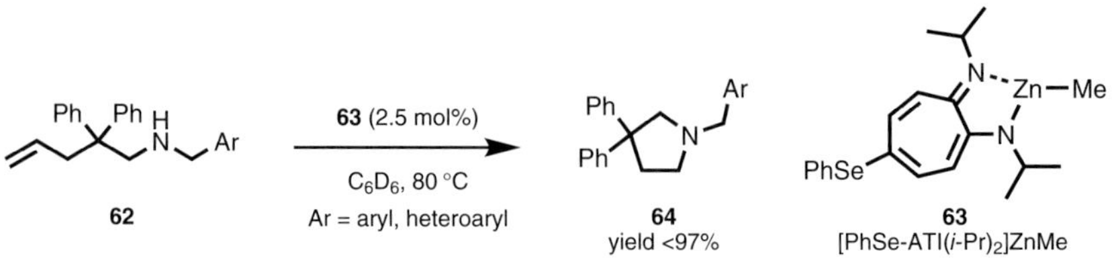

Scheme 8.19

8.2.3
Selenium-Ligands in Organocatalytic Asymmetric Aldol Reactions

In 2008, Braga and his coworkers synthesized cysteine and selenocysteine-derived prolinamides **66** and **67**, which were used as catalysts in enantioselective aldol reactions of various aromatic aldehydes **51** with acetone **65**. The aldol products **68** were obtained in moderate yields with up to 94% enantiomeric excess using cysteine-derived prolinamide as catalyst while selenocysteine-derived prolinamide **67** was found less effective and the products **68** were obtained with up to 85% enantiomeric excess (Scheme 8.20) [42].

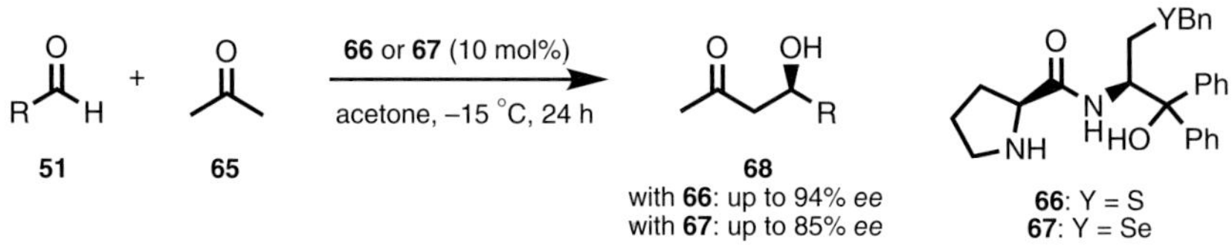

Scheme 8.20

8.2.4
Selenium-Ligands in Stereoselective Darzens Reactions

Recently, selenium-mediated stereoselective Darzens reactions have been reported. Watanabe and colleagues have developed an enantioselective synthesis of chiral epoxides **71** by a Darzens reaction of phenacyl bromide **69** with aldehydes **51** using a novel Lewis acid/Brønsted base catalyst **70** formed by the C_2 symmetric chiral selenide bearing isoborneol skeletons. The chiral epoxides **71** were obtained in good yields with up to 62% enantiomeric excess as shown in Scheme 8.21 [43]. Both aromatic and aliphatic aldehydes can be employed in this reaction, but the enantioselectivity was very low with aliphatic aldehydes.

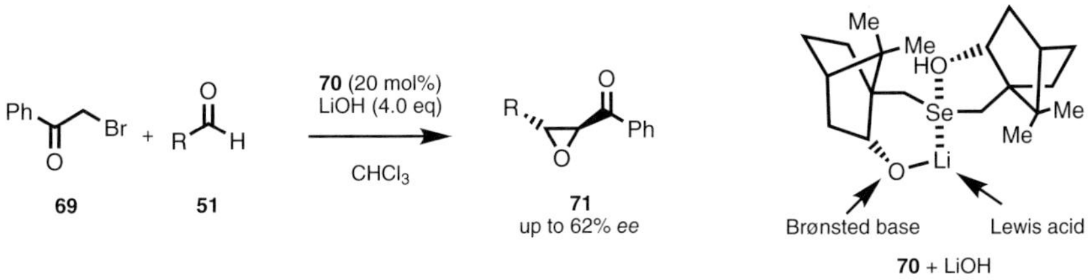

Scheme 8.21

8.2.5
Selenium-Catalyzed Carbonylation Reactions

Selenium-catalyzed carbonylation reactions have been intensively studied because of their cheap and easily available catalyst, mild reaction conditions, and their phase-transfer like properties. The study of these reactions was started in the 1970s by Sonoda and coworkers [7, 44]. After these studies, Lu and Liu reported the monoreduction of dinitro compounds **72** with catalytic amounts of elemental selenium, carbon monoxide, and water using sodium acetate as a base in DMF under atmospheric pressure. The reaction proceeded with good to excellent yields using *m*-dinitrobenzene and homologous substituted *m*-dinitrobenzenes **72** as substrates (Scheme 8.22) [45]. The catalyst and base can be recovered and reused with no loss of catalytic activity over at least three cycles. However, *o*- and *p*-dinitrobenzenes cannot be reduced under these conditions due to their electron deficiency.

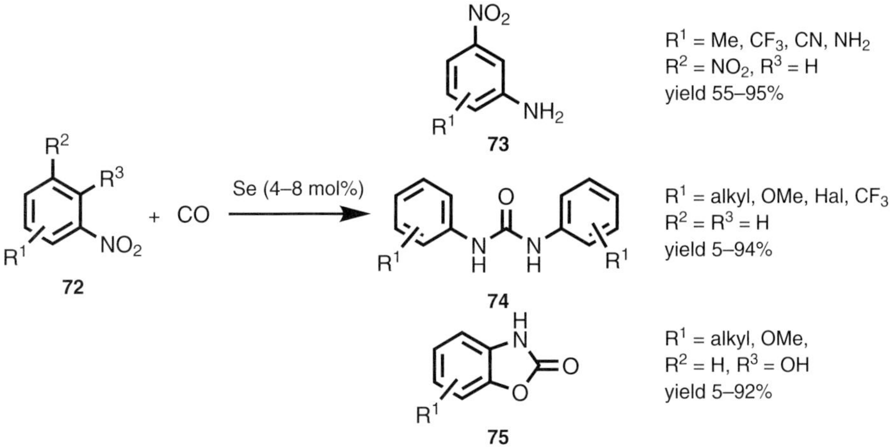

R^1 = Me, CF_3, CN, NH_2
R^2 = NO_2, R^3 = H
yield 55–95%

R^1 = alkyl, OMe, Hal, CF_3
R^2 = R^3 = H
yield 5–94%

R^1 = alkyl, OMe,
R^2 = H, R^3 = OH
yield 5–92%

Scheme 8.22

1,3-Diaryl ureas **74** can be obtained from nitroarenes **72** with KOH as base under similar reaction conditions [46]. In 2006, Lu and coresearchers additionally reported the same reaction under solvent-free conditions in high yields [47]. The selenium-catalyzed reductive carbonylation of 2-nitrophenols with 8 mol% of selenium in the presence of organic or inorganic bases depending on solvent produces 2-benzoxazolones **75** in good yields (Scheme 8.22) [48].

8.2.6
Selective Reduction of α,β-Unsaturated Carbonyl Compounds

The selective reduction of α,β-unsaturated carbonyl compounds **76** was achieved with catalytic amounts of selenium (20 mol%) with carbon monoxide and water in dimethylformamide as shown in Scheme 8.23. The reducing agent hydrogen selenide is formed *in situ* and reduces unsaturated carbonyl compounds **76** to corresponding saturated ketones **77** in high yields (76–99%) [49]. However, aromatic aldehydes can be reduced and selenated to symmetrical diselenides. The authors found implications that isolated carbon–carbon double bonds are unreactive under these conditions.

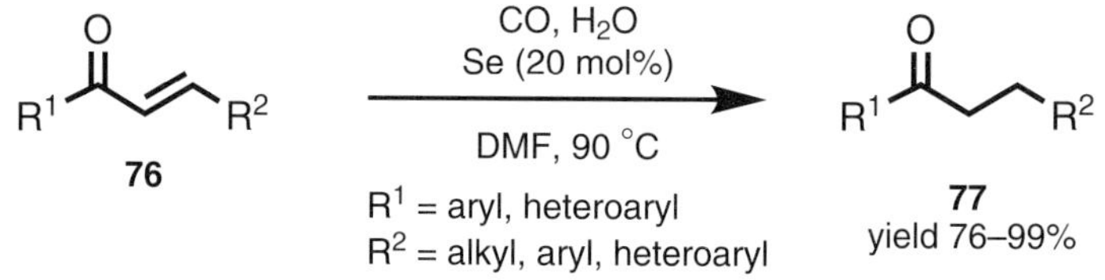

Scheme 8.23

8.2.7
Selenium-Catalyzed Halogenations and Halocyclizations

In 1979, Sharpless and Hori investigated the synthetic utility of arylselenenyl chlorides and aryl diselenides as catalysts in the nonradical chlorination of olefins with N-chlorosuccinimide (NCS) [50]. Later, Tunge and coworkers reported selenocatalytic allylic chlorination of allylic acids, esters, arenes, and nitriles **78** by using a similar approach [51]. The reaction products **79** were obtained with high selectivity since the electron-withdrawing groups push the reaction toward the formation of the rearranged allylic chlorides as shown in Scheme 8.24 [51].

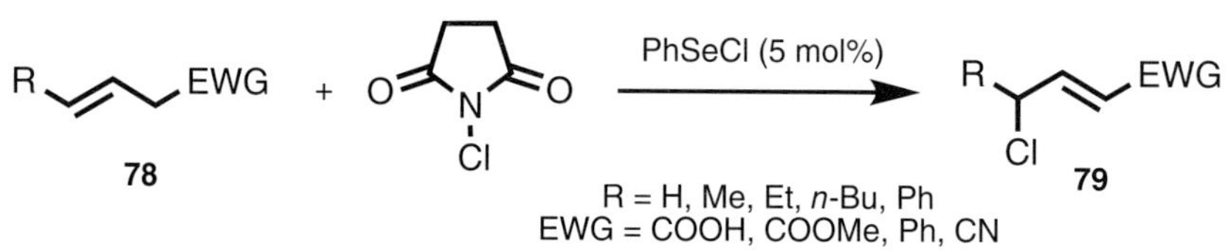

Scheme 8.24

The allylic chlorination of polyprenoids was developed by a comparable approach using NCS and a catalytic amount of a polymer-supported selenenyl bromide [52]. The method showed high regioselectivity toward the functionalization of the terminal double bond and is also compatible with the presence of different functional groups in the starting material.

Phenylselenenyl chloride has been used as an effective catalyst for the α-halogenation of ketones in the presence of NCS as halogen source. The mechanistic studies suggest that the reaction involves an oxidative addition of NCS to selenium which activates the chlorine atom for the nucleophilic attack by the enol or enolate.

Recently, Tunge and Mellegaard reported a diphenyl diselenide-catalyzed halolactonization of unsaturated acids **80** mediated by N-bromosuccinimide (NBS). The halogenated lactones **81** were obtained in good yields [53]. On the basis of experimental studies, a possible catalytic cycle was proposed (Scheme 8.25). According to this catalytic cycle, NBS undergoes nucleophilic attack by the diselenide to produce the cationic selenium complex **A** with a succinimide counterion. The resulting succinimide anion is expected to rapidly deprotonate the carboxylic acid to form the carboxylate. Furthermore, the formation of lactone **81** proceeds through the nucleophilic displacement of a selenium-coordinated bromonium ion **B** [53].

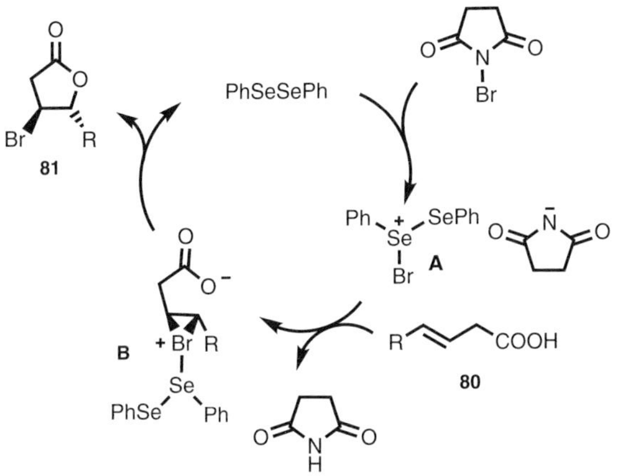

Scheme 8.25

8.2.8
Selenium-Catalyzed Staudinger–Vilarrasa Reaction

Vilarrasa and coworkers developed a selenium-catalyzed Staudinger–Vilarrasa reaction for the direct ligation of carboxylic acids **82** and azides **83**. These reactions were achieved under mild reaction conditions using 20 mol% of dipyridyl diselenide and trimethyl phosphine in toluene at room temperature. The substituted amides **84** were obtained in excellent yields (Scheme 8.26) [54].

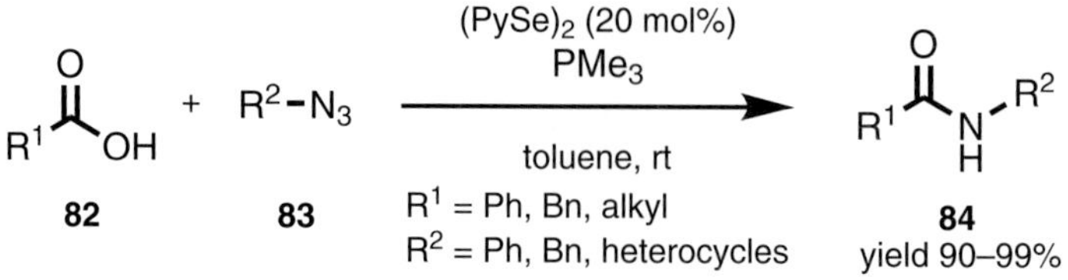

Scheme 8.26

R^1-C(O)-OH (82) + R^2-N_3 (83) →[(PySe)$_2$ (20 mol%), PMe$_3$, toluene, rt] R^1-C(O)-NH-R^2 (84)

R^1 = Ph, Bn, alkyl
R^2 = Ph, Bn, heterocycles
yield 90–99%

On the basis of NMR studies of reaction intermediates, the possible catalytic cycle for the Staudinger–Vilarrasa reaction of a carboxylic acid with azides in the presence of trimethyl phosphine and dipyridyl diselenide has been proposed as shown in Scheme 8.27.

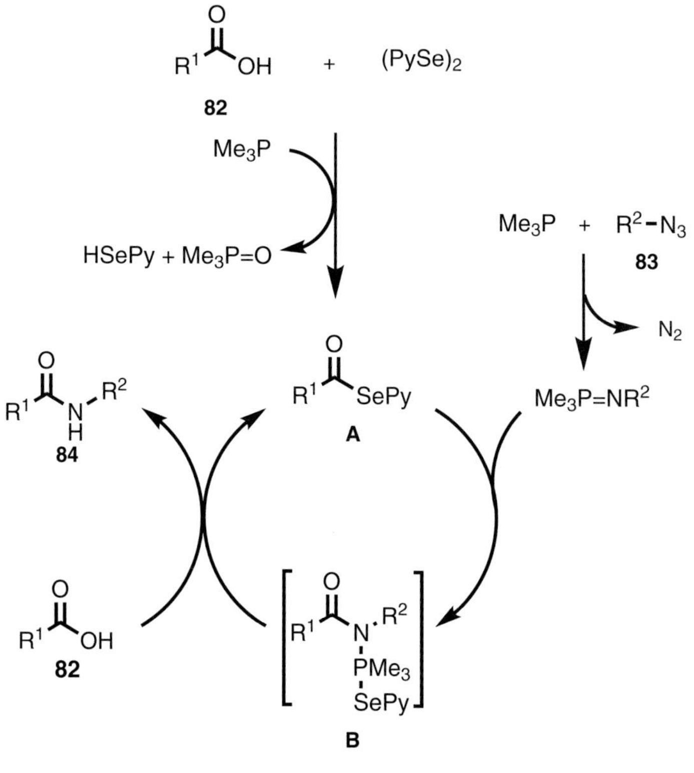

Scheme 8.27

8.2.9
Selenium-Catalyzed Elimination Reactions of Diols

The fluorous selenide mediated elimination reactions of vicinal diols have been investigated by Crich and his research group [55]. They have achieved the conversion of vicinal dimesylates **85** to the corresponding alkenes **86** using sodium borohydride as stoichiometric reagent and a catalytic quantity of bis(4-perfluorohexylphenyl) diselenide in excellent yields (74–99%) in ethanol at reflux temperature (Scheme 8.28). The diselenide is recovered in up to 88% by continuous fluorous extraction.

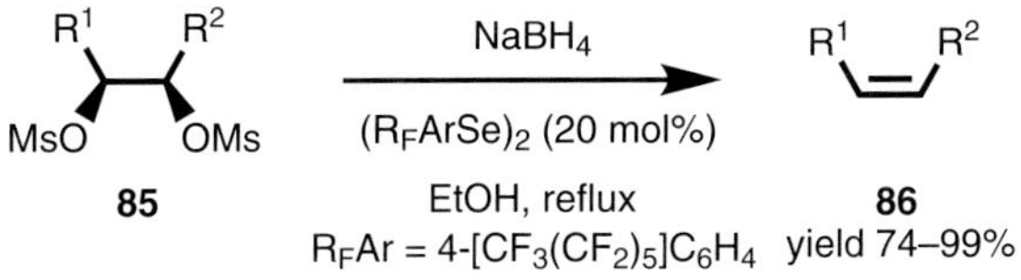

Scheme 8.28

The reaction starts with a S_N2 displacement of one mesyl group by the areneselenide anion, followed by formation of the cyclic episelenenium ion **B**, and final selenophilic attack by the second equivalent of selenide anion to liberate the product and regenerate the catalyst (Scheme 8.29) [55].

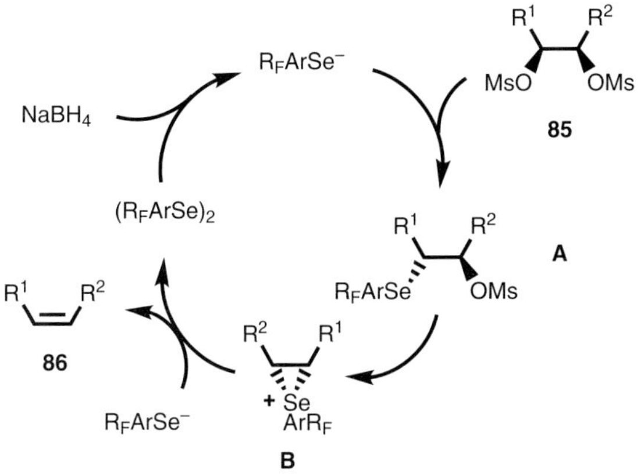

Scheme 8.29

8.2.10
Selenium-Catalyzed Hydrostannylation of Alkenes

Schiesser and collogues demonstrated the efficient homolytic hydrostannylation of various electron-donating alkenes **87** with tin hydrides **88** by using catalytic amount of diphenyl diselenide and 2,2′-azobis[2-*iso*-butyronitrile] (AIBN). The hydrostannylated products **89** were isolated in 78–96% yield (Scheme 8.30) [56]. This reaction presumably benefits from the increased rate of H-atom transfer of the *in situ* generated polarity-reversed catalyst benzeneselenol.

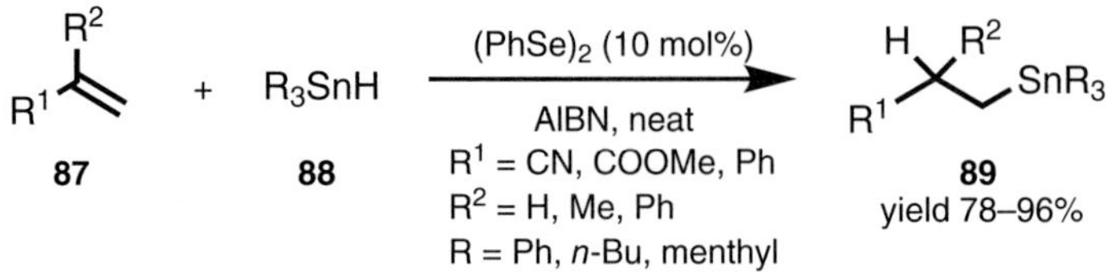

Scheme 8.30

8.2.11
Selenium-Catalyzed Radical Chain Reactions

In later 1990s, Crich and his colleagues reported the synthesis of a minimally fluorous (52% F) diaryl diselenide. On reduction *in situ* with tributylstannane, this diselenide provides a fluorous selenol which is effective in inhibiting a range of stannane-mediated radical rearrangements, including a cyclopropylcarbinyl ring opening [57].

In 2000, a novel design was reported by Pandey and coworkers to initiate radical-based chemistry in a catalytic fashion. The design of the concept was based on the phenylselenyl group transfer reaction from alkyl phenyl selenides **90** by utilizing PhSeSiR$_3$ as a catalytic reagent. The reaction products **91** were obtained in good yields (Scheme 8.31) [58].

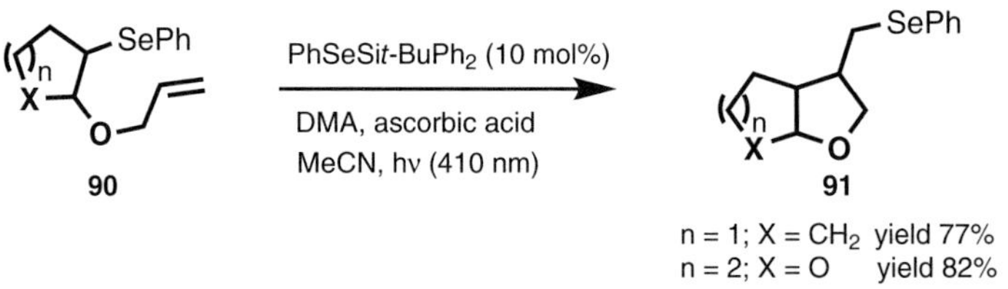

Scheme 8.31

Recently, Clive and coworkers have developed a selenium-catalyzed radical cyclization of iodo *O*-trityl oximes **92** in the presence of diphenyl diselenide (20 mol%) and Hünig's base using THF as a solvent and 1,1′-azobis(cyclohexanecarbonitrile) (ABC) as initiator. Five- and six-membered carbocycles **93** were obtained up to 96% yield as shown in Scheme 8.32 [59].

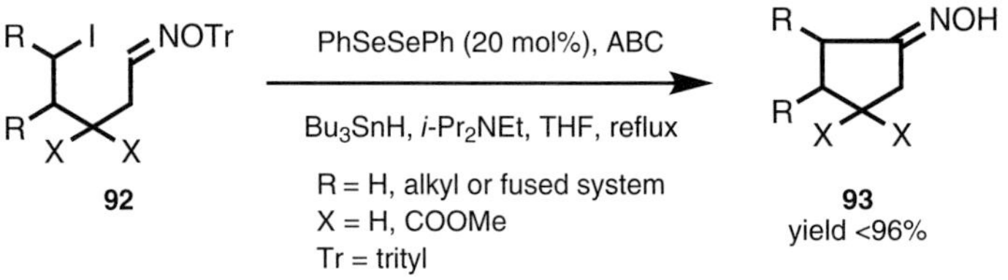

Scheme 8.32

8.2 Selenium-Catalyzed Reactions | 341

Crich and coworkers investigated the benzeneselenol-catalyzed radical addition of *ortho*-functionalized aryl iodides **94** to benzene in the presence of AIBN and tributyltin hydride. The addition products, aryl cyclohexadienes **95**, were isolated in 41–54% yield (Scheme 8.33) [60]. The catalyst benzeneselenol was prepared *in situ* by the reduction in diphenyl diselenide using tributyltin hydride. Despite the poor yields, aryl cyclohexadienes **95** are important synthetic intermediates and can be further cyclized into tetrahydrodibenzofurans and tetrahydrocarbazoles **96** in good yields using phenylselenenyl bromide (Scheme 8.33).

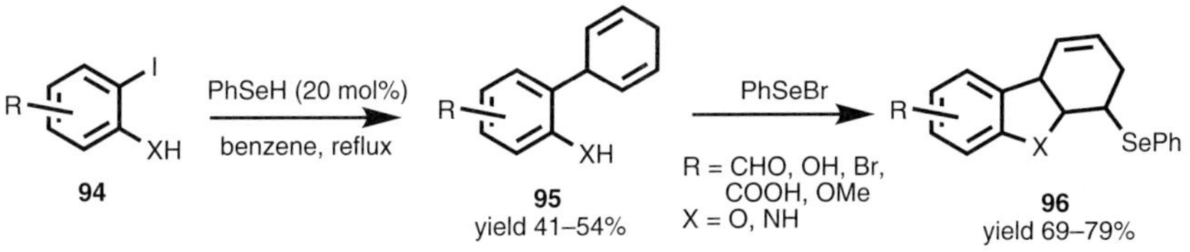

Scheme 8.33

The same research group reported the total synthesis of natural product carbazomycin B **102** by using a benzeneselenol-catalyzed intramolecular radical addition to benzene, followed by selenocyclization, deselenation, rearomatization, and saponification. The synthesis of carbazomycin B **102** started from 2-methoxy-3,4-dimethyl-5-nitrophenol **97**, which was further converted into compound **98** by using iodination, followed by reduction and reaction with methyl chloroformate (Scheme 8.34) [61].

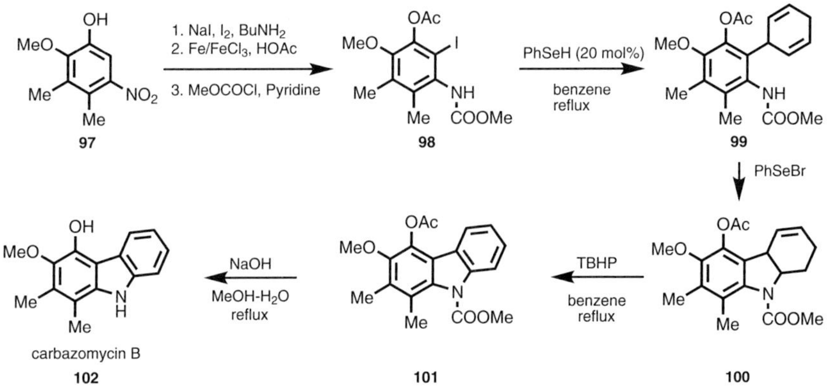

Scheme 8.34

Crich and coworkers also developed the dearomatizing radical arylation of five-membered heterocycles **104** by a benzeneselenol-catalyzed radical addition of functionalized aryl iodides **103** to furan and thiophene using 2,2′-azo(4-methoxy-2,4-dimethyl-4-methoxyvaleronitrile) (V-70) as radical starter. The adduct radicals are trapped by the selenol to give the 2-aryl-dihydro heterocyclic products **105** and **106** in up to 45% and 35% yields, respectively (Scheme 8.35) [62].

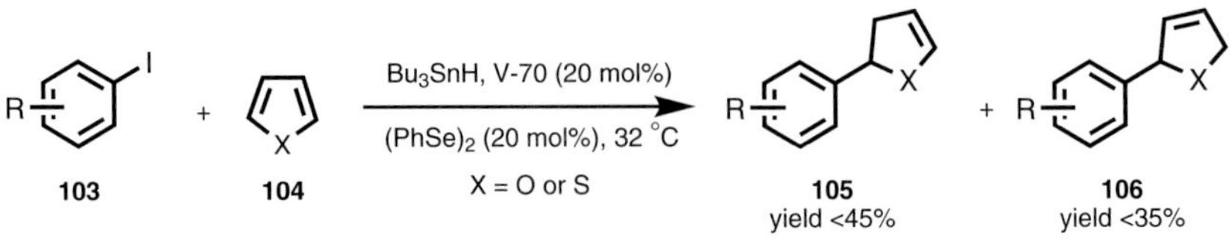

Scheme 8.35

When the iodide is an *o*-iodophenol, a cyclization follows the radical addition and provides bridged bicyclic acetals (Scheme 8.36). Unlike furan and thiophene, pyrrole provides complex reaction mixtures from which fully aromatized 2-aryl pyrroles could only be obtained in low yield [62].

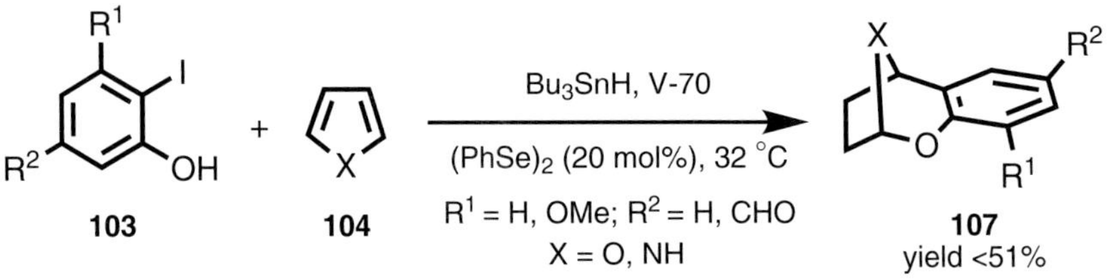

Scheme 8.36

8.2.12
Selenium-Catalyzed Oxidation Reactions

Organic selenium reagents have been recognized as powerful oxidants [63] since long time, but they have received the main attention after their use as catalyst with Sharpless oxidant, *tert*-butyl hydroperoxide [64]. Taylor and Flood have been used polystyrene-bound phenylseleninic acid as catalyst efficiently for the oxidation of olefins, ketones, and alcohols in good to excellent yields [65]. After this discovery, several selenium compounds have been explored as catalyst for the oxidation of various organic functionalities such as olefins, alcohols, alkynes, amines, and carbonyl compounds.

8.2.12.1 Selenium-Catalyzed Epoxidation of Alkenes
The first selenium-catalyzed epoxidation of olefins was reported by Sharpless and his coworkers during late 1970s [66, 67]. In 1999, Knochel and coworkers prepared 2,4-bisperfluorooctylphenyl butylselenide as a catalyst for the epoxidation of various alkenes **108** with hydrogen peroxide in a fluorous biphasic system. The

epoxidation products **109** were isolated in good to excellent yields (Scheme 8.37) [68]. The catalyst was soluble in perfluorinated solvents and can easily be recovered simply by phase separation. Furthermore, the catalyst can be reused several times without any influence on yield and reaction time. Later on, Detty and coworkers used aryl benzyl selenoxides as catalysts for the epoxidation of various cyclic and acyclic olefinic substrates [69].

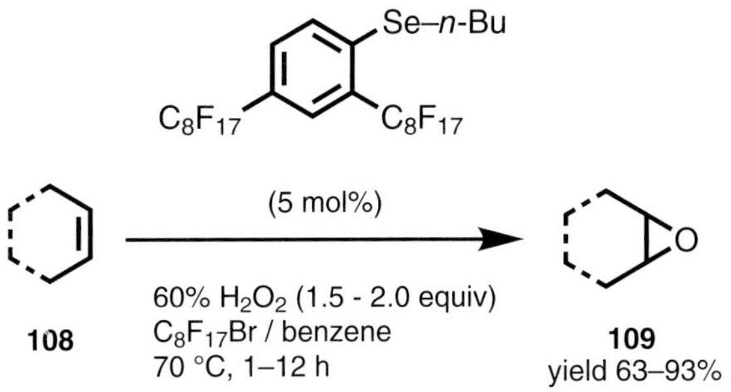

Scheme 8.37

Recently, Arends and coworkers reported the bis[3,5-bis(trifluoromethyl)-diphenyl] diselenide catalyzed epoxidation of olefins with hydrogen peroxide as an oxidant in good yields using glycerol-based solvents as green reaction media [70]. In the catalytic cycle, the diselenide is oxidized by hydrogen peroxide to give the corresponding arylseleninic acid which is the actual catalytic species for epoxidation (Scheme 8.38) [71, 72].

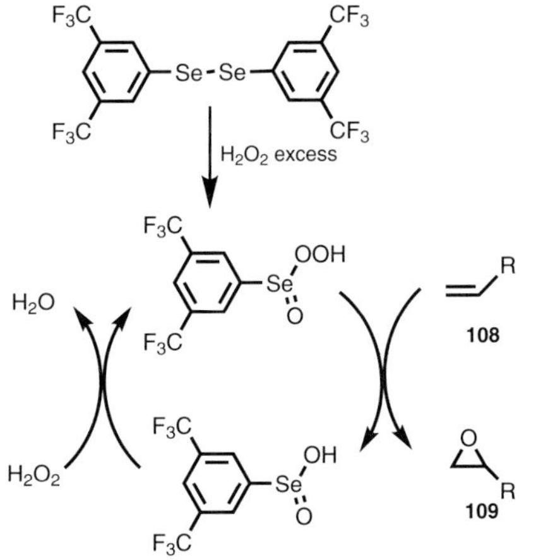

Scheme 8.38

8.2.12.2 Selenium-Catalyzed Dihydroxylation of Alkenes

For the first time, selenium-catalyzed oxidations of alkenes into *trans*-diols were observed by Knochel and coworkers in 1999 [68]. In 2007, Gogoi and Konwar developed a new transition metal-free approach for the *trans*-dihydroxylation of alkenes **110** catalyzed by selenium dioxide using hydrogen peroxide as oxidant. The *trans*-diols **111** were isolated as reaction products in 50–88% yield (Scheme 8.39) [73]. The oxidation does not affect other functionalities (alcohol, ester, alkyl, ether, halogens) in the substrate. However, aliphatic olefins exhibited better results in comparison to aromatic olefins and sterically hindered double bonds that showed poor yields compared to less hindered ones.

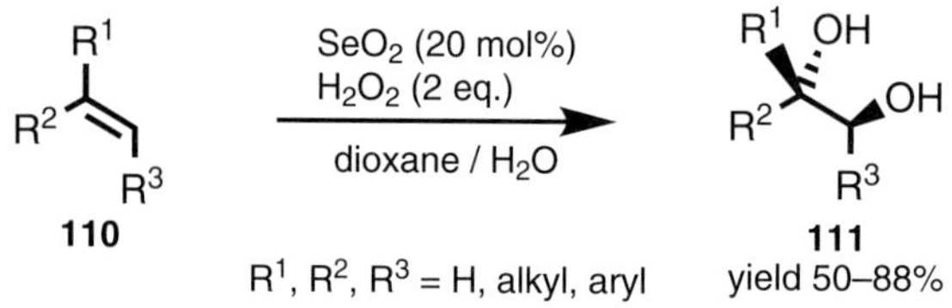

Scheme 8.39

Santi and coworkers developed the first general and efficient dihydroxylation of alkenes **110** in good yields catalyzed by diphenyl diselenide using hydrogen peroxide as an oxidant (Scheme 8.40). The chiral diols **111** were obtained with high selectivity but the reactions were found to be extremely slow [74].

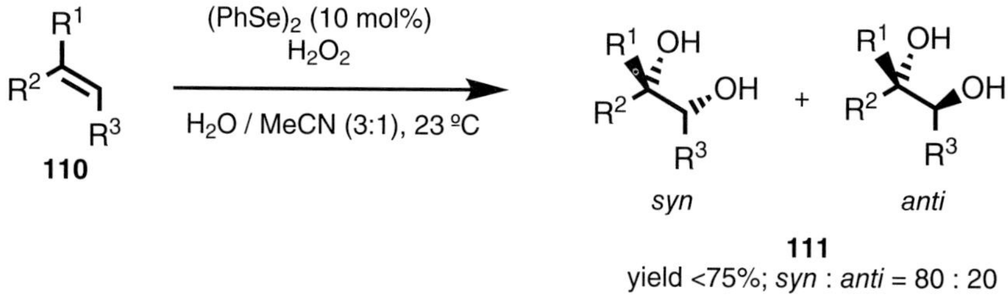

Scheme 8.40

The syn/anti-selectivities could be explained on the basis of the mechanism proposed in Scheme 8.41. The reaction is initiated by a peroxy acid catalyzed epoxidation [71, 72]. Furthermore, the intermediate **A** can be rapidly converted into the corresponding diol either through a S_N2 mechanism (path A, Scheme 8.41), or through a carbocation **B** (path B), depending on the electronic and steric properties of the starting alkene.

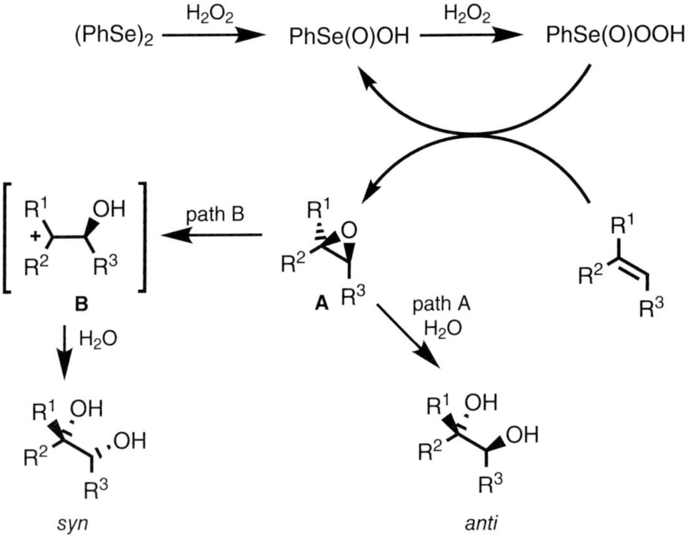

8.2 Selenium-Catalyzed Reactions | 345

Scheme 8.41

Recently, Bois and Brodsky reported the hydroxylation of unactivated tertiary C–H bond of adamantane by a two-stage catalytic process using 20 mol% of benzoxathiazine as oxygen-transfer catalyst, 1 mol% of diaryl diselenide, and four equivalents of urea-hydrogen peroxide (UHP). The hydroxylated reaction product, 1-adamantanol **112**, was isolated in 80% yield (Scheme 8.42). After the success of this reaction, hydroxylations of unactivated tertiary C–H bonds of various scaffolds have been performed [75].

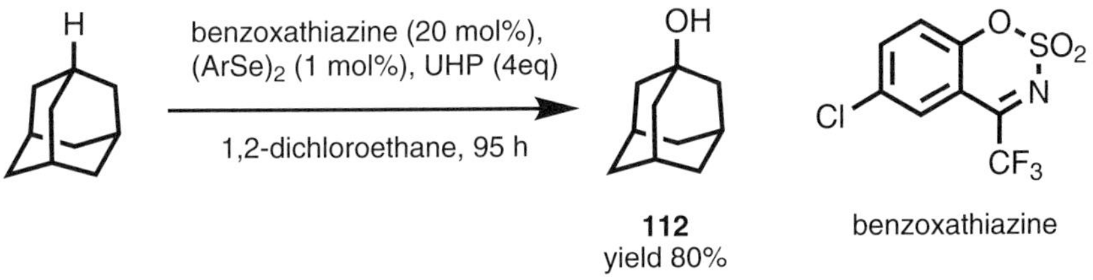

Scheme 8.42

The selenium-catalyzed synthesis of iodohydrins has been achieved by the oxidation of cyclic and acyclic olefins using *N*-iodosuccinimide (NIS) in a water–acetonitrile solvent mixture in good to moderate yields with up to 95% *trans*-selectivity (Scheme 8.43) [76].

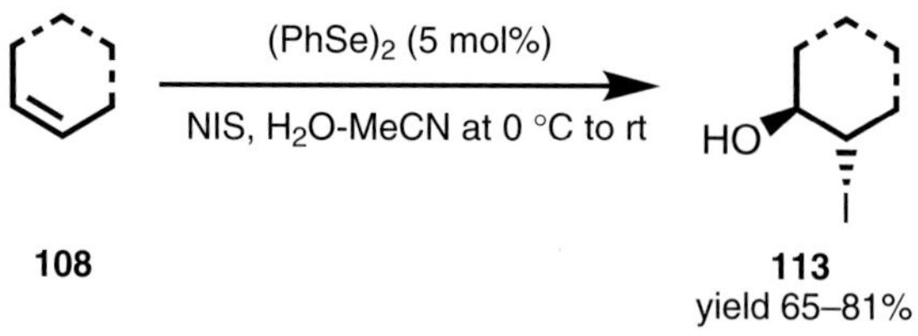

Scheme 8.43

8.2.12.3 Selenium-Catalyzed Oxidation of Alcohols

The oxidations of alcohols are highly relevant reactions in synthetic organic chemistry [77]. Although oxidation of alcohols with the stoichiometric amount of selenium-cored oxidants are known since late 1970s [63], but the first selenium-catalyzed oxidation of alcohols were reported in 1996 by Onami and coworkers [78]. They have achieved the oxidation of alcohols **114** into corresponding carbonyl compounds **115** catalyzed by dimethyl 2,2′-diselanediyldibenzoate using N-chloro-4-chlorobenzenesulfonamide sodium as oxidant. The same research group prepared bis[2-(2-pyridyl)phenyl] diselenide as catalyst in oxidations of alcohols with the same sodium salt as oxidant (Scheme 8.44) [79].

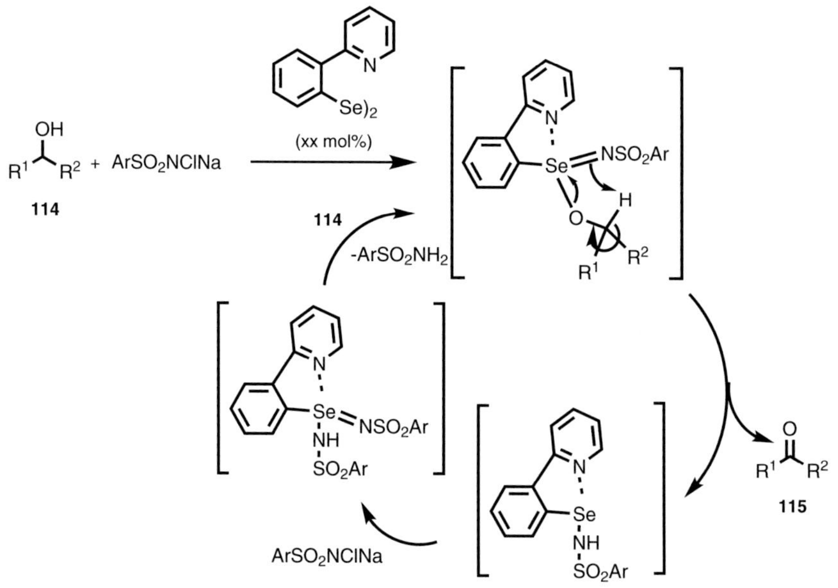

Scheme 8.44

Arends and his coworkers reported the diphenyl diselenide catalyzed oxidation of alcohols using *tert*-butyl hydroperoxide as terminal oxidant (Scheme 8.45) [80].

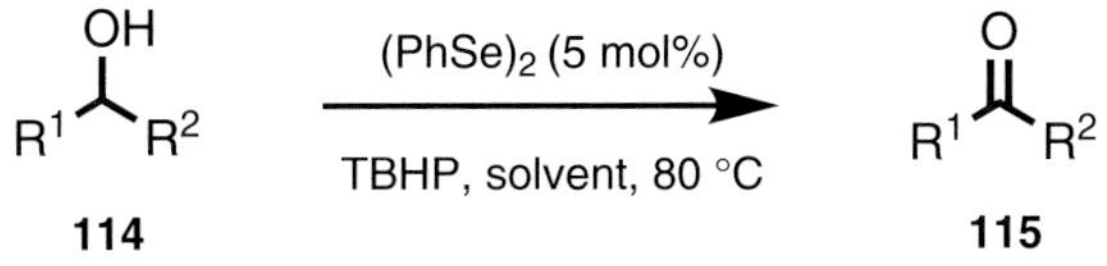

Scheme 8.45

$$R^1R^2C(OH)H \xrightarrow[\text{TBHP, solvent, 80 °C}]{(PhSe)_2 \ (5 \ \text{mol\%})} R^1C(O)R^2$$

114 → **115**

Diphenyl diselenide is a precatalyst in this reaction and the first step of the reaction sequence is the formation of an intermediate benzeneseleninic acid anhydride by the oxidation of *tert*-butyl hydroperoxide (TBHP). The benzeneseleninic acid anhydride initiates the catalytic cycle and forms intermediate **A**, which reacts to the carbonyl compound and to benzeneseleninic acid [63]. Finally, benzeneseleninic acid is oxidized by another molecule of TBHP to continue the catalytic cycle (Scheme 8.46).

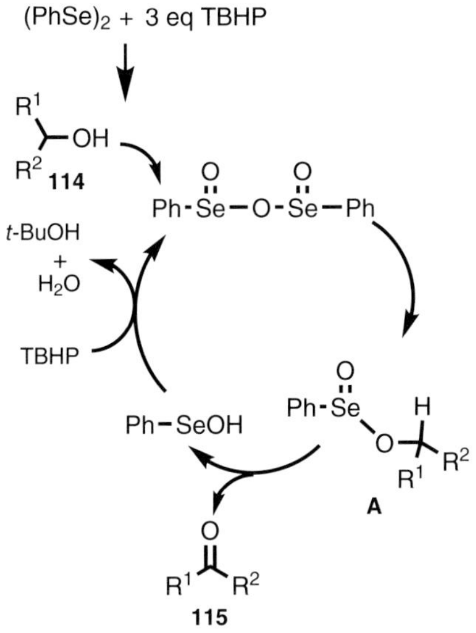

Scheme 8.46

Recently, Singh *et al.* have developed piano-stool type complexes of ruthenium with arenes and N-[2-(arylchalcogeno)ethyl]morpholines, which have been used as catalyst for the oxidation of primary and secondary alcohols with different oxidants such as N-methylmorpholine N-oxide, *tert*-butyl hydroperoxide, and sodium periodate. These complexes have been shown tremendous catalytic activity because only 0.001 mol% of them was sufficient to catalyze the reaction [81].

8.2.12.4 Baeyer–Villiger Oxidation

Sheldon and colleagues developed the catalytic utility of organoselenium compounds in Baeyer–Villiger oxidations [82]. In 2001, they reported the bis[3,5-bis(trifluoromethyl)phenyl] diselenide catalyzed Baeyer–Villiger oxidation of various carbonyl compounds **115** with 60% hydrogen peroxide in excellent yields as shown in Scheme 8.47.

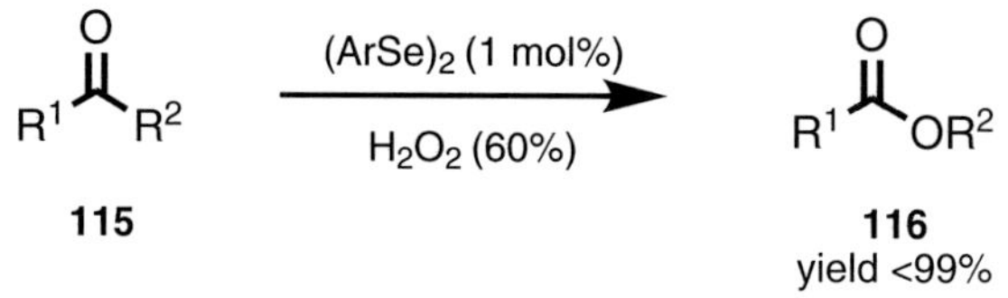

Scheme 8.47

The catalytic cycle is initiated by the *in situ* formation of peroxyseleninic acid from the corresponding diselenide (Scheme 8.48) [82].

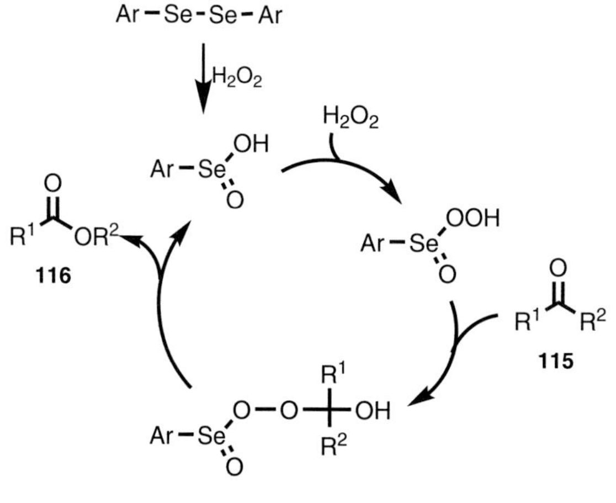

Scheme 8.48

Uemura and coworkers reported an example of an asymmetric Baeyer–Villiger reaction in 2002 [83]. Diselenides **117** and **118** bearing a chiral oxazoline moiety in the *ortho*-position (Figure 8.6) have been synthesized and tested in this reaction, but unfortunately only low enantioselectivities (up to 19% *ee*) have been observed.

Later, Ichikawa, and coworkers explored also diselenide catalyzed Baeyer–Villiger oxidations with various cyclohexanones using 30% hydrogen peroxide. Lactones were isolated in good to moderate yields [84]. Detty and coworkers have used

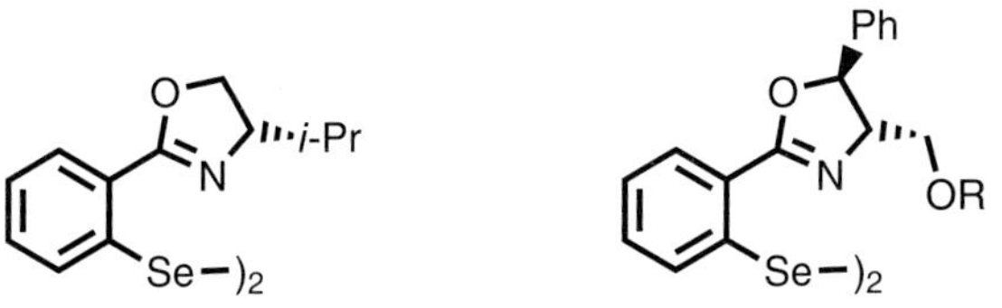

Figure 8.6 Chiral diselenides **117** and **118** bearing an oxazoline moiety in the *ortho*-position.

various selenoxides as catalysts in Baeyer–Villiger reactions using hydrogen peroxide as stoichiometric oxidant [69]. In the beginning of 2000, oxidation of aldehydes into corresponding carboxylic acids has been achieved using stoichiometric amounts of *tert*-butyl hydroperoxide in the presence of 5 mol% ebselen [85].

8.2.12.5 Selenium-Catalyzed Allylic Oxidation of Alkenes

In 2004, Crich and Zou prepared highly electron-deficient new fluorous seleninic acid and used as catalyst for allylic oxidation reactions [86]. They have achieved the allylic oxidation of alkenes **119** to the corresponding enones **120** with 10 mol% fluorous seleninic acid and three equivalents iodoxybenzene in trifluoromethylbenzene. The products were obtained in excellent yields (Scheme 8.49) [86]. The fluorous seleninic acid can be synthesized in two steps from perfluorooctyl iodide and dibutyl diselenides to afford perfluorooctylbutyl selenide, which oxidizes after treatment with hydrogen peroxide to perfluorooctylseleninic acid.

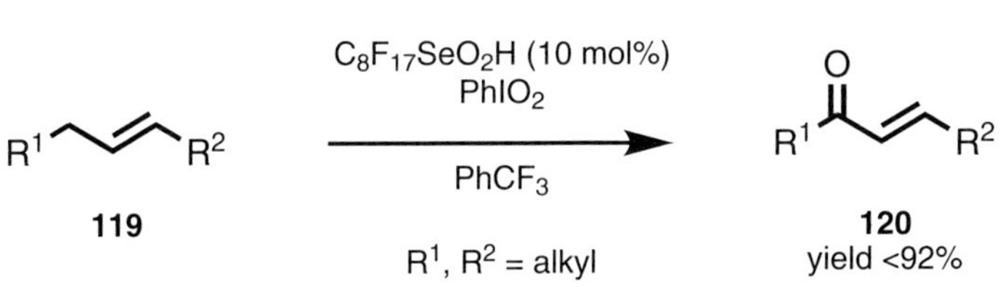

Scheme 8.49

The reaction is possibly initiated with formation of an intermediate **A** via an ene reaction followed by a 2,3-sigmatropic rearrangement to form selenoxide intermediate **B**. Intermediate **B** then undergoes oxidation with iodoxybenzene to yield the enone and seleninic acid which re-enters the catalytic cycle (Scheme 8.50).

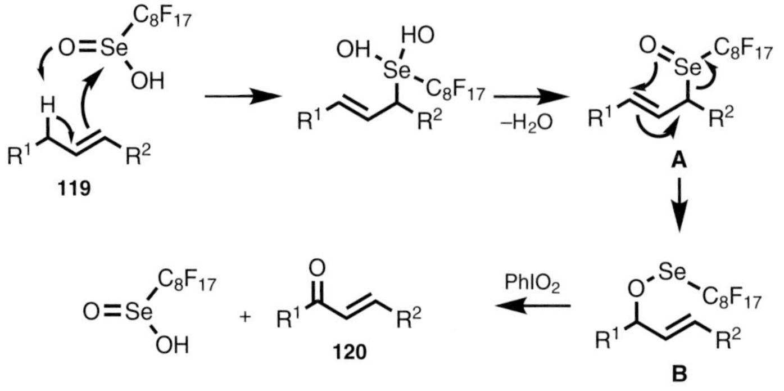

Scheme 8.50

8.2.12.6 Selenium-Catalyzed Oxidation of Aryl Alkyl Ketones

Later on, the same research group reported an efficient oxidation of aryl alkyl ketones **121** into the corresponding ketoacids **122** catalyzed by fluorous seleninic acid using iodoxybenzene as oxidant (Scheme 8.51). Benzylic methylene groups have been also oxidized by using catalyst–oxidant combination to the corresponding ketones. Fluorous diselenide can be used in these reactions as catalyst precursors [87].

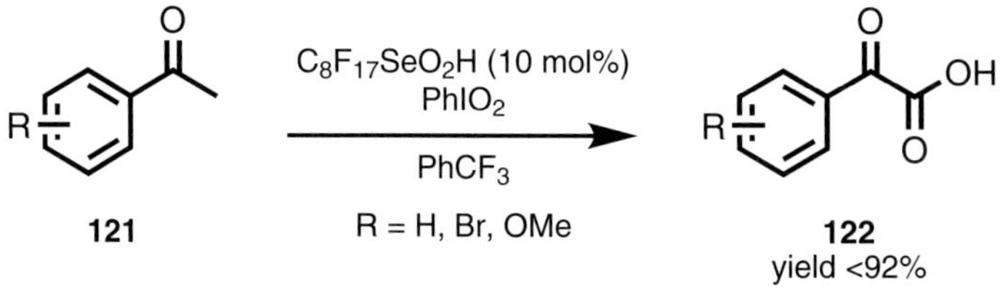

Scheme 8.51

8.2.12.7 Selenium-Catalyzed Oxidation of Primary Aromatic Amines

In 2007, Bäckvall and coworkers developed the selective oxidation of aniline **123** into corresponding nitroso compound **124** using 10 mol% of diphenyl diselenide with hydrogen peroxide as oxidant in high yields. Furthermore, the generated nitroso compounds were subsequently transferred into variety of oxazines **125** in good yields using hetero-Diels–Alder reactions (Scheme 8.52) [88].

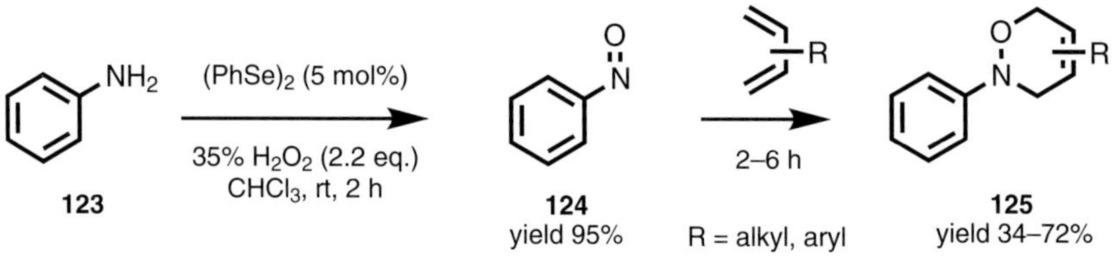

Scheme 8.52

Rück-Braun and her colleagues also developed the selenium-catalyzed oxidation of various anilines **123** into azoxyarenes **126** using 10 mol% of selenium dioxide with three equivalents of 30% hydrogen peroxide in methanol at room temperature in high yields (Scheme 8.53). 4-Alkoxy-*N*-(4-nitrophenyl)anilines can be obtained under the same reaction conditions in reasonable yields. By using nonpolar solvents (*n*-heptane) and establishing heterogeneous reaction conditions, the formation of nitrosoarenes is favored (up to 95% yield) [89].

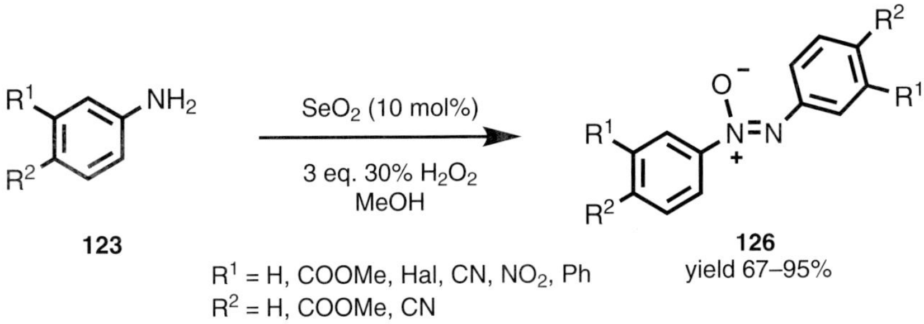

Scheme 8.53

8.2.12.8 Selenium-Catalyzed Oxidation of Alkynes

Recently, Santi and coworkers developed selenium-catalyzed oxidations of alkynes. They have reported the oxidation of alkynes **127** into corresponding 1,2-unprotected dicarbonyl compounds **128** using catalytic amounts of diphenyl diselenide with ammonium persulfate as oxidant. The reaction products were obtained with up to 85% yields but reactions suffered from long reaction times (Scheme 8.54). Interestingly, hemiacetals were isolated by the oxidation of terminal alkynes under the same reaction conditions [90].

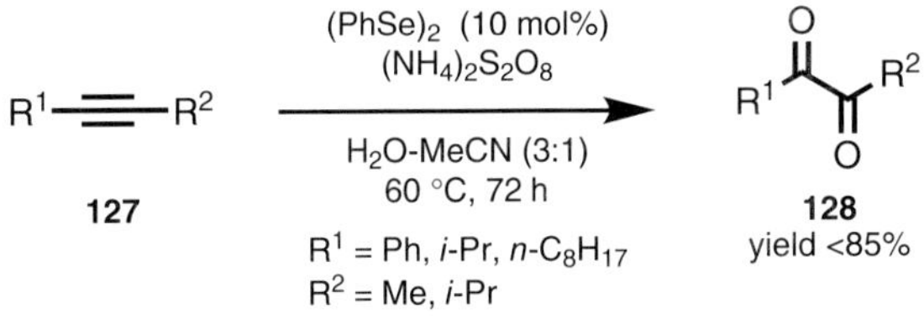

Scheme 8.54

The proposed catalytic cycle is depicted in Scheme 8.55. The reaction is initiated from seleninic acid, which is formed by the reaction of diphenyl diselenide with ammonium persulfate. This electrophile, in the presence of water, promotes a hydroxyselenenylation on the triple bond leading to the enol **A** that exists in a

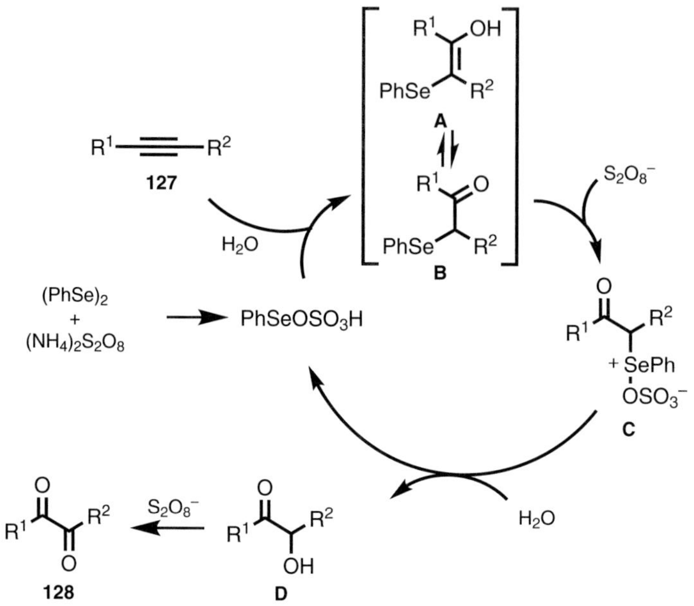

Scheme 8.55

tautomeric equilibrium with the ketone **B**. The excess of ammonium persulfate activates the phenylselenium moiety to a nucleophilic substitution by a molecule of water to form an intermediate **D**, which leads to 1,2-diketone **128** on oxidation [90].

8.2.12.9 Selenium-Catalyzed Oxidation of Halide Anions

Recently, Detty, and coworkers have investigated the application of arylseleninic acids [91] and selenoxides [92] as catalysts for the oxidation of bromide into hypobromite using hydrogen peroxide as oxidant. Furthermore, the generated hypobromite species can be utilized for the bromocyclization of unsaturated alcohols or acids **129** (Scheme 8.56).

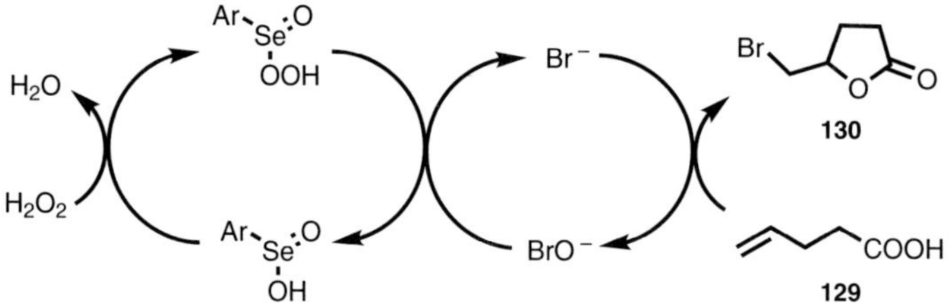

Scheme 8.56

The same research group has synthesized dendrimeric polyphenyl selenides and used these compounds as catalysts in the oxidation of bromide with hydrogen peroxide for subsequent reactions with alkenes [93]. A dendrimer with 12 PhSe-groups showed an autocatalytic effect between the selenium moieties and resulted in turnover numbers of >60,000. The reaction is initiated by the bromonium cation generated via the uncatalyzed background reaction [94]. Furthermore, the same research group developed xerogel-sequestered selenoxide as catalyst for brominations with hydrogen peroxide and sodium sromide in aqueous environment [95]. The reaction with this catalyst is 23 times faster in comparison to the homogeneous reaction and the catalyst can be recovered and recycled without apparent loss of activity for at least four cycles.

The organoselenides-catalyzed oxidation of halides has been also developed by using hydrogen peroxide as an oxidant. Furthermore, the selenides catalyze the transfer of oxidized halogens from N-halosuccinimides to alkenes and ketones. Thus, organoselenides catalyze oxidative halogenation reactions including halolactonization, α-halogenation of ketones, and allylic halogenation [96].

8.2.13
Stereoselective Catalytic Selenenylation–Elimination Reactions

Chiral selenium reagents have been used as catalyst in stereoselective selenenylation–elimination reactions. These catalytic reactions provide double-bond transpositioned allylic ethers or allylic alcohols from the corresponding alkenes. In these catalytic reactions, different oxidants have been used in stoichiometic amounts to activate the chiral selenium electrophiles. Wirth and coworkers developed a sequence of methoxyselenylation and oxidative β-hydride elimination of β-methylstyrene using only catalytic amounts of chiral diselenide **131** (Scheme 8.57). The reaction products were obtained in low yields and with up to 75% enantiomeric excess [97].

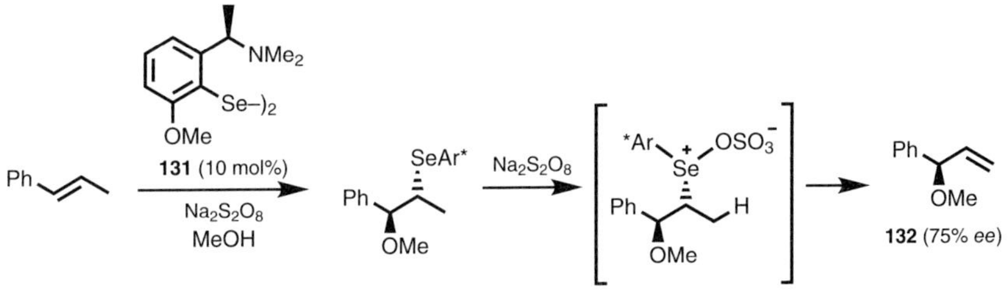

Scheme 8.57

Furthermore, Wirth and coworkers developed the catalytic cyclization of β,γ-unsaturated carboxylic acids **133** to butenolides **135** using catalytic amounts of diphenyl diselenide and [bis(trifluoroacetoxy)iodo]benzene as stoichiometric oxidant. The efforts were also made toward the stereoselective synthesis of butenolides **135** using 5 mol% of chiral diselenides **134**. The chiral butenolides **135** were obtained in low yields and poor selectivity with maximum enantiomeric excess of 22% (Scheme 8.58) [98].

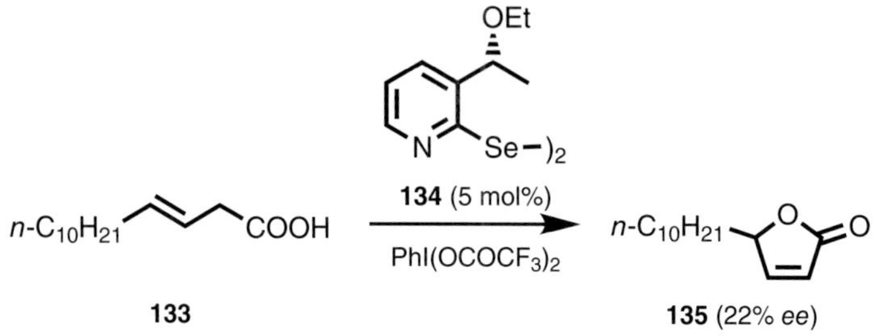

Scheme 8.58

The proposed catalytic cycle is shown in Scheme 8.59 [98]. The reaction is initiated by the oxidation of diphenyl diselenide by the hypervalent iodine reagent PhI(OCOCF$_3$)$_2$ to form phenylselenenyl trifluoroacetate **136**. Reagent **136** then reacts with the β,γ-unsaturated carboxylic acid **133** in a cyclization to yield compound **137**. The selenide in lactone **137** can then be activated for elimination by [bis-(trifluoroacetoxy)iodo]benzene. Finally the catalytic cycle proceeds mainly via the intermediate **138** to the butenolide **135** in an elimination process while regenerating the selenium electrophile **136**.

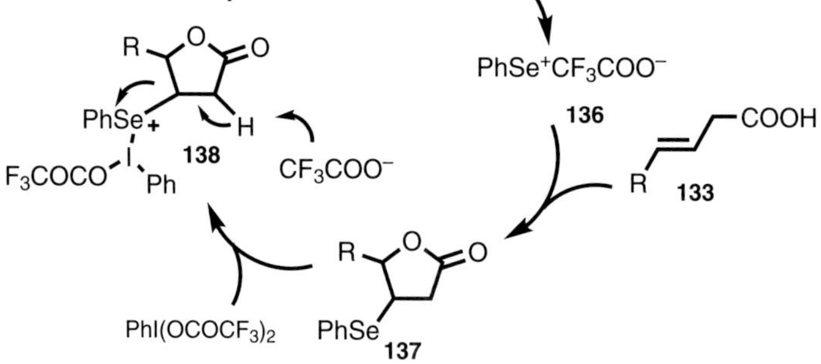

Scheme 8.59

Recently, Wirth and his coworkers further explored the same catalytic approach and developed the selenium-mediated cyclization of stilbenecarboxylic acids **139** to the corresponding isocoumarin derivatives **140** using diselenide or disulfide reagents as catalyst and [bis(trifluoroacetoxy)iodo]benzene as stoichiometric oxidant. The reaction products, isocoumarins **140**, were isolated in excellent yields (Scheme 8.60) [99].

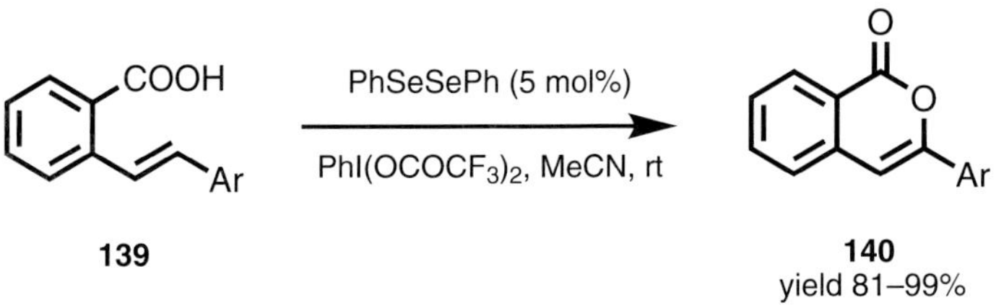

Scheme 8.60

8.2.14
Selenium-Catalyzed Diels–Alder Reactions

Lenardao and coworkers reported the synthesis of various selenonium and telluronium salts in excellent yields. These compounds behave as acidic ionic liquids at room temperature. Phenyl butyl ethyl selenonium tetrafluoroborate **142** has been employed as catalyst in hetero-Diels–Alder reactions of aryl imines derived from citronellal **141**. The cyclized octahydroacridines **143** were obtained in good yields (Scheme 8.61) [100].

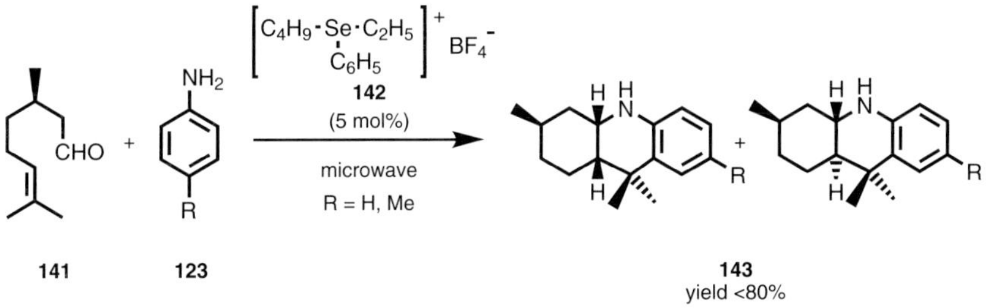

Scheme 8.61

8.2.15
Selenium-Catalyzed Synthesis of Thioacetals

Later on, the selenium-catalyzed synthesis of thioacetals **145** was achieved by the reaction of carbonyl compounds **115** with thiols using same ionic liquid **142** as catalyst (15 mol%) under solvent-free conditions in good to excellent yields (Scheme 8.62) [101].

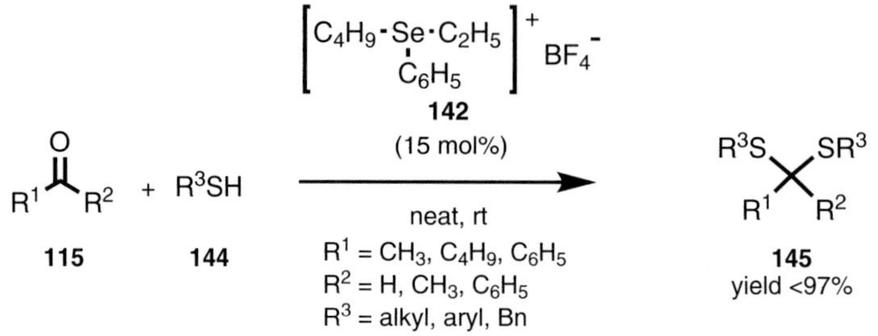

$$R^1\text{C(O)}R^2 + R^3SH \xrightarrow[\text{neat, rt}]{\begin{bmatrix} C_4H_9\text{-}Se\text{-}C_2H_5 \\ | \\ C_6H_5 \end{bmatrix}^+ BF_4^-\\ \mathbf{142}\\ (15\text{ mol\%})} R^1R^2C(SR^3)_2$$

115 + **144**

$R^1 = CH_3, C_4H_9, C_6H_5$
$R^2 = H, CH_3, C_6H_5$
$R^3 = $ alkyl, aryl, Bn

145 yield <97%

Scheme 8.62

8.2.16
Selenium-Catalyzed Baylis–Hillman Reaction

Recently, the same research group developed selenium catalyzed Baylis–Hillman reactions of aldehydes **51** and electron-deficient olefins **146** using the same ionic liquid **142** as catalyst and DABCO as a base in acetonitrile at room temperature. The Baylis–Hillman adducts **147** were obtained in moderate to good yields and in relatively short reaction times under mild reaction conditions (Scheme 8.63) [102].

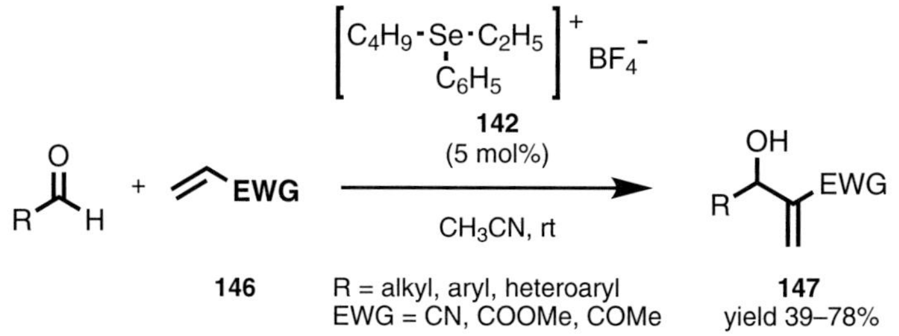

$$R\text{CHO} + \text{CH}_2\text{=CH-EWG} \xrightarrow[\text{CH}_3\text{CN, rt}]{\begin{bmatrix} C_4H_9\text{-}Se\text{-}C_2H_5 \\ | \\ C_6H_5 \end{bmatrix}^+ BF_4^-\\ \mathbf{142}\\ (5\text{ mol\%})} R\text{CH(OH)C(=CH}_2)\text{EWG}$$

146

R = alkyl, aryl, heteroaryl
EWG = CN, COOMe, COMe

147 yield 39–78%

Scheme 8.63

References

1. (a) Huguet, J.L. (1967) *Adv. Chem. Ser.*, **76**, 345; (b) Jones, D.N., Mundy, D., and Whitehouse, R.D. (1970) *J. Chem. Soc., Chem. Commun.*, 86; (c) Walter, R. and Roy, J. (1971) *J. Org. Chem.*, **36**, 2561.

2. (a) Wirth, T. (2000) *Angew. Chem.*, **112**, 3890; (2000) *Angew. Chem., Int. Ed.*, **39**, 3740; (b) Freudendahl, D.M., Shahzad, S.A., and Wirth, T. (2009) *Eur. J. Org. Chem.*, 1649; (c) Browne, D.M. and Wirth, T. (2006) *Curr. Org. Chem.*, **10**,

1893; (d) Wirth, T. (1999) *Tetrahedron*, **55**, 1.

3 (a) Santi, C., Santoro, S., and Battistelli, B. (2010) *Curr. Org. Chem.*, **14**, 2442; (b) Braga, A.L., Lüdtke, D.S., Vargas, F., and Braga, R.C. (2006) *Synlett*, 1453.

4 (a) Back, T.G. (1999) *Organoselenium Chemistry*, Oxford University Press, Oxford; (b) Wirth, T. (2000) *Organoselenium Chemistry: Modern Developments in Organic Synthesis, Top. Curr. Chem.*, vol. 208, Springer, Berlin; (c) Liotta, D. (1987) *Organoselenium Chemistry*, John Wiley & Sons, Inc., New York; (d) Paulmier, C. (1986) *Selenium Reagents and Intermediates in Organic Synthesis*, Pergamon Press, Oxford; (e) Patai, S., and Rappoport, Z. (1986). *The Chemistry of Organic Selenium and Tellurium Compounds*, vol. 1 (eds S. Patai and Z. Rappoport), Wiley, New York.; (f) Patai, S. (1987) *The Chemistry of Organic Selenium and Tellurium Compounds*, vol. 2, Wiley, New York, ; (g) Nicolaou, K.C. and Petasis, N.A. (1984) *Selenium in Natural Products Synthesis*, CIS, Philadelphia; (h) Krief, A. and Hevesi, L. (1988) *Organoselenium Chemistry I*, Springer, Berlin; (i) Klayman, D.L. and Günther, W.H.H. (1973) *Organic Selenium Compounds: Their Chemistry and Biology*, Wiley, New York.

5 (a) Beaulieu, P.L. and Déziel, R. (1999) *Organoselenium Chemistry: A Practical Approach*, (ed. T.G. Back), Oxford University Press, Oxford, pp. 35–66; (b) Nishibayashi, Y. and Uemura, S. (2000) *Top. Curr. Chem.*, **208**, 201.

6 (a) Mukherjee, A.J., Zade, S.S., Singh, H.B., and Sunoj, R.B. (2010) *Chem. Rev.*, **110**, 4357; (b) Guillena, G. and Ramon, D.J. (2006) *Tetrahedron Asymmetry*, **17**, 1465; (c) Tiecco, M., Testaferri, L., Marini, F., Bagnoli, L., Santi, C., Temperini, A., Sternativo, S., and Tomassini, C. (2005) *Phosphorus Sulfur*, **180**, 729; (d) Uemura, S. (1998) *Phosphorus Sulfur*, **136–138**, 219; (e) Wessjohann, L.A. and Sinks, U. (1998) *J. Prakt. Chem.*, **340**, 189; (f) Nishibayashi, Y. and Uemura, S. (1996) *Rev. Heteroatom Chem.*, **14**, 83; (g) Tiecco, M., Testaferri, L., Tingoli, M., Bagnoli, L., Marini, F., Santi, C., and Temperini, A. (1996) *Gazz. Chim. Ital.*, **126**, 635; (h) Reich, H.J. and Wollowitz, S. (1993) *Org. React.*, **44**, 1; (i) Reich, H.J. (1979) *Acc. Chem. Res.*, **12**, 22; (j) Sharpless, K.B., Gordon, K.M., Lauer, R.F., Patrick, D.W., Sinder, S.P., and Young, M.W. (1975) *Chem. Scr.*, **8A**, 9; (k) Gosselck, J. (1963) *Angew. Chem.*, **75**, 831; (1963) *Angew. Chem., Int. Ed. Engl.*, **2**, 660; (l) Petragnani, N., Stefani, H.A., and Valduga, C.J. (2001) *Tetrahedron*, **57**, 1411; (m) Tiecco, M. (2000) *Top. Curr. Chem.*, **208**, 7; (n) McGarrigle, E.M., Myers, E.L., Illa, O., Shaw, M.A., Riches, S.L., and Aggarwal, V.K. (2007) *Chem. Rev.*, **107**, 5841.

7 Sonoda, N., Yamamoto, G., Natsukawa, K., Kondo, K., and Murai, S. (1975) *Tetrahedron Lett.*, **24**, 1969.

8 Nishibayashi, Y., Singh, J.D., Segawa, K., Fukuzawa, S.-I., and Uemura, S. (1994) *J. Chem. Soc., Chem. Commun.*, 1375.

9 Noyori, R. and Kitamura, M. (1991) *Angew. Chem.*, **103**, 34; (1991) *Angew. Chem., Int. Ed. Engl.*, **30**, 49.

10 Pu, L. and Yu, H.-B. (2001) *Chem. Rev.*, **101**, 757.

11 (a) Wirth, T. (1995) *Tetrahedron Lett.*, **36**, 1849; (b) Wirth, T., Kulicke, K.J., and Fragale, G. (1996) *Helv. Chim. Acta*, **79**, 1957; (c) Santi, C. and Wirth, T. (1999) *Tetrahedron Asymmetry*, **10**, 1019.

12 (a) Braga, A.L., Schneider, P.H., Paixão, M.W., Deobald, A.M., Peppe, C., and Bottega, D.P. (2006) *J. Org. Chem.*, **71**, 4305; (b) Braga, A.L., Paixão, M.W., Ludtke, D.S., Silveira, C.C., and Rodrigues, O.E.D. (2003) *Org. Lett.*, **5**, 2635; (c) Braga, A.L., Galetto, F.Z., Rodrigues, O.E.D., Silveira, C.C., and Paixão, M.W. (2008) *Chirality*, **20**, 839.

13 Braga, A.L., Rodrigues, O.E.D., Paixão, M.W., Appelt, H.R., Silveira, C.C., and Bottega, D.P. (2002) *Synlett*, 2338.

14 Bolm, C., Kesselgruber, M., Grenz, A., Hermanns, N., and Hildebrand, J. (2001) *New J. Chem.*, **25**, 13.

15 Schwab, R.S., Soares, L.C., Dornelles, L., Rodrigues, O.E.D., Paixão, M.W., Godoi, M., and Braga, A.L. (2010) *Eur. J. Org. Chem.*, 3574.

16 Nishibayashi, Y., Segawa, K., Singh, J.D., Fukuzawa, S., Ohe, K., and Uemura, S. (1996) *Organometallics*, **15**, 370.
17 Nishibayashi, Y., Singh, J.D., Arikawa, Y., Uemura, S., and Hidai, M. (1997) *J. Organomet. Chem.*, **531**, 13.
18 Börner, C., Dennis, M.R., Sinn, E., and Woodward, S. (2001) *Eur. J. Org. Chem.*, 24352.
19 Braga, A.L., Silva, S.J.N., Ludtke, D.S., Drekener, R.L., Silveira, C.C., Rocha, J.B.T., and Wessjohann, L.A. (2002) *Tetrahedron Lett.*, **43**, 7329.
20 Shi, M., Wang, C., and Zhang, W. (2004) *Chem. Eur. J.*, **10**, 5507.
21 (a) Trost, B.M. and Crawley, M.L. (2003) *Chem. Rev.*, **103**, 2921; (b) Trost, B.M. and van Vranken, D.L. (1996) *Chem. Rev.*, **96**, 395; (c) Uenishi, J. and Hamada, M. (2001) *Tetrahedron Asymmetry*, **12**, 2999.
22 Sprinz, J., Kiefer, M., and Helmchen, G. (1994) *Tetrahedron Lett.*, **35**, 1523.
23 Hiroi, K., Suzuki, Y., and Abe, I. (1999) *Tetrahedron Asymmetry*, **10**, 1173.
24 You, S.-L., Hou, X.-L., and Dai, L.-X. (2000) *Tetrahedron Asymmetry*, **11**, 1495.
25 (a) Hou, X.-L., Wu, X.-W., Dai, L.-X., Cao, B.-X., and Sun, J. (2000) *Chem. Commun.*, 1195; (b) Keehn, P.M., and Rosenfeld, S.M. (1983) *Cyclophanes*, vol. 71, Academic Press, New York; (c) Richards, C.J., Damalidis, T., Hibbs, D.E., and Hursthouse, M.B. (1995) *Synlett*, 74; (d) Sammakia, T. and Latham, H.A. (1995) *J. Org. Chem.*, **60**, 6002.
26 (a) Zielinska-Błajet, M., Siedlecka, R., and Skarzewski, J. (2007) *Tetrahedron Asymmetry*, **18**, 131; (b) Sehnem, J.A., Milani, P., Nascimento, V., Andrade, L.H., Dorneles, L., and Braga, A.L. (2010) *Tetrahedron Asymmetry*, **21**, 997.
27 Braga, A.L., Paixão, M.W., and Marin, G. (2005) *Synlett*, 1975.
28 Vargas, F., Sehnem, J.A., Galetto, F.Z., and Braga, A.L. (2008) *Tetrahedron*, **64**, 392.
29 Yao, Q., Kinney, E.P., and Zheng, C. (2004) *Org. Lett.*, **6**, 2997.
30 Mingzhong, C., Jun, Z., Hong, Z., and Caisheng, S. (2002) *React. Func. Polym.*, **50**, 191.
31 Cai, M., Liu, G., and Zhou, J. (2005) *J. Mol. Catal. A*, **227**, 107.
32 Jones, R.C., Canty, A.J., Gardiner, M.G., Skelton, B.W., Tolhurst, V.-A., and White, A.H. (2010) *Inorg. Chim. Acta*, **363**, 77.
33 Das, D., Rao, G.K., and Singh, A.K. (2009) *Organometallics*, **28**, 6054.
34 Wallner, O.L. and Szabo, K.J. (2005) *J. Org. Chem.*, **70**, 9215.
35 Sebelius, S., Olsson, V.J., and Szabo, K.J. (2005) *J. Am. Chem. Soc.*, **127**, 10478.
36 Aydin, J., Selander, N., and Szabo, K.J. (2006) *Tetrahedron Lett.*, **47**, 8999.
37 Yao, Q. and Sheets, M. (2006) *J. Org. Chem.*, **71**, 5384.
38 Aydin, J., Kumar, K.S., Eriksson, L., and Szabó, K.J. (2007) *Adv. Synth. Catal.*, **349**, 2585.
39 Nishibayashi, Y., Imajima, H., Onodera, G., Hidai, M., and Uemura, S. (2004) *Organometallics*, **23**, 26.
40 Nishibayashi, Y., Imajima, H., Onodera, G., Inada, Y., Hidai, M., and Uemura, S. (2004) *Organometallics*, **23**, 5100.
41 Dochnahl, M., Lohnwitz, K., Luhl, A., Pissarek, J.-W., Biyikal, M., Roesky, P.W., and Blechert, S. (2010) *Organometallics*, **29**, 2637.
42 Schwab, R.S., Galetto, F.Z., Azeredo, J.B., Braga, A.L., Lüdtke, D.S., and Paixão, M.W. (2008) *Tetrahedron Lett.*, **49**, 5094.
43 Watanabe, S., Hasebe, R., Ouchi, J., Nagasawa, H., and Kataoka, T. (2010) *Tetrahedron Lett.*, **51**, 5778.
44 Miyata, T., Kondo, K., Murai, S., Hirashima, T., and Sonoda, N. (1980) *Angew. Chem.*, **92**, 1040; (1980) *Angew. Chem., Int. Ed. Engl.*, **19**, 1008.
45 Liu, X.-Z. and Lu, S.-W. (2003) *Chem. Lett.*, **32**, 1142.
46 Wang, X., Lu, S., and Yu, Z. (2004) *Adv. Synth. Catal.*, **346**, 929.
47 Wang, X., Li, P., Yuan, X., and Lu, S. (2006) *J. Mol. Catal. A*, **253**, 261.
48 Wang, X., Ling, G., Xue, Y., and Lu, S. (2005) *Eur. J. Org. Chem.*, 1675.
49 Tian, F. and Lu, S. (2004) *Synlett*, 1953.
50 Hori, T. and Sharpless, K.B. (1979) *J. Org. Chem.*, **44**, 4204.
51 Tunge, J.A. and Mellegaard, S.R. (2004) *Org. Lett.*, **6**, 1205.

52 Barrero, A.F., Quílez del Moral, J.F., Herrador, M.M., Cortés, M., Arteaga, P., Catalán, J.V., Sánchez, E.M., and Arteaga, J.F. (2006) *J. Org. Chem.*, **71**, 5811.
53 Mellegaard, S.R. and Tunge, J.A. (2004) *J. Org. Chem.*, **69**, 8979.
54 Bures, J., Martın, M., Urpi, F., and Vilarrasa, J. (2009) *J. Org. Chem.*, **74**, 2203.
55 Crich, D., Neelamkavil, S., and Sartillo-Piscil, F. (2000) *Org. Lett.*, **2**, 4029.
56 Ford, L., Wille, U., and Schiesser, C.H. (2006) *Helv. Chim. Acta*, **89**, 2306.
57 Crich, D., Hao, X., and Lucas, M. (1999) *Tetrahedron*, **53**, 14261.
58 Pandey, G., Rao, K.S.S.P., and Rao, K.V.N. (2000) *J. Org. Chem.*, **65**, 4309.
59 Clive, D.L.J., Pham, M.P., and Subedi, R. (2007) *J. Am. Chem. Soc.*, **129**, 2713.
60 Crich, D. and Sannigrahi, M. (2002) *Tetrahedron*, **58**, 3319.
61 Crich, D. and Rumthao, S. (2004) *Tetrahedron*, **60**, 1513.
62 Crich, D. and Patel, M. (2005) *Org. Lett.*, **7**, 3625.
63 Barton, D.H.R., Brewster, A.G., Hui, A.H.F., Lester, D.J., Ley, S.V., and Back, T.G. (1978) *J. Chem. Soc., Chem. Commun.*, 952.
64 Umbreit, M.A. and Sharpless, K.B. (1977) *J. Am. Chem. Soc.*, **99**, 5526.
65 Taylor, R.T. and Flood, L.A. (1983) *J. Org. Chem.*, **48**, 5160.
66 Sharpless, K.B. and Young, M.W. (1975) *J. Org. Chem.*, **40**, 947.
67 Hori, T. and Sharpless, K.B. (1978) *J. Org. Chem.*, **43**, 1689.
68 Betzemeier, B., Lhermitte, F., and Knochel, P. (1999) *Synlett*, 489.
69 Goodman, M.A. and Detty, M.R. (2006) *Synlett*, 1100.
70 Garcia-Marin, H., van der Toorn, J.C., Mayoral, J.A., Garcia, J.I., and Arends, I.W.C.E. (2009) *Green Chem.*, **11**, 1605.
71 Syper, L. and Mlochowski, J. (1984) *Synthesis*, 747.
72 Syper, L. and Mlochowski, J. (1987) *Tetrahedron*, **43**, 207.
73 Gogoi, P., Sharma, S.D., and Konwar, D. (2007) *Lett. Org. Chem.*, **4**, 249.
74 Santoro, S., Santi, C., Sabatini, M., Testaferri, L., and Tiecco, M. (2008) *Adv. Synth. Catal.*, **350**, 2881.
75 Brodsky, B.H. and Bois, J.D. (2005) *J. Am. Chem. Soc.*, **127**, 15391.
76 Carrera, I., Brovetto, M.C., and Seoane, G.A. (2006) *Tetrahedron Lett.*, **47**, 7849.
77 Backvall, J.-E. (2004) *Modern Oxidation Methods*, Wiley, New York.
78 Onami, T., Ikeda, M., and Woodard, S.S. (1996) *Bull. Chem. Soc. Jpn.*, **69**, 3601.
79 Ehara, H., Noguchi, M., Sayama, S., and Onami, T. (2000) *J. Chem. Soc., Perkin Trans. 1*, 1429.
80 van der Toorn, J.C., Kemperman, G., Sheldon, R.A., and Arends, I.W.C.E. (2009) *J. Org. Chem.*, **74**, 3085.
81 Singh, P., and Singh, A.K. (2010) *Eur. J. Inorg. Chem.*, 4187.
82 Brink, G., Vis, J.-M., Arends, I.W.C.E., and Sheldon, R.A. (2001) *J. Org. Chem.*, **66**, 2429.
83 Miyake, Y., Nishibayashi, Y., and Uemura, S. (2002) *Bull. Chem. Soc. Jpn.*, **75**, 2233.
84 Ichikawa, H., Usami, Y., and Arimoto, M. (2005) *Tetrahedron Lett.*, **46**, 8665.
85 Wójtowicz, H., Brzaszcz, M., Kloc, K., and Mlochowski, J. (2001) *Tetrahedron*, **57**, 9743.
86 Crich, D. and Zou, Y. (2004) *Org. Lett.*, **6**, 775.
87 Crich, D. and Zou, Y. (2005) *J. Org. Chem.*, **70**, 3309.
88 Zhao, D., Johansson, M., and Bäckvall, J.-E. (2007) *Eur. J. Org. Chem.*, 4431.
89 Gebhardt, C., Priewisch, B., Irran, E., and Rück-Braun, K. (2008) *Synthesis*, 1889.
90 Santoro, S., Battistelli, B., Gjoka, B., Si, C.-W.S., Testaferri, L., Tiecco, M., and Santi, C. (2010) *Synlett*, 1402.
91 Drake, M.D., Bateman, M.A., and Detty, M.R. (2003) *Organometallics*, **22**, 4158.
92 Goodman, M.A. and Detty, M.R. (2004) *Organometallics*, **23**, 3016.
93 Francavilla, C., Drake, M.D., Bright, F.V., and Detty, M.R. (2001) *J. Am. Chem. Soc.*, **123**, 57.
94 Drake, M.D., Bright, F.V., and Detty, M.R. (2003) *J. Am. Chem. Soc.*, **125**, 12558.

95 Bennett, S.M., Tang, Y., McMaster, D., Bright, F.V., and Detty, M.R. (2008) *J. Org. Chem.*, **73**, 6849.

96 Mellegaard-Waetzig, S.R., Wang, C., and Tunge, J.A. (2006) *Tetrahedron*, **62**, 7191.

97 (a) Wirth, T., Häuptli, S., and Leuenberger, M. (1998) *Tetrahedron Asymmetry*, **9**, 547; (b) Browne, D.M., Niyomura, O. and Wirth, T. (2008) *Phosphorus Sulfur*, **183**, 1026.

98 (a) Browne, D.M., Niyomura, O., and Wirth, T. (2007) *Org. Lett.*, **9**, 3169; (b) Freudendahl, D.M., Santoro, S., Shahzad, S.A., Santi, C., and Wirth, T. (2009) *Angew. Chem.*, **121**, 8559; (2009) *Angew. Chem., Int. Ed.*, **48**, 8409.

99 Shahzad, S.A., Venin, C., and Wirth, T. (2010) *Eur. J. Org. Chem.*, 465.

100 Lenardao, E.J., Mendes, S.R., Ferreira, P.C., Perin, G., Silveira, C.C., and Jacob, R.G. (2006) *Tetrahedron Lett.*, **47**, 7439.

101 Lenardao, E.J., Borges, E.L., Mendes, S.R., Perin, G., and Jacob, R.G. (2008) *Tetrahedron Lett.*, **49**, 1919.

102 Lenardao, E.J., Feijo, J.O., Thurow, S., Perin, G., Jacob, R.G., and Silveira, C.C. (2009) *Tetrahedron Lett.*, **50**, 5215.

9
Biological and Biochemical Aspects of Selenium Compounds
Bhaskar J. Bhuyan and Govindasamy Mugesh

9.1
Introduction

Selenium was discovered by Swedish scientist Jöns Jakob Berzelius in the year 1818 and was named after the Greek goddess of moon, Selene [1]. Selenium was considered as poison for a long time as it was known to cause major health problems such as livestock disease [2] and intoxication in experimental animals [3–7]. Selenium biochemistry gained some appreciation in 1950s with the discovery by Pinsent that some bacteria grew faster in selenium-fortified media [8]. After years of empirical studies on selenium-deficiency syndrome in experimental animals led to the discovery in 1973 that two bacterial enzymes formate dehydrogenase [9] and glycine reductase [10] contain selenium. The biochemical role of selenium in mammalian system was clearly established almost subsequently with the discovery that the antioxidant enzyme glutathione peroxidase (GPx) [11, 12] contains a selenocysteine residue at the active site. The number of selenoproteins identified has grown substantially in recent years (Table 9.1) [13, 14]. In prokaryotes, formate dehydrogenases [15], hydrogenases [16–18], and glycine reductase [19] are a few representative examples in which selenocysteine [20, 21] has been identified as the selenium moiety. In contrast, selenium is bound to a cysteine residue in CO dehydrogenase, where it forms a redox active center with cofactor-bound molybdenum [22]. In eukaryotes, iodothyronine deiodinases (IDs) [23–26], thioredoxin reductases (TrxRs) [27–30], selenophosphate synthetase [31], and selenoprotein P [32, 33] represent important classes of selenoenzymes in addition to the well-known GPx [11, 12, 34–36].

9.2
Biological Importance of Selenium

Selenium is a main group element and its natural abundance is only ~0.05% relative to sulfur. However, selenium resembles sulfur in many of their physio-chemical properties [37]. The redox potential of selenium compounds is generally

Organoselenium Chemistry: Synthesis and Reactions, First Edition. Edited by Thomas Wirth.
© 2012 Wiley-VCH Verlag GmbH & Co. KGaA. Published 2012 by Wiley-VCH Verlag GmbH & Co. KGaA.

Table 9.1 Selenocysteine-containing enzymes and their biological functions[a].

Enzyme	Reaction
Glutathione peroxidases (GPx)	$H_2O_2 + 2GSH \rightarrow H_2O + GSSG$
Phospholipid-hydroperoxide-GPx	$ROOH + 2GSH \rightarrow R\text{-}OH + H_2O + GSSG$
Type 1 iodothyronine deiodinase	L-Thyroxine + $2e^- + H^+ \rightarrow$ 3,5,3'-triiodothyronine + I^-
Thioredoxin reductase	$NADPH + Trx_{ox} \rightarrow NADP^+ + Trx_{red}$
Formate dehydrogenases	$HCOOH \rightarrow CO_2 + 2H^+ + 2e^-$
NiFeSe hydrogenases	$H_2 \rightarrow 2H^+ + 2e^-$
Glycine reductase	$Gly + 2e^- + 4H+ + ADP + P_i \rightarrow$ acetete + NH_4^+ + ATP
Selenophosphate synthatase	$HSe^- + ATP \rightarrow HSe\text{-}PO_3H_2 + AMP + P_i$
Selenoprotein W	?
Selenoprotein P	Antioxidant?

a) GSH, reduced glutathione; ROOH, lipid hydroperoxide; Trx, thioredoxin.

Table 9.2 Tolerable upper intake levels for selenium for infants, children, and adults [38].

Age	Males and females (µg/day)
0–6 months	45
7–12 months	60
1–3 years	90
4–8 years	150
9–13 years	280
14 years and above	400

lower, and therefore, selenium compounds are more reactive than their sulfur analogues. Selenium is an essential trace element required for normal functioning of human body. The daily tolerable intake level of selenium depends on the age group (Table 9.2) [38]. The Institute of Medicine of the National Academy of Sciences, United States has prescribed the upper limit of selenium intake by a normal individual of the age of 14 and above as 400 µg per day. Selenium deficiency leads to the development of heart diseases, hypothyroidism, and weakened immune system [39, 40].

9.3
Selenocysteine: The 21st Amino Acid

Selenium is found *in vivo* in the form of selenocysteine and selenomethionine. However, selenomethionine does not have any major biological role and

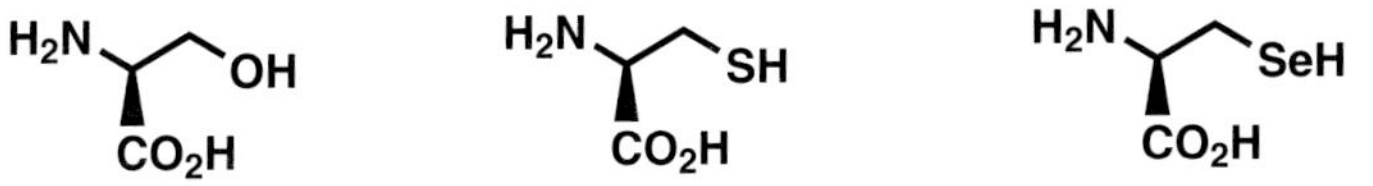

Figure 9.1 Chemical structures of serine, cysteine, and selenocysteine residues. According to the CIP rules, L-Cys and L-Sec have R-configuration.

no regulated incorporation has been reported for this amino acid. The major biological form of selenium is, therefore, represented by the amino acid L-selenocysteine (**3**, Sec, one letter code "U"). Its chemical structure is similar to that of L-serine (**1**) or L-cysteine (**2**) with the side chain containing a selenol moiety in place of the hydroxyl or thiol group. According to the "Cahn–Ingold–Prelog priority rules" [41, 42], due to the presence of a thiol or selenol moiety, there is an alteration in the order of priorities in Cys and Sec amino acid residues. Therefore, Sec and Cys are the only two natural amino acid residues having an R-configuration (Figure 9.1) at the α-carbon for the naturally occurring L-isomer. The presence of a Sec residue in the active center of an enzyme instead of a Cys confers a dramatic catalytic advantage. The lower pK_a (5.2) of the selenol group in the active site as compared to a thiol moiety (8.5) may account for this catalytic advantage [43]. Therefore, the selenol group is fully dissociated at physiological pH and the dissociated selenolate in enzyme's active site is a much better nucleophile than the corresponding thiolate. The significant difference in redox potential between selenocysteine/selenocystine (−710 mV versus standard silver reference electrode (SSE), i.e., −488 mV versus normal hydrogen reference electrode (NHE)) and cysteine/cystine (−455 mV versus SSE, −233 mV versus NHE) which is about 250 mV confers considerably more reducing properties to selenocysteine [44]. The unique redox behavior of selenium makes the Sec residues more reactive than Cys and, therefore, the Sec residues in selenoenzymes can be termed as "superreactive cysteines."

9.4
Biosynthesis of Selenocysteine

In contrast to the other naturally occurring amino acids, Sec is not directly coded in the genetic code. It is coded with an UGA codon, which normally stops translation. The UGA codon recognizes the signal for selenocysteine with a special mRNA secondary structure known as selenocysteine insertion sequence [45]. The details of the Sec incorporation sequence vary from one species to the other. However, in all the selenocysteine-recognizing organisms, Sec is not incorporated directly as selenocysteine to the respective tRNA but involves a

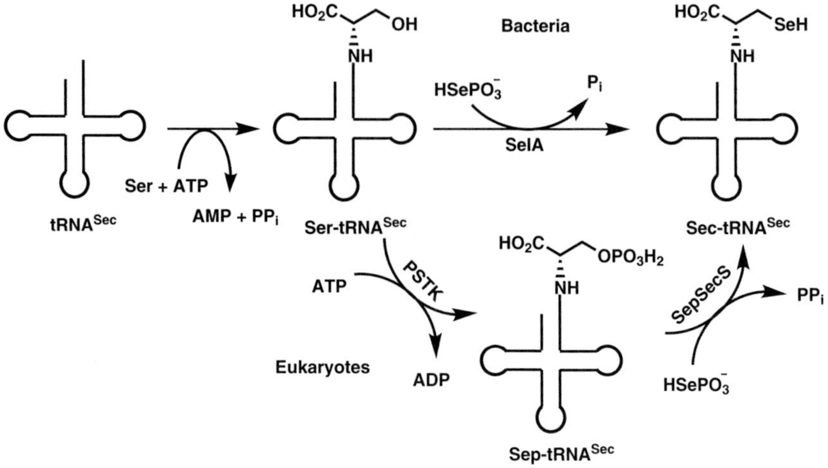

Scheme 9.1 Schematic representation of selenocysteine synthesis in prokaryotes and eukaryotes. No tRNA synthetase is known till date to directly incorporate selenocysteine residues. Selenocysteine is synthesized indirectly in all the organisms involving incorporation of serine followed by selenium insertion. In bacteria, selenium is inserted directly to Ser-tRNA^{Sec}, whereas in eukaryotes, phosphorylation of serine takes place followed by selenium insertion. Sec-tRNA^{Sec} is then used by ribosomes to synthesize selenocysteine. This figure is modified from Söll and coworkers [51].

number of steps. This is mainly because of two reasons: (i) Because of the cellular abundance of Cys as compared to Sec, there is an inhibition of Sec incorporation [46]. (iivc) If Cys is incorporated instead of Sec, there is a dramatic alteration in the enzymatic behavior [47]. Therefore, selenocysteine biosynthesis begins with the insertion of serine to tRNA^{Sec} (Ser-tRNA^{Sec}) [48–50]. In bacteria, Ser-tRNA^{Sec} is then converted to Sec-tRNA^{Sec} by selenocysteine synthetase (SelA). In *Archaea* and eukaryotes, Ser-tRNA^{Sec} is first phosphorylated at serine-OH to form Sep-tRNA^{Sec} by *O*-phosphoseryl-tRNA kinase which is then converted to Sec-tRNA^{Sec} by Sep-tRNA:Sec-tRNA synthetase (SepSecS) as shown in Scheme 9.1 [51–54]. The eukaryotic evolution in selenocysteine incorporation from Sep-tRNA^{Sec} is advantageous over the prokaryotic selenocysteine biosynthesis as the phosphoryl group is a better leaving group as compared to a hydroxyl group, and therefore, is a minimum energy-consuming process [54]. In both selenocysteine incorporation mechanisms, selenophosphate is used as selenium source. Sec-tRNA^{Sec} is then used by the ribosomal proteins to synthesize selenocysteine residues.

The crystal structure of human Sec-tRNA^{Sec} in complex with SepSecS has been solved recently by Söll and coworkers [55]. SepSecS enzyme consists of four domains (Figure 9.2), which can potentially bind four tRNA^{Sec}. However, it is known that the SepCysS, the closest homologue of SepSecS, functions as a dimer

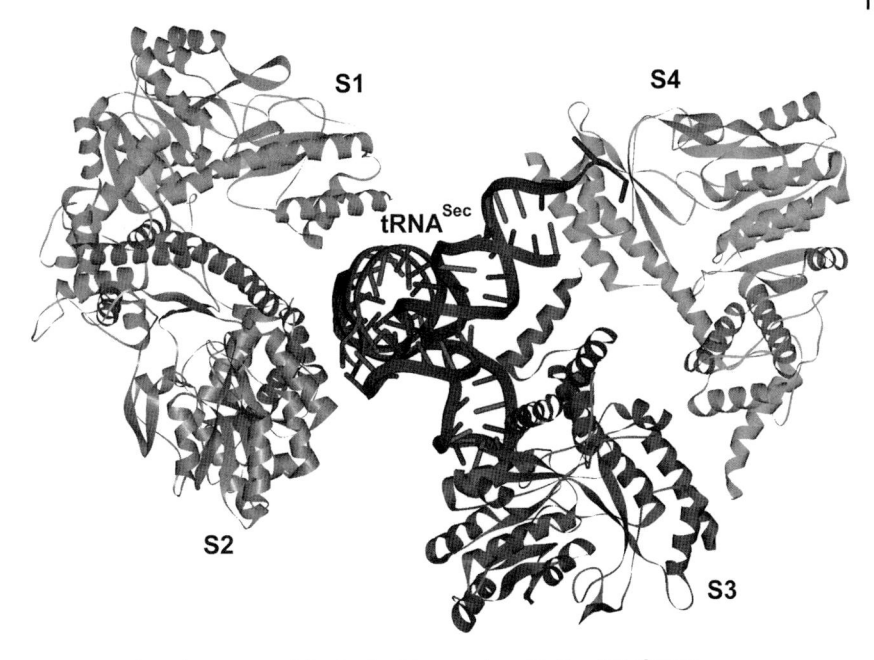

Figure 9.2 Crystal structure of the SepSecS in complex with Sep-tRNA^{Sec} [55].

[56]. The presence of two additional subunits in selenocysteine synthetase enzyme (SepSecS) is structurally important in addition to the two domains that are significant for the catalytic activity. The noncatalytic domain of SepSecS plays a crucial role in the reorientation of the selenophosphate and Sep-tRNA^{Sec} in close proximity for the reaction to take place. The binding of Sep-tRNA^{Sec} to SepSecS leads to a conformational change that allows the o-phosphoserine covalently attached to tRNA to react with selenophosphate to form Sec-tRNA^{Sec}.

It is known that the conversion of Sep-tRNA^{Sec} to Sec-tRNA^{Sec} is mediated by a prosthetic group called pyridoxal phosphate (PLP) bound to SepSecS enzyme. Based on the crystal structure of SepSecS and theoretical studies, Söll and coworkers proposed the mechanism of conversion of Sep-tRNA^{Sec} to Sec-tRNA^{Sec} as given in Scheme 9.2. The reaction begins with movement of the Sep-tRNA^{Sec} to the close proximity of the Schiff base between PLP and Lys284 of SepSecS. The amine of Sep-tRNA^{Sec} attacks at the Schiff base to form an external aldimine. The Lys284 residue reorients itself to abstract one proton from the α-carbon of the Sep moiety. Electron delocalization along the pyridine ring results in the elimination of the phosphate group to produce dehydroalanine-tRNA^{Sec} intermediate. Selenophosphate then binds to this intermediate with the release of the inorganic phosphate to produce a Sec-tRNA^{Sec}–PLP complex. An attack of the Lys284 residue at the Schiff base between Sec-tRNA^{Sec} and PLP leads to the regeneration of the SepSecS–PLP complex with the release of the tRNA^{Sec}.

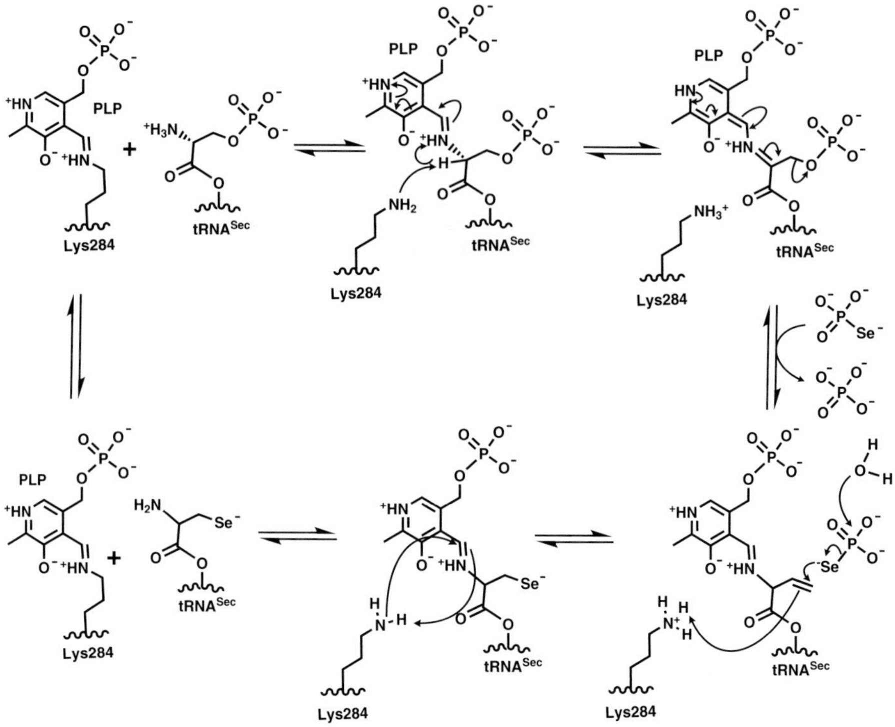

Scheme 9.2 Schematic representation of various steps involved in the biosynthesis of Sec-tRNA^{Sec} [55].

9.5
Chemical Synthesis of Selenocysteine

As the biosynthesis of selenocysteine is very complex, chemical synthesis of selenocysteine has been found to be very useful for various mutagenic experiments. A number of peptides containing Sec residue have been synthesized by solid phase peptide synthetic (SPPS) routes, using both Fmoc- and Boc-protection strategies. The problems associated with the synthesis of Sec are mainly due to the facile oxidation of the selenol to the diselenide as compared to thiol or hydroxyl groups of Cys and Ser, respectively. The synthesis of selenocysteine was first reported by Fredga in the year 1936 [57]. However, the synthesis of selenocysteine in an optically pure form was first reported by Soda and coworkers [58, 59]. Stocking *et al.* improved their procedure to obtain L-Sec from Boc-protected β-iodo-L-alanine-methylester by treating with dilithium diselenide (Scheme 9.3) [60]. Siebum *et al.* followed the Mitsunobo procedure to introduce different isotopes to selenocysteine such as ^{13}C, ^{15}N, or ^{77}Se for various NMR studies [61] (Scheme 9.4).

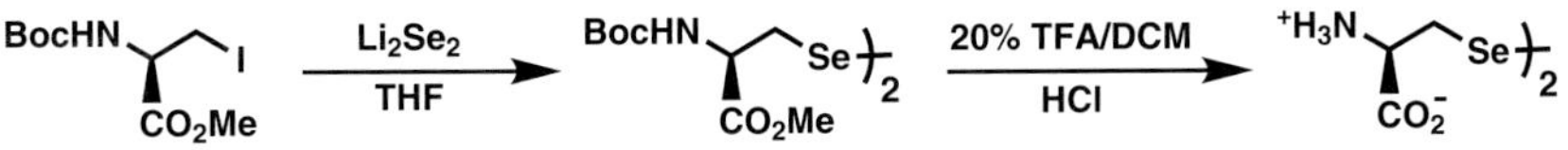

9.6 Chemical Synthesis of Sec-Containing Proteins and Peptides

Scheme 9.3 Synthesis of selenocysteine in optically pure form [60].

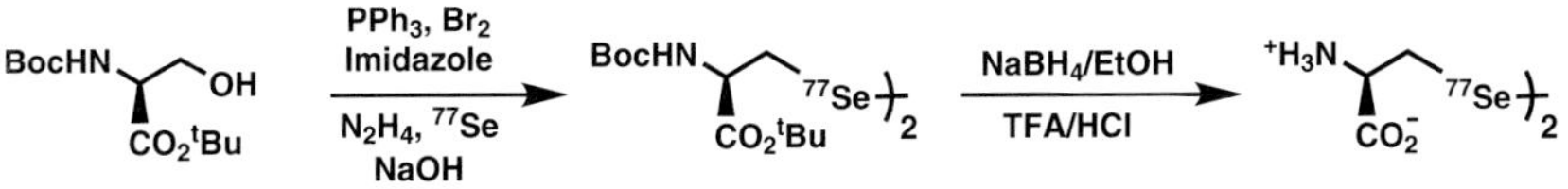

Scheme 9.4 Synthesis of selenocysteine with ^{77}Se enrichment [61].

The first example of an optically pure Sec-benzyl derivative was reported by Walter and coworkers in late 1960s [62]. Nucleophilic displacement of O-tosyl moiety of L-serine derivatives (Fmoc-protected) by benzyl selenolate anion afforded the Sec derivative. This methodology was applied for the synthesis of various other derivatives such as Fmoc-Sec(Bn)-OH, Fmoc-Sec(Ph)-OH, and Fmoc-Sec(PMB)-OH (Schemes 9.5 and 9.6) [63–65].

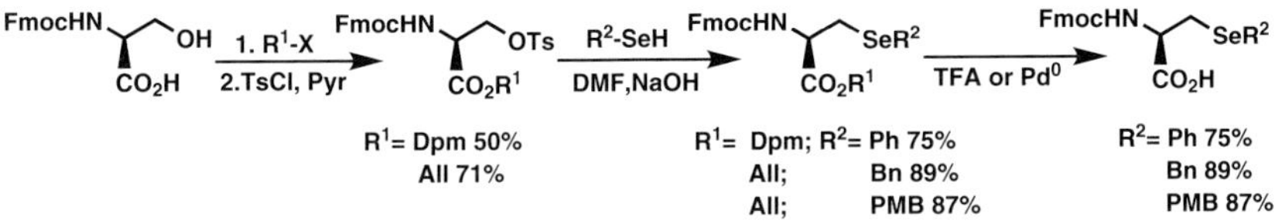

Scheme 9.5 Synthesis of various substituted selenocysteine derivatives [63]. (Dpm, diaminopimelate; All, allyl; Ph, phenyl; Bn, benzyl; PMB, para-methoxybenzyl).

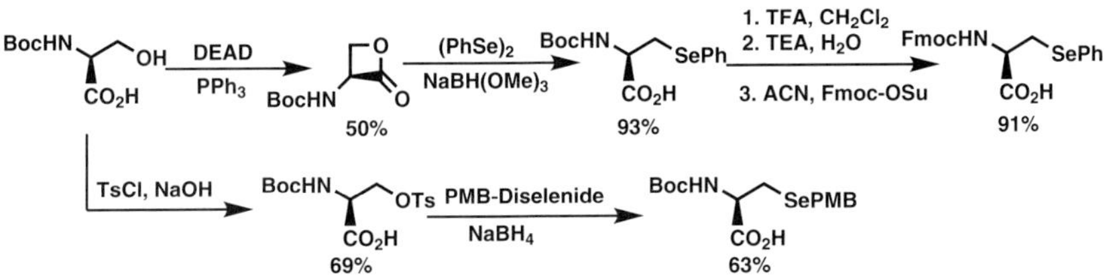

Scheme 9.6 An alternate approach for the synthesis of selenocysteine building blocks [65].

9.6
Chemical Synthesis of Sec-Containing Proteins and Peptides

Chemical synthesis of proteins and peptides containing a selenocysteine residue can be achieved by SPPS (Scheme 9.7). A combination of SPPS and native chemical ligation [66–71] is considered as a very effective tool for the synthesis of peptides and proteins. In general, Cys and Sec are incorporated in a similar manner and the Fmoc-protected Sec derivative, Fmoc-Sec (PMB)-OH, provides the peptides in good yield. Because of the higher reactivity of Sec as compared to Cys, better coupling efficiency and smooth chain assembly are observed during the synthesis of Sec-peptides. In the native chemical ligation of Sec-containing peptides, the thioester derivative of a peptide is coupled with Sec-containing peptide having free selenol group as shown in Scheme 9.7. A nucleophilic attack of the selenol group at the thioester bond and subsequent selenium-to-nitrogen acyl shift produces the Sec-containing peptides.

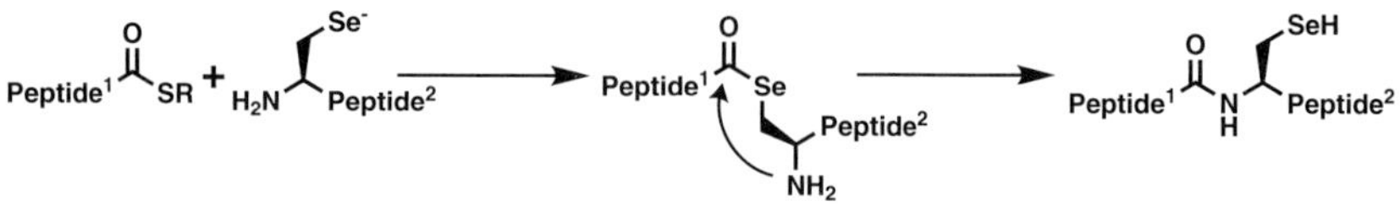

Scheme 9.7 Synthesis of Sec-containing peptides by native chemical ligation.

Expressed protein ligation [69, 72] is another tool for the synthesis of Sec-containing peptides. This is an extension of native chemical ligation that was first reported in 1998. With the success of native chemical ligation in the synthesis of Sec-containing peptides, it became apparent that Sec can be used at the point of ligation in the expressed protein ligation as shown in Scheme 9.8. A target protein truncated at the C-terminus is overexpressed in *Escherichia coli* as a fusion to an intein domain affinity tag (either a hexa His tag or chitin-binding domain). The cell lysate is then passed through an affinity column to bind the target protein through the tag to separate it from the cellular constituents. The intein domain mediates a self-catalyzed rearrangement to form an internal thioester bond to the

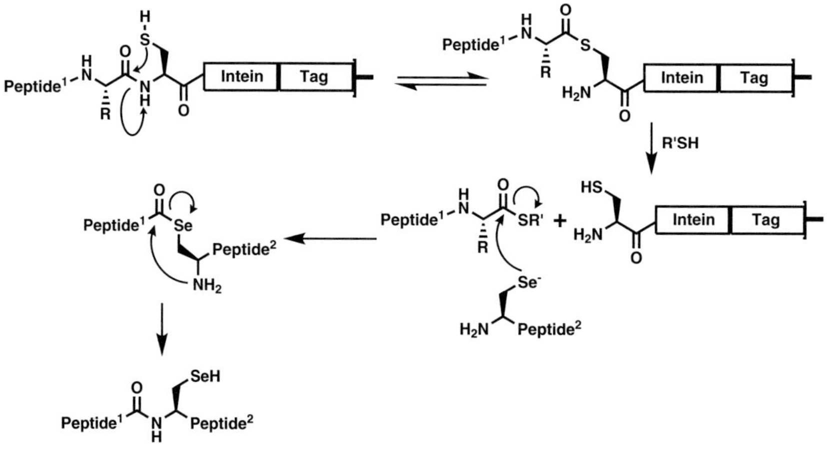

Scheme 9.8 Synthesis of Sec-containing peptides by expressed protein ligation [71].

Cys residue. When an external thiol is added to the resin, a trans-esterification takes place that leads to the cleavage of the truncated protein from the resin. Addition of the peptide-containing Sec with free selenol results in the formation of the desired mutant protein containing Sec residue.

9.7
Selenoenzymes

9.7.1
Glutathione Peroxidases

GPx is an antioxidant selenoenzyme that protects various organisms from reactive oxygen species (ROS) such as hydroperoxides at the expense of the cellular thiol glutathione (GSH). The GPx superfamily contains four types of enzymes, the cytosolic GPx (cGPx), phospholipid hydroperoxide GPx (PHGPx), plasma GPx (pGPx), and gastrointestinal GPx (giGPx), all of which require Sec in the form of selenol for their catalytic activity [36, 69, 72–76]. The catalytic activity of these enzymes differs considerably depending on the hydroperoxides and thiol cosubstrate. The classical GPx utilizes mainly GSH as reducing agent for the reduction of H_2O_2 and some organic hydroperoxides such as cumene hydroperoxide (CumOOH) and tert-butyl hydroperoxide (TBHP) (tBuOOH). The PHGPx also uses GSH as physiological reducing substrate, but the hydroperoxide substrate specificity is broader for this enzyme. For example, this enzyme is active on all phospholipid hydroperoxides, fatty acid hydroperoxides, cumene hydroperoxide, TBHP, cholesterol hydroperoxides, and H_2O_2 [77]. In contrast to cGPx and PHGPx, the hydroperoxide substrate specificity of the plasma enzyme is highly restricted. Although pGPx can reduce H_2O_2 and organic hydroperoxides, it is approximately one order of magnitude less active than the cGPx. Furthermore, GSH is a poor reducing substrate for this enzyme. As the concentration of GSH in its reduced form is very low in human plasma, it is quite unlikely that GSH is the reducing substrate for the plasma enzyme [78].

The catalytic cycle of GPx enzymes involves the oxidation of the selenol moiety (E-SeH) to produce the corresponding selenenic acid (E-SeOH), which upon reaction with GSH generates a selenenyl sulfide (E-SeSG) intermediate. The attack of a second GSH at the selenenyl sulfide regenerates the active site selenol as shown in Scheme 9.9. The antioxidant activity of GPx is generally measured using GR–GSSG coupled assay in which the glutathione disulfide (GSSG) is reduced back to GSH by glutathione reductase (GR) [79]. In the overall process, 2 equivalents of GSH are oxidized to the corresponding disulfide (GSSG), whereas the hydroperoxide is reduced to corresponding alcohol or water [11, 80]. The reduction of GSSG by GR requires NADPH as cofactor. As NADPH exhibits UV absorption at 340 nm, the GPx activity can be conveniently determined by spectrophotometric methods.

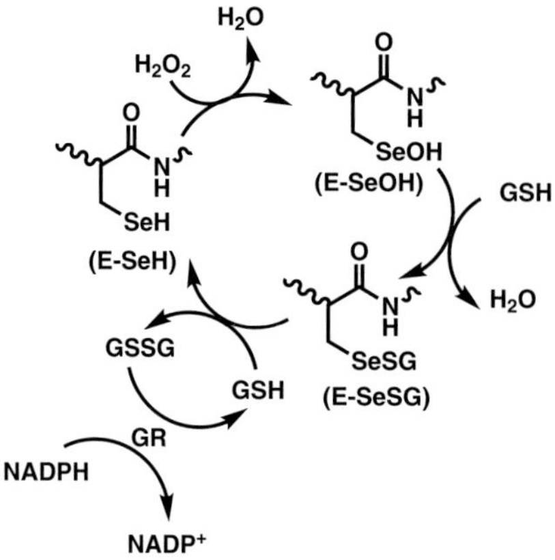

Scheme 9.9 Proposed mechanism for the reduction of hydroperoxides by glutathione peroxidase and reduction of glutathione disulfide by glutathione reductase.

When GSH is depleted in the reaction mixture, the selenenic acid (E-SeOH) undergoes an overoxidation to produce the seleninic acid (E-SeO$_2$H). The crystal structure of bovine erythrocyte GPx (Figure 9.3a) confirms the formation of a seleninic acid [81]. Furthermore, the crystal structure of the semisynthetic enzyme selenosubtilisin indicates the formation of a seleninic acid, indicating that the seleninic acid is probably the most stable form in air. However, the seleninic acid may lie off the main catalytic cycle in the presence of thiol cofactor as the selenenic acid is rapidly converted to the selenenyl sulfide by GSH. It has been reported that the Sec residue forms a "catalytic triad" with two other amino acid residues, tryptophan (Trp) and glutamine (Gln), by noncovalent interactions (Figure 9.3b) [82]. These interactions activate the selenol moiety to remain in the catalytically active selenol form for an efficient reduction of hydroperoxides.

The importance of GPx in maintaining the cellular balance of antioxidant led to the development of several synthetic compounds that functionally mimic the GPx activity. Furthermore, it has been reported that a protein scaffold is not essential for GSH to function as a cofactor in reducing peroxides. Most of the synthetic GPx mimics reported in the literature can be classified into two major categories. The first type consists of compounds having heteroatoms such as nitrogen, oxygen, or sulfur directly bonded to the selenium center. The second category of GPx mimics comprises of the compounds that do not have the heteroatom directly bonded to selenium center, but placed in close proximity to it. In this type of compounds, weak intramolecular nonbonded selenium–heteroatom interactions are observed [83]. Among a number of synthetic GPx mimics, 2-phenyl-1,2-benzoisoselenazol-3-(2H)-one (**1**, ebselen), one of the most promising synthetic antioxidants that exhibits numerous biological activities both *in vitro* and *in vivo* systems [83–87], uses GSH for its catalytic activity. Ebselen is an excellent scavenger of ROS such as peroxynitrite (PN) and the rate of the reaction between ebselen and PN is about three orders of magnitude higher than that of naturally occurring small molecules

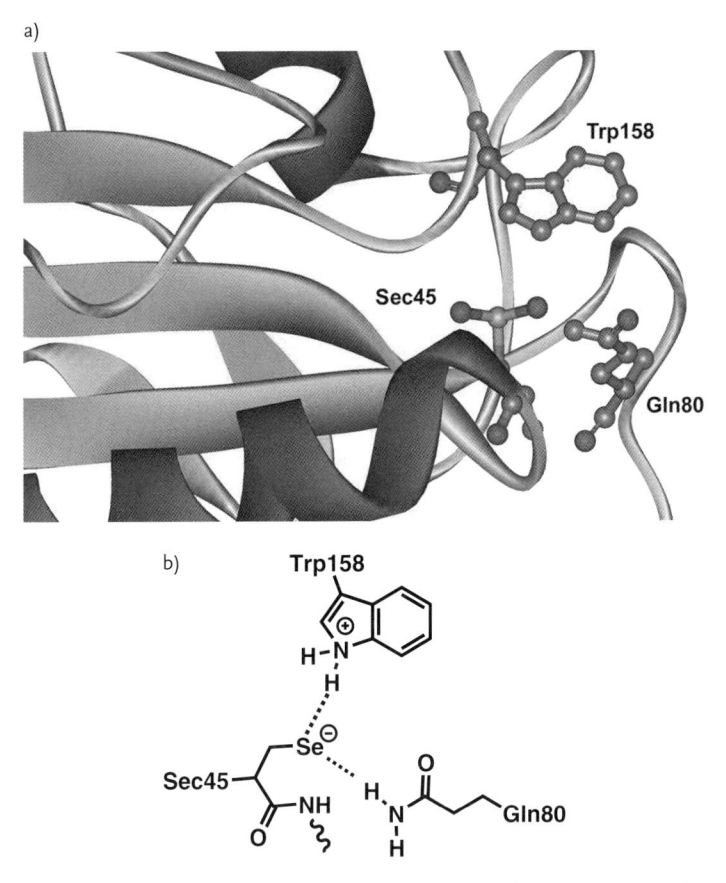

Figure 9.3 (a) X-ray crystal structure of the active site of glutathione peroxidase (PDB Code 1GP1) [81]. (b) The selenol group of Sec45 is highly activated by a catalytic triad involving Gln80 and Trp158 residues.

such as ascorbate, cysteine, and methionine [88]. Ebselen and related derivatives effectively protect against lipid peroxidation induced by transition metals [84, 89]. Furthermore, ebselen has been shown to inhibit a number of enzymes that include nitric oxide synthase, the enzymes involved in inflammatory diseases such as lipoxygenase and cycloxygenase, NADPH oxidase, protein kinase C, glutathione-S-transferase, cytochrome P-450 and b-5 reductases, H^+/K^+-ATPase, the cysteine proteases such as papain, and prostaglandin H synthetase [90–92]. Ebselen also exhibits significant antitumor and immunomodulating activities and has been suggested to have a potential to protect the ROS-mediated brain damage. Interestingly, most of the biological activities of ebselen have been associated with its ability to mimic the enzymatic properties of GPx [11–13].

Ebselen was first synthesized by Lesser and Weiss in 1924 [93]. Its ability to mimic GPx *in vitro* was first demonstrated by Müller *et al.* [84] and Wendel *et al.* [85] in 1984. After this discovery, extensive research has been carried out to identify efficient methods for the synthesis of ebselen and its analogues. A simple method reported by Engman and Hallberg involves *ortho*-lithiation of benzanilide, followed by the selenium insertion [94]. Cyclization of the Se–Li compound was carried out in the presence of cupric bromide as shown in Scheme 9.10.

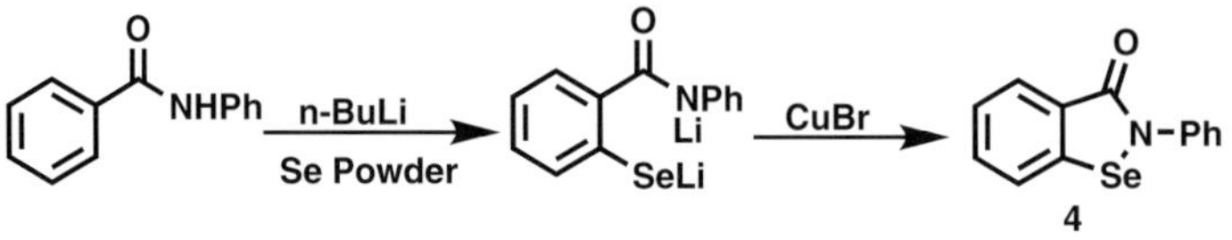

Scheme 9.10 Synthesis of ebselen (1) from benzanilide [94].

The anti-inflammatory, antiatherosclerotic, and cytoprotective properties of ebselen (4) have led to the design and synthesis of new GPx mimics for potential therapeutic applications. Several synthetic organoselenium compounds are known to act as antioxidants by reducing H_2O_2 and PN (ONOO$^-$) and also by preventing lipid peroxidation. Some representative examples of synthetic GPx mimics are given in Figure 9.4 [95–105]. These include the ebselen analogue (5),

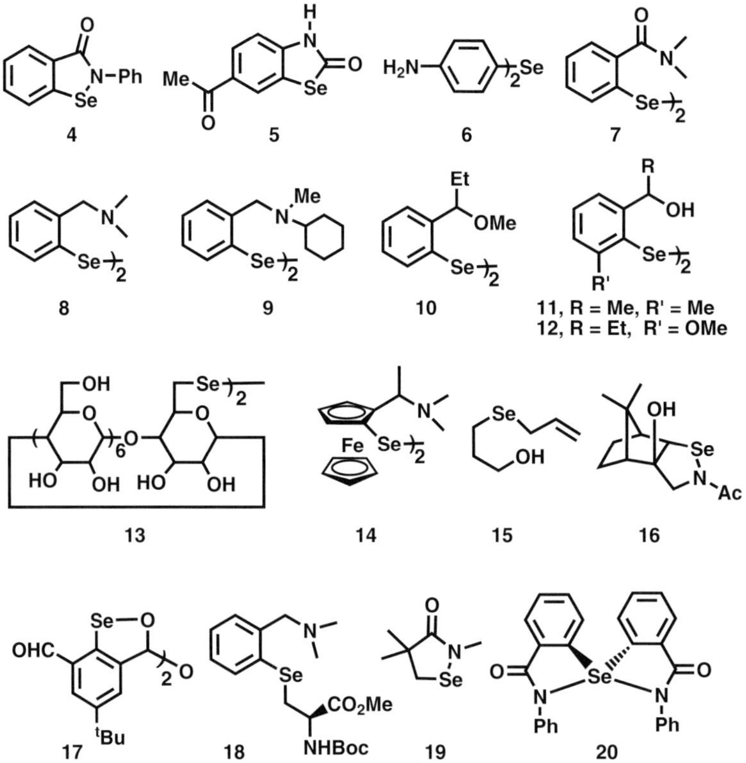

Figure 9.4 Chemical structures of some GPx mimics reported in the literature.

benzoselenazolinones, a variety of substituted diaryl selenides such as **6** and **7**, diselenides such as **8–14** and related derivatives (**15–19**), spirodiazaselenurane (**20**), and the semisynthetic selenoenzyme selenosubtilisin [106, 107]. It was found that the diselenides having very strong Se···N intramolecular interactions were less active, whereas the diselenides that contain a built-in basic amino group but exhibit weak Se···N interactions showed excellent GPx activity. The ferrocenyl-based diselenides such as **14** have been shown to display much higher activities as compared to the phenyl-based diselenides.

Although the anti-inflammatory effect of ebselen has been studied extensively, the catalytic cycle for the reduction of hydroperoxides by ebselen is controversial [86, 108–110]. This is probably due the different thiols and peroxides used in different assays. Initially, it was proposed that ebselen first reacts with one equivalent of thiol to produce the selenenyl sulfide **21**, which is then converted to selenol **22** at the expense of another equivalent of thiol cofactor. The selenol **22** was believed to be the active species that reduces peroxide to form the selenenic acid **23** as shown in Scheme 9.11. The catalytic cycle is completed with the release of a water molecule to regenerate ebselen. In presence of an excess thiol, the selenenic acid (**23**) reacts with the thiol to produce the corresponding selenenyl sulfide (**21**).

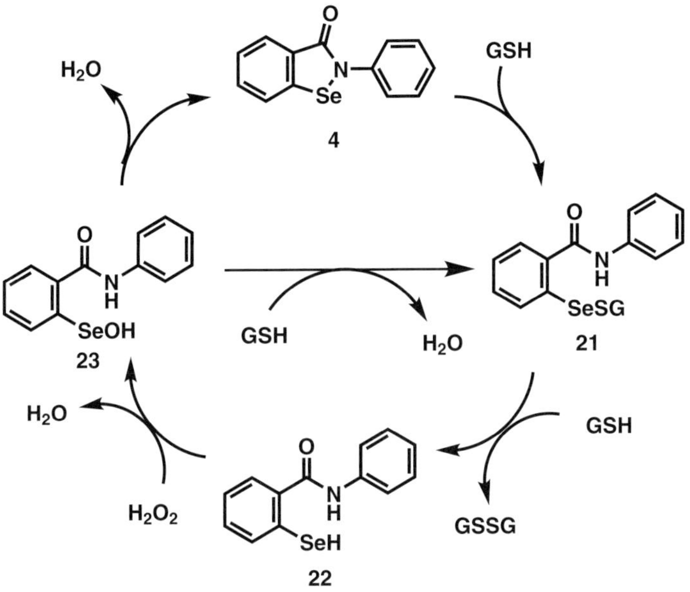

Scheme 9.11 Initially proposed catalytic mechanism of ebselen.

Recent studies have shown that ebselen is a poor catalyst in reducing peroxides when aryl thiols such as PhSH or BnSH are used as thiol cofactors [111, 112]. Our group has studied the GPx activity of various ebselen derivatives to understand the reason for the relatively poor activity of ebselen analogues in aromatic thiol assays [112–114]. It was observed that the reaction of ebselen with aromatic thiols does not produce the selenol. This is due to extensive thiol exchange reactions that take place at the selenenyl sulfide intermediate step. The attack of the incoming thiol can take place either at the selenium center or at the sulfur center as shown in Scheme 9.12. If the thiol attacks the sulfur center, the selenol is produced with the elimination of a disulfide. However, when the incoming thiol attacks at the selenium center, it leads to the generation of another selenenyl sulfide intermediate. In the catalytic cycle of ebselen, the selenium center is more electrophilic than the sulfur due to a strong Se···O interaction, which favors the attack of thiol at the selenium center [112]. As a result, the selenol required for the reduction of peroxides is not formed in sufficient quantities, which accounts for the poor GPx-like activity of ebselen analogues in aromatic thiol assays.

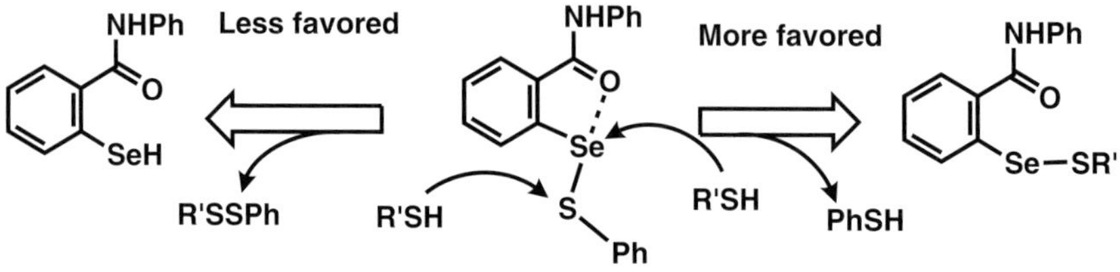

Scheme 9.12 Thiol exchange reactions taking place at the selenenyl sulfide intermediate. An attack of the incoming thiol at the selenium center is more favored due to strong Se–O/N interactions.

Compounds **24–34** having different substituents at the nitrogen have been studied as GPx mimics to understand the effect of the substituents on GPx activity (Figure 9.5) [113]. These studies indicate that the carbonyl oxygen invariably interacts with selenium in the selenenyl sulfide intermediates irrespective of the substituents attached to the nitrogen. Therefore, all the selenenyl sulfides derived from **24–34** undergo thiol exchange reactions when aromatic thiols are used in the assay. These studies also reveal that the nature of thiol, but not the peroxide, plays an important role in the catalytic mechanism. The reactivity of thiol toward the selenenyl sulfide intermediate may also modulate the thiol exchange reactions.

As discussed previously, strong Se···O/N interactions reduce the GPx-like activity of ebselen derivatives due to extensive thiol exchange reactions. Therefore, weakening of such interactions is considered to be beneficial for the catalytic efficiency. This could be achieved by introducing a substituent at the thiol that can interact with the sulfur instead of selenium. For example, compound **33** having both Se···O and S···N interactions overcomes the thiol exchange reaction to a large extent (Scheme 9.13) [112]. The Se···N interaction considerably reduces the Se···O interaction, which leads to a decrease in the eletrophilic reactivity of selenium. This facilitates an attack of the external thiol at the sulfur center, generating selenol **22** and disulfide **37**. These observations suggest that the introduction of

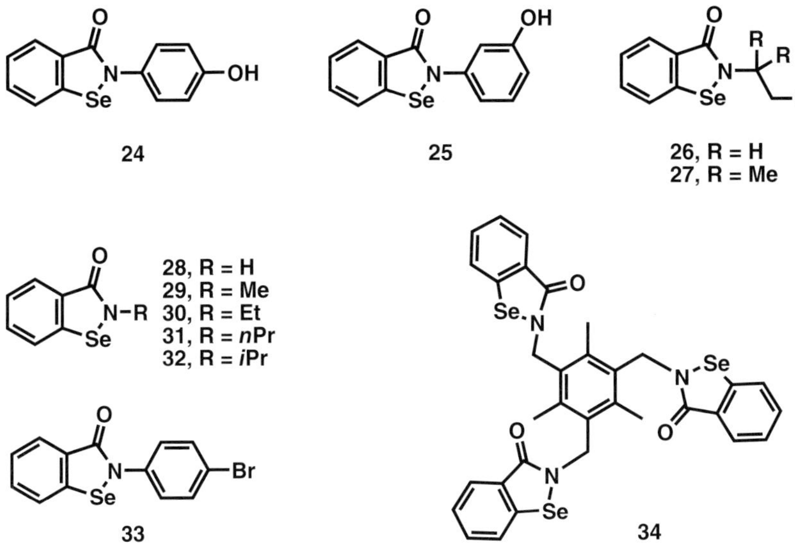

Figure 9.5 Chemical structures of some GPx mimics reported in the literature.

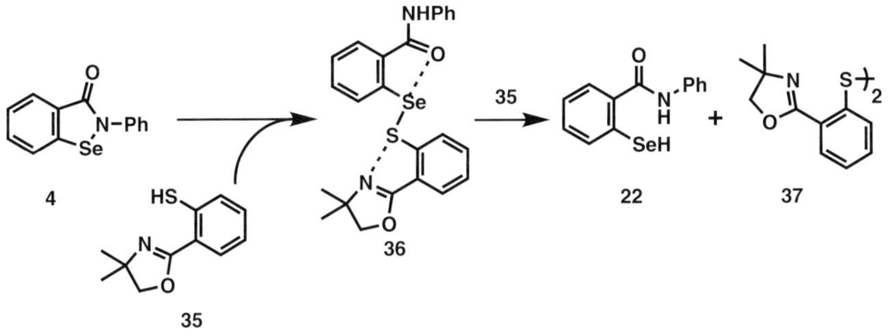

Scheme 9.13 Weakening of Se⋯O interactions in the selenenyl sulfide intermediate by introducing S⋯N interactions.

groups that can reduce the undesired thiol exchange reactions can enhance the GPx-like activity of ebselen derivatives.

In an attempt to understand the catalytic mechanism of ebselen, it was observed that ebselen can readily react with peroxides even in the absence of thiol to produce the seleninic acid **38**. Treatment with an excess amount of thiol converts the seleninic acid to the corresponding selenenic acid **23**, which upon reaction with another equivalent of thiol produces the selenenyl sulfide **39** (Scheme 9.14). In the absence of thiol, compound **23** eliminates a water molecule to regenerate ebselen (**4**). The disproportionation of the selenenyl sulfide **39** to diselenide **40** is the rate-determining step and the rate of disproportionation depends on the nature of thiol employed for the assay. In this reaction, GSH was found to have higher reactivity than PhSH.

Scheme 9.14 Revised catalytic mechanism of ebselen analogues.

As the diselenide **37** was found to be a key intermediate in the catalytic mechanism of ebselen, a number of *sec*- and *tert*-amide-based compounds (**41–49**) have been studied as GPx mimics (Figure 9.6) [115]. These diselenides are inactive toward PhSH, and therefore, their reaction with H_2O_2 is important for their catalytic activity. The *tert*-amide-based diselenides exhibit much better GPx activity than the corresponding *sec*-amide-based compounds [115]. Although the reactivity toward thiols cannot be altered by the introduction of *tert*-amide groups, the Se···O interactions in the selenenyl sulfides derived from **46–49** were found to be much weaker than the interaction observed in the corresponding selenenyl sulfides obtained from compounds **41–45**. This may account for the higher GPx

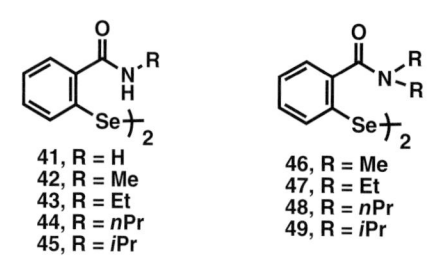

41, R = H
42, R = Me
43, R = Et
44, R = nPr
45, R = iPr

46, R = Me
47, R = Et
48, R = nPr
49, R = iPr

Figure 9.6 *Sec*- and *tert*-amide-based diselenides as GPx mimics.

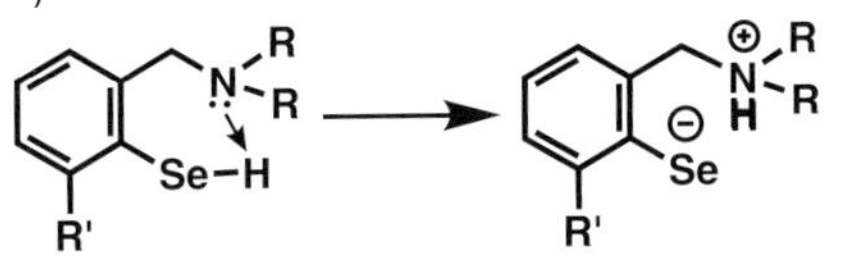

Figure 9.7 (a) Amine-based diselenides as GPx mimics. (b) Activation of selenol moiety by basic amino group.

a)

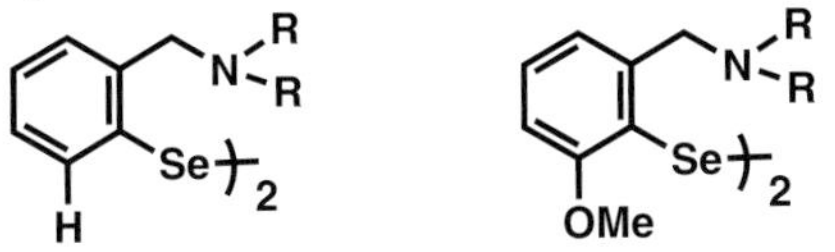

8, R = Me
50, R = Et
51, R = nPr
52, R = iPr

53, R = Me
54, R = Et
55, R = nPr
56, R = iPr

activity of the *tert*-amide-based compounds as compared to the *sec*-amide diselenides.

As discussed previously, the Sec, Trp, and Gln residues form a "catalytic triad" in all enzymes from the GPx superfamily, which activates the selenol moiety at the active site through hydrogen bonding. This concept led to the development of diaryl diselenides such as **8**, **50–56** (Figure 9.7a) having basic amino functionality near the selenium center as GPx mimics [116]. For example, reduction of the diselenide bonds in compounds **8** and **9** produces the catalytically active selenols, which are activated by hydrogen bonding with the basic amino groups to form more reactive selenolate species (Figure 9.7b). Interestingly, the introduction of a methoxy group at the 6-position of the aromatic ring significantly enhances the catalytic activity of the parent compounds. The methoxy substituent blocks the attack of a thiol at the selenium center. Therefore, the thiol preferably attacks at the sulfur center of the selenenyl sulfide to generate the catalytically active selenol species.

The catalytic efficiency of synthetic GPx mimics can be modulated by using various substituents on the aromatic ring or the nitrogen atom. The Se···O interactions in the selenenyl sulfide intermediate are responsible for the poor GPx-like activity of many synthetic compounds including ebselen due to extensive thiol exchange reactions. The substituents that can weaken the Se···O interactions are expected to increase the GPx activity (Figure 9.8). The *tert*-amide- and amine-based compounds appear to be potential candidates for further pharmacological studies as they exhibit good *in vitro* GPx-like activities [117, 118].

The GPx-like antioxidant activity of some aliphatic compounds (**15, 57–64**) has been studied extensively by Back and Moussa [119, 120]. In contrast to many aromatic selenides and diselenides, these aliphatic compounds are very reactive toward peroxides and readily produce the corresponding selenoxides. It has been

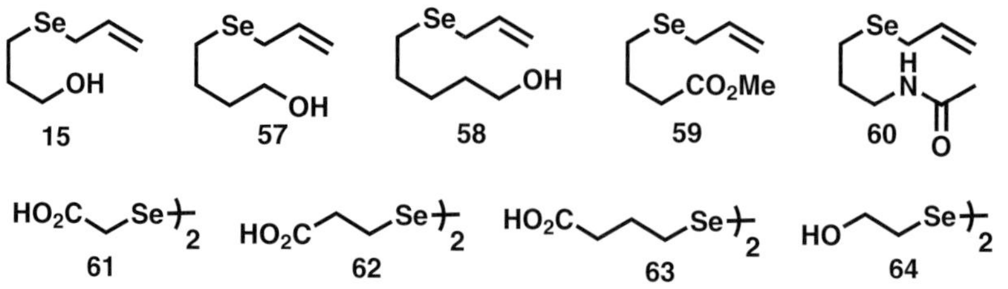

Figure 9.8 Aliphatic selenium compounds as GPx mimics [120].

shown that the compounds having an allylic chain and an alcohol-containing arm attached to the selenium center exhibits excellent activity in the presence of TBHP as the substrate and benzyl thiol as the thiol cofactor. In this series of compounds, the selenide **15** having the spacer link of three carbon atoms between the selenium center and alcohol moiety was found to be the most efficient mimic of GPx. Compound **57** having an amide side chain has also been shown to be a good GPx mimic.

The higher activities of allylic monoselenides having an *n*-propyl alcohol or *n*-propyl amide substituent can be ascribed to their facile reaction with peroxide. For example, compounds **15** reacts rapidly with peroxide to produce the selenoxide **65**, which undergoes a facile [2,3] sigmatropic rearrangement to produce the selenelate ester **66** (Scheme 9.15). Oxidation of the selenium center in compound **66** leads to the formation of an unstable intermediate **68**, which undergoes a rapid cyclization to produce the cyclic seleninate ester **69**. Overall, the selenide **15** undergoes a series of oxidation and [2,3]-sigmatropic rearrangement steps, which lead to the formation of the cyclic seleninate ester **69**. Formation of cyclic product **67** was not observed due to overoxidation at the selenium center. While compound **15** acts as a procatalyst, the seleninate ester **69** has been shown to be the true catalyst in the process. In summary, the study by Back and coworkers suggest that GPx-like activity of allyl selenides is generally much higher than that of aliphatic diselenides due to a novel catalytic pathway.

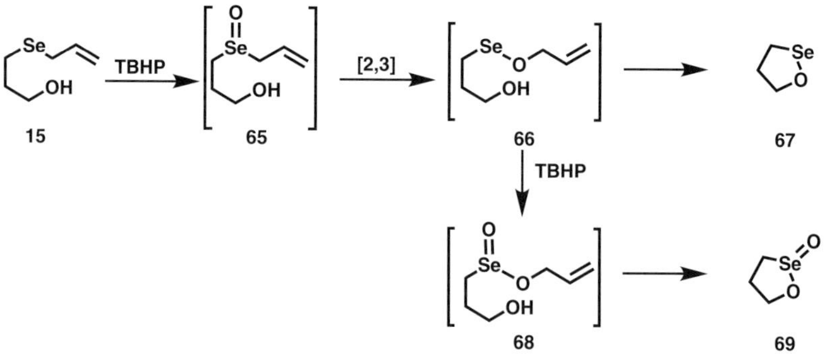

Scheme 9.15 Reaction of compound **15** with peroxide produces the selenoxide that undergoes [2,3] sigmatropic rearrangement to produce the cyclic compound **69** [120].

Singh and coworkers have reported the GPx activity of the selenenate ester **16**, which undergoes redox reactions in the presence of PhSH and H_2O_2. In the catalytic cycle, compound **16** reacts rapidly with H_2O_2 to the corresponding seleninate **70**, which upon reaction with PhSH produces the thioseleninate **71**. Addition of PhSH to compound **71** produces the corresponding selenenic acid **72**. Elimination of a water molecule then regenerates the selenenate ester **16**. Similar to the catalytic cycle of ebselen, the cyclization of the selenenic acid to cyclic compounds may protect the selenium moiety from overoxidation. The facile isolation of compound **16** indicates the importance of Se···O interactions in the stabilization and isolation of unstable species. These results are in agreement with the reports of Back and coworkers that compounds containing Se–O bonds can be equally effective catalysts as the commonly studied Se–N derivatives (Scheme 9.16).

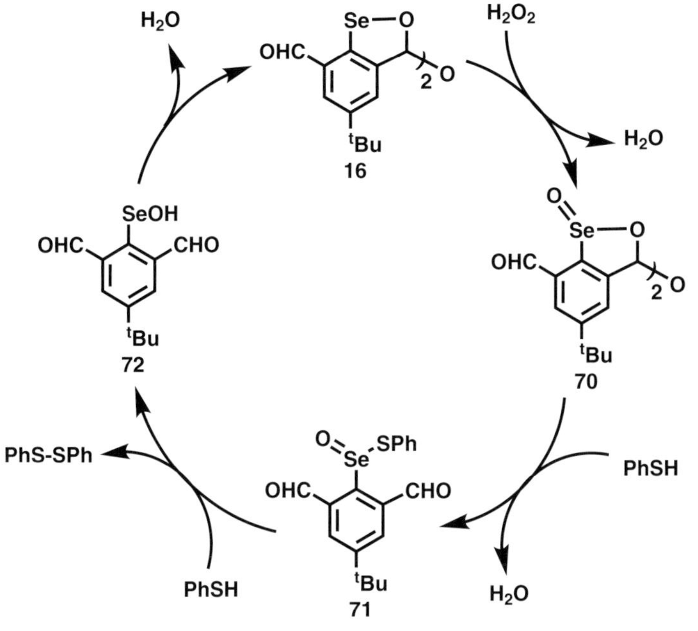

Scheme 9.16 GPx-like catalytic cycle of compound **13** [121].

9.7.2
Iodothyronine Deiodinase

Thyroid gland is one of the largest and most important organs in the endocrine system. The main biological function of the thyroid gland is the production of thyroid hormones such as thyroxine (T4) and 3,5,5′-triiodothyronine (T3), apart from maintaining the normal functions of the cardiovascular system and the central nervous system. T4 is a prohormone, which is produced *in vivo* by thyroid peroxidase that catalyzes the conversion of L-tyrosine to T4 by iodination followed by phenolic coupling reactions. The triiodo derivative T3 is the active thyroid hormone, which is produced from T4 by an outer-ring deiodination by iodothyronine deiodinases (ID-1 or ID-2, Scheme 9.17) [122–128]. The phenolic ring is referred to as outer ring and the tyrosyl ring is referred to as inner ring. From the crystal structure of T3, it is observed that the two aromatic rings are almost perpendicular with an angle between the two planes being 82° [129].

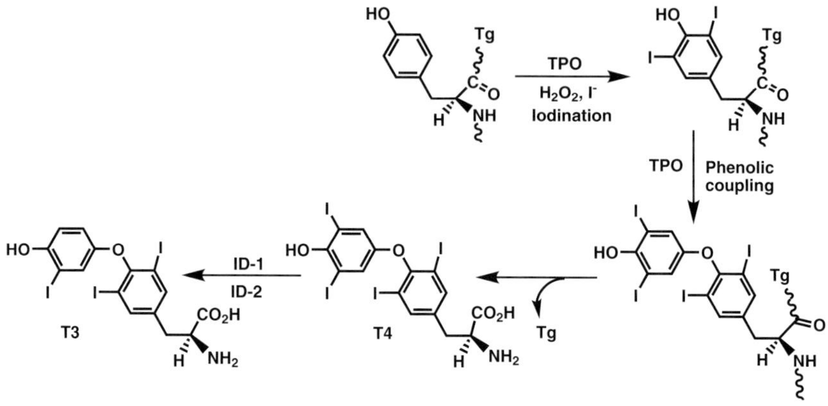

Scheme 9.17 Biosynthesis of thyroid hormone *in vivo*. Iodination of tyrosine residue of thyroglobulin by TPO followed by a phenolic coupling reaction results in the formation of prohormone T4. Outer-ring deiodination of T4 by iodothyronine deiodinases leads to the formation of active hormone T3.

IDs catalyze the regioselective deiodination of the various iodothyronines. These enzymes can be classified as ID-1, ID-2, or ID-3 depending on the position of deiodination and their selectivity for inner- versus outer-ring deiodination. ID-1 removes iodine from both outer and inner-rings and the enzyme activity is effectively inhibited by thiourea-based compounds such as 6-*n*-propyl-2-thiouracil (PTU) [130]. In contrast, ID-2 removes iodine selectively from the outer ring of T4 and the type 3 enzyme (ID-3) is selective to the removal of inner-ring iodines [131, 132]. In contrast to ID-1, the other two enzymes are less sensitive to the PTU and other related inhibitors. All the three types of deiodinases constitute a group of dimeric integral membrane proteins containing a Trx-type fold [133–135]. The outer-ring deiodination of T4 by ID-1 generates the active hormone T3 and inner-ring deiodination leads to the formation of reverse-T3 (rT3). ID-1 also removes iodine from the inner ring of rT3 to generate T2 (Scheme 9.18). It is known that rT3 is a better substrate for ID-1 than T4 or T3. Therefore, ID-1 is not a kinetically efficient enzyme. In contrast, ID-2 catalyzes the removal of iodine from the outer ring of T4 and rT3 to generate T3 and T2, respectively. As T4 is a better substrate for ID-2 as compared to rT3, ID-2 generally controls the production of the active hormone T3. In contrast to ID-2, ID-3 catalyzes the deiodination of inner-ring iodines of T4 and T3 to produce rT3 and T2, respectively. For ID-3, T3 has been

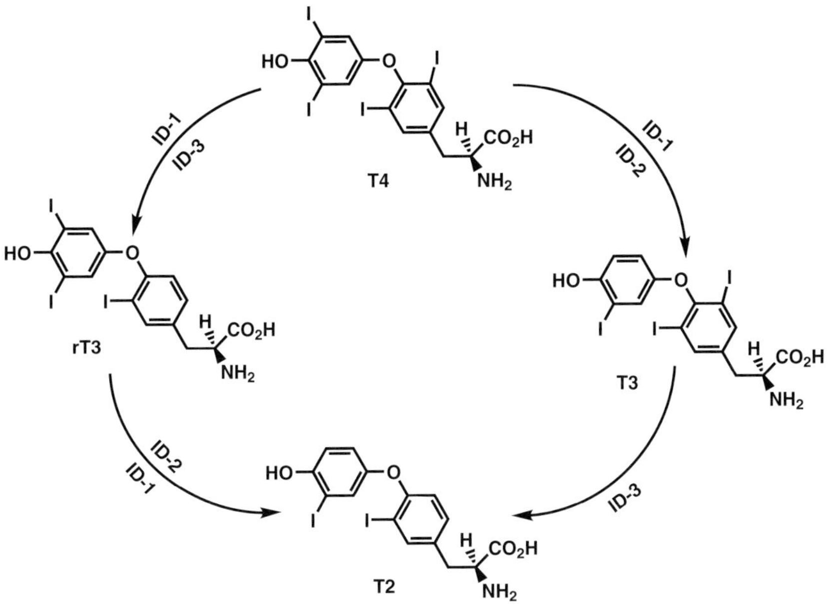

Scheme 9.18 Regioselective deiodination reactions catalyzed by three different iodothyronine deiodinases.

shown to be a better substrate than T4. Therefore, ID-3 is mainly responsible for the inactivation of the thyroid hormones. ID-2 and ID-3 together maintain the cellular balance in the production of thyroid hormones. Various properties and the substrate specificity for all the three types of deiodinases are summarized in Table 9.3.

All the three types of IDs are integral membrane-bound enzymes. Because of difficulties in purification of membrane-bound proteins and expression of selenocysteine-containing proteins, the crystal structures of IDs are not available. However, analyses of the amino acid sequences of these enzymes indicate that all three enzymes contain a selenocysteine residue at the active site and the amino acid residues near the active site are conserved in different species. The key amino acid residues near the active site are summarized in Table 9.4. Callebaut et al. [135]

Table 9.3 Summary of properties of the iodothyronine deiodinases [131, 136, 137].

	ID-1	ID-2	ID-3
Molecular mass (Da)	29 000	30 500	31 500
Tissue	Liver, kidney, thyroid	CNS, brown fat, pituitary	CNS, skin, placenta
Site of deiodination	Inner and outer	Outer	Inner
Substrate	rT3 > T4 > T3	T4 ≥ rT3	T3 > T4
Active site	Sec	Sec	Sec

and Dentice et al. [138] have carried out extensive sequence alignment experiments of the ID enzymes and proposed a three-dimensional active site. According to their model studies, the deiodinases contain a Trx-type folding motif as shown in Figure 9.9.

Table 9.4 Conserved amino acid residues in ID enzymes [135].

ID-1	ID-2	ID-3
F121	F128	F139
S123	S130	S141
Sec126	Sec133	Sec144
S128	P135	P146
F129	F136	F147
E155	D162	E173
E156	E163	E174
H158	H165	H176
S160	S167	S178
W163	W170	W181
H174	H185	H193
S212	S226	Y231
E214	E228	E233
H158	H165	H176

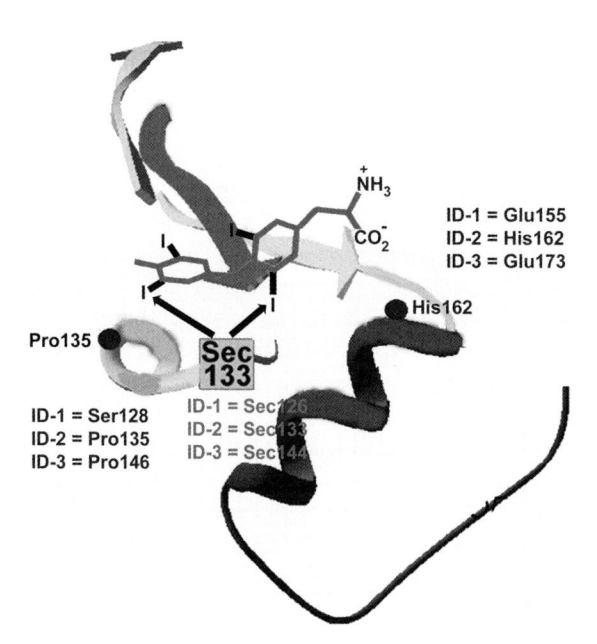

Figure 9.9 Schematic representation of the putative active site of deiodinases deduced from sequence alignment considering the ID-2 enzyme as basic structure. Position for P135 in ID-2 is substituted by S128 in ID-1 and H165 in ID-2 is replaced by E158 and E176, respectively, in ID-1 and ID-3. Figure is modified from Callebaut et al. [135].

The production of thyroid hormones is controlled *in vivo* by hypothalamus and pituitary gland. When low levels of thyroid hormones are detected in the blood by the hypothalamus and the pituitary, thyroid releasing hormone (TRH) is released, which stimulates the pituitary to release the thyroid-stimulating hormone (TSH). Consequently, increased levels of TSH stimulate the thyroid to produce more thyroid hormone, thereby returning the level of thyroid hormone in the blood back to normal. The three glands and the hormones they produce make up the hypothalamic–pituitary–thyroid control system (Figure 9.10). When high levels of thyroid hormones are detected in the blood, there is a negative control of the TRH and TSH, which leads to a decrease in the production of the thyroid hormones. When the TSH receptors are activated by certain auto-antibodies, the thyroid gland starts producing the hormones without any signal from the hypothalamus–pituitary system, leading to an overproduction of thyroid hormones. In addition, these auto-antibodies also stimulate ID-1 and probably ID-2, which then together produce T3. As these antibodies are not under pituitary feedback control system, there is no negative influence on thyroid activity, and therefore, the uncontrolled production of thyroid hormones leads to "hyperthyroidism," which is generally treated by antithyroid drugs such as methimazole, PTU, and carbimazole that

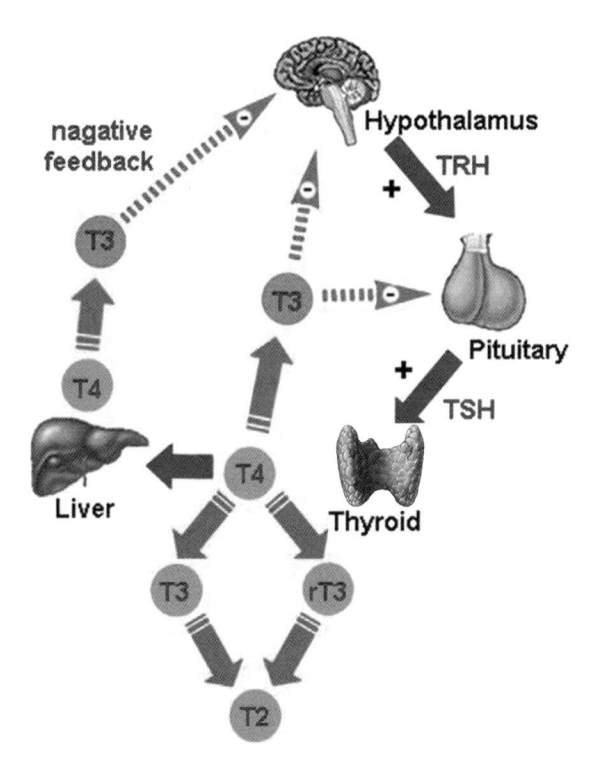

Figure 9.10 Schematic representation of the regulation of thyroid hormone production by the hypothalamic–pituitary–thyroid control system.

either block the thyroid hormone biosynthesis or reduce the conversion of T4 to T3 [122–128, 139, 140]. Similarly, low level of thyroid hormone production leads to "hypothyroidism."

The 5′-deiodination catalyzed by ID-1 is a ping-pong, bisubstrate reaction in which the selenol (or selenolate) group of the enzyme (E–SeH or E–Se$^-$) first reacts with thyroxine (T4) to produce an enzyme–selenenyl iodide intermediate. Subsequent reaction of the selenenyl iodide with an yet unidentified intracellular cofactor (dithiothreitol *in vitro*) completes the catalytic cycle and regenerates the enzyme active site [141, 142]. It is known that the antithyroid drug, PTU, inhibits the activity of the enzyme probably by reacting with the selenenyl iodide intermediate, and that gold-containing drugs such as gold thioglucose inhibit the deiodinase activity by reacting with the selenol (or selenolate) group of the native enzyme. According to the mechanism shown in Scheme 9.19, the complete cycle requires two substrates: thyroxine (T4) and a cellular thiol or another cofactor.

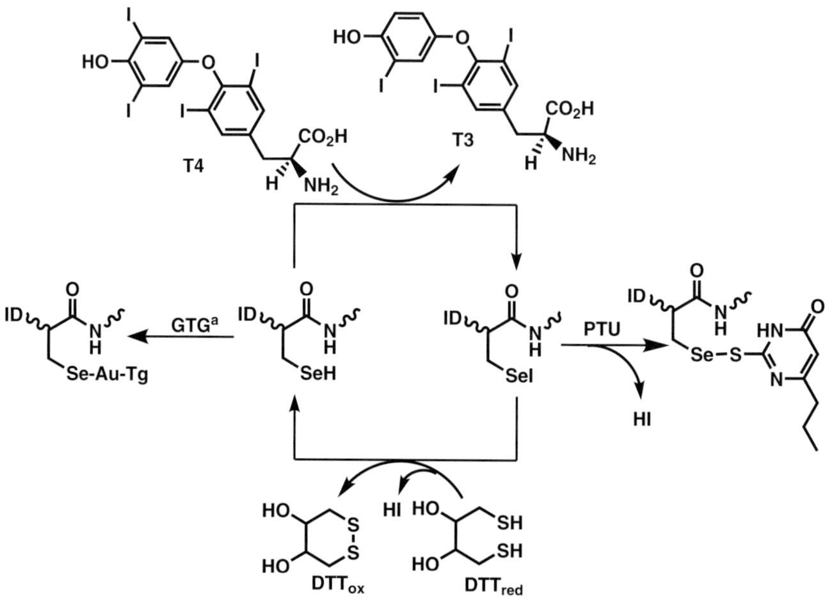

Scheme 9.19 Proposed mechanism for the inhibition of deiodinase activity of type-I iodothyronine deiodinase by gold thioglucose with the formation of a protein–gold–selenolate complex.

9.7.3
Synthetic Mimics of IDs

In recent years, synthetic mimics of IDs gained interest due to their potential pharmaceutical applications. Initial attempts to functionally mimic the deiodinases have seen only limited success. In one of the initial studies, Reglinski and coworkers have employed benzeneselenol (PhSe$^-$) as the deiodinating reagent to remove iodine from a few diiodotyrosine derivatives by refluxing in ethanol [143]. However, it has been found that benzeneselenolate can deiodinate only the activated iodo compound **75** to generate compound **78** (Scheme 9.20).

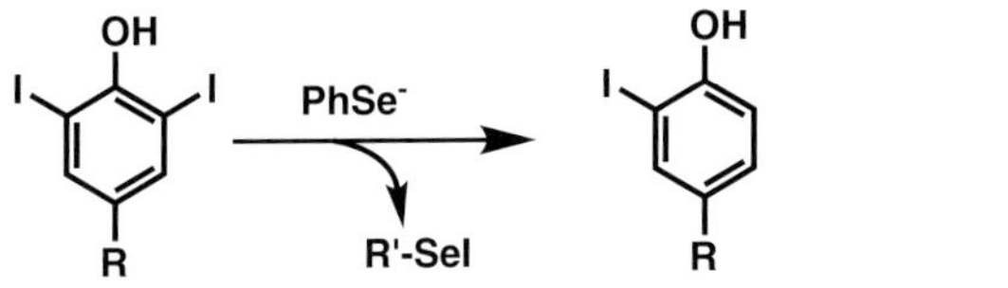

73, R = CH$_2$CH(NH$_2$)CO$_2$H
74, R = NH$_2$
75, R = NO$_2$

76, R = CH$_2$CH(NH$_2$)CO$_2$H
77, R = NH$_2$
78, R = NO$_2$

Scheme 9.20 Deiodination of various diiodophenol derivatives by benzeneselenolate [143].

Compounds **73** and **74** cannot react under these reaction conditions. Engman and coworkers have demonstrated that phenyltellurolate (PhTe$^-$) is capable of removing iodine from the phenolic rings of various model compounds [144]. When the reactions were carried out with various sulfur, selenium, and tellurium-containing nucleophiles such Na$_2$Te, NaHTe, PhTe$^-$ Na$_2$Se, and Na$_2$S, only PhTe$^-$ exhibited some selectivity in the deiodination. This compound has been shown to remove one of the iodines from the aromatic ring of compound **79** to generate compound **80**. In contrast, removal of both the iodines from compound **79** to generate **81** was observed when Na$_2$S was used as the deiodinating agent (Scheme 9.21).

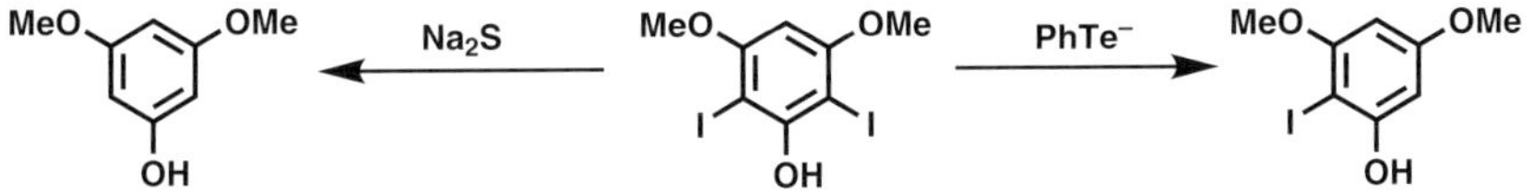

Scheme 9.21 Deiodination of model compounds by sulfur and tellurium reagents [144].

Recently, Goto and coworkers have demonstrated that *N*-butyrylthyroxine methyl ester (**82**), a thyroxine derivative, can be converted to the corresponding triiodo derivative **83** by the sterically hindered selenol **84** (Scheme 9.22) [145]. In this study, the selenium nucleophile has been shown to attack at one of the outer-ring iodines to produce the corresponding selenenyl iodide **85**, which was isolated and analyzed by X-ray crystallographic studies. According to the mechanism proposed for this transformation, the T4 derivative first undergoes a tautomerization to produce the corresponding keto derivative. The nucleophilic attack of the selenol at the positively charged iodine leads to the formation of the selenenyl iodide and compound **83** (Scheme 9.23).

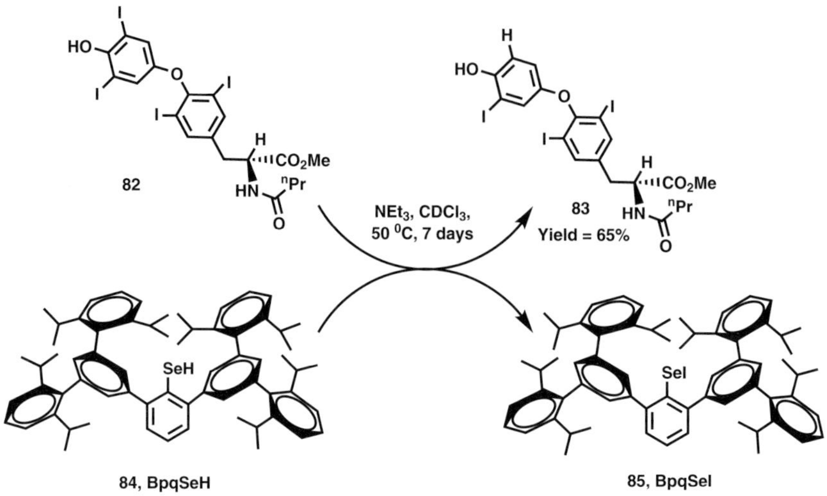

Scheme 9.22 Deiodination of the thyroxine derivative **82** by Bpq-selenol to produce the corresponding selenenyl iodide **85** and T3 derivative **83**.

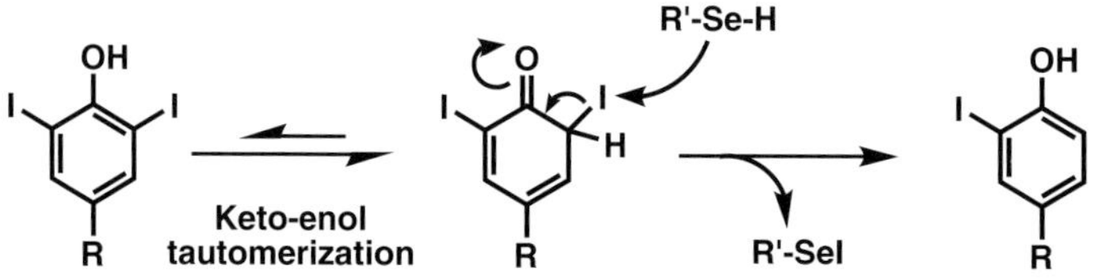

Scheme 9.23 Mechanism of deiodination via enol–keto tautomerism as proposed by Goto et al. [145].

This study provided an experimental evidence for the formation of a selenenyl iodide (R–Se–I) in the deiodination of a thyroxine derivative by an organoselenol.

From the above model studies, it appears that the outer-ring deiodination of T4 derivatives is more favored than the inner-ring deiodination. We have recently reported the first chemical model for the inner-ring deiodination of T4 and T3 to produce rT3 and T2, respectively, under physiologically relevant reaction conditions [146]. The naphthyl-based selenols and thiols (compounds **86–88**, Figure 9.11) were used as ID-3 mimics. This study indicates that the selenol **86** is more effective in removing iodine from the inner ring of T4 than the corresponding thiol (**87**). This study also suggests that the presence of a selenol is important for the deiodination activity as compound **88** having a phenyl ring attached to the selenium center does not show any noticeable activity. The presence of a thiol at the 8-position of the naphthyl ring is also important for the activity as compound **90** is found to be inactive. This is in agreement with the report by Sun et al. that the presence of a highly conserved Cys residue (Cys124) at the active site of ID-1 (Sec126, Table 9.4) plays an important role in enhancing both outer- and inner-ring deiodination reactions [147]. The lower activity of dithiol (**87**) as compared to

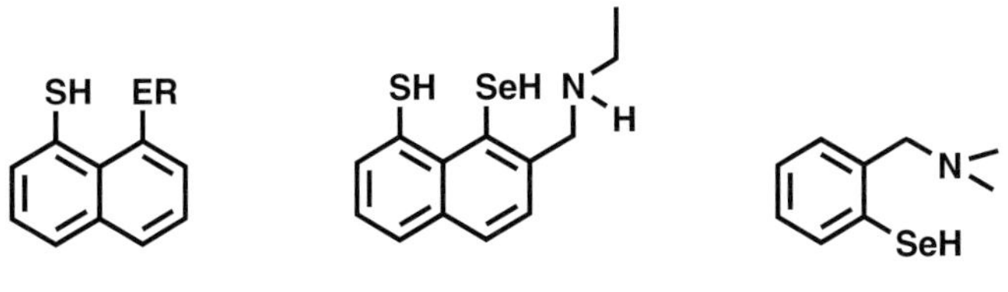

86, E = Se; R = H
87, E = S; R = H
88, E = Se; R = Ph

89

90

Figure 9.11 Model compounds used as ID-3 mimics [146].

compound **86** is also in agreement with the mutation studies, which demonstrated that the replacement of Sec by Cys reduces the substrate turnover number of T4 and T3 by six- and twofold, respectively.

9.7.4
Thioredoxin Reductase

Glutathione and Trx systems play key roles in the cellular redox regulation in almost all organisms, and therefore, are targets for a number of clinical conditions. Glutathione is a tripeptide containing a γ-glutamate linkage (Figure 9.12a) to the cysteine thiol, whereas Trxs are a class of small peptides (~10–12 kDa) containing redox reactive disulfide bonds. Typically, Trx contains a conserved -Trp-Cys-Gly-Pro-Cys-Lys- moiety at the active site. The human Trx contains 105 amino acid residues with the active site Cys residues at 32- and 35-positions (Figure 9.12b). Trx acts as cofactor in various oxidoreduction reactions *in vivo*. In a typical reaction performed by Trx, the active site cysteines undergo oxidation to form an internal disulfide bond [148–151]. Therefore, the oxidation of Trx represents the last step in the reactions involving this peptide as a cofactor. The catalytic site of Trx is regenerated by TrxR. The reduction of disulfide linkage in Trx by TrxR is very similar to the reduction of GSSG GR.

TrxR is a member of the pyridine-disulfide oxidoreductase family [152]. The enzymes of this family such as GR, lipoamide dehydrogenase, and trypanothione reductases form homodimers and each of the subunits contains a redox active intramolecular disulfide and a tightly bound FAD molecule. TrxR catalyzes the reduction of disulfide bond in Trx by using NADPH as a cofactor. The mammalian TrxR contains two redox active centers, one at the N-terminal (Cys59/Cys64) near the FAD moiety and the other at the highly flexible C-terminal (Cys497/Sec498). The Sec is the penultimate amino acid residue in TrxR [149, 150, 153]. This structural motif in TrxR provides the environment for extensive thiol-disulfide-selenenyl sulfide exchange reactions. The currently accepted model for the catalytic mechanism of these enzymes involves the transfer of electrons from the reduced nicotinamide ring of NADPH via FAD to the N-terminal cysteines (Cys59 and Cys64). The electron is further transferred to the Cys497/Sec498 redox pair (Scheme 9.24), where the Trx is reduced back to the dithiol form.

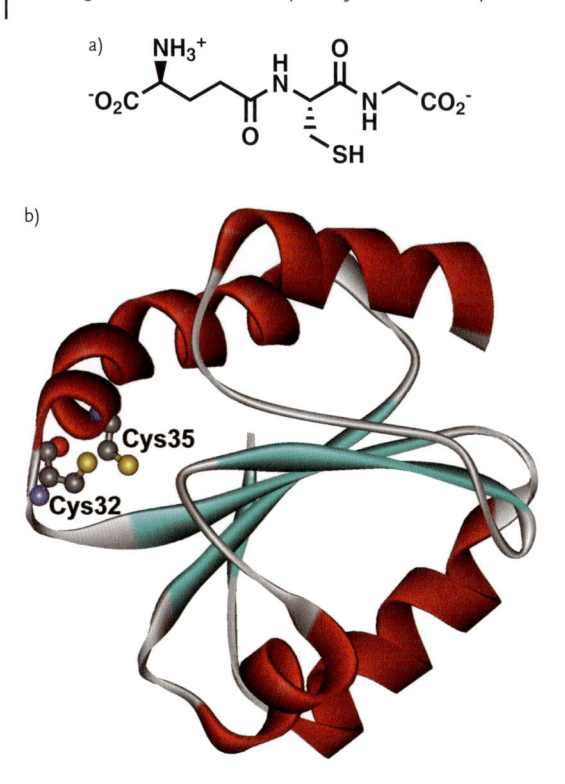

Figure 9.12 (a) Structure of glutathione showing γ-glutamate linkage to cysteine. (b) Crystal structure of human thioredoxin with active site thiols in reduced form (PDB code: 1AIU) [148].

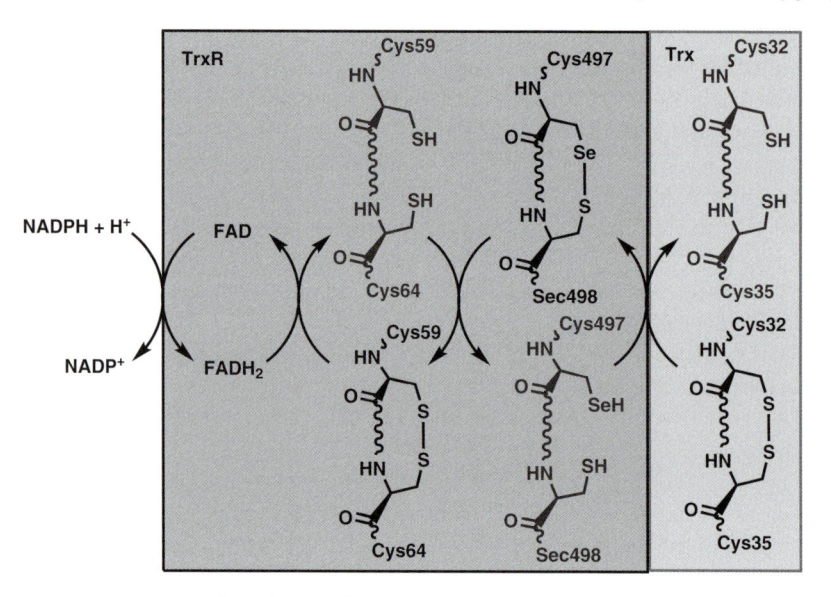

Scheme 9.24 Proposed mechanism for the reduction of Trx by TrxR.

Several mutational studies have been carried out to understand the role of selenium in TrxR and the crystal structure for the Sec to Cys mutant has been reported [154–158]. However, until recently, the crystal structure of native TrxR has not been reported due to the facile oxidation of the selenol moiety. Recently, Cheng *et al.* have reported the crystal structure of the native TrxR1 from rat [153], which confirms the homodimeric nature of the mammalian TrxR (Figure 9.13). The crystal structure indicates that the N-terminal redox pair is positioned near the C-terminal redox pair of the other domain. While the structure of one domain indicates that the Cys59–Cys64 redox pair is situated away from the Cys497–Sec498 redox pair, the structure of the active site indicates that the Cys497–Sec498 is located very close to the Cys59–Cys64 pair of the neighboring domain. The electron transfer takes place from N-terminal redox pair to C-terminal redox pair via FAD. The aromatic ring of Tyr116 participates in the electron transfer process as a tremendous loss of activity was observed on mutation of the crucial Tyr116.

The proposed mechanism for the catalytic reduction of Trx by TrxR is shown in Scheme 9.25. According to this mechanism, the reduction of FAD moiety by NADPH generates the EH_2 state of the enzyme. Electron transfer from FAD to the Cys59/Cys64 pair leads to the cleavage of the disulfide bond to produce the reduced cysteine residues. The proton transfer is facilitated by His472 and Glu477 residues present at the active site. The FAD moiety is oriented near the Cys59/Cys64 pair and a charge transfer from Cys64 to FAD leads to the stabilization of the dithiol. This leads to the formation of the selenenyl sulfide intermediate. The electron transfer from dithiol to Cys497/Sec498 pair is greatly influenced by Tyr116 moiety. Further reduction of the selenenyl sulfide intermediate by thiol leads to the generation of active site selenol in the EH_4 state of the enzyme. An attack of the selenol at the disulfide linkage of Trx produces a -Se–S–S-Trx species, which upon reaction with free thiol group of TrxR generates the dithiol species. During the reduction of disulfide bond in Trx, the Cys497/Sec498 pair forms a selenenyl sulfide bond, which upon reduction by another thiol regenerates the active site. This catalytic mechanism indicates that the reduction of Trx by TrxR takes place via a number of steps and thiol exchange reactions play important roles. However, in contrast to the GPx mechanism, the thiol exchange reactions take place internally. Brandt *et al.* have proposed that His472, Glu477, and Sec498 form a "catalytic triad" at the active site, which is similar to the catalytic triad proposed for GPx [159].

9.8
Summary

After several years of extensive research on biochemical roles of selenium, this element is now considered as an essential micronutrient for mammals. The major biological form of selenium is represented by the amino acid selenocysteine (Sec, U). Although Sec is a rare amino acid, Sec shares many structural and functional

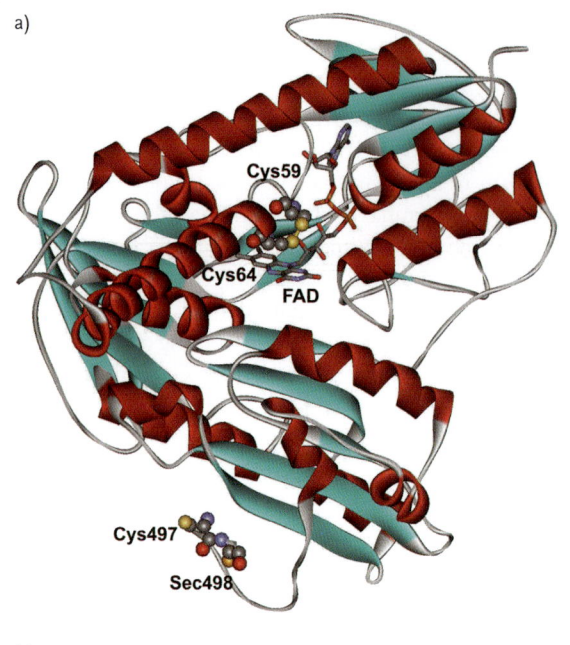

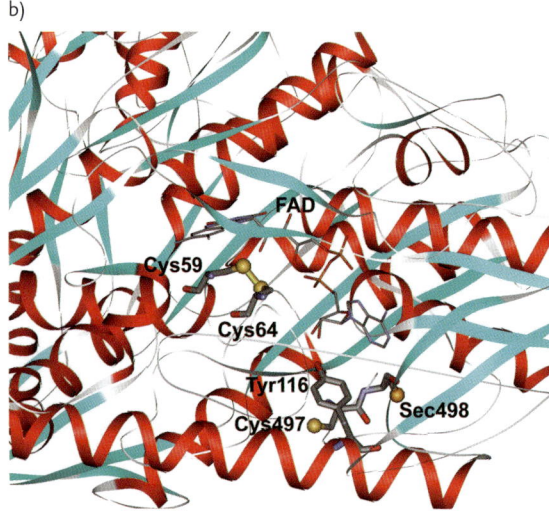

Figure 9.13 X-ray crystal structure of the active site of thioredoxin reductase. (a) Cys59–Cys64 forms an internal disulfide bond and Cys497–Sec498 pair is flexible at the C-terminal. (b) Tyr116 plays a crucial role in the electron transfer via FAD from Cys59–Cys64 to Cys497–Sec498 redox pairs. (TrxR, PDB Code 3EAN) [153].

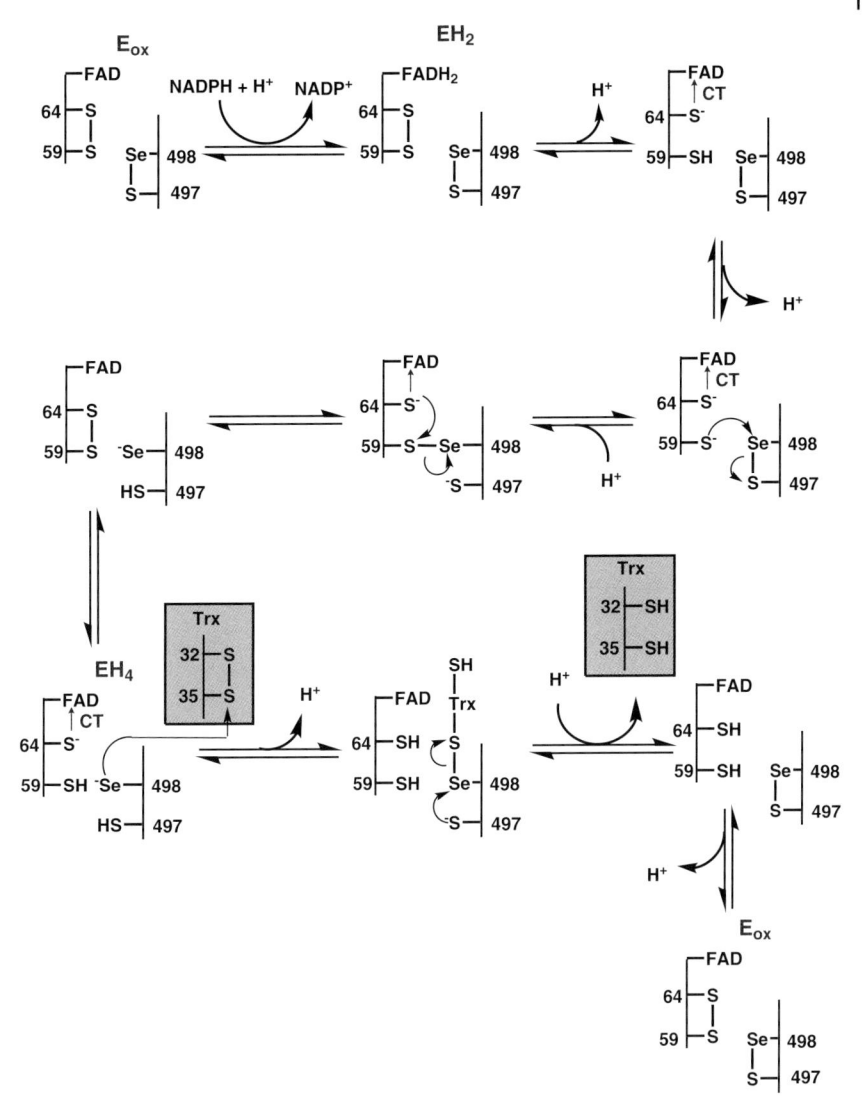

Scheme 9.25 The catalytic mechanism of thioredoxin reductase. This figure is modified from Cheng et al. [153].

similarities with cysteine (Cys, C). However, due to its interesting redox properties, many oxidoreductase enzymes utilize Sec for their biological activity. GPx represents an important mammalian oxidoreductase that contains a Sec residue at the active site. GPx functions as an antioxidant enzyme by reducing harmful hydroperoxides, which is important for the maintenance of cellular concentration of ROS *in vivo*. GPx mimics such as ebselen are in clinical trial as potential candidates for the treatment of a number of disease states including inflammation. In addition to the antioxidant properties, Sec plays an important role in the activation and inactivation of thyroid hormones. While the Sec-containing IDs control the thyroid hormone level, synthetic mimics of these enzymes may become important in the treatment of thyroid-related problems such as hyperthyroidism and hypothyroidism. The flavoprotein, TrxR, is involved in protein folding, redox control, and possibly detoxification. This enzyme catalyzes the reduction of Trx using NADPH as a reducing agent. It should be noted that the Trx system mediates the final step in the electron-transfer mechanism for the reduction of nucleoside diphosphate. The design and synthesis of compounds that mimic the biological function of TrxR may become an active area of research in future.

References

1 Berzelius, J.J. (1818) *Afhandl. Fys. Kemi Mineralog.*, **6**, 42.
2 Hutton, J.G. (1931) *J. Am. Chem. Soc.*, **23**, 1076.
3 Franke, K.W. (1934) *J. Nutr.*, **8**, 597.
4 Franke, K.W. (1934) *J. Nutr.*, **8**, 609.
5 Franke, K.W. and Potter, V.R. (1934) *J. Nutr.*, **8**, 615.
6 Franke, K.W. (1935) *J. Nutr.*, **10**, 223.
7 Franke, K.W. and Painter, E.P. (1935) *J. Nutr.*, **10**, 599.
8 Pinsent, J. (1954) *Biochem. J.*, **57**, 10.
9 Andreesen, J.R. and Ljungdahl, L. (1973) *J. Bacteriol.*, **116**, 867.
10 Turner, D.C. and Stadtman, T.C. (1973) *Arch. Biochem. Biophys.*, **154**, 366.
11 Flohé, L., Günzler, E.A., and Schock, H.H. (1973) *FEBS Lett.*, **32**, 132.
12 Rotruck, J.T., Pope, A.L., Ganther, H.E., Swason, A.B., and Hoekstra, D.G. (1973) *Science*, **179**, 588.
13 Böck, A. (1994) Selenium proteins containing selenocysteine, in *Encyclopedia of Inorganic Chemistry*, vol. 8, (ed. R.B. King), John Wiley Ltd, Chichester, UK, p. 370.
14 Flohé, L., Andreesen, J.R., Brigelius-Flohé, R., Maiorino, M., and Ursini, F. (2000) *IUBMB Life*, **49**, 411.
15 Boyington, J.C., Gladyshev, V.N., Khangulov, S.V., Stadtman, T.C., and Sun, P.D. (1997) *Science*, **275**, 1305.
16 Wilting, R., Schorling, S., Persson, B.C., and Böck, A. (1977) *J. Mol. Biol.*, **266**, 637.
17 Garcin, E., Vernede, X., Hatchikian, E.C., Volbeda, A., Frey, M., and Fontecilla-Camps, J.C. (1999) *Structure*, **7**, 557.
18 Pfeiffer, M., Bingemann, R., and Klein, A. (1998) *Eur. J. Biochem.*, **256**, 447.
19 Wagner, M., Sonntag, D., Grimm, R., Pich, A., Eckerskorn, C., Söhling, B., and Andreesen, J.R. (1999) *Eur. J. Biochem.*, **260**, 38.
20 Böck, A., Forchhammer, K., Heider, J., Leinfelder, W., Sawers, G., Veprek, B., and Zinoni, F. (1991) *Mol. Microbiol.*, **5**, 515.
21 Stadtman, T.C. (1996) *Annu. Rev. Biochem.*, **65**, 83.
22 Dobbek, H., Gremer, L., Meyer, O., and Huber, R. (1999) *Proc. Natl. Acad. Sci. U.S.A.*, **96**, 8884.
23 Behne, D., Kyriakopoulos, A., Meinhold, H., and Köhrle, J. (1990) *Biochem. Biophys. Res. Commun.*, **173**, 1143.
24 Arthur, J.R., Nicol, F., and Beckett, G.J. (1990) *Biochem. J.*, **272**, 537.
25 Davey, J.C., Becker, K.B., Schneider, M.J., Germain, G.L., and Galton, V.A. (1995) *J. Biol. Chem.*, **270**, 26786.
26 Croteau, W., Whittemore, S.K., Schneider, M.J., and Germain, D.L. (1995) *J. Biol. Chem.*, **270**, 16569.
27 Lescure, A., Gautheret, D., Carbon, P., and Krol, A. (1999) *J. Biol. Chem.*, **274**, 38147.
28 Tamura, T., and Stadtman, T.C. (1996) *Proc. Natl. Acad. Sci. U.S.A.*, **93**, 1006.
29 Lee, S.R., Kim, J.R., Kwon, K.S., Yoon, H.W., Leveine, R.L., Ginsburg, A., and Rhee, S.G. (1999) *J. Biol. Chem.*, **274**, 4722.

30 Watabe, S., Makino, Y., Ogawa, K., Hiroi, T., Yamamoto, Y., and Takahashi, S.Y. (1999) *Eur. J. Biochem.*, **264**, 74.
31 Mustacich, D. and Powis, G. (2000) *Biochem. J.*, **346**, 1.
32 Mills, G.C. (1957) *J. Biol. Chem.*, **229**, 189.
33 Motsenbocker, M.A. and Tappel, A.L. (1984) *J. Nutr.*, **114**, 279.
34 Ursini, F., Maiorino, M., Valente, M., Ferri, L., and Gregolin, C. (1982) *Biochim. Biophys. Acta*, **710**, 197.
35 Takahasi, K., Avissar, N., Whittin, J., and Cohen, H. (1987) *Arch. Biochem. Biophys.*, **256**, 677.
36 Chu, F.-F., Doroshow, J.H., and Esworthy, R.S. (1993) *J. Biol. Chem.*, **268**, 2571.
37 Gromer, S., Eubel, J.K., Lee, B.L., and Jacob, J. (2005) *Cell. Mol. Life Sci.*, **62**, 2414.
38 Dietary Supplement Fact Sheet: Selenium. National Institutes of Health; Office of Dietary Supplements, http://ods.od.nih.gov/factsheets/selenium/ (accessed 5 January 2009).
39 Combs, G.F. (2000) *Biofactors*, **12**, 39.
40 Zimmerman, M.B., and Kohrle, J. (2002) *Thyroid*, **12**, 867.
41 Cahn, R.S., Ingold, C.K., and Prelog, V. (1966) *Angew. Chem. Int. Ed.*, **5**, 385.
42 Prelog, V. and Helmchen, G. (1982) *Angew. Chem. Int. Ed.*, **21**, 567.
43 Huber, R.E. and Criddle, R.S. (1967) *Arch. Biochem. Biophys.*, **122**, 164.
44 Jacob, C., Giles, G.I., Giles, N.M., and Sies, H. (2003) *Angew. Chem. Int. Ed.*, **42**, 4742.
45 Walczak, R., Westhof, E., Carbon, P., and Krol, A. (1996) *RNA*, **2**, 367.
46 Sliwkowski, M.X., and Stadtman, T.C. (1985) *J. Biol. Chem.*, **260**, 3140.
47 Young, P.A. and Kaiser, I.I. (1975) *Arch. Biochem. Biophys.*, **171**, 483.
48 Leinfelder, W., Zehelein, E., Mandrand-Berthelot, M.A., and Böck, A. (1988) *Nature*, **331**, 723.
49 Gladyshev, V.N. and Hatfield, D.L. (1999) *J. Biomed. Sci.*, **6**, 151.
50 Kryukov, G.V., Castellano, S., Novoselov, S.V., Lobanov, A.V., Zehtab, O., Guigo, R., and Gladyshev, V.N. (2003) *Science*, **300**, 1439.
51 Sheppard, K., Yuan, J., Hohn, M.J., Jester, B., Devine, K.M., and Söll, D. (2008) *Nucleic Acid Res.*, **36**, 1813.
52 Forchhammer, K., Leinfelder, W., Boesmiller, K., Veprek, B., and Böck, A. (1991) *J. Biol. Chem.*, **266**, 6318.
53 Yuan, J., Palioura, S., Salazar, J.C., Su, D., O'Donoghue, P., Hohn, M.J., Cardoso, A.M., Whitman, W.B., and Söll, D. (2006) *Proc. Natl. Acad. Sci. U.S.A.*, **103**, 18923.
54 Carlson, B.A., Xu, X.M., Kryukov, G.V., Rao, M., Berry, M.J., Gladyshev, V.N., and Hatfield, D.L. (2004) *Proc. Natl. Acad. Sci. U.S.A.*, **101**, 12848.
55 Palioura, S., Sherrer, R.L., Steitz, T.A., Söll, D., and Simonović, M. (2009) *Science*, **325**, 321.
56 Fukunaga, R. and Yokoyama, S. (2007) *J. Mol. Biol.*, **370**, 128.
57 Fredga, A. (1936) *Svensk Kemisk Tidskrift*, **48**, 160.
58 Chocat, P., Esaki, N., Tanaka, H., and Soda, K. (1985) *Anal. Bioanal. Chem.*, **148**, 485.
59 Tanaka, H. and Soda, K. (1987) *Meth. Enzymol.*, **143**, 240.
60 Stocking, E.M., Schwarz, J.N., Senn, H., Salzmann, M., and Silks, L.A. (1997) *J. Chem. Soc., Perkin Trans. 1*, (25) 2443.
61 Siebum, A.H.G., Woo, W.S., Raap, J., and Lugtenburg, J. (2004) *Eur. J. Org. Chem.*, **13**, 2905.
62 Theodoropoulos, D., Schwartz, I.L., and Walter, R. (1967) *Biochemistry*, **6**, 3927.
63 Gieselman, M.D., Xie, L., and van Der Donk, W.A. (2001) *Org. Lett.*, **3**, 1331.
64 Muttenthaler, M. and Alewood, P.F. (2008) *J. Pept. Sci.*, **14**, 1223.
65 Bhuyan, B.J. and Mugesh, G. (2011) *Org. Biomol. Chem.*, **9**, 1356.
66 Wieland, T., Bokelmann, E., Bauer, L., Lang, H.U., and Lau, H. (1953) *Liebigs Ann. Chem.*, **583**, 129.
67 Dawson, P.E., Muir, T.W., Clark-Lewis, I., and Kent, S.B. (1994) *Science*, **266**, 776.
68 Dawson, P.E. and Kent, S.B. (2000) *Annu. Rev. Biochem.*, **69**, 923.
69 Hondal, R.J., Nilsson, B.L., and Raines, R.T. (2001) *J. Am. Chem. Soc.*, **123**, 5140.
70 Roelfes, G. and Hilvert, D. (2003) *Angew. Chem. Int. Ed.*, **42**, 2275.

71 Roy, G., Sarma, B.K., Phadnis, P.P., and Mugesh, G. (2005) *J. Chem. Sci.*, **117**, 287.
72 Quaderer, R., Sewing, A., and Hilvert, D. (2001) *Helv. Chim. Acta*, **84**, 1197.
73 Maiorino, M., Aumann, K.-D., Brigelius-Flohé, R., Doria, R.D., van den Heuvel, L., McCarthy, J., Rovery, A., Ursini, F., and Flohé, L. (1995) *Biol. Chem. Hoppe-Seyler*, **376**, 651.
74 Rocher, C., Lalanne, J.-L., and Chaudiére, J. (1992) *Eur. J. Biochem.*, **205**, 955.
75 Maddipati, K.R. and Marnett, L.J. (1987) *J. Biol. Chem.*, **262**, 17398.
76 Brigelius-Flohé, R. (1999) *Free Radic. Biol. Med.*, **27**, 951.
77 Maiorino, M., Gregolin, C., and Ursini, F. (1990) *Meth. Enzymol.*, **186**, 448.
78 Björnstedt, M., Xue, J., Huang, W., Åkesson, B., and Holmgren, A. (1994) *J. Biol. Chem.*, **269**, 29382.
79 Maiorino, M., Roveri, A., Coassin, M., and Ursini, F. (1988) *Biochem. Pharmacol.*, **37**, 2267.
80 Flohé, L. (1982) *Free Radicals in Biology*, vol. 5, (ed. W.A. Pryor), Academic Press, New York, p. 223.
81 Epp, O., Ladenstein, R., and Wendel, A. (1983) *Eur. J. Biochem.*, **133**, 51–69.
82 Ursini, F., Maiorino, M., Brigelius-Flohé, R., Aumann, K.-D., Roveri, A., Schomburg, D., and Flohé, L. (1995) *Meth. Enzymol.*, **252**, 38.
83 Mugesh, G., and Singh, H.B. (2000) *Chem. Soc. Rev.*, **29**, 347.
84 Müller, A., Cadenas, E., Graf, P., and Sies, H. (1984) *Biochem. Pharmacol.*, **33**, 3235.
85 Wendel, A., Fausel, M., Safayhi, H., Tiegs, G., and Otter, R. (1984) *Biochem. Pharmacol.*, **33**, 3241.
86 Sies, H. (1986) *Angew. Chem. Int. Ed.*, **25**, 1058.
87 Mugesh, G., du Mont, W.-W., and Sies, H. (2001) *Chem. Rev.*, **101**, 2125.
88 Masumoto, H., Kissner, R., Koppenol, W.H., and Sies, H. (1996) *FEBS Lett.*, **398**, 179.
89 Noguchi, N., Yoshida, Y., Kaneda, H., Yamamoto, Y., and Niki, E. (1992) *Biochem. Pharmacol.*, **44**, 39.
90 Zembowicz, A., Hatchett, R.J., Radziszewski, W., and Gryglewski, R.J. (1993) *J. Pharmacol. Exp. Ther.*, **267**, 1112.
91 Hattori, R., Inoue, R., Sase, K., Eizawa, H., Kosuga, K., Aoyama, T., Masayasu, H., Kawai, C., Sasayama, S., and Yui, Y. (1994) *Eur. J. Pharmacol.*, **267**, R1.
92 Nikawa, T., Schuch, G., Wagner, G., and Sies, H. (1994) *Biochem. Pharmacol.*, **47**, 1007.
93 Lesser, R. and Weiss, R. (1924) *Dtsch. Chem. Ges.*, **57**, 1077.
94 Engman, L. and Hallberg, A. (1989) *J. Org. Chem.*, **54**, 2964.
95 Zade, S.S., Panda, S., Tripathi, S.K., Singh, H.B., and Wolmershaeuser, G.E. (2004) *Eur. J. Org. Chem.*, **18**, 3857.
96 Galet, V., Bernier, J.-L., Hénichart, J.-P., Lesieur, D., Abadie, C., Rochette, L., Lindenbaum, A., Chalas, J., Renaud de la Faverie, J.-F., Pfeiffer, B., and Renard, P. (1994) *J. Med. Chem.*, **37**, 2903.
97 Andersson, C.M., Hallberg, A., Brattsand, R., Cotgreave, I.A., Engman, L., and Persson, J. (1993) *Bioorg. Med. Chem. Lett.*, **3**, 2553.
98 Wilson, S.R., Zucker, P.A., Huang, R.-R.C., and Spector, A. (1989) *J. Am. Chem. Soc.*, **111**, 5936.
99 Iwaoka, M. and Tomoda, S. (1994) *J. Am. Chem. Soc.*, **116**, 2557.
100 Jauslin, M.L., Wirth, T., Meier, T., and Schoumacher, F. (2002) *Hum. Mol. Genet.*, **11**, 3055.
101 Wirth, T. (1998) *Molecules*, **3**, 164.
102 Sun, Y., Mu, Y., Ma, S., Gong, P., Yan, G., Liu, J., Shen, J., and Luo, G. (2005) *Biochim. Biophys. Acta*, **1743**, 199.
103 Mugesh, G., Panda, A., Singh, H.B., Punekar, N.S., and Butcher, R.J. (2001) *J. Am. Chem. Soc.*, **123**, 839.
104 Sarma, B.K., Manna, D., Minoura, M., and Mugesh, G. (2010) *J. Am. Chem. Soc.*, **132**, 5364.
105 Wu, Z.-P. and Hilvert, D. (1989) *J. Am. Chem. Soc.*, **111**, 4513.
106 Wu, Z.-P. and Hilvert, D. (1990) *J. Am. Chem. Soc.*, **112**, 5647.
107 Engman, L., Stern, D., Cotgreave, I.A., and Andersson, C.M. (1992) *J. Am. Chem. Soc.*, **114**, 9737.
108 Bell, I.M. and Hilvert, D. (1993) *Biochemistry*, **32**, 13969.
109 Back, T.G. and Dyck, B.P. (2079) *J. Am. Chem. Soc.*, **1997**, 119.

110 Fisher, H. and Dereu, N. (1987) *Bull. Soc. Chim. Belg.*, **96**, 757.
111 Kice, J.L. and Purkiss, D.W. (1987) *J. Org. Chem.*, **52**, 3448.
112 Sarma, B.K. and Mugesh, G. (2005) *J. Am. Chem. Soc.*, **127**, 11477.
113 Bhabak, K.P. and Mugesh, G. (2007) *Chem. Eur. J.*, **13**, 4594.
114 Sarma, B.K. and Mugesh, G. (2008) *Chem. Eur. J.*, **14**, 10603.
115 Bhabak, K.P. and Mugesh, G. (2009) *Chem. Asian J.*, **4**, 974.
116 Bhabak, K.P. and Mugesh, G. (2008) *Chem. Eur. J.*, **14**, 8640.
117 Bhabak, K.P. and Mugesh, G. (2010) *Acc. Chem. Res.*, **43**, 1408.
118 Sarma, B.K. and Mugesh, G. (2008) *Org. Biomol. Chem.*, **6**, 965.
119 Back, T.G. and Moussa, Z. (2002) *J. Am. Chem. Soc.*, **124**, 12104.
120 Back, T.G. and Moussa, Z. (2003) *J. Am. Chem. Soc.*, **125**, 13455.
121 Zade, S.S., Singh, H.B., and Butcher, R.J. (2004) *Angew. Chem. Int. Ed.*, **43**, 4513.
122 Werner, S.C. and Ingbar, S. (1991), *The Thyroid: A Fundamental and Clinical Text*, (eds L.E. Braverman and R.D. Utiger), Lippincott, Philadelphia.
123 Darras, V.M., Hume, R., and Visser, T.J. (1999) *Mol. Cell Endocrinol.*, **151**, 37.
124 Carayon, P., and Ruf, J. (1990) *Thyroperoxidase and Thyroid Autoimmunity*, John Libbey, London.
125 Dunford, H.B. (1999) *Heme Peroxidases*, John Wiley & Sons, Inc., New York.
126 Visser, T.J., Leonard, J.L., Kaplan, M.M., and Larsen, P.R. (1982) *Proc. Natl. Acad. Sci. U.S.A.*, **79**, 5080.
127 Berry, M.J., Banu, L., and Larsen, P.R. (1991) *Nature*, **349**, 438.
128 Köhrle, J. (2005) *Thyroid*, **15**, 841.
129 Camerman, N. and Camerman, A. (1971) *Science*, **175**, 764.
130 Oppenheimer, J.H., Schwartz, H.L., and Surks, M.I. (1972) *J. Clin. Invest.*, **51**, 2493.
131 Bianco, A.C., Salvatore, D., Gereben, B., Berry, M.J., and Larsen, P.R. (2002) *Endocr. Rev.*, **23**, 38.
132 Köhrle, J. (1999) *Mol. Cell. Endocrinol.*, **151**, 103.
133 Bianco, A.C. and Larsen, P.R. (2005) *Thyroid*, **15**, 777.
134 Leonard, J.L., Visser, T.J., and Leonard, D.M. (2001) *J. Biol. Chem.*, **276**, 2600.
135 Callebaut, I., Morelli, C.C., Mornon, J.-P., Gereben, B., Buettner, C., Huang, S., Castro, B., Fonseca, T.L., Harney, J.W., Larsen, P.R., and Bianco, A.C. (2003) *J. Biol Chem.*, **278**, 36887.
136 Köhrle, J., Jakob, F., Contempré, B., and Dumont, J.E. (2005) *Endocr. Rev.*, **26**, 944.
137 Bianco, A.C. and Kim, B.W. (2006) *J. Clin. Invet.*, **116**, 2571.
138 Dentice, M., Bandyopadhyay, A., Gereben, B., Callebaut, I., Christoffolete, M.A., Kim, B.W., Nissim, S., Mornon, J.-P., Zavacki1, A.M., Zeöld, A., Capelo, L.P., Morelli, C.C., Ribeiro, R., Harney, J.W., Tabin, C.J., and Bianco, A.C. (2005) *Nat. Cell Biol.*, **7**, 698.
139 Roy, G., Nethaji, M., and Mugesh, G. (2004) *J Am. Chem. Soc.*, **126**, 2712.
140 Roy, G. and Mugesh, G. (2005) *J Am. Chem. Soc.*, **127**, 15207.
141 Berry, M.J., Kieffer, J.D., Harney, J.W., and Larsen, P.R. (1991) *J. Biol. Chem.*, **266**, 14155.
142 Köhrle, J. (2002) *Meth. Enzymol.*, **347**, 125.
143 Beck, C., Jensen, S.B., and Reglinski, J. (1994) *Bioorg. Med. Chem. Lett.*, **4**, 1353.
144 Vasil'ev, A.A. and Engman, L. (1998) *J. Org. Chem.*, **63**, 3911.
145 Goto, K., Sonoda, D., Shimada, K., Sase, S., and Kawashima, T. (2010) *Angew. Chem. Int. Ed.*, **49**, 545–547.
146 Manna, D. and Mugesh, G. (2010) *Angew. Chem. Int. Ed.*, **49**, 9246.
147 Sun, B.C., Harney, J.W., Berry, M.J., and Larsen, P.R. (1997) *Endocrinology*, **138**, 5452.
148 Andersen, J.F., Sanders, D.A.R., Gasdaska, J.R., Weichsel, A., Powis, G., and Montfort, W.R. (1997) *Biochemistry*, **36**, 13979.
149 Freemerman, A.J., Gallegos, A., and Powis, G. (1999) *Cancer Res.*, **59**, 4090.
150 Oblong, J.E., Berggren, M., Gasdaska, P.Y., and Powis, G. (1994) *J. Biol. Chem.*, **269**, 11714.
151 Sarma, B.K. and Mugesh, G. (2006) *Inorg. Chem.*, **45**, 5307.
152 Sandalova, T., Zhong, L., Lindqvist, Y., Holmgren, A., and Schneider, G. (2001) *Proc. Natl. Acad. Sci. U.S.A.*, **98**, 9533.

153 Cheng, Q., Sandalova, T., Lindqvist, Y., and Arnér, E.S.J. (2009) *J. Biol. Chem.*, **284**, 3998.
154 Deponte, M., Urig, S., Arscott, L.D., Fritz-Wolf, K.R.R., Herold-Mende, C., Koncarevic, S., Meyer, M., Davioud-Charvet, E., Ballou, D.P., Williams, C.H., Jr., and Becker, K. (2005) *J. Biol. Chem.*, **280**, 20628.
155 Schröder, D., Schwarz, H., Hrušák, J., and Pyykkö, P. (1998) *Inorg. Chem.*, **37**, 624.
156 Filipovska, A., Kelso, G.F., Brown, S.E., Beer, S.M., Smith, R.A., and Murphy, M.P. (2005) *J. Biol. Chem.*, **280**, 24113.
157 Isab, A.A., Hormann, A.L., Coffer, M.T., and Shaw, C.F. (1988) *J. Am. Chem. Soc.*, **110**, 3278.
158 Coffer, M.T., Shaw, C.F., Hormann, A.L., Mirabelli, C.K., and Crooke, S.T. (1987) *J. Inorg. Biochem.*, **30**, 177.
159 Brandt, W. and Wessjohann, L.A. (2005) *ChemBioChem*, **6**, 386.

⁷⁷Se NMR Values

Compiled with the help of all chapter authors and Dr. Sohail A. Shazad

Compound	Chapter/No.	Chemical shift (ppm)	Solvent	Reference
(cyclohexyl-Se)$_2$ with isopropyl	1/	290	CDCl$_3$	Scianowski, J., Rafinski, Z., and Wojtczak, A. (2006) Eur. J. Org. Chem., 3216–3225.
cyclohexyl-SePh with Me	1/	305	CDCl$_3$	Duddeck, H., Wagner, P., and Biallaß, A. (1991) Magn. Reson. Chem., 29, 248–259.
(cyclohexyl-Se)$_2$ with isopropyl	1/	336	CDCl$_3$	Scianowski, J., Rafinski, Z., and Wojtczak, A. (2006) Eur. J. Org. Chem., 3216–3225.
cyclohexyl-SePh with Me	1/	345	CDCl$_3$	Duddeck, H., Wagner, P., and Biallaß, A. (1991) Magn. Reson. Chem., 29, 248–259.
bicyclic (Se)$_2$	1/	351	CDCl$_3$	Scianowski, J., Rafinski, Z., and Wojtczak, A. (2006) Eur. J. Org. Chem., 3216–3225.
bicyclic (Se)$_2$	1/	352	CDCl$_3$	Scianowski, J., Rafinski, Z., and Wojtczak, A. (2006) Eur. J. Org. Chem., 3216–3225.
cyclohexyl-SePh with Me	1/	358	CDCl$_3$	Duddeck, H., Wagner, P., and Biallaß, A. (1991) Magn. Reson. Chem., 29, 248–259.

Structure	Ratio	δ (ppm)	Solvent	Reference
2-(SeMe)-... with SMe, OMe	1/24	366	CDCl₃	Tiecco, M., Testaferri, L., Santi, C., Tomasini, C., Marini, F., Bagnoli, L., and Temperini, A. (2002) *Chem. Eur. J.*, **8**, 1118–1124.
(Se)₂ cyclohexane derivative	1/	367	CDCl₃	Scianowski, J., Rafinski, Z., and Wojtczak, A. (2006) *Eur. J. Org. Chem.*, 3216–3225.
SePh cyclohexyl	1/	400		Duddeck, H., Wagner, P., and Biallaß, A. (1991) *Magn. Reson. Chem.*, **29**, 248–259.
SePh (Me-cyclohexyl)	1/	411	CDCl₃	Duddeck, H., Wagner, P., and Biallaß, A. (1991) *Magn. Reson. Chem.*, **29**, 248–259.
N-cyclohexyl benzyl (Se)₂	1/	430		Tomoda, S., Fujita, K., and Iwaoka, M. (1997) *Phosphorus Sulfur*, **67**, 247–252.
SMe (Se)₂	1/23	431		Tiecco, M., Testaferri, L., Santi, C., Tomasini, C., Santoro, S., Marini, F., Bagnoli, L., Temperini, A., and Costantino, F. (2006) *Eur. J. Org. Chem.*, 4867–4873.
NMe₂ (Se)₂	1/	444		Santi, C., Fragale, G., and Wirth, T. (1998) *Tetrahedron: Asymmetry*, **9**, 3625–3628.

(*Continued*)

Compound	Chapter/No.	Chemical shift (ppm)	Solvent	Reference
(camphor-Se)₂	1/	444	CDCl₃	Scianowski, J., Rafinski, Z., and Wojtczak, A. (2006) *Eur. J. Org. Chem.*, 3216–3225.
Ph-imidazolidine-(o-C₆H₄Se)₂	1/19	446		Santi, C. and Wirth, T. (1999) *Tetrahedron: Asymmetry*, **10**, 1019–1023.
o-(PhHNOC)C₆H₄(Se)₂	1/	450		Fisher, H. and Dereu, N. (1987) *Bull. Soc. Chim. Belg.*, **96**, 757–768.
SiMe₃/NMe₂ aryl (Se)₂	1/	463		Santi, C., Fragale, G., and Wirth, T. (1998) *Tetrahedron: Asymmetry*, **9**, 3625–3628.
(norbornyl-Se)₂	1/	464	CDCl₃	Scianowski, J., Rafinski, Z., and Wojtczak, A. (2006) *Eur. J. Org. Chem.*, 3216–3225.
(PhSe)₂	1	481		Odom, O. A., Vernon, W. D., and Dechter, J. J. (1978) *J. Magn. Reson.*, **32**, 19–21.
O₂N-aryl-(pyrrolidinyl)(Se)₂	1/	483		Santi, C., Fragale, G., and Wirth, T. (1998) *Tetrahedron: Asymmetry*, **9**, 3625–3628.

Structure	Ratio	δ	Solvent	Reference
Ar-SMe, SeI	1/	557		Tiecco, M., Testaferri, L., Santi, C., Tomassini, C., Santoro, S., Marini, F., Bagnoli, L., Temperini, A., and Costantino, F. (2006) *Eur. J. Org. Chem.*, 4867–4873.
Ar-N(Ph)(Se)₂	1/20	624	CDCl₃	Tiecco, M., Testaferri, L., Santi, C., Tomassini, C., Bagnoli, L., Marini, F., and Temperini, A. (2000) *Tetrahedron: Asymmetry*, **11**, 4645–4650.
Ar-N(Ph)(Se)₂	1/	626		Tiecco, M., Testaferri, L., Santi, C., Tomassini, C., Bagnoli, L., Marini, F., and Temperini, A. (2000) *Tetrahedron: Asymmetry*, **11**, 4645–4650.
Ar-SMe, SeBr	1/29	751	CDCl₃	Tiecco, M., Testaferri, L., Santi, C., Tomassini, C., Santoro, S., Marini, F., Bagnoli, L., Temperini, A., and Costantino, F. (2006) *Eur. J. Org. Chem.*, 4867–4873.
Ph-SeNEt₂	1/	764		Roustesuo, P., Hakkinen, A.M., Liias-Lepisto, R., and Sallminen, U. (1988) *Spectrochim. Acta*, **44A**, 1105–1108.
Ar-SMe, SeCl	1/28	797	CDCl₃	Tiecco, M., Testaferri, L., Santi, C., Tomassini, C., Santoro, S., Marini, F., Bagnoli, L., Temperini, A., and Costantino, F. (2006) *Eur. J. Org. Chem.*, 4867–4873.
Ar-SeBr	1/	832	CDCl₃	Tiecco, M., Testaferri, L., Santi, C., Tomassini, C., Santoro, S., Marini, F., Bagnoli, L., Temperini, A., and Costantino, F. (2006) *Eur. J. Org. Chem.*, 4867–4873.

(*Continued*)

Compound	Chapter/No.	Chemical shift (ppm)	Solvent	Reference
PhSeOMe	1/	832		Nakanishi, W., Ikeda, Y., and Iwamura, H. (1982) *Org. Magn. Reson.*, **22**, 117–122.
PhSeBr	1/	867		Tiecco, M., Testaferri, L., Santi, C., Tomasini, C., Santoro, S., Marini, F., Bagnoli, L., Temperini, A., and Costantino, F. (2006) *Eur. J. Org. Chem.*, 4867–4873.
Ph₂Se⁺ ⊖BF₄	1/	976		Sandor, P. and Radics, L. (1981) *Org. Magn. Reson.*, **16**, 148–151.
(sec-Bu)C₆H₄SeCl	1/	1004	CDCl₃	Tiecco, M., Testaferri, L., Santi, C., Tomasini, C., Santoro, S., Marini, F., Bagnoli, L., Temperini, A., and Costantino, F. (2006) *Eur. J. Org. Chem.*, 4867–4873.
PhSeCl	1/	1044		Tiecco, M., Testaferri, L., Santi, C., Tomasini, C., Santoro, S., Marini, F., Bagnoli, L., Temperini, A., and Costantino, F. (2006) *Eur. J. Org. Chem.*, 4867–4873.
Me₃SnSePh	2/	11.3	CCl₄	Kennedy, J. D. and McFarlane, W. (1973) *J. Chem. Soc., Dalton Trans.*, 2134.
Li⁺ ⁻Se-C≡C-butyl	2/50	−15.1	C₆D₆	Pietschnig, R., Merz, K., and Schäfer, S. (2005) *Heteroatom Chem.*, **16**, 169.
Mes-N(Se)C=N-H (imidazole-2-selone)	2/5b	17	CDCl₃	Landry, V. K., Minoura, M., Pang, K., Buccella, D., Kelly, B. V., and Parkin, G. (2006) *J. Am. Chem. Soc.*, **128**, 12490.
(2,4,6-Me₃C₆H₂)SeH	2/	26.9	C₆D₆	Briand, G. G., Decken, A., and Hamilton, N. S. (2010) *Dalton Trans.*, 3833.

Compound	Ratio	δ	Solvent	Reference
(2,4,6-Me$_3$C$_6$H$_2$)SeInMe$_2$	2/	31.7	CDCl$_3$	Briand, G. G., Decken, A., and Hamilton, N. S. (2010) *Dalton Trans.*, 3833
EtSeH	2/	39	CDCl$_3$	Odom, J. D., Dawson, W. H., and Ellis, P. D. (1979) *J. Am. Chem. Soc.*, **101**, 5815.
PhSeZnCl	2/	−41.6	THF-d_8	Santi, C., Santoro, S., Battistelli, B., Testaferri, L., and Tiecco, M. (2008) *Eur. J. Org. Chem.*, 5387.
⟨alkynyl⟩−Se⁻ Li⁺	2/50	−114.6	THF-d_8	Pietschnig, R., Merz, K., and Schäfer, S. (2005) *Heteroatom Chem.*, **16**, 169.
Pb(SePh)$_3^-$	2/	127	CD$_3$OD	Arsenault, J. J. I. and Dean, P. A. W. (1983) *Can. J. Chem.*, **61**, 1516.
tBuSeNa	2/	129	H$_2$O	McFarlane, W. and Wood, R. J. (1972) *J. Chem. Soc. (A)*, 1397.
MeSeH	2/	−130	CDCl$_3$	Odom, J. D., Dawson, W. H., and Ellis, P. D. (1979) *J. Am. Chem. Soc.*, **101**, 5815.
PhSeH	2/	145	neat	McFarlane, W. and Wood, R. J. (1972) *J. Chem. Soc. (A)*, 1397.
EtSeNa	2/	−150	H$_2$O	McFarlane, W. and Wood, R. J. (1972) *J. Chem. Soc. (A)*, 1397.
Sn(SePh)$_3^-$	2/	164	CD$_3$OD	Arsenault, J. J. I. and Dean, P. A. W. (1983) *Can. J. Chem.*, **61**, 1516.
As(SePh)$_3$	2/	266	CDCl$_3$	Arsenault, J. J. I. and Dean, P. A. W. (1983) *Can. J. Chem.*, **61**, 1516.
tBuSeH	2/	278	neat	McFarlane, W. and Wood, R. J. (1972) *J. Chem. Soc. (A)*, 1397.
Bi(SePh)$_3$	2/	311	CDCl$_3$	Arsenault, J. J. I. and Dean, P. A. W. (1983) *Can. J. Chem.*, **61**, 1516.

(*Continued*)

Compound	Chapter/No.	Chemical shift (ppm)	Solvent	Reference
pyridine-2(1H)-selone (N-H)	2/4b	314.0	CDCl$_3$	Laube, J., Jäger, S., and Thöne, C. (2001) *Eur. J. Inorg. Chem.*, 1983.
Ph$_3$SnSeMe	2/	−320	CCl$_4$	Kennedy, J. D. and McFarlane, W. (1973) *J. Chem. Soc., Dalton Trans.*, 2134.
MeSeNa	2/	−330	H$_2$O	Odom, J. D., Dawson, W. H., and Ellis, P. D. (1979) *J. Am. Chem. Soc.*, **101**, 5815
Sb(SePh)$_3$	2/	397	CDCl$_3$	Arsenault, J. J. I. and Dean, P. A. W. (1983) *Can. J. Chem.*, **61**, 1516.
[(pyridine-2-selenolate)(18-crown-6)K]$^+$	2/11	441.8	CD$_3$OD	Laube, J., Jäger, S., and Thöne, C. (2001) *Eur. J. Inorg. Chem.*, 1983.
Me$_3$Si–Se–Cu(PEt$_2$Ph)(PEt$_2$Ph)	2/13	−546	CD$_2$Cl$_2$ (−60°C)	Borecki, A. and Corrigan, J. F. (2007) *Inorg. Chem.*, **46**, 2478.
[Li$_2$(nBuSe)$_2$(TMEDA)$_2$]	2/49	−660	C$_6$D$_6$	Clegg, W., Davies, R. P., Snaith, R., and Wheatley, A. E. H. (2001) *Eur. J. Inorg. Chem.*, 1411.
[WSe$_4$]$^{2-}$ 2Et$_4$N$^+$	2/24	1235	DMF	Wardle, R. W. M., Mahler, C. H., Chau, C.-N., and Ibers, J. A. (1988) *Inorg. Chem.*, **27**, 2790.

Structure		Value	Solvent	Reference
Ph-CH(OH)-CH2-SeMe	4	30.6	CDCl3	Costa, C. E., Clososki, G. C., Barchesi, H. B., Zanotto, S. P., Graça Nascimento, M., and Comasseto, J. V. (2004) *Tetrahedron: Asymmetry*, **15**, 3945–3954.
Ph-CH(OAc)-CH2-SeMe	4	64.3	CDCl3	Costa, C. E., Clososki, G. C., Barchesi, H. B., Zanotto, S. P., Graça Nascimento, M., and Comasseto, J. V. (2004) *Tetrahedron: Asymmetry*, **15**, 3945–3954.
CH(OH)-CH2-SePh (butyl chain)	4	234.2	CDCl3	Costa, C. E., Clososki, G. C., Barchesi, H. B., Zanotto, S. P., Graça Nascimento, M., and Comasseto, J. V. (2004) *Tetrahedron: Asymmetry*, **15**, 3945–3954.
CH(OH)-CH2-SePh (hexyl chain)	4	237.0	CDCl3	Costa, C. E., Clososki, G. C., Barchesi, H. B., Zanotto, S. P., Graça Nascimento, M., and Comasseto, J. V. (2004) *Tetrahedron: Asymmetry*, **15**, 3945–3954.
Aryl-CH(OMe)-, ortho-SeBu	4	229.8	CDCl3	Piovan, L., Alves, M. F. M., Juliano, L., Brömme, D., Cunha, R. L. O. R., and Andrade, L. H. (2010) *J. Braz. Chem. Soc.*, **21**, 2108–2118.
Aryl-CH2-OMe, ortho-SeBu	4	240.0	CDCl3	Piovan, L., Alves, M. F. M., Juliano, L., Brömme, D., Cunha, R. L. O. R., and Andrade, L. H. (2010) *J. Braz. Chem. Soc.*, **21**, 2108–2118.
iPr-CH(OH)-CH2-SePh	4	240.8	CDCl3	Costa, C. E., Clososki, G. C., Barchesi, H. B., Zanotto, S. P., Graça Nascimento, M., and Comasseto, J. V. (2004) *Tetrahedron: Asymmetry*, **15**, 3945–3954.

(*Continued*)

Compound	Chapter/No.	Chemical shift (ppm)	Solvent	Reference
OH, SePh (isopropyl)	4	240.8	CDCl$_3$	Costa, C. E., Clososki, G. C., Barchesi, H. B., Zanotto, S. P., Graça Nascimento, M., and Comasseto, J. V. (2004) *Tetrahedron: Asymmetry*, **15**, 3945–3954.
OH, SePh, Ph	4	251.9	CDCl$_3$	Costa, C. E., Clososki, G. C., Barchesi, H. B., Zanotto, S. P., Graça Nascimento, M., and Comasseto, J. V. (2004) *Tetrahedron: Asymmetry*, **15**, 3945–3954.
OAc, SePh	4	260.3	CDCl$_3$	Costa, C. E., Clososki, G. C., Barchesi, H. B., Zanotto, S. P., Graça Nascimento, M., and Comasseto, J. V. (2004) *Tetrahedron: Asymmetry*, **15**, 3945–3954.
OAc, SePh (hexyl)	4	264.0	CDCl$_3$	Costa, C. E., Clososki, G. C., Barchesi, H. B., Zanotto, S. P., Graça Nascimento, M., and Comasseto, J. V. (2004) *Tetrahedron: Asymmetry*, **15**, 3945–3954.
OAc, SePh	4	264.3	CDCl$_3$	Costa, C. E., Clososki, G. C., Barchesi, H. B., Zanotto, S. P., Graça Nascimento, M., and Comasseto, J. V. (2004) *Tetrahedron: Asymmetry*, **15**, 3945–3954.
OAc, SePh	4	265.4	CDCl$_3$	Costa, C. E., Clososki, G. C., Barchesi, H. B., Zanotto, S. P., Graça Nascimento, M., and Comasseto, J. V. (2004) *Tetrahedron: Asymmetry*, **15**, 3945–3954.
SeMe, Cl vinyl	4	267.2	CDCl$_3$	Comasseto, J. V., Menezes, P. H., Stefani, H. A., Zeni, G., and Braga, A. L. (1996) *Tetrahedron*, **52**, 9687–9702.
SeMe, Cl vinyl	4	267.3	CDCl$_3$	Comasseto, J. V., Menezes, P. H., Stefani, H. A., Zeni, G., and Braga, A. L. (1996) *Tetrahedron*, **52**, 9687–9702.

Structure	Value	Solvent	Reference
CF₃, Ph, OMe, O, O, SePh	271.0	4 CDCl₃	Costa, C. E., Clososki, G. C., Barchesi, H. B., Zanotto, S. P., Graça Nascimento, M., and Comasseto, J. V. (2004) *Tetrahedron: Asymmetry*, **15**, 3945–3954.
CF₃, Ph, OMe, O, O, SePh	269.9	4 CDCl₃	Costa, C. E., Clososki, G. C., Barchesi, H. B., Zanotto, S. P., Graça Nascimento, M., and Comasseto, J. V. (2004) *Tetrahedron: Asymmetry*, **15**, 3945–3954.
SeMe, Br	296.2	4 CDCl₃	Comasseto, J. V., Menezes, P. H., Stefani, H. A., Zeni, G., and Braga, A. L. (1996) *Tetrahedron*, **52**, 9687–9702.
SeMe, Br	295.6	4 CDCl₃	Comasseto, J. V., Menezes, P. H., Stefani, H. A., Zeni, G., and Braga, A. L. (1996) *Tetrahedron*, **52**, 9687–9702.
SeMe, I	335.4	4 CDCl₃	Comasseto, J. V., Menezes, P. H., Stefani, H. A., Zeni, G., and Braga, A. L. (1996) *Tetrahedron*, **52**, 9687–9702.
OH, SePh, SePh	355.0, 357.9	4 CDCl₃	Costa, C. E., Clososki, G. C., Barchesi, H. B., Zanotto, S. P., Graça Nascimento, M., and Comasseto, J. V. (2004) *Tetrahedron: Asymmetry*, **15**, 3945–3954.
OAc, SePh, SePh	374.7, 382.4	4 CDCl₃	Costa, C. E., Clososki, G. C., Barchesi, H. B., Zanotto, S. P., Graça Nascimento, M., and Comasseto, J. V. (2004) *Tetrahedron: Asymmetry*, **15**, 3945–3954.

(Continued)

Compound	Chapter/No.	Chemical shift (ppm)	Solvent	Reference
SnBu₃, SePh (naphthalene)	4	383.3		Toledo, F. T., Comasseto, J. V., and Raminelli, C. (2010) *J. Braz. Chem. Soc.*, **21**, 2164–2168.
SeMe, I (alkene)	4	426.3	CDCl₃	Comasseto, J. V., Menezes, P. H., Stefani, H. A., Zeni, G., and Braga, A. L. (1996) *Tetrahedron*, **52**, 9687–9702.
SePh, SnBu₃, F (benzene)	4	437.2		Toledo, F. T., Comasseto, J. V., and Raminelli, C. (2010) *J. Braz. Chem. Soc.*, **21**, 2164–2168.
SePh, Cl (alkene)	4	467.4	CDCl₃	Comasseto, J. V., Menezes, P. H., Stefani, H. A., Zeni, G., and Braga, A. L. (1996) *Tetrahedron*, **52**, 9687–9702.
OMe, Se-Br, Br (benzene)	4	483.6	CDCl₃	Piovan, L., Alves, M. F. M., Juliano, L., Brömme, D., Cunha, R. L. O. R., and Andrade, L. H. (2010) *J. Braz. Chem. Soc.*, **21**, 2108–2118.
SePh, Br (alkene)	4	493.9	CDCl₃	Comasseto, J. V., Menezes, P. H., Stefani, H. A., Zeni, G., and Braga, A. L. (1996) *Tetrahedron*, **52**, 9687–9702.
Ph, SePh, Cl (alkene)	4	501.4	CDCl₃	Comasseto, J. V., Menezes, P. H., Stefani, H. A., Zeni, G., and Braga, A. L. (1996) *Tetrahedron*, **52**, 9687–9702.

4	508.1	CDCl₃	Piovan, L., Alves, M. F. M., Juliano, L., Brömme, D., Cunha, R. L. O. R., and Andrade, L. H. (2010) *J. Braz. Chem. Soc.*, **21**, 2108–2118.
4	528.0	CDCl₃	Comasseto, J. V., Menezes, P. H., Stefani, H. A., Zeni, G., and Braga, A. L. (1996) *Tetrahedron*, **52**, 9687–9702.
4	538.5	CDCl₃	Comasseto, J. V., Menezes, P. H., Stefani, H. A., Zeni, G., and Braga, A. L. (1996) *Tetrahedron*, **52**, 9687–9702.
4	549.4	CDCl₃	Piovan, L., Alves, M. F. M., Juliano, L., Brömme, D., Cunha, R. L. O. R., and Andrade, L. H. (2010) *J. Braz. Chem. Soc.*, **21**, 2108–2118.
4	555.0	CDCl₃	Comasseto, J. V., Menezes, P. H., Stefani, H. A., Zeni, G., and Braga, A. L. (1996) *Tetrahedron*, **52**, 9687–9702.
4	554.9	CDCl₃	Comasseto, J. V., Menezes, P. H., Stefani, H. A., Zeni, G., and Braga, A. L. (1996) *Tetrahedron*, **52**, 9687–9702.
4	569.6	CDCl₃	Comasseto, J. V., Menezes, P. H., Stefani, H. A., Zeni, G., and Braga, A. L. (1996) *Tetrahedron*, **52**, 9687–9702.
4	584.1	CDCl₃	Comasseto, J. V., Menezes, P. H., Stefani, H. A., Zeni, G., and Braga, A. L. (1996) *Tetrahedron*, **52**, 9687–9702.

(Continued)

Compound	Chapter/No.	Chemical shift (ppm)	Solvent	Reference
SePh / Br	4	584.7	CDCl$_3$	Comasseto, J. V., Menezes, P. H., Stefani, H. A., Zeni, G., and Braga, A. L. (1996) *Tetrahedron*, **52**, 9687–9702.
SePh / I	4	633.0	CDCl$_3$	Comasseto, J. V., Menezes, P. H., Stefani, H. A., Zeni, G., and Braga, A. L. (1996) *Tetrahedron*, **52**, 9687–9702.
SePh	4	633.3	CDCl$_3$	Comasseto, J. V., Menezes, P. H., Stefani, H. A., Zeni, G., and Braga, A. L. (1996) *Tetrahedron*, **52**, 9687–9702.
(OH)...Se)$_2$	5/	206.8		Ścianowski, J., Rafiński, Z., Wojtczak, A., and Burczyński, K. (2009) *Tetrahedron: Asymmetry*, **20**, 2871–2879.
(...Se)$_2$	5/	290.4		Ścianowski, J., Rafiński, Z., and Wojtczak, A. (2006) *Eur. J. Org. Chem.*, 3216–3225.
(...Se)$_2$	5/	336.3		Ścianowski, J., Rafiński, Z., and Wojtczak, A. (2006) *Eur. J. Org. Chem.*, 3216–3225.

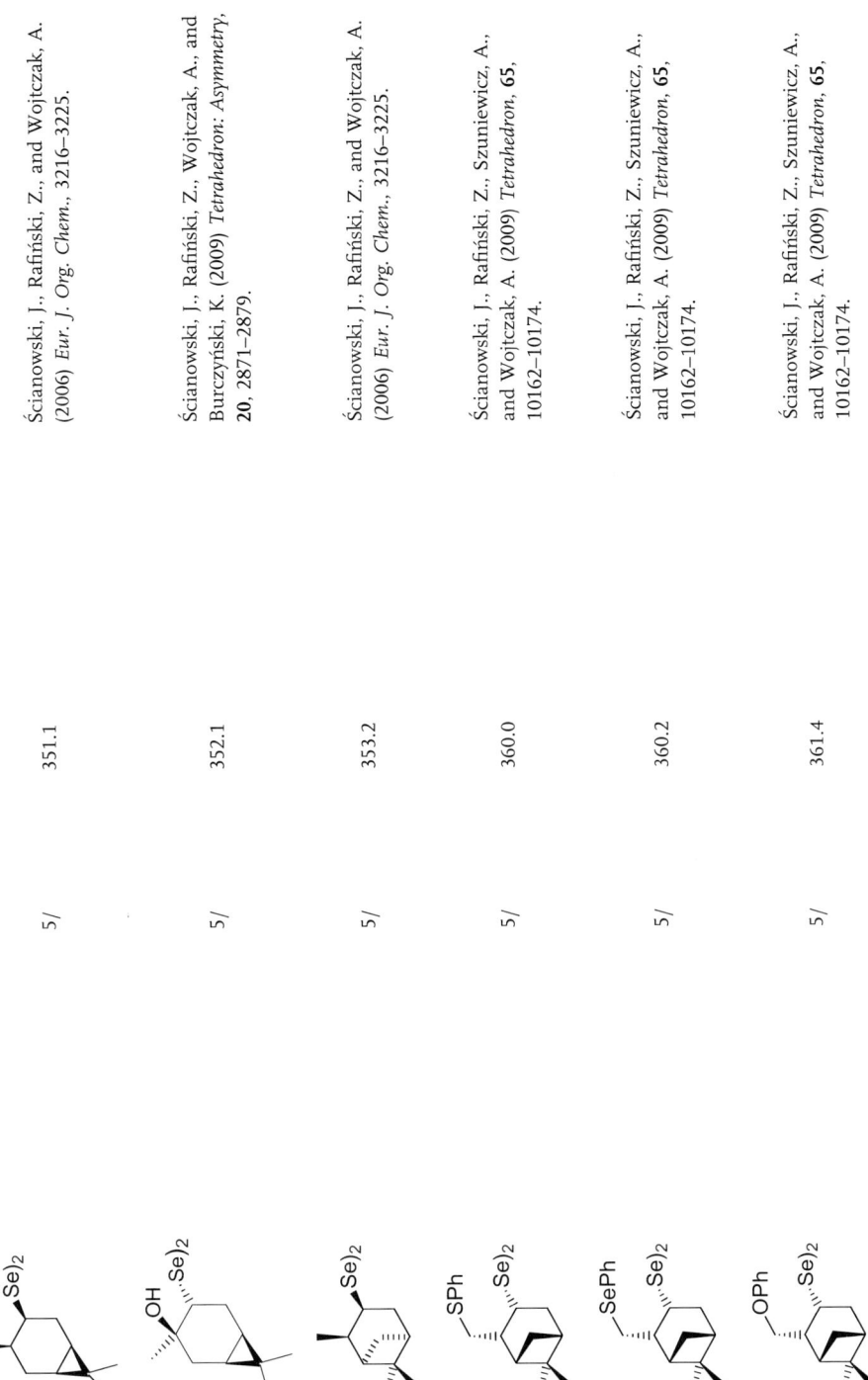

5/	351.1	Ścianowski, J., Rafiński, Z., and Wojtczak, A. (2006) *Eur. J. Org. Chem.*, 3216–3225.
5/	352.1	Ścianowski, J., Rafiński, Z., Wojtczak, A., and Burczyński, K. (2009) *Tetrahedron: Asymmetry*, **20**, 2871–2879.
5/	353.2	Ścianowski, J., Rafiński, Z., and Wojtczak, A. (2006) *Eur. J. Org. Chem.*, 3216–3225.
5/	360.0	Ścianowski, J., Rafiński, Z., Szuniewicz, A., and Wojtczak, A. (2009) *Tetrahedron*, **65**, 10162–10174.
5/	360.2	Ścianowski, J., Rafiński, Z., Szuniewicz, A., and Wojtczak, A. (2009) *Tetrahedron*, **65**, 10162–10174.
5/	361.4	Ścianowski, J., Rafiński, Z., Szuniewicz, A., and Wojtczak, A. (2009) *Tetrahedron*, **65**, 10162–10174.

(*Continued*)

Compound	Chapter/No.	Chemical shift (ppm)	Solvent	Reference
Ph—Se⁺—Et BF$_4^-$ / Bu	5/72a	420.0		Lenardão, E.J., Mendes, S. R., Ferreira, P.C., Perin, G., Silveira, C. C., and Jacob, R. G. (2006) *Tetrahedron Lett*, **47**, 7439–7442
Ph—≡—SePh$_2^+$ CF$_3$SO$_3^-$	5/35	469.6		Kataoka, T., Watanabe, S., Yamamoto, K., Yoshimatsu, M., Tanabe G., and Muraoka, O. (1998) *J. Org. Chem*., **63**, 6382–6386.
(structure with SePh$_2$, N⁺, S, Ph, Ph$_2$Se, 2BPh$_4^-$)	5/256	695.5		Aucott, S.M., Bailey, M.R., Elsegood, M.R.J., Gilby, L.M., Holmes, K.E., Kelly, P.F., Papageorgiou, M.J., and Pedron-Haba, S. (2004) *New. J. Chem*., **28**, 959–966.
(dibenzo structure with Se–Cl)	5/79d	726.7		Kataoka, T., Iwamura, T., Tsutsui, H., Kato, Y., Banno, Y., Aoyama, Y., and Shimizu, H. (2001) *Heteroatom Chem*., **12**, 317–326.
O=Se(COOH)(cyclohexyl)	5/232	756.6		Iwaoka, M. and Kumakura, F. (2008) *Phosphorus Sulfur*, **183**, 1009–1017.
(benzoxaselenole with Ph, Se–Cl)	5/79b	797.2		Kataoka, T., Iwamura, T., Tsutsui, H., Kato, Y., Banno, Y., Aoyama, Y., and Shimizu, H. (2001) *Heteroatom Chem*., **12**, 317–326.
(benzoxaselenole with Ph, Se–Cl)	5/79a	814.9		Kataoka, T., Iwamura, T., Tsutsui, H., Kato, Y., Banno, Y., Aoyama, Y., and Shimizu, H. (2001) *Heteroatom Chem*., **12**, 317–326.

5/80e	858.5	Kataoka, T., Iwamura, T., Tsutsui, H., Kato, Y., Banno, Y., Aoyama, Y., and Shimizu, H. (2001) *Heteroatom Chem.*, **12**, 317–326.
5/217	865.4	Iwaoka, M. and Kumakura, F. (2008) *Phosphorus Sulfur*, **183**, 1009–1017.
5/80b	983.3	Kataoka, T., Iwamura, T., Tsutsui, H., Kato, Y., Banno, Y., Aoyama, Y., and Shimizu, H. (2001) *Heteroatom Chem.*, **12**, 317–326.
5/80a	1009.2	Kataoka, T., Iwamura, T., Tsutsui, H., Kato, Y., Banno, Y., Aoyama, Y., and Shimizu, H. (2001) *Heteroatom Chem.*, **12**, 317–326.
5/288	1182.1	Yu, S.-C., Borchert, A., Kuhn, H., and Ivanov, I. (2008) *Chem. Eur. J.*, **14**, 7066–7071.
5/219a	1294.0	Nakashima, Y., Shimizu, T., Hirabayashi, K., Yasui, M., Nakazato, M., Iwasaki, F., and Kamigata, N. (2005) *Bull. Chem. Soc. Jpn.*, **78**, 710–714.
5/219c	1316.0	Nakashima, Y., Shimizu, T., Hirabayashi, K., Yasui, M., Nakazato, M., Iwasaki, F., and Kamigata, N. (2005) *Bull. Chem. Soc. Jpn.*, **78**, 710–714.

(*Continued*)

Compound	Chapter/No.	Chemical shift (ppm)	Solvent	Reference
Cyclohexyl-Se(=O)-OH	5/219d	1322.0		Nakashima, Y., Shimizu, T., Hirabayashi, K., Yasui, M., Nakazato, M., Iwasaki, F., and Kamigata, N. (2005) *Bull. Chem. Soc. Jpn.*, **78**, 710–714.
2,4,6-tri-tert-butylphenyl–N=Se=N–2,4,6-tri-tert-butylphenyl	5/137f	1844.0		Maaninen, T., Tuononen, H.M., Kosunen, K., Oilunkaniemi, R., Hiitola, J., Lattinen, R., and Chivers, T. (2004) *Z. Anorg. Allg. Chem.*, **630**, 1947–1954.
Leu, BocHN–CH–N=C=Se	6/	−357.6		Chennakrishnareddy, G., Nagendra, G., Hemantha, H. P., Das, U., Row, T. N. G., and Sureshbabu, V. V. (2010) *Tetrahedron*, **66**, 6718.
Ala, ZHN–CH–N=C=Se	6/	−356.6		Chennakrishnareddy, G., Nagendra, G., Hemantha, H. P., Das, U., Row, T. N. G., and Sureshbabu, V. V. (2010) *Tetrahedron*, **66**, 6718.
R¹–C(=Se)–SeCH₂CH₂SiMe₃	6/	863–955, 1652–1847 (169–175, 222–226)		Tani, K., Murai, T., and Kato, S. (2002) *J. Am. Chem. Soc.*, **124**, 5960.
R¹–N(C(=Se))–C(Se)(R²)(R³) (cyclic)	6/	531–747, 920–1033		Koketsu, M., Yamamura, Y., and Ishihara, H. (2006) *Synthesis*, 2738.
R¹–C(=Se)–SR²	6/	1423–1634		Murai, T., Kakami, K., Hayashi, A., Komuro, T., Takada, H., Fujii, M., Kanda, T., and Kato, S. (1997) *J. Am. Chem. Soc.*, **119**, 8592.

Structure		Range	Solvent	Reference
Ar-C(=N-N(Ph))-SeR (triazole)	6/	158–358		Koketsu, M., Yamamura, Y., and Ishihara, H. (2006) *Heterocycles*, **68**, 1191.
Ar-C(Se)-NH$_2$	6/	529–716		Li, Y., Hua, G.-X., Slawin, A. M. Z., and Woollins, J. D. (2009) *Molecules*, **14**, 884.
R-C(=O)-N(H)-C(=Se)-TeBu	6/	960–1120		Kanda, T., Aoki, H., Mizoguchi, K., Shiraishi, S., Murai, T., and Kato, S. (1996) *Organometallics*, **15**, 5753.
R-C(Se$^-$)=S $^+$NR$_4$	6/	1110–1257 (201–209)	THF-d$_6$ or CD$_3$CN	Tani, K., Murai, T., and Kato, S. (2002) *J. Am. Chem. Soc.*, **124**, 5960.
R-C(Se$^-$)=Se $^+$NMe$_4$	6/	1362–1493 (208–2015)	CD$_3$CN	Tani, K., Murai, T., and Kato, S. (2002) *J. Am. Chem. Soc.*, **124**, 5960.
R^1-C(=Se)-N(H)-R^2	6/	516–622 (206–215)		Mutoh, Y. and Murai, T. (2004) *Organometallics*, **23**, 3907.
Se=C(N-R^1)-CH(R^2)-C(=O)- (cyclic)	6/	574–679 1051–1226		Garud, D. R., Tanahashi, N., Ninomiya, M., and Koketsu, M. (2009) *Tetrahedron*, **65**, 4775.
Cyclic Se-N-C(=O)-CH$_2$-CH(R)-	6/	295–384 (55.2, 178.0)		Kanoh, K., Ishihara, H., and Koketsu, M. (2007) *Synthesis*, 2617.

(*Continued*)

Compound	Chapter/No.	Chemical shift (ppm)	Solvent	Reference
(Se / Se / R¹ / R² / HN / O)	6/	627–715 1043–1057		Koketsu, M., Yamamura, Y., and Ishihara, H. (2006) *Synthesis*, 2738.
(S / R / SeMe)	6/	696–747 (158–170)		Murai, T., Kamoto, T., and Kato, S. (2000) *J. Am. Chem. Soc.*, **122**, 9850.
(Se / O / Ar / N / N,R / H / H)	6/	320–357		Koketsu, M., Yamamura, Y., Aoki, H., and Ishihara, H. (2006) *Phosphorus Sulfur*, **181**, 2699.
(Me–Se⁺ / R / N / Ph / OTf⁻)	6/	382–417 (162–172)		Mutoh, Y. and Murai, T. (2004) *Organometallics*, **23**, 3907.
(Se / Ph–N / O / R)	6/	256–287		Koketsu, M., Takahashi, A., and Ishihara, H. J. (2007) *Heterocyclic Chem.*, **44**, 79.
(Se / H / O / R / N / N-Ph / H)	6/	293–310		Koketsu, M., Yamamura, Y., and Ishihara, H. (2006) *Heterocycles*, **68**, 1191.
(Se / O / Ar / N / NR₂ / H)	6/	463–476		Koketsu, M., Yamamura, Y., Aoki, H., and Ishihara, H. (2006) *Phosphorus Sulfur*, **181**, 2699.
(Me-N / Se / N-Me imidazole)	6/	–6		Roy, G. and Mugesh, G. (2008) *Phosphorus Sulfur*, **183**, 908.

Structure		Value	Solvent	Reference
(diketone-Se-N-R² with R¹ groups)	6/	751–757		Koketsu, M., Yamamura, Y., Ando, H., and Ishihara, H. (2006) *Heterocycles*, **68**, 1267.
(Mes-N Se NH)	6/	−5 222		Landry, V. K., Minoura, M., Pang, K., Buccella, D., Kelly, B. V., and Parkin, G. (2006) *J. Am. Chem. Soc.*, **128**, 12490.
(Se N-R, R = CH₂C₆H₅ = CH₂C₆H₄-OMe)	6/	−3 −0.23		(a) Roy, G. and Mugesh G. (2008) *Phosphorus Sulfur Silicon*, 183, 908; (b) Bhabak, K. P. and Mugesh, G. (2010) *Chem. Eur. J.*, **16**, 1175.
(R-N Se, HN-N-Ph)	6/	87–90	DMSO-d₆	Koketsu, M., Yamamura, Y., and Ishihara, H. (2006) *Heterocycles*, **68**, 1191.
(Se⁻ NBu₄⁺, N-CH₂Ph)	6/	222.6 (179.9)	CD₃CN	Mutoh, Y. and Murai, T. (2004) *Organometallics*, **23**, 3907.
(Ph-Se-C(Ph)(H)-NHBn)	6/	246	Acetone-d6	Andaloussi, M. B. D. and Mohr, F. (2010) *J. Organomet. Chem.*, **695**, 1276.
(Leu-BocHN... Se, C(=Se)-NH-C₆H₄-Cl)	6/	261.1		Chennakrishnareddy, G., Nagendra, G., Hemantha, H. P., Das, U., Row, T. N. G., and Sureshbabu, V. V. (2010) *Tetrahedron*, **66**, 6718.

(*Continued*)

Compound	Chapter/No.	Chemical shift (ppm)	Solvent	Reference
MeS, ZHN, H, N(H), C(=O)OMe, Se	6/	262.2		Chennakrishnareddy, G., Nagendra, G., Hemantha, H. P., Das, U., Row, T. N. G., and Sureshbabu, V. V. (2010) *Tetrahedron*, **66**, 6718.
Me–Se, Ph–C(=N–CH2Ph)	6/	262.9 (138.9)		Mutoh, Y. and Murai, T. (2004) *Organometallics*, **23**, 3907.
Ph–N(H)–C(=Se)–O–CH2CH=CH2	6/	278.4		Garud, D. R., Makimura, M., Ando, H., Ishihara, H., and Koketsu, M. (2007) *Tetrahedron Lett.*, **48**, 7764.
Ph3C–N(H)–C(=Se)–N(H)–NH2	6/	319	Acetone-d6	Andaloussi, M. B. D. and Mohr, F. (2010) *J. Organomet. Chem.*, **695**, 1276.
Mes–N(N)–Se)2	6/	400 (163)		Landry, V. K., Minoura, M., Pang, K., Buccella, D., Kelly, B. V., and Parkin, G. (2006) *J. Am. Chem. Soc.*, **128**, 12490.
Ph–C(=N–CH(iPr)–CH(iPr))–O–Se	6/	394.8		Koketsu, M., Kiyokuni, T., Sakai, T., Ando, H., and Ishihara, H. (2006) *Chem. Lett.*, **35**, 626.
Ph–C(=N–CH(iPr)–CH(iPr))–O–Se	6/	405.9		Koketsu, M., Kiyokuni, T., Sakai, T., Ando, H., and Ishihara, H. (2006) *Chem. Lett.*, **35**, 626.
Ar–N=C(Se)–CH(iPr)–CH2–NH, Ar = 4-CH3C6H4	6/	411.5		Koketsu, M., Kiyokuni, T., Sakai, T., Ando, H., and Ishihara, H. (2006) *Chem. Lett.*, **35**, 626.

Structure		Value	Solvent	Reference
Me-C(=O)-SeH	6/	434.7 (neat)		Castano, J. A. G., Romano, R. M., Beckers, H., Willner, H., Boese, R., and Védova, C. O. D. (2008) *Angew. Chem. Int. Ed.*, **47**, 10114.
BnO¹³CH₂ — N(Se)¹³C — O — CH(iPr) (oxazoline)	6/	447 (241)		Olivault-Shiflett, M., Kimball, D. B., and Silks L. A. (2004) *J. Org. Chem.*, **69**, 5150.
Ph-C(Se⁻)=O⁻ NMe₄⁺	6/	442.9 (192.1)	CD₃CN	Murai, T., Kamoto, T., and Kato, S. (2000) *J. Am. Chem. Soc.*, **122**, 9850.
Cyclohexyl-N(iPr)-C(=Se)-N(Me)	6/	453		Sivapriya, K., Suguna, P., Banerjee, A., Saravanan, V., Rao, D. N., and Chandrasekaran, S. (2007) *Bioorg. Med. Chem. Lett.*, **17**, 6387.
p-CH₃C₆H₄—(Se/N ring with OEt, OH, CO₂H)	6/	563.8		(a) Yoshimatsu, M., Yamamoto, T., Sawa, A., Kato, T., Tanabe, G., and Muraoka, O. (2009) *Org. Lett.*, **11**, 2952; (b) Yoshimatsu, M., Matsui, M., Yamamoto, T., and Sawa, A. (2010) *Tetrahedron*, **66**, 7975.
H-C(=Se)-NBn₂	6/	569.5		Murai, T., Hori, R., Maruyama, T., and Shibahara, F. (2010) *Organometallics*, **29**, 2400.
p-CH₃C₆H₄—(Se/N ring with CO₂Me, =O)	6/	568.8		(a) Yoshimatsu, M., Yamamoto, T., Sawa, A., Kato, T., Tanabe, G., Muraoka, O. (2009) *Org. Lett.*, **11**, 2952; (b) Yoshimatsu, M., Matsui, M., Yamamoto, T., and Sawa, A. (2010) *Tetrahedron*, **66**, 7975.

(*Continued*)

Compound	Chapter/No.	Chemical shift (ppm)	Solvent	Reference
Se=CH-N(pyrrolidine)	6/	631.9		Saravanan, V., Mukherjee, C., Das. S., and Chandrasekaran, S. (2004) *Tetrahedron Lett.*, **45**, 681.
t-Bu-C(=Se)-NBn₂	6/	664.5		Murai, T., Hori, R., Maruyama, T., and Shibahara, F. (2010) *Organometallics*, **29**, 2400.
Bn-N(H)-C(=Se)-NHBn	6/	667.8		Saravanan, V., Mukherjee, C., Das. S., and Chandrasekaran, S. (2004) *Tetrahedron Lett.*, **45**, 681.
Z-**127** (Ph-C(=O)-C=C(Se)-Se-C(=C)-C(=O)-Ph)	6/	736		Amosova, S. V., Elokhina, V. N., Nakhmanovich, A. S., Larina, L. I., Martynov, A. V., Steele, B. R., and Potatov, V. A. (2008) *Tetrahedron Lett.*, **49**, 974.
Se-thiazole	6/	728.9		Below, H., Pfeiffer, W.-D., Geisler, K., Lalk, M., and Langer, P. (2005) *Eur. J. Org. Chem.*, 3637.
CH₃-C(=Se)-N(piperidin-4-one)	6/	744.5		Saravanan, V., Mukherjee, C., Das. S., and Chandrasekaran, S. (2004) *Tetrahedron Lett.*, **45**, 681.
Me₃Ge-C(=Se)-NBn₂	6/	785.2		Murai, T., Hori, R., Maruyama, T., and Shibahara, F. (2010) *Organometallics*, **29**, 2400.

	813	6/	Amosova, S. V., Elokhina, V. N., Nakhmanovich, A. S., Larina, L. I., Martynov, A. V., Steele, B. R., and Potatov, V. A. (2008) *Tetrahedron Lett.*, **49**, 974.
	827.1	6/	Murai, T., Hori, R., Maruyama, T., and Shibahara, F. (2010) *Organometallics*, **29**, 2400.
	893.9 (215.5)	6/	Murai, T., Kamoto, T., and Kato, S. (2000) *J. Am. Chem. Soc.*, **122**, 9850.
	955.9	6/	Murai, T., Kakami, K., Hayashi, A., Komuro, T., Takada, H., Fujii, M., Kanda, T., and Kato, S. (1997) *J. Am. Chem. Soc.*, **119**, 8592.
	993	6/	Amosova, S. V., Elokhina, V. N., Nakhmanovich, A. S., Larina, L. I., Martynov, A. V., Steele, B. R., and Potatov, V. A. (2008) *Tetrahedron Lett.*, **49**, 974.
	1016.5	6/	Kanda, T., Aoki, H., Mizoguchi, K., Shiraishi, S., Murai, T., and Kato, S. (1996) *Organometallics*, **15**, 5753.
	1926	6/	Okuma, K., Kojima, K., Kaneko, I., Tsujimoto, Y., Ohta, H., and Yokomori, Y. (1994) *J. Chem. Soc., Perkin Trans. 1*, 2151.

(*Continued*)

Compound	Chapter/No.	Chemical shift (ppm)	Solvent	Reference
R-substituted Se=CH-Ar, R = 2,4,6-[CH(SiMe$_3$)$_2$]$_3$	6	2075		Takeda, N., Tokitoh, N., and Okazaki, R. (1996) *Angew. Chem. Int. Ed. Engl.*, **35**, 660.
C$_6$H$_5$–SeH	8	145	neat	McFarlane, W. and Wood, R., (1972) *J. Chem. Soc. A*, 1397.
BocHN-CH(Bn)-CH$_2$-Se)$_2$ **4**	8/4	314.4	CDCl$_3$	Braga, A. L., Paixao, M. W., Ludtke, D. S., Silveira, C. C., and Rodrigues, O. E. D. (2003) *Org. Lett.*, **5**, 2635.
[C$_4$H$_9$–Se–C$_2$H$_5$ / C$_6$H$_5$]$^+$ BF$_4^-$ **142**	8/142	420	CDCl$_3$	Lenardão, E. J., Mendes, S. R., Ferreira, P. C., Perin, G., Silveira, C. C., and Jacob, R. G. (2006) *Tetrahedron Lett.*, **47**, 7439.
Ar-CH(Me)(pyrrolidinyl), Se)$_2$ **1**	8/1	429.8	CDCl$_3$	Poleschner, H. and Seppelt, K. (2004) *Chem. Eur. J.*, **10**, 6565.
Ar-CH(NMe$_2$)(n-Pr), Se)$_2$ **3c**	8/3c	444.5	CDCl$_3$	Santi, C. and Wirth, T. (1999) *Tetrahedron: Asymmetry*, **10**, 1019.

Compound	δ (^{77}Se)	Solvent	Reference
8/3b	445.8	CDCl$_3$	Santi, C., Fragale, G., and Wirth, T. (1998) *Tetrahedron: Asymmetry*, **9**, 3625.
8/11	451	CDCl$_3$	Mugesh, G., Panda, A., Singh, H. B., Punekar, N. S., and Butcher, R. J. (2001) *J. Am. Chem. Soc.*, **123**, 839.
8/3d	463.5	CDCl$_3$	Santi, C. and Wirth, T. (1999) *Tetrahedron: Asymmetry*, **10**, 1019.
8/18	473.1	CDCl$_3$	Wirth, T. and Fragale, G. (1997) *Chem. Eur. J.*, **3**, 1894.
8/134	474.8	CDCl$_3$	Wirth, T. and Fragale, G. (1997) *Chem. Eur. J.*, **3**, 1894.

(*Continued*)

Compound	Chapter/No.	Chemical shift (ppm)	Solvent	Reference
Ph-(Se)₂	8	481	CDCl₃	Odom, O. A., Vernon, W. D., and Dechter, J. (1978) *J. Magn. Reson.*, **32**, 19.
O₂N-C₆H₃(-(Se)₂)-CH(CH₃)-N(pyrrolidine) **3a**	8/3a	483.0	CDCl₃	Santi, C. and Wirth, T. (1999) *Tetrahedron: Asymmetry*, **10**, 1019.
Ph-SeBr	8	867	CDCl₃	Tiecco, M., Testaferri, L., Santi, C., Tomasini, C., Santoro, S., Marini, F., Bagnoli, L., Temperini, A., and Costantino, F. (2006) *Eur. J. Org. Chem.*, 4867.
Ph-SeCl	8	1044	CDCl₃	Tiecco, M., Testaferri, L., Santi, C., Tomasini, C., Santoro, S., Marini, F., Bagnoli, L., Temperini, A., and Costantino, F. (2006) *Eur. J. Org. Chem.*, 4867.
Me-N(imidazoline-Se)-NH **MeSeI**	9/	−5	CDCl₃	Roy, G., Nethaji, M., and Mugesh, G. (2005) *J. Am. Chem. Soc.*, **127**, 15207.
C₆H₄(CH₂-N=)(SeH) **90**	9/90	3	MeOD	Manna, D. and Mugesh, G. (2010) *Angew. Chem. Int. Ed.*, **49**, 9246.

Structure	Coupling	δ (ppm)	Solvent	Reference
89	9/89	108	MeOD	Manna, D. and Mugesh, G. (2010) Angew. Chem. Int. Ed., **49**, 9246.
SeMe imidazole	9	117	CDCl₃	Roy, G., Nethaji, M., and Mugesh, G. (2005) J. Am. Chem. Soc., **127**, 15207.
84	9/84	135	CDCl₃	Goto, K., Sonoda, D., Shimada, K., Sase, S., and Kawashima, T. (2010) Angew. Chem. Int. Ed., **49**, 545–547.
86	9/86	162	MeOD	Manna, D. and Mugesh, G. (2010) Angew. Chem. Int. Ed., **49**, 9246.
X = Cl; X = Br	9	200; 244	CDCl₃; CDCl₃	Potapov, V. A., Kurkutov, E. O., Musalov, M. V., and Amosova, S. V. (2010) Tetrahedron Lett., **51**, 5258.

(Continued)

Compound	Chapter/No.	Chemical shift (ppm)	Solvent	Reference
(structure with NHBoc, Se, OMe; HOOC)	9	213	CDCl$_3$	Bhuyan, B. J. and Mugesh, G. (2011) *Org. Biomol. Chem.*, DOI:10.1039/C0OB00823K
(structure 18, with Se, NMe$_2$, CO$_2$Me, NHBoc)	9/18	216	CDCl$_3$	Phadnis, P. P. and Mugesh, G. (2005) *Org. Biomol. Chem.*, **3**, 2476.
(structure with NHBoc, Se, OMe, pyrrolidine, MeO$_2$C)	9	222	CDCl$_3$	Bhuyan, B. J. and Mugesh, G. (2011) *Org. Biomol. Chem.*, DOI:10.1039/C0OB00823K
(structure 22, benzamide with SeH)	9/22	232	CDCl$_3$	Sarma, B. K. and Mugesh, G. (2005) *J. Am. Chem. Soc.*, **127**, 11477.
(structure Se–Au–PR$_3$, R = Me, Et, Ph)	9	87, 76, 74	CDCl$_3$	Bhabak, K. P. and Mugesh, G. (2009) *Inorg. Chem.*, **48**, 2449.
(structure with Se, OMe, pyrrolidine, HO$_2$C)	9	247	CDCl$_3$	Bhuyan, B. J. and Mugesh, G. (2011) *Org. Biomol. Chem.*, DOI:10.1039/C0OB00823K

9	270 (X=Cl) 287 (X=Br)	CDCl$_3$ CDCl$_3$	Potapov, V. A., Kurkutov, E. O., Musalov, M. V., and Amosova, S. V. (2010) *Tetrahedron Lett.*, **51**, 5258.
9	280	CDCl$_3$	Roy, G., Nethaji, M., and Mugesh, G. (2005) *J. Am. Chem. Soc.*, **127**, 15207.
9	305	CDCl$_3$	Bhuyan, B. J. and Mugesh, G. (2011) *Org. Biomol. Chem.*, DOI:10.1039/C0OB00823K
9	309	CDCl$_3$	Bhuyan, B. J. and Mugesh, G. (2011) *Org. Biomol. Chem.*, DOI:10.1039/C0OB00823K
9	345 77	CDCl$_3$	Hua, G., Fuller, A. L., Slawin, A. M. Z., and Woollins, J. D. (2010) *Eur. J. Org Chem.*, 2607.

(Continued)

Compound	Chapter/No.	Chemical shift (ppm)	Solvent	Reference
	9	353 67	CD_2Cl_2	Hua, G., Fuller, A. L., Slawin, A. M. Z., and Woollins, J. D. (2010) *Eur. J. Org Chem.*, 2607.
	9	383	$CDCl_3$	Garud, D. R., Toyoda, Y., and Koketsu, M. (2009) *Tetrahedron Lett.*, **50**, 3035.
	9	390	$CDCl_3$	Denmark, S. E. and Edwards, M. G. (2006) *J. Org. Chem.*, **71**, 7293.
	9	397	$CDCl_3$	Roy, G., Nethaji, M., and Mugesh, G. (2005) *J. Am. Chem. Soc.*, **127**, 15207.
53, R = Me 54, R = Et 55, R = nPr 56, R = iPr	9/53 9/54 9/55 9/56	418 417 416 415	$CDCl_3$ $CDCl_3$ $CDCl_3$ $CDCl_3$	Bhabak, K. P., Mugesh, G. (2008) *Chem. Eur. J.*, **14**, 8640.

9/50	424	CDCl₃	Bhabak, K. P. and Mugesh, G. (2008) *Chem. Eur. J.*, **14**, 8640.
9/51	420	CDCl₃	
9/52	406	CDCl₃	
9/46	433	CDCl₃	Bhabak, K. P. and Mugesh, G. (2009) *Chem. Asian J.*, **4**, 974.
9/47	425	CDCl₃	
9/48	423	CDCl₃	
9/49	412	CDCl₃	
9/71	435	CDCl₃	Zade, S. S., Singh, H. B., and Butcher, R. J. (2004) *Angew. Chem. Int. Ed.*, **43**, 4513.
9/42	439	DMSO-D6	Bhabak, K. P. and Mugesh, G. (2009) *Chem. Asian J.*, **4**, 974.
9/43	440	DMSO-D6	
9/44	439	DMSO-D6	
9/45	440	DMSO-D6	

50, R = Et
51, R = nPr
52, R = iPr

46, R = Me
47, R = Et
48, R = nPr
49, R = iPr

71

42, R = Me
43, R = Et
44, R = nPr
45, R = iPr

(*Continued*)

Compound	Chapter/No.	Chemical shift (ppm)	Solvent	Reference
40	9/40	440	CDCl$_3$	Sarma, B. K. and Mugesh, G. (2005) *J. Am. Chem. Soc.*, **127**, 11477.
14	9/14	451	CDCl$_3$	Mugesh, G., Panda, A., Singh, H. B., Punekar, N. S., and Butcher, R. J. (2001) *J. Am. Chem. Soc.*, **123**, 839.
85	9/85	465	CDCl$_3$	Goto, K., Sonoda, D., Shimada, K., Sase, S., and Kawashima, T. (2010) *Angew. Chem. Int. Ed.*, **49**, 545–547.
21	9/21	545	CDCl$_3$	Sarma, B. K. and Mugesh, G. (2008) *Chem. Eur. J.*, **14**, 10603.

9/36	547	CDCl$_3$	Sarma, B. K. and Mugesh, G. (2005) *J. Am. Chem. Soc.*, **127**, 11477.
9/20	570	CDCl$_3$	Sarma, B. K., Manna, D., Minoura, M., and Mugesh, G. (2010) *J. Am. Chem. Soc.*, **132**, 5364.
9/39	588	CDCl$_3$	Sarma, B. K. and Mugesh, G. (2008) *Chem. Eur. J.*, **14**, 10603.
9/28	794	DMSO-D6	Bhabak, K. P. and Mugesh, G. (2007) *Chem. Eur. J.*, **13**, 4594.

(*Continued*)

432 | *77Se NMR Values*

Compound	Chapter/No.	Chemical shift (ppm)	Solvent	Reference
19	9/19	819	CDCl$_3$	Reich, H. J. and Jasperse, C. P. (1987) *J. Am. Chem. Soc.*, **109**, 5549.
34	9/34	853	DMSO-D6	Bhabak, K. P. and Mugesh, G. (2007) *Chem. Eur. J.*, **13**, 4594.
	9	857	CDCl$_3$	Bhabak, K. P. and Mugesh, G. (2009) *Inorg. Chem.*, **48**, 2449.
16	9/16	885	CDCl$_3$	Back, T. G. and Dyck, B. P. (1997) *J. Am. Chem. Soc.*, **119**, 2079.

9/24	912	DMSO-D6	Bhabak, K. P. and Mugesh, G. (2007) Chem. Eur. J., **13**, 4594.
9/29	913	CDCl$_3$	Bhabak, K. P. and Mugesh, G. (2009) Chem. Eur. J., **15**, 9846.
9/30	876	CDCl$_3$	
9/31	883	CDCl$_3$	
9/32	820	CDCl$_3$	
9/4a	960	CDCl$_3$	Fischer, H. and Dereu, N. (1987) Bull. Soc. Chim. Belg., **96**, 757.
9/33	965	CDCl$_3$	Bhabak, K. P. and Mugesh, G. (2007) Chem. Eur. J., **13**, 4594.
9/25	975	DMSO-D6	Bhabak, K. P. and Mugesh, G. (2007) Chem. Eur. J., **13**, 4594.
9/38	1122	CDCl$_3$	Sarma, B. K. and Mugesh, G. (2008) Chem. Eur. J., **14**, 10603.

(Continued)

Compound	Chapter/No.	Chemical shift (ppm)	Solvent	Reference
70 (OHC-aryl-tBu with Se=O cyclic)	9/70	1318	$CDCl_3$	Zade, S. S., Singh, H. B., and Butcher, R. J. (2004) *Angew. Chem. Int. Ed.*, **43**, 4513.
69 (cyclic Se=O with O)	9/69	1341	CD_3NO_2	Back, T. G. and Moussa, Z. (2003) *J. Am. Chem. Soc.*, **125**, 13455.
72 (SeOH, CHO, tBu aryl)	9/72	1399	$CDCl_3$	Zade, S. S., Singh, H. B., and Butcher, R. J. (2004) *Angew. Chem. Int. Ed.*, **43**, 4513.
17 (OHC-aryl-tBu with Se-O cyclic)	9/17	1401	$CDCl_3$	Zade, S. S., Singh, H. B., and Butcher, R. J. (2004) *Angew. Chem. Int. Ed.*, **43**, 4513.

Index

a

acetaldehyde, oxidation   238–9
acetic acid   239
– selenoacetic acid   258
acetoxyselenenylation   11
acetyl radicals   113
actinide selenolate   73
activation energies   287–8
– selenocarbonyls   259
acyclic selenium-stabilized carbanions   176
acyl radicals   119–22, 131
acyl selenide precursors   119–22, 131, 137
N-acylaziridines   241
adamantane   345
addition reactions
– addition–cyclization   136
– aldehydes   322–4
– carbonyl compounds   180–4
– chiral diselenides   5–6, 195
– double bonds   11–30
– enantioselective   8–9, 10, 186
– enones, organometallic addition to   325–6
– epoxides   53
– nucleophiles
– – carbon-centered   26–8
– – chiral   28–30
– – nitrogen-centered   22–6
– – oxygen-centered   11–22
– organometallics to vinyl/alkynylselenides   158–61
– radical addition   341–2
– stereoselective   8, 26, 176, 322–4
– Z-symmetric alkenes   14–15
alcohols
– β-amino alcohols   24
– benzyl alcohols   203, 204
– nucleophilic selenium reactions   53, 54, 62
– oxidation   346–7
– vinyl alcohol   216, 217
– see also allylic alcohols
aldehydes
– acetaldehyde oxidation   238–9
– alkyl selenide cyclization onto   117
– allylation   331–2
– aromatic, benzylidenation   205–6
– diorganozinc addition to   322–4
– selenoacetals from   157
– in selenoaldehyde synthesis   263
– α,β-unsaturated aldehydes   15–16
– see also selenoaldehydes
aldol reactions   276
– organocatalytic   334
aliphatic compounds
– GPx mimics   377–8
– see also alkenes; alkynes
alkenes
– allylic oxidation   349–50
– azidoselenenylation   24
– dihydroxylation   344–6
– dihydroxyselenenylation   20, 213
– epoxidation   342–3
– hydrostannylation   339–40
– methoxyhydroxyselenenylation   213
– selenium electrophiles reaction with   8–11
– seleno group transfer onto   134–7
– selenocyclizations   31
– selenodiimide reactions with   223
– selenofunctionalization   1
– from selenoxide elimination   225–6, 287, 293
– Z-symmetric   14–15
– see also double bonds
alkenyl selenides   125
alkenylselenonium salts   199–200
alkenylation, enolates   200
alkoxychloroselenuranes   203

Organoselenium Chemistry: Synthesis and Reactions, First Edition. Edited by Thomas Wirth.
© 2012 Wiley-VCH Verlag GmbH & Co. KGaA. Published 2012 by Wiley-VCH Verlag GmbH & Co. KGaA.

alkoxyselenonium salts 203–4
(Z)-β-alkoxyvinylsulfones 197, 198
alkyl-acyl selenides 122
alkyl cyclization 116–17
(Z)-O-alkyl enol ethers 199
alkyl radicals 112, 133
– from selenide precursors 115–19
– in seleno-alkylation 137
– in seleno-selenation 135–6
alkyl selenides 115–19
alkylation
– asymmetric allylic 326–8
– enolates 196
– selenium-stabilized carbanions 177–80
– seleno-alkylation 137
– α-selenocarbonyl compounds 167
2-alkylaziridines 116
alkynes
– *one-pot* oxidation 20–1
– oxidation 351–2
– seleno group transfer onto 134–7
alkynylselenides 158–61
alkynylselenonium salts 197–8
1-alkynyl(diphenyl)selenonium salts 197
allenecarboxylic esters 295–6
allenes
– hydroxyselenenylation 21
– selenocyclizations 33
– from selenoxide elimination 293–5
allenylselenonium salts 201
O-allyl hydroxylamines 36–7
O-allyl oximes 37
allylation, aldehydes 331–2
allylic alcohols 168, 170
– chiral 297–305
– kinetic resolution 15
– [2,3]-sigmatropic rearrangement 287, 288
allylic alkylation 326–8
allylic chiral amines 309–11
allylic chlorides 336–7
allylic oxidation 349–50
allylic selenimides 305–11
allylic selenium ylides 311–17
allylic selenoxides 297–305
allylic sulfoxides 297
aluminum selenolates 95
amides 225
– nucleophilic selenium reactions 62, 63
– *sec-* and *tert-*amide diselenides 376–7
– selenocyclization 41
– *see also* lithium diisopropylamide (LDA); selenoamides; selenoformamides; sulfonamides

β-amido selenides 241
amidoselenenylation 22–3
amines
– amine-based diselenides as GPx mimics 377
– chiral allylic 309–11
– PhSe group reactions 114
– primary aromatic, oxidation 350–1
– secondary amines 36, 223
– in selenocyclizations 35–6
amino acid residues
– Cys and Sec 363
– ID enzymes 381, 382
– thioredoxins 387, 388
β-amino alcohols 24
α-aminocarbene 273
ammonium persulfate 19, 20
ammonium selenolates 67
anilines 350–1
anisole 28
anti-inflammatories 371, 372, 373
anti-Markovnikov adduct 10
anti-selenoxide elimination 293
anticancer compounds 114, 121, 130, 178, 371
antifungals 186
Antillatoxin 185
antimony selenolates 101
antioxidant activity 369, 370, 372, 377
antithyroid drugs 383–4
L-Arabinose 29
arenesulfonyl azetidines 241
aromatic homolytic substitution 116, 139–40
arsenic selenolates 100–1
aryl alkyl ketones 350
aryl bromides 329
aryl methyl ketones 212
aryl radicals 132–3, 141, 142
aryl selenides 121, 125, 141, 142
aryl-substituted epoxides 204–5
β-arylselenoazides 25–6
(Z)-β-arylthio-α-functionalized ethenes 200
β-arylthiovinylselenonium salts 198, 199, 200
Asperdiol 178
asymmetric reactions
– aldol 334
– allylic alkylation 326–8
– amidoselenenylation 23
– azidoselenenylation 25
– carbocyclization 44
– carboselenenylation 28
– hydroxyselenenylation 17

– *one-pot* selenenylation–deselenenylation 18–19
– selenocyclization 34
– selenofunctionalizations 35
– *Z*-symmetric alkenes 14–15
auto-antibodies 383
Avenaciolide 186
axial organolithiums 173–5
azides
– β-arylselenoazides 25–6
– carboxylic acid reaction 337–8
– selenium tetraazide 249
azidoselenenylation 24–6, 135
aziridine 26
– *N*-acylaziridines 241
– 2-alkylaziridines 116
– derivatives 211
2,2′-azobis[2-*iso*-butyronitrile] (AIBN) 115

b

Baeyer–Villiger oxidation 234–5, 347–9
barium selenolates 70
Baylis–Hillman reaction 202, 356
benzanilide 372
benzeneseleninic acid 239, 347
benzeneselenol (PhSeH) 55, 114, 120, 341
– in radical reactions 138–41
benzeneselenolate 385
benzimidazole 129
benzophenones 204
benzoselenohydroximates 124
benzoselenophenes 134
benzyl alcohols 203, 204
benzyl ethers 204
benzylidenation, aldehydes 205–6
benzylideneacetophenone 214–15
bicyclic compounds, synthesis 45
bicyclo[2.2.2]octenones 119
bimolecular S_H2 reactions 112–15
binuclear iron complexes 78–9
bismuth selenolates 101–2
bis(phenylseleno) acetals 154
Boc-protection 366, 367
boron selenolates 95
bromides
– aryl 329
– oxidation 352–3
– phenylselenenyl bromide 3
– solid-phase selenyl bromide 128–9
N-bromosuccinimide 337
butenolides 22, 179, 353–4
tert-butyl hydroperoxide *see* Sharpless oxidant
n-butyl lithium 148, 150, 151, 155–8, 173–5

t-butyl lithium 173–4
butyrolactones 180

c

C-nor-D-homosteroids 131
cadmium selenide 259
cadmium selenolates 93–4
Cahn–Ingold–Prelog priority rules 363
calcium selenolates 70–1
calothrixin B 121
camphor-derived selenenyls 14
camphor-derived selenonium ylides 206, 207–8
camphor selenenyl sulfate 16–17, 19–20
carbamoyl radicals 125, 126
carbanions, selenium-stabilized
– acyclic 176
– cyclic 173–6
– preparation 147–61
– reactivity with electrophiles 161–8, 169–72
– stereochemistry 168, 173–6
– total synthesis 176–86
carbazomycin B 140, 141, 341
carbocyclization 42–4
carbohydrate derivatives 28–9
carbon–carbon bond formation, stereospecific 27
carbon monoxide loss 120, 121
carbon nucleophiles
– addition reactions 26–8
– selenocyclizations 42–4
carbonyl compounds
– α-selenenylation 2
– Baeyer–Villiger oxidation 234–5, 348–9
– carbocyclization 43
– labeling 281
– oxycarbonyl radicals 125, 126
– selenium-stabilized carbanions addition to 180–4
– α,β-unsaturated 167, 185–6, 336
– *see also* aldehydes; ketones; selenocarbonyls
carbonyl diselenide 259
carbonyl selenide 259
carbonylation 335–6
carboselenenylation 28
carboxylic acids
– acyl selenides from 120
– azide reaction 337–8
(+)-2-carene 220, 221
catalysis
– polarity reversal catalysis 138–9, 141
catalysts, selenium 321–56
– Baylis–Hillman reaction 356

- carbonylation reactions 335–6
- Diels–Alder reactions 355
- dihydroxylation of alkenes 344–6
- diol elimination reactions 338–9
- halogenations and halocyclizations 336–7
- hydrostannylation of alkenes 339–40
- organocatalytic aldol reactions 334
- oxidation reactions 342–53
- – alcohols 346–7
- – alkenes 349–50
- – alkynes 351–2
- – aryl alkyl ketones 350
- – halide anions 352–3
- – primary aromatic amines 350–1
- radical chain reactions 340–2
- Staudinger–Vilarrasa reaction 337–8
- stereoselective selenenylation–elimination 353–5
- thioacetal synthesis 355–6
- transition metal-catalyzed reactions 324–34

"catalytic triad" 370, 371, 377, 389
cesium selenolate 69
chalcogenolates 267
chiral allenecarboxylic esters 295–6
chiral allenes 293–5
chiral allylic alcohols 297–305
chiral allylic amines 309–11
chiral cyclohexylidenemethyl ketones 294–5
chiral cyclopropanes 206–9
chiral diselenides, in electrophilic addition 5–6, 195
chiral epoxides 334–5
chiral nucleophiles, addition reactions 28–30
"chiral pool" approach 30
chiral selenimides 307–9
chiral selenonium salts 196
chiral selenonium ylides 204–8
chiral selenoxides, preparation and properties 288–91
chirality transfer 303–4, 306, 312, 314
chloptosin 38–9
chloramine–T 305–6, 308, 309
chloramine–T hydrate 231
chlorides
- allylic 336–7
- phenylselenenyl chloride 21, 22, 337
chloroselenenylation reaction 10
chloroselenurane 301–2, 311, 315–16
N-chlorosuccinimide 306, 336, 337
chlorotrimethylsilane 174
chromium carbenes 266

cinnamyl selenides 298–301, 309, 315–16
cinnamyl sulfides 315–16
CO dehydrogenases 361
cobalt selenolates 81–2
condensation reactions
- palladium-catalyzed 332–3
- selenium dioxide catalyzed 216
copper compounds
- copper oxide nanopowder 64–5
- in organometallic addition to enones 325–6
- selenolates 90–1, 97
counterion, facial selectivity 14
crossover experiment 10
CuOTf 308, 309
α-Cuparenone 180
cyclic selenium-stabilized carbanions 173–6
cyclization
- alkyl cyclization 116–17
- dienes 42, 136
- double cyclization 44–5
- halocyclization 336–7
- iodocyclizations 275, 276
- using selenide building blocks 127–8
- see also radical cyclization; selenocyclizations
cyclofunctionalizations 35, 36, 40
cyclohexadienes 140, 341
cyclohexene 17
cyclohexenyl disulfonamides 222
cyclohexylidenemethyl ketones 294–5
cyclopropane derivatives 206–10, 242
cysteine (Cys) 56, 64, 249, 363, 386–7
- in glutathione 388, 389, 390
cytosolic GPx (cGPx) 369

d

Darzens reactions 334–5
dasycarpidone 121
Davis oxidant 290, 291, 294, 295, 297–8
deiodination reactions 380, 381, 385, 386
density functional theory (DFT) 6, 14, 70
deoxygenation
- selenones 243
- selenoxides 229
deprotonation, selenides 149–54
deselenenylation
- nitrogen-centered nucleophiles 23
- *one-pot* selenenylation–deselenenylation 18–19
- reductive 167, 169
detosylative cyclization 40

diastereoselective reactions
– alkenylation of enolates 200
– methoxyselenenylation of styrene 12–14
– oxidation of selenides 290, 291
– selenoxide elimination 295–6
– [2,3]-sigmatropic rearrangement 299–305, 310–11, 315–17
– see also enantioselective reactions
diazoles, cyclization onto 127
dichlorodimethyl selane 244
dicoordinated tetravalent selenonium salts 191, 211–24
dicoordinated trivalent selenonium salts 191, 192–4
Diels–Alder adducts 257, 263, 269
Diels–Alder reactions 119, 270, 355
dienes
– carbocyclization 42
– cyclization 136
– selenenylation 11
– selenodiimide reactions with 222, 223
– synthesis 182
diethyl ether 157, 158, 167
diethylzinc addition 322–3
diferrocenyl diselenide 324, 325
dihydroxylation, alkenes 344–6
dihydroxyselenenylation, alkenes 20, 213
diisobutyl aluminum hydride (DIBAL-H) 160, 161, 166
dilithium carborane-1,2-diselenolate 81–2
dilithium diselenide 366
dimethyl(4-methylphenyl)selenonium trifluoromethanesulfonate 201
dimethylacetal intermediate 15–16
dimethyldioxirane (DMDO) 234
dimethylselenoxide 228–9
dinitrobenzenes 335
diols
– elimination reactions 338–9
– trans-diols 344
1,4-dioxanes 29
diphenyl diselenide 5
– in carbocyclization 42
– carbon–carbon triple bond oxidation 20
– reaction with mercury 94
– reaction with zinc 92
diphenylzinc addition 323–4
diphosphine oxide 115
dipyridylurea derivatives 217, 218
1,3-diselenanes 175
diselenides
– chiral, addition reactions 5–6, 195
– cleavage of Se–Se bond 65
– for diethylzinc addition 322–3

– disodium diselenide 60
– as GPx mimics 373, 375, 376–7
– samarium selenolates from 71–2
– sec- and tert-amide diselenides 376–7
– X-ray crystallography 7
– see also diphenyl diselenide
diselenoic acid esters 257, 258
1,3-disenole-2-selones 269
disilanyl selenides 61
disodium diselenide 60
disulfide bonds 387, 389, 390
dithioacetals 202
dithiotreitol (DDT) 226–7, 230
domino reactions 122, 130–1
double bonds
– addition reactions 11–30
– selenol attachment to 57
double cyclization 44–5

e

ebselen 370–6
electron transfer mechanism (S_N2) 4
electronegativity, selenium 56–7
electrophiles
– carbanions reactivity with 161–8, 169–72
– selenium, alkene reaction with 8–11
electrophilic selenium reagents 1–45
– addition reactions to double bonds 11–30
– reactivity 1
– reactivity and properties 7–11
– selenocyclizations 30–45
– synthesis 3–7
element-lithium exchange 154–8
elimination reactions
– diols 338–9
– selenenylation–elimination 353–5
– see also selenoxide elimination
ellipticine 121, 124, 130
enantioselective reactions
– addition 8–9, 10, 186
– oxidation prochiral selenides 290, 291
– selenoxide elimination 293–5
– [2,3]-sigmatropic rearrangement 297–9, 309–10, 312, 313–15
– see also diastereoselective reactions; stereoselective reactions
endo-transition state 300, 311
ene reaction, (+)-2-carene 220, 221
enol-keto tautomerism 386
enolates
– alkenylation 200
– alkylation 196
– carbocyclization 43

enones
- formation 349
- organometallic addition to 325–6
epoxides/epoxidation
- alkenes 342–3
- chiral 334–5
- formation 168, 170–1
- nucleophilic addition 53
- opening reaction 179
- from selenonium ylides 204–6, 209, 211
- styrene epoxide 8–9
epoxy triflates 170
eprosartan 134
equatorial organolithiums 173–5
erythro-selenide 292
Eschenmoser coupling reaction 272
Escherichia coli 368
esters
- allenecarboxylic esters 295–6
- β-keto ester 44
- seleninate ester 378, 379
- seleninic acid esters 225, 234
- selenoic acid esters 257, 258
- selenonic esters 240
- α,β-unsaturated esters 117–18, 303
ethers
- (Z)-O-alkyl enol ethers 199
- benzyl ethers 204
- diethyl ether 157, 158, 167
- selenoethers 29, 31–2
eukaryotes, selenocysteine synthesis 364
exo-transition state 300
expressed protein ligation 368–9

f
facial selectivity 15, 25
- counterion 14
FAD, in TrxR catalysis 387–9, 391
ferrocene-bearing selenuranes 247–8
ferrocenyl selenides 310, 316, 327
- diferrocenyl diselenide 324, 325
ferrocenyl selenolate platinum(II) complex 89–90
ferrocenyl selenoxides 300
Fe_2Se_2 cluster 78, 79
Fe_4Se_4 cluster 79, 80
Fe_2SeY cluster 78
Fischer carbene complexes 266
fluorous seleninic acid 349
Fmoc-protection 366, 367, 368
formate dehydrogenases 361, 362
formic acid 238
free selenolate 66
Frulanolide 177–8

g
gallium selenolates 96
gastrointestinal GPx (giGPx) 369
geranyl selenides 301
germanium, tributylgermanium radicals 112, 115
germanium selenolate 99
glutamine (Gln) 370, 371
glutathione 387, 388
glutathione disulfide (GSSG) 369, 370
glutathione peroxidases (GPx) 92, 227, 239, 361, 369–79
- biological functions 362
glutathione reductase (GR) 369, 370, 391
glutathione thiol (GSH) 369–70
glycine reductase 361, 362
glyoxal 238, 239
glyoxylic acid 238
gold selenolates 91–2
gold thioglucose 384
groups 1-15 elements, selenolates 67–102
Grubbs catalyst 127

h
halide anions, oxidation 352–3
halides
- nucleophilic selenium reactions 53, 54, 55, 62
- organo, phenylselenenylation 330
- selenenyl 3, 6
- in S_H2 reactions 113
- *see also* bromides; chlorides
halocyclization 336–7
halogen–lithium exchange 155
halogenation, selenium-catalyzed 336–7
halogens, in selenocarbonyl compounds 257, 258
haloselenenylation 10
haloselenuranes 244
hemiacetals 16, 351
heptemerone G 2
heteroarenes, PhSe group reactions 114
hexanuclear palladium cluster 55
hexavalent hexacoordinated compounds 244, 250–1
hexavalent pentacoordinated compounds 244, 249–50
hexavalent tetracoordinated compounds 191, 240–3
homolytic substitution
- aromatic 116, 139–40
- selenium 111–26, 132–4
horsfiline 115
hydrazine derivatives 39, 40

hydroalumination, alkynylselenides 160, 161
hydroaminations, intramolecular 334
hydrogen peroxide 20, 213
hydrogenases 361
hydroperoxides 369, 370, 373
hydrosilylation, ketones 324–5
hydrostannylation, alkenes 339–40
hydroxy carbamic acids 41, 42
hydroxyselenenylation
– alkynes 21
– allenes 21
– dihydroxyselenenylation 20, 213
– *one-pot* reaction 18, 19
– from selenenyl sulfates 16–17
hydroxyselenides
– in aromatic carbon–carbon formation 27
– β-hydroxyselenides 167, 168, 169–72, 180
– carbocyclization 42–3
– enantioselective addition 8–9
hydrozirconation, alkynylselenides 160, 161
hyperthyroidism 383
hypervalent derivatives 192, 244–51
hypothalamic–pituitary–thyroid control system 383
hypothyroidism 384

i

ID-1 deiodinase 380, 381, 382, 383, 384, 386
ID-2 deiodinase 380, 381, 382, 383
ID-3 deiodinase 380, 381, 382, 386, 387
imidazo-2-selone 261
imidazol-2-ylidene complexes 84–5, 97
imidazole 262
imidazoline-2-selones 279–80
imidoyl radicals 123–4, 130
imidoyl selenides 123–4, 130
imines
– cyclization onto 127
– PhSe group reactions 114, 116
– selenocyclization 38
– sulfonimines 332, 333
iminium salts 37, 38, 265, 267
indium selenolates 66, 96–7
indole alkaloids 121
indoles, from imidoyl selenides 123–4
inorganic nucleophilic selenium reagents 59–65
intramolecular hydroaminations 334
iodocyclizations 275, 276
iodothyronine deiodinase 362, 379–84
– synthetic mimics 384–7

ionic liquids 202–3
iridium selonates 82–4
iron compounds
– ferrocene-bearing selenuranes 247–8
– ferrocenylselenolate platinum(II) complex 89–90
– Fe_2Se_2 cluster 78, 79
– Fe_4Se_4 cluster 79, 80
– selenolates 78–80
– *see also* ferrocenyl selenides
isocoumarins 355
isocyanates 268
isocyanoacetates 332, 333
isoselenocyanates 259, 268
– reactions 277–8
isoxazolidines 37
isoxazolines 36–7

j

juruenolide C 125, 126

k

K-selectride 123
keto acids 212, 350
β-keto ester 44
ketodiphenylselenonium ylide 211
ketones
– aryl alkyl, oxidation 350
– aryl methyl ketones 212
– hydrosilylation 324–5
– α,β-unsaturated 210, 235, 293–5
– *see also* selenoketones
Kharasch reaction 134–5
Khusimone 184

l

lactones 177–8, 179–80
– *Lauraceae* lactones 182–3
– sesquiterpene 213–14
– solandelactone E 303, 304
lanthanide benzeneselenolates 72–3
Lauraceae lactones 182–3
lead selenolates 100
$LiAlH_4$ 61, 266
LiAlHSeH 61–2
ligands, selenium
– allylation of aldehydes 331–2
– asymmetric allylic alkylation 326–8
– condensation reactions 332–3
– intramolecular hydroaminations 334
– Mizoroki–Heck reactions 328–30
– organometallic addition to enones 325–6
– phenylselenenylation of organohalides 330

– stereoselective Darzens reactions 334–5
– stereoselective hydrosilylation of ketones 324–5
– substitution reactions 331, 333
LiN(i-Bu)$_2$ 155
lipid peroxidation 371, 372
lithium
– axial/equatorial organolithiums 173–5
– *n*-butyl lithium 148, 150, 151, 155–8, 173–5
– element-lithium exchange 154–8
– LiAlH$_4$ 61, 266
– LiAlHSeH 61–2
– selenium–lithium exchange 147, 148, 154–5, 157, 173
lithium 4,4′-di-*tert*-butylbiphenyl group 161
lithium alkyl selenolate 67
lithium benzeneselenolate 67, 68
lithium cobalt selenide complex 81–2
lithium diisopropylamide (LDA) 148, 149–52, 154, 175–6
lithium eneselenolates 68
α-lithium selenide 148, 159
lithium selenolates 67–9
Lyngbya majuscule 185
Lys284 residue 365, 366

m

magnesium 70, 229, 243
maltol 262–3
manganese selenolates 76–7, 97
Markovnikov adducts 10, 11, 97
mercury selenolates 94–5
metal selonates 55, 66
methaneseleninic acid 239
methaneselenolate 59
bis(*p*-methoxyphenyl) selenide 207
methoxyselenenylation
– alkenes 213
– *one-pot* reaction 18, 19
– styrene 12–14
– α,β-unsaturated aldehydes 15–16
– 2-vinylperhydro-1,3-benzoxazines 30
methyl cinnamate 14
β-methylstyrene 353
microwave-assisted reactions 212, 215, 216
milfasartan 134
mimics
– GPx 370, 372, 374–8
– IDs 384–7
mitomycin 114–15
Mizoroki–Heck reaction 201, 328–30

molybdenum cobalt selenide complex 81
molybdenum selenolates 75, 76

n

NADPH 369, 370, 387–9, 391
naphthyl-based selenols and thiols 386, 387
NaSeH 265–6
native chemical ligation 56, 367, 368
nickel dithiolene–diselenolene 84
nickel selenolates 84–6, 97
NiFeSe hydrogenases 362
ninhydrin 216
niobium selenolate 74–5
nitriles, aromatic 264
nitrogen nucleophiles
– addition reactions 22–6
– selenocyclizations 35–40
– vs oxygen in selenocyclizations 40–2
normal hydrogen reference electrode (NHE) 363
nuclear magnetic resonance (NMR) 6, 398–434
– haloselenenylation of styrene derivatives 10
– seleniranium salts 195
– selenocarbonyls 278
– selenocysteine studies 366, 367
– selenolates
– – ammonium 67
– – lithium 67–8
– – molybdenum 75
– selenomethionine 227
– selenones 241
– selone form 59
nuclear Overhauser effect (n.O.e.) 6
nucleophiles
– carbon, selenocyclizations 42–4
– carbon-centered, addition reactions 26–8
– chiral, in addition reactions 28–30
– nitrogen, selenocyclizations 35–40
– nitrogen-centered, in addition reactions 22–6
– oxygen, selenocyclizations 31–5
– oxygen-centered, in addition reactions 11–22
– oxygen vs nitrogen in selenocyclizations 40–2
nucleophilic selenium reagents
– development of 53–4
– inorganic 59–65
– organic 65–102
– recent applications 54–6

nucleophilic substitution 53, 54
– intramolecular and intermolecular 241, 242
– palladium-catalyzed 331
– ruthenium-catalyzed 333
nucleophilicity, selenolates 58–9
C-nucleoside 125

o

octanuclear silver cluster 55
olefins see alkenes
one-pot azidoselenenylation of alkenes 24
one-pot oxidation of alkynes 20–1
one-pot selenenylation–deselenenylation 18–19
optically active selenium ylides 312–13
organic nucleophilic selenium
– preparation 65–6
– structure 66–7
– see also selenolates
organocatalytic aldol reactions 334
organohalides, phenylselenenylation 330
organometallics
– addition to enones 325–6
– conjugate addition to vinyl- and alkynylselenides 158–61
– coordination polymers 64
– selenocarbonyls 278–80
– selenonium salt reactions with 203–4
osmium 80, 228
oxalic acid 238
1,2-oxaselenolane Se-oxide 233
1,3-oxaselenolanes 242
oxazines 350
1,3-oxazolidin-2-ones 242
oxazolines 37, 262, 303, 308, 327
– isoxazolines 36–7
– 4,5-oxazolines 23–4
– ring-opening 328
oxidation reactions
– acetaldehyde 238–9
– alcohols 346–7
– alkenes 349–50
– alkynes 351–2
– aryl alkyl ketones 350
– Baeyer–Villiger oxidation 234–5, 347–9
– halide anions 352–3
– primary aromatic amines 350–1
– selenides 225–6, 290, 291
– selenium-catalyzed 342–53
– using selenium dioxide 212–15, 218
oxidoreductases 54, 387, 391
oxycarbonyl radicals 125, 126

oxygen nucleophiles
– addition reactions 11–22
– selenocyclizations 31–5
– vs nitrogen in selenocyclizations 40–2
oxyselenenylation 11, 29

p

palladium reagents
– in allylation of aldehydes 331–2
– in asymmetric allylic alkylation 326–8
– in condensation reactions 332–3
– in Mizoroki–Heck reactions 328–30
– in phenylselenenylation of organohalides 330
– selenolates 86–7, 100, 201
– in substitution reactions 331
[2.2]paracyclophanyl selenide 300, 327
Pederin 185
pentacoordinated hexavalent compounds 244, 249–50
pentacoordinated systems 239, 240
pentalenene 122
d,l-Pentalenene 184
pentavalent compounds 191, 239–40
peptides
– polypeptides 56
– Sec-containing 367–9
perfluoroalkyl selenoxides 229–30
permethyl cyclohexane 180–1
peroxide
– GPx reaction with 373, 374, 378
– see also hydrogen peroxide; hydroperoxides
peroxynitrite (PN) 370
perselenuranes 250–1
phellandrene 221
phenanthridone 139
phenyl butyl ethyl selenonium tetrafluoroborate ([pbeSe]BF$_4$) 202–3
phenyl sulfide radical 113
phenyl tributylstannyl selenide 55, 99
phenyl vinyl selenide 119
phenyl vinyl sulfide 119
phenylselenenyl bromide 3
phenylselenenyl chloride 21, 22, 337
phenylselenenyl halides, reaction with zinc 93
phenylselenenyl sulfate (PSS) 3
phenylselenenylation, organohalides 330
phenylselenide (PhSe$^-$) anion 114, 120, 123, 141–2
phenylselenide (PhSe) group 111–19
– abstraction rate 112, 113, 120
– building blocks in radical synthesis 126–8

– in domino reactions   130–1
– transfer onto alkenes and alkynes   134–7
phenylseleno arylsulfonamides   241
2-(phenylseleno)enones   159
phenylselenomethyllithium   185
phenyltellurolate   385
phosphoindigo   217
phosphole selenides/sulfides   270
phospholipid hydroperoxide GPx   362, 369
phosphorus selenolates   100
phosphorus–selenium compounds   63, 100
photochemical degradation, selenoxides   226
PhSe see phenylselenide (PhSe) group
PhSeH see benzeneselenol (PhSeH)
PhSeMgBr   70
PhSeSiMe$_3$ 76, 97
pincer-type complexes   330, 331, 332
pK_a, selenols   57, 363
plasma GPx (pGPx)   369
platinum selenolates   87–9
polarity reversal catalysis (PRC)   138–9
polymer-supported selenium reagents   4–5
polypeptides   56
polyselenolate dianions   58
polyselenolates   58
polyselenols   58
polystyrene-bound phenylseleninic acid   342
potassium diisopropylamide (KDA)   154, 175
potassium selenocyanate (KSeCN)   262, 267
potassium selenolates   68
primary amines
– oxidation   350–1
– in selenocyclizations   35–6
primary selenoamides   261, 271
prochiral selenides
– enantioselective imidation   308
– enantioselective oxidation   290, 291
prokaryotes, selenocysteine synthesis   364
prolinamides   334
6-n-propyl-2-thiouracil (PTU)   380
proteins
– Sec-containing   367–9
– selenoproteins   361, 362
– see also selenoenzymes
pseudocodeine   303, 304
pyramidal inversion   307, 312
pyrazole   128–9
pyridine-2-selenol   57
pyridoxal phosphate (PLP)   365, 366
bis(2-pyridyl) diselenide   59
pyrroles, cyclization onto   127

pyrrolidin-3-ones   118
pyrrolizidines   36

q
Quadragel®   128–9
quinolines   123–4

r
radical addition   341–2
radical chain reactions   340–2
radical clock reactions   138
radical cyclization   115, 124, 133, 137, 340–1
– selenium-stabilized carbanions   184
– α,β-unsaturated esters   117–18
radical precursors   111–26
– domino reactions   130–1
radical protective groups   113, 114, 127
radical reactions
– PhSeH in   138–41
– selenide building blocks   126–8
radical reagents   115
radical reduction   118
radioactive organoselenium compounds   55–6
radioisotopes, selenocysteine studies   366, 367
reaction constants, bimolecular S$_H$2 reactions   112
reactive oxygen species (ROS)   369, 370, 371
redox potential   361–2, 363
redox reactivity   54
reduction
– radical reduction   118
– α,β-unsaturated carbonyl compounds   336
regioselective reactions
– deiodination   381
– selenoxide elimination   292
– see also enantioselective reactions; stereoselective reactions
reverse-T3 (rT3)   380, 386
rhenium selenolate   77
rhodium reagents   324, 325
– selonates   82–4
rotational barriers, selenocarbonyls   260
ruthenium reagents   324
– selenolates   80
– in substitution reactions   333

s
salinosporamide A   33–4
samarium selenolates   60, 71–2

(+)-Samin 2
Schiff base 365
Se–N interactions 7, 374
Se–O interactions 7, 374, 375, 376, 377, 379
Se–S interactions 6
Sec *see* selenocysteine (Sec)
secondary amines 223
– in selenocyclizations 36
selenadiadines 271–2
selenadienes 274
selenafluorene 246
selenenate ester 379
selenenic acid 369, 370, 373, 375, 379
selenenyl halides 3, 6
selenenyl iodide 384, 385, 386
selenenyl sulfates 16–17, 19–20
selenenyl sulfide 369, 370, 373, 374, 375, 377, 389
selenenyl triflates 25, 28, 37
selenenylation–deselenenylation, *one-pot* reaction 18–19
selenenylation–elimination, stereoselective 353–5
selenides
– alkyl 115–19
– alkynyl 158–61
– aryl 121, 125, 141, 142
– deprotonation 149–54
– oxidation 225–6, 290, 291
– *see also* diselenides; *individual groups of selenides*
selenimides 225, 230–2
– allylic 305–11
seleninate ester 378, 379
seleninic acid 225, 232, 234, 237–8, 370, 375
– enantiomers 239
– fluorous 349
seleninic acid esters 225, 234
seleninic anhydrides 225, 237
seleniranium ion 8, 9–10, 17
seleniranium salts 194, 195
selenite, radioactive 55–6
selenium
– biological importance 361–2
– electronegativity 56–7
– homolytic substitution 111–26, 132–4
– intermolecular S_H2 onto 132
– intramolecular S_H2 cyclization 132–4
– in selenocarbonyl synthesis 262, 265, 266, 268
– tolerable upper intake levels 362
selenium-catalyzed reactions 321–56

selenium dioxide 211, 212–20
selenium–heteroatom nonbonding interactions 6
selenium-ligated transition metal-catalyzed reactions 324–34
selenium–lithium exchange 147, 148, 154–5, 157, 173
selenium tetraazide 249
selenium ylides 311–17
seleno-alkylation 137
seleno-selenation 135–6
seleno-sulfonation 136–7
selenoacetals, carbanions from 155–8, 173, 174
selenoacetic acid 258
selenoacylsilanes 263
selenoaldehydes 257, 258, 263, 269
– manipulation of 270
selenoamides 258, 264, 267
– manipulations 271–4
– molecular structure 261
– synthesis 277
selenoazoles 271
selenocarbamate 276, 280
selenocarbamyllithium 274
selenocarbonates 276
selenocarbonyls 257–81
– manipulation 270–8
– molecular structure 261
– organometallics 278–80
– α-selenocarbonyls 2, 167
– synthetic procedures 261–70
– theoretical aspects 259–61
selenocyanates 7, 219–20
selenocyclizations 30–45
– carbon nucleophiles 42–4
– double cyclization 44–5
– nitrogen nucleophiles 35–40
– oxygen nucleophiles 31–5
– oxygen vs nitrogen nucleophiles 40–2
selenocysteine (Sec) 56, 64, 249, 363, 390–2
– biosynthesis 363–6
– chemical synthesis 366–7
– enzymes containing 361, 362
– in GPx catalytic cycle 369, 370
– proteins and peptides 367–9
– Sec-tRNA[Sec] 364, 365, 366
– 21st amino acid 362–3
– in TrxR catalytic cycle 389, 390
selenodiimides 211, 220–4
selenodithiocarbonates 259
selenoenzymes 54, 361, 369–89
– biological functions 362
– *see also* glutathione peroxidases (GPx)

selenoetherifications 31–2
selenoethers 29
selenoformaldehyde 257, 258
selenoformamides 263, 264, 271, 274
selenoic acid esters 257, 258
selenoimidates 268, 272, 275
selenoimidazole 57
selenoiminium salts 272
selenoisocyanates 262, 267, 277
selenoketones 257, 258, 270–1
– metal complexes 278
selenol(s)
– in GPx catalysis 369–70, 371
– in GPx mimic catalysis 373, 374, 377
– naphthyl-based 386, 387
– pK_a 57, 363
– polyselenols 58
– synthesis 97
– tautomerism 57–8
selenolactam 270–1, 272
selenolactonizations 31–2
μ_2 selenolate 66
μ_3 selenolate 66
μ_4 selenolate 66
selenolates
– ammonium 67
– Groups 1-15 elements 67–102
– nucleophilicity 58–9
– structures 66
selenomethionine 227–8, 246, 362
selenones 240
– deprotonation 153
– synthetic applications 241–3
selenonic acids 234–5, 240
selenonic amides 240
selenonic esters 240
selenonium ionic liquids 202–3
selenonium salt oxides 239–40
selenonium salts 355
– dicoordinated tetravalent 191, 211–24
– dicoordinated trivalent 191, 192–4
– five-membered or larger 194, 195
– tricoordinated tetravalent 191, 225–39
– tricoordinated trivalent 191, 194–211
selenonium ylides 194, 195, 204–11
selenophenes 134, 219
selenophosphate 364, 365
selenophosphate synthatase 362
selenoproteins 361, 362
selenopyranoquinolines 273
selenopyrazines 273
3-selenoquinoline 265
selenosemicarbazones 278–9
selenosubtilisin 370
selenothioc acid esters 257, 258

selenothymidines 269–70
α-selenotocopherol 132
selenoureas 259
– manipulations 271, 274–5
– metal complexes 278
– synthesis 264, 266, 267, 268, 277
selenoureido products 268
selenouronium salts 194
selenoxide elimination 292–6
– alkenes from 225–6, 287, 293
– chiral allenes and α,β-unsaturated ketones from 293–5
– *syn* elimination 167, 168, 169, 182, 292
selenoxides 225–30
– allylic 297–305
– chiral, preparation and properties 288–91
– conversion to selenimides 307
– deprotonation 153
– enantio- and diastereomerically pure 289, 290
– [2,3]sigmatropic rearrangement 378
– α,β-unsaturated compounds synthesis 2
selenoximines 240
selenurane oxides 245, 249–50
selenuranes 227–8, 244–9
– [2,3]-sigmatropic rearrangement 301–2, 311, 315–16
selonates, metal clusters 55
selones 57, 59, 192, 193
– imidazoline-2-selones 279–80
– synthesis 265, 266, 269
Sep-tRNA^{Sec} 364, 365
SepCysS enzyme 364
SepSecS enzyme 364–5
serine 363, 367
– Ser-tRNA^{Sec} 364
Se–Se bond cleavage 65
sesquiterpene lactones 213–14
S_H2 reactions *see* homolytic substitution
Sharpless oxidant 290, 291, 294, 295, 299, 342, 347
[2,3]-sigmatropic rearrangement 220–1, 287, 288
– selenoxides 378
– via allylic selenimides 305–11
– via allylic selenium ylides 311–17
– via allylic selenoxides 297–305
silica-supported polysiloxane complexes 329, 330
silicon selenolates 97–8
– bis(trimethylsilyl) selenide 60–1, 97
silver selenolates 90–1
silver triflate 3
single electron transfer (SET) 141–3
S–N interactions 375

sodium benzeneselenolate 68
sodium hydrogen selenide 59–60
sodium selenolates 67, 68
solandelactone E 303, 304
solid phase peptide synthetic (SPPS) routes 366, 367
solid phase selenyl bromide 128–9
solid-phase synthesis 128–30, 136
solid-state structure, potassium selenolates 68
solvent polarity, rotational barriers 260
spirodienyl radicals 139, 140
standard silver reference electrode (SSE) 363
Staudinger–Vilarrasa reaction 337–8
stereoselective reactions
– addition 26
– – diorganozinc reagents to aldehydes 322–4
– – selenium electrophile–alkene reactions 8
– – selenium-stabilized carbanion 176
– Darzens 334–5
– hydrosilylation of ketones 324–5
– organoselenium compounds 288
– selenenylation–elimination 353–5
– see also diastereoselective reactions; enantioselective reactions
stereospecific carbon–carbon bond formation 27
styrene
– haloselenenylation 10
– methoxyselenenylation 12–14
styrene epoxide 8–9
(E)-β-styrylselenonium triflate 209–10
substitution radical-nucleophilic, unimolecular ($S_{RN}1$) reactions 141–3
sulfenamides 113
sulfimides 307
sulfonamides 221–4, 241
sulfonimines 332, 333
sulfonium salts 216, 217
sulfonyl radicals 136–7
sulfur ylides 311, 312, 314, 315
"superreactive enzymes" 363
Swern reaction conditions 113
syn elimination 167, 168, 169, 182, 292

t
T2 thyroid hormone 380, 381, 386
tantalum selenolate 74–5
tautomerism
– enol-keto tautomerism 386
– selenocarbonyls 259
– selenols 57–8

taxane model system 182
tellurium–lithium exchange 156
telluroamides 272
telluronium salts 355
terminal selenolate 66
terpene-based electrophilic selenium 7, 14, 30
terpenes 131
tetrabutylammonium hydroxide (TBAH) 194
tetracoordinated hexavalent compounds 191, 240–3
tetracoordinated tetravalent compounds 191, 244–9
tetraethylammonium tetraselenotungstate 62–3, 267
tetrahydrofuran 157, 158, 167
tetrahydrofuran-3-ones 117, 118
tetrahydrofuranols 117
tetraphosphorus decaselenide 63
tetrathiafulvalene (TTF) derivatives 192, 259
tetravalent compounds
– dicoordinated 191, 211–24
– tetracoordinated 244–9
– tricoordinated 191, 225–39
thiazolines 24
thioacetals 355–6
thioimides 265
thiol cofactor 370, 373, 384
thiol exchange reactions 374
thiols, naphthyl-based 386, 387
thioredoxin reductase (TrxR) 362, 387–9, 390, 391, 392
thioredoxins 387, 388
thioureas 267
threo-selenide 292
thyroid peroxidase (TPO) 379–80
thyroid releasing hormone (TRH) 383
thyroid-stimulating hormone (TSH) 383
thyroxine (T4) 379–80, 381, 384, 385, 386
tin compounds
– hydrostannylation of alkenes 339–40
– selenolates 99–100
– tin–lithium exchange 156
– see also tributyltin hydride; tributyltin radicals
titanium selenolates 73
tolerable upper intake levels 362
bis-(p-toluenesulfonyl)selenodiimide 220, 221, 222, 223
N-tosylamides 231
N-tosylselenimides 230–1
total synthesis, selenium-stabilized carbanions in 176–86

toxicity 80, 249, 279
transition metals
– catalysis 324–34
– *see also* organometallics; *individual metals*
2,4,6-tri-*tert*-butylphenyl (TTBPSe) group 26–7
tributylgermanium radicals 112, 115
tributyltin hydride 111, 112, 119, 126, 128
tributyltin radicals 111–12, 119, 120
– PhSeH interaction with 138–9
tricoordinated tetravalent selenonium salts 191, 225–39
tricoordinated trivalent selenonium salts 191, 194–211
triethylborane 115
3,5,5′-triiodothyronine (T3) 379, 380, 381, 383, 386
2,4,6-triisopropylphenyl group (TIPP) 160, 161
trilobacin 34
trimethylselenonium hydroxide 195
trimethylsilyl radicals 126
bis(trimethylsilyl) selenide 60–1, 97
tris(trimethylsilyl)silane 111, 112
tris(trimethylsilyl)silane radical 113, 119, 120
triphenylphosphine 244
"*tripod*Co^0" series 82
2,3,4-trisubstituted morpholines 30
O-trityl oximes 141
trivalent compounds
– dicoordinated 191, 192–4
– tricoordinated 191, 194–211
tRNA^{Sec} 364, 366
tryptophan (Trp) 370, 371
tungsten carbenes 266
tungsten selenolates 75–6
Tyr116 389, 390

u

UGA codon 363
α,β-unsaturated acyl selenide precursors 122
α,β-unsaturated aldehydes 15–16
α,β-unsaturated carbonyl compounds 167, 185–6
– selective reduction 336
α,β-unsaturated compounds, from selenoxides 2
α,β-unsaturated esters 117–18, 303
α,β-unsaturated ketones 210, 235, 293–5

urea derivatives 217, 218
urea hydrogen peroxide (UHP) 213

v

vanadium 74, 303
Vernomenin 177
vinyl alcohol 216, 217
vinyl radicals 134
vinyl selenides 125, 293–4
– conjugate addition of organometallics to 158–61
– deprotonation 154
vinyl sulfides 216, 217
2-vinylperhydro-1,3-benzoxazines 30
vinyl(phenyl)selenide 159
(*E*)-vinylselenonium salt 209–10
vinylselenonium tetrafluoroborates 209

w

Wang solid phase resin 129
withasomnine 128–9
Wittig-type conversions 204
Woollins' reagent 63–4, 264
Wurtz reaction 64

x

X-ray analysis/crystallography
– diselenides 7
– glutathione peroxidase 371
– lithium alkyl selenolate 67
– metal selonates 66
– molybdenum areneselenolates 75
– Se–S interaction in selenenyl halides 6
– selenocarbonyls 261, 266, 278
– thioredoxin reductase 389, 390

y

2-ylidene-1,3-diselenones 193, 194
ylides
– allylic selenium 311–17
– imidazol-2-ylidene complexes 84–5, 97
– selenonium 194, 195, 204–11

z

Z-symmetric alkenes, asymmetric addition 14–15
zinc reagents
– diethylzinc 322–3
– diphenylzinc 323–4
– in intramolecular hydroaminations 334
– selenolates 65, 92–3